RESEARCH IN SURFACE FORCES

ISSLEDOVANIYA V OBLASTI POVERKHNOSTNYKH SIL

ИССЛЕДОВАНИЯ В ОБЛАСТИ ПОВЕРХНОСТНЫХ СИЛ

RESEARCH IN SURFACE FORCES

Proceedings of a Series of Symposia Sponsored by
the Institute of Physical Chemistry of the Academy of
Sciences of the USSR

Edited by Academician B. V. Deryagin

RESEARCH IN SURFACE FORCES

Edited by Academician B. V. Deryagin
Director, Laboratory of Surface Phenomena
Institute of Physical Chemistry
Academy of Sciences of the USSR

Volume 3

Translated from Russian by
J. E. S. Bradley
Senior Lecturer in Physics
University of London

 Springer Science+Business Media, LLC · 1971

The Russian text, which comprises the Proceedings of the Third Conference on Surface Forces sponsored by the Institute of Physical Chemistry of the Academy of Sciences of the USSR, and which originally was printed for the Institute by Nauka Press in Moscow in 1967, has been extensively corrected and up-dated by the editor for this edition. The English translation is published under an agreement with Mezhdunarodnaya Kniga, the Soviet book export agency.

The transliteration of the editor's name imprinted on the title page and throughout this volume is in accord with the British Standard System. However, the name is also frequently spelled *Deryaguin* in the Western literature.

Library of Congress Catalog Card Number 62-15549

ISBN 978-1-4899-5668-2   ISBN 978-1-4899-5666-8 (eBook)
DOI 10.1007/978-1-4899-5666-8

© 1971 Springer Science+Business Media New York
Originally published by Consultants Bureau, New York
Softcover reprint of the hardcover 1st edition 1971

CONTENTS

SECTION V

KINÉTIC EFFECTS IN THIN FILMS OF LIQUID

SECTION VI

TRANSPORT EFFECTS IN DISPERSE SYSTEMS AND POROUS BODIES

SECTION VII

SURFACE FORCES IN ADHESION, COHESION, AND FRICTION

SECTION VIII

SURFACE FORCES IN GASES

SECTION I

ANOMALOUS PROPERTIES
OF LIQUID BOUNDARY LAYERS

THE POLYMORPHIC FORMS OF LIQUIDS
IN CONDENSATION ON HYDROPHILIC SURFACES

B. V. Deryagin, N. N. Fedyakin, and M. V. Talaev

Surface Phenomena Division, Institute of Physical Chemistry,
Academy of Sciences of the USSR

Condensation of unsaturated vapors of polar liquids has been observed in quartz capillaries of radius
1-20 μ. The equilibrium vapor pressure of the liquid is deduced from the condensation rate (water
0.93, acetic acid 0.95, acetone 0.95, methanol 0.94) and is found to be independent of the radius.
The liquid is thus the bulk phase of a structural modification of the corresponding liquid. Viscosity
measurements in the case of water give a value more than ten times that for bulk water.

Introduction

It has been shown [1-3] that the surface of a solid or liquid can alter the structure-sensitive properties of adjacent layers of polar liquids 10^{-6} to 10^{-5} cm thick, and so it is assumed
that boundary layers with altered structure have the same thickness. This large radius of action of an interface is difficult to explain in terms of direct action of van der Waals forces,
since these decrease rapidly with distance and at distances of 10^{-6}-10^{-5} cm produce molecular
energies much less than thermal ones.

Although a liquid lacks long-range order, we thus have to suppose that an interface has
its effect transmitted from layer to layer via some form of epitaxial mechanism. The rel-
evant elementary forces may be various intermolecular forces, in particular hydrogen bonding,
which can determine the structure of a framework of molecules. The most uncertain feature
is what restricts the range of action of the surface on the structure of the liquid.

Here particular interest attaches to the observation [4] that an unsaturated vapor of a polar
liquid in certain instances will condense in a glass capillary to give columns whose physical
properties differ from those of the bulk liquid. The special structure must here extend out to
distances of the order of the capillary radius, which in certain cases is many microns.

This effect might have a mechanism different from that of formation of a boundary layer,
but the two undoubtedly involve structural change in the liquid.

TABLE 1. Electrical Conductivity of Columns of Anomalous Water

Liquid	r, μ	l, mm	R, mΩ	$\rho \cdot 10^4$, $\Omega^{-1} \cdot cm^{-1}$	c, % $\cdot 10^2$
Distilled water left in capillary 1 h	65	2.38	20	3.52	0.12
Distilled water after 7 h	171	2.1	1	9.1	3.4
20% suspension of ground glass after 20 days	72	4.2	2	53	18.5
Anomalous column after 6 months	82	2.5	1.55	15.3	5.3
Anomalous column after 7 days	100	2.1	2	7.5	2.62
Tabulated values for 0.4% solution	–	–	–	116	40

TABLE 2. Vapor Pressures of Anomalous Liquid Columns

Substance	Capillary	Radius, μ	p_0/p_s
Water	Quartz	12	0.93
"	"	21	0.93
"	"	5	0.93
"	No. 46 and No. 23 glasses	1–5	0.93
Acetone	Quartz	3	0.95
"	"	8	0.95
"	"	15	0.95
Methanol	"	19	0.94
The same	"	6	0.94
Acetic acid	"	23	0.95

Polymorphism of Liquids Produced by Condensation in Glass Capillaries

The first experiments were done with glass capillaries, so the effect might be ascribed to leaching (the liquid would then be an alkali solution). In fact, small volumes of water act on glass as do alkalis [5-7], i.e., the surface layers are completely disrupted.

However, it has been shown [8] that even a column of alkali in a glass capillary does not give the effects observed with a column of liquid having these novel properties. We have measured the electrical conductivity on these anomalous columns; Table 1 gives the results, in which r is radius, l is distance between electrodes, R is resistance, ρ is electrical conductivity, and c is the alkali concentration corresponding to ρ.

It is clear that normal water in a capillary differs only slightly in alkali concentration from anomalous water, the differences being due to the state of the surface and the time spent in the capillary. The equivalent alkali concentrations are less than those needed to explain the reduced vapor pressure by more than a factor 10^4, so leaching cannot explain the liquid of anomalous structure, although it may have an effect on column growth over long periods. It is therefore of especial interest to examine anomalous liquids in quartz capillaries.

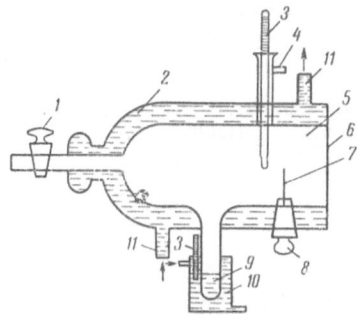

Fig. 1. Apparatus for study-ing condensation of vapor in capillaries: 1) stopcock; 2) thermostatic jacket; 3) ther-mometer; 4) to pressure gauge; 5) chamber; 6) obser-vation window; 7) capillary; 8) plug; 9) sidearm; 10) ther-mostat; 11) tube for connec-tion to thermostat.

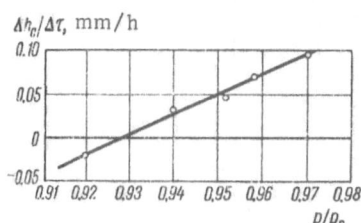

Fig. 2. Column growth rate as a function of p/p_s for water.

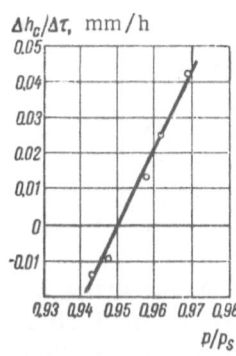

Fig. 3. Column growth rate as a function of p/p_s for acetone.

Polymorphism on Condensation in Quartz Capillaries

We subsequently used quartz capillaries with various liquids in the apparatus of Fig. 1. The thermostatic cham-ber 5 contains a quartz capillary 7 open at one end. The chamber is evacuated, and a vapor pressure p less than the saturation pressure p_s is set up, p being adjusted by means of sidearm 9 fitted with its own thermostat 10, which is set at a temperature below the chamber temperature. Stop-cock 1 connects the chamber to the vacuum line. A column of liquid is formed in capillary 7 after shutting the chamber off from the vacuum line. A cathetometer is used to measure the length of the column to 1 μ and the rate of growth for various p/p_s. If p is low enough, the column shortens by evaporation and the rate of growth is negative. So the rate of growth can be plotted as a function of p/p_s.

Figures 2-4 show curves which are linear within the experimental error, the intercept on the abscissa giving the equilibrium values p_0/p_s which are tabulated in Table 2 for 20°C.

It was found that p_0/p_s was independent of capillary radius, at least within the limits 1-20 μ.

Viscosity of Liquid Condensed in Capillaries

The viscosity is a structure-sensitive property, so viscosity measurement can indicate whether a special struc-ture is involved. The method has previously been described [4, 9] for glass capillaries and has been applied [10] to quartz ones. Poiseuille's equation applies to the velocity v produced by a pressure difference ΔP in a capillary of radius r

$$v = \frac{1}{\eta} \cdot \frac{r^2}{8\Delta l} \cdot \Delta P, \qquad (1)$$

in which η is viscosity and Δl is column length.

It was found that the column does not move as a solid but has a parabolic velocity profile, except at the ends; this was established by observing the motion of small dust par-ticles.

The first step was to measure the viscosity of bulk water drawn into the column. Here we must note that a given velocity in a capillary with dry walls requires a ΔP hundreds of times that predicted by Poiseuille's law, on ac-count of wetting hysteresis; but one or two runs up and down the column will dispose of this effect.

TABLE 3. Viscosity η of Ordinary Water in Quartz Capillaries

r, μ	$\Delta l,$ mm	$t, °C$	$\eta \cdot 10^2,$ g/cm·sec (meas)	$\eta_0 \cdot 10^2,$ g/cm·sec (meas)	r, μ	$\Delta l,$ mm	$t, °C$	$\eta \cdot 10^2,$ g/cm·sec (table)	$\eta_0 \cdot 10^2,$ g/cm·sec (table)
5.0	0.300	22.0	0.91	0.960	3.2	0.092	18.0	1.12	1.060
4.1	0.120	22.0	1.02	1.005	8.6	0.340	21.4	1.06	0.981

TABLE 4. Viscosity η of Water Condensed in Quartz Capillaries

r, μ	$\Delta l,$ mm	$t, °C$	$\eta \cdot 10^2,$ g/cm·sec	η/η_0	r, μ	$\Delta l,$ mm	$t, °C$	$\eta \cdot 10^2,$ g/cm·sec	η/η_0
3.2	0.092	18.0	17.4	16.03	7.5	0.26	20.3	18.6	18.60
4.0	0.127	22.5	16.5	16.83	8.6	0.34	24.4	19.0	21.20
4.1	0.12	20.0	22.0	21.90					

TABLE 5. Viscosity η of Water Condensed in Glass Capillaries

Liquid	r, μ	$\Delta l,$ mm	$\eta \cdot 10^2,$ g/cm·sec	η/η_0	Liquid	r, μ	$\Delta l,$ mm	$\eta \cdot 10^2,$ g/cm·sec	η/η_0
Water	1.06	0.39	12.5	12.5	Water	1.85	0.61	12.5	12.5
»	1.78	0.63	15.0	15.0	»	2.17	1.57	11.0	11.0
»	2.13	1.5	18.4	18.4	Acetone	1.4	4.2	0.52	1.6
»	2.64	0.98	8.0	8.0	»	1.45	1.64	0.58	1.7
»	1.37	0.69	12.5	12.5	»	1.85	6.58	0.55	1.7

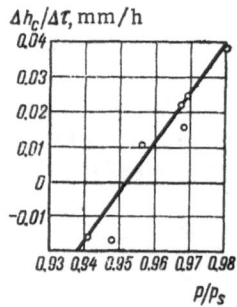

Fig. 4. Column growth rate as a function of p/p_s for acetic acid.

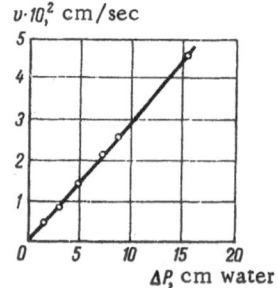

Fig. 5. Relation of speed v to pressure difference ΔP for a column of normal water.

Figure 5 shows the results from one experiment. The straight line passes through the origin, and the viscosity is close to the tabulated value, which shows that (1) is applicable and that there was no hysteresis. Table 3 gives results for ordinary water in quartz capillaries.

The second step was to measure η for water condensed in quartz capillaries at p/p_s of 0.95-0.98.

The columns were produced in the usual way in the chamber of Fig. 1 and were run twice along the capillary before measurement, as for ordinary water. Figure 6 shows the results

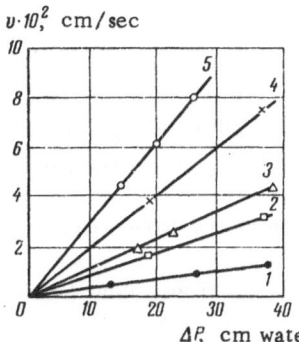

Fig. 6. Relation of speed
v to pressure difference
ΔP for a column of anom-
alous water run along the
capillary for the following
number of times: 1) third;
2) fourth; 3) fifth; 4) sixth;
5) seventh.

from one experiment; here again, the relation of v to ΔP is
linear, and the straight line passes through the origin; but the
slopes are far lower than for ordinary water, though they increase with each cycle.

Table 4 gives results [9] for η for condensed water, as
calculated from the first run (neglecting the two preliminary
runs). Table 5 gives analogous results [4, 9] for glass capillaries. The last column in each of these tables gives η / η_0,
in which η_0 is the value for ordinary water at the same temperature. These two sets of results differ only slightly.

The viscosity thus differs from that of ordinary water by
more than an order of magnitude for capillaries of radius 1–10 μ
or so, whereas the ratio for acetone is much smaller. The fall
in η on repeated displacement of the column is sometimes
considerable, but a halt always allows the initial η to be re-
covered. The following conclusions may be drawn:

1. The initial state (high η) is the more stable one.

2. The fall in η on movement shows that the anomalous
water has a special structure.

3. Movement of an anomalous column is not sufficient to convert it to ordinary liquid.†

Discussion

Previous papers [11–14] have represented a deviation from van der Waal's views on the
nature of the liquid state. It may now be assumed that a liquid has a short-range order,
but much remains to be done to elucidate this. New mechanical properties have recently been
observed in liquids [15–17], and these do not fit into the framework of existing concepts. Our
results require even greater changes in views on the structure and properties of liquids.

We have observed changes only for polar liquids (water, acetone, acetic acid, methanol),
the molecules of which have OH groups.

It may be that these columns are formed by condensation under the influence of hydrogen
bonds to the OH groups in the surface of the capillary. Initially, a polymolecular layer with a
special structure is produced, which subsequently extends during condensation via an auto-
epitaxial mechanism previously observed only for crystals (solid and liquid). Unsaturation
in the vapor may favor this because liquid with the normal structure cannot be produced. Also,
an anomalous column is not formed in a capillary previously wetted with ordinary water if the
walls are not allowed to dry out [18].

Thixotropic gel-sol transformation by stirring is an example of conversion of a more
stable state by flow into a less stable one (though still far from the state of a normal liquid),
with recovery on standing [19]. Here we have only an analogy, because the anomalous liquid
is not a colloidal system.

There is no doubt that an unsaturated vapor of certain polar compounds condenses in a
glass or quartz capillary to give a liquid whose vapor pressure and other properties distinguish
it from the bulk ordinary liquid.

---

† This is not produced even by heating to 380°C (with subsequent cooling) [10].

References

1. B. V. Deryagin, Zh. Fiz. Khim., 1:29 (1932).
2. B. V. Deryagin and V. V. Karasev, Dokl. AN SSSR, 62(4):761 (1948); Wear, 1:277 (1958); Proc. 2nd Int. Congress of Surface Activity, Vol. 3 (1957), p. 389; Kolloidn. Zh., 15:365 (1953); Zh. Tekhn. Fiz., 27(5):1076 (1957); Zh. Fiz. Khim., 33:100 (1959); B. V. Deryagin, N. N. Zakhavaeva, and S. V. Andreev, Inzh.-Fiz. Zh., 5(5):22 (1962).
3. Z. M. Zorin, in: Research in Surface Forces, Vol. 2, Consultants Bureau, New York (1966), p. 76.
4. N. N. Fedyakin, in: Research in Surface Forces, Vol. 2, Consultants Bureau, New York (1966), p. 126.
5. V. S. Molchanov and O. S. Molchanova, Trudy Gos. Opt. Inst., 24(145):23 (1956).
6. Yu. A. Gastev, Tr. Vses. Nauchn.-Issled. Inst.-Stekla, No. 32, 37 (1953).
7. Handbook on the Production of Glass, Vol. 1, Moscow, Gos. Izd. Lit. po Str., Arkhitekt., i Stroit. Mat. (1963), p. 219.
8. B. V. Deryagin, N. N. Fedyakin, A. V. Novikova, and M. V. Talaev, Dokl. AN SSSR, 165(4):878 (1965).
9. B. V. Deryagin and N. N. Fedyakin, Dokl. AN SSSR, 147(2):403 (1962).
10. B. V. Deryagin, N. N. Fedyakin, and M. V. Talaev, Dokl. AN SSSR, 167(2):376 (1966).
11. A. Z. Golik, in: Structure and Physical Properties of Materials in the Liquid State, Izd. Kiev. Univ. (1953), p. 17.
12. A. Z. Golik and D. N. Karlikov, Dokl. AN SSSR, 114:361 (1957).
13. F. Zernicke and J. Prins, Z. Phys., 41:184 (1927).
14. P. Debye and H. Menke, Ergeb. der Tech. Röntg., 11:256 (1931).
15. U. B. Bazaron, B. V. Deryagin, and A. V. Bulgadaev, this volume, p. 36; Zh. Éksp. Teor. Fiz., 51(10):970 (1966).
16. N. F. Bondarenko and S. V. Nerpin, this volume, p. 345.
17. B. V. Deryagin, S. V. Nerpin, and M. A. Arutyunyan, this volume, p. 9.
18. N. N. Fedyakin, Kolloidn. Zh., 24:497 (1962).
19. B. V. Derjaguin (Deryagin), Discussion of the Faraday Society on the Stability of Colloids in Aqueous and Nonaqueous Media; Dokl. AN SSSR, 170(4):876 (1966); 172(5):1121 (1967).

CONDENSATION OF WATER VAPOR
IN GLASS CAPILLARIES

B. V. Deryagin, S. V. Nerpin,
and M. A. Arutyunyan

Institute of Physical Chemistry, Academy of Sciences of the USSR;
Agrophysical Institute, Lenin All-Union Academy of Agricultural Sciences;
Deserts Institute, Academy of the Turkmen SSR

It is shown that condensate columns are formed in capillaries of radius 1-50 μ from unsaturated water vapor only in the absence of a film of ordinary water on the capillary walls.

Condensation of water vapor on radiatively cooled soils and sands is of considerable practical importance in localities with a hot dry climate. It is usually considered that the water vapor must become saturated for condensation to occur, but it has been shown [1-5] that condensation can begin in glass and quartz capillaries at about 93% relative humidity. We have examined some aspects of this phenomenon.

The tests were done with cylindrical glass capillaries (50 μ > r > 1 μ), some of which had been wetted internally with water droplets. Figure 1 shows the apparatus, which consisted of the chamber 2 containing the capillary 1, the microscope 3, the light source 5, and the heat filter 4. The capillary was imaged on a screen. It was found that water columns with anomalous properties were formed in capillaries of radius up to several microns only if the capillary had not been wetted with bulk water. It was subsequently found that condensation would also occur in such wetted capillaries if these were first carefully dried.

These results indicate that a previously applied film differs in structure from the film formed by condensation of an unsaturated vapor, with a consequent difference between the saturation vapor pressures over the surfaces. The condensate has a lower saturation vapor pressure, while the reduction in vapor pressure over a film of bulk water is very small and is due solely to the curvature of the surface. Thickening of the condensate film and formation of columns indicate that the ordered structure is retained as the film grows.

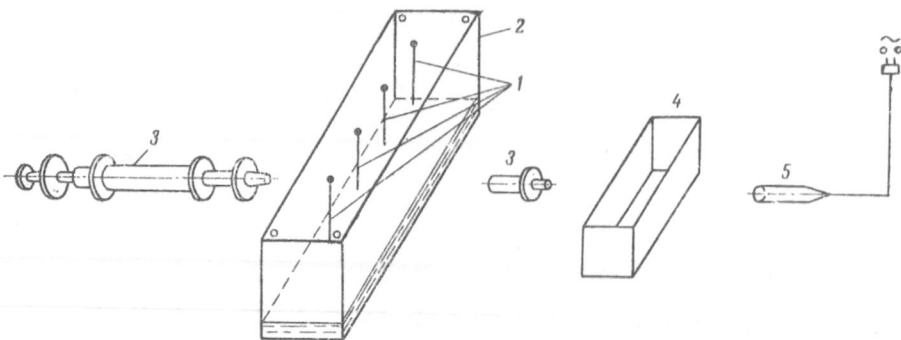

Fig. 1. Apparatus for studying condensation and evaporation in glass capillaries.

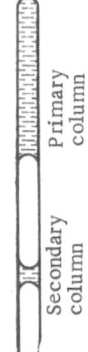

Fig. 2. Disposition of the primary and secondary water columns in a capillary during evaporation.

These effects might be due to salts dissolved from the glass, not to a different structure. As the salt concentration in a film might be high (while the film is thinner than a film of bulk water), the vapor pressure over the film should be reduced, and so the rate of condensation on such a film should be higher than that on a film of bulk water. This factor should continue to act until the films become of the same thickness, at which point the rates should become identical for the different films. However, this is not observed. An ordered structure in condensate columns is also indicated by evaporation tests performed with the apparatus of Fig. 1.

Chamber 2 was kept at a relative humidity well below saturation, and capillaries 1 were arranged with a secondary condensate column of water between the open end and the primary (ordinary) column (Fig. 2). The small column does not evaporate in such a system until the main column has evaporated completely. If the secondary condensate column is replaced by a column produced in some other way, e.g., by passing the open end of a capillary filled with water through the flame of a spirit burner (which causes the column to split into two), the column near the mouth evaporates first, and then the main column. If several columns are produced in the same way, the first column evaporates first, then the second, and so on.

This evaporation via a condensate column can be explained if the more ordered structure results in a lower vapor pressure over the secondary column. The resulting pressure gradient causes transfer of water between the two columns. The liquid can also migrate as a film along the wall as a result of the gradient in the capillary pressure set up by the vapor-pressure gradient. Retention of the secondary column indicates that these two modes of flow balance the loss of vapor from the secondary column to the mouth of the capillary.

Conclusions

1. It has been found that condensation producing secondary columns at vapor pressures near the saturation point occurs only in capillaries whose walls have not been wetted with bulk water.

2. If a secondary condensate column lies between the mouth of the capillary and the main column of bulk liquid, the main column shortens on evaporation, and the secondary column evaporates only after the main column has vanished.

3. The results indicate that the film formed by condensation does not have the structure of the bulk liquid and that the order persists as the film thickens and when a column is formed.

References

1. N. N. Fedyakin, Kolloidn. Zh., 24:497 (1962).
2. B. V. Deryagin and N. N. Fedyakin, Dokl. AN SSSR, 147(2):403 (1962).
3. B. V. Deryagin, M. V. Talaev, and N. N. Fedyakin, Dokl. AN SSSR, 165(3):597 (1965).
4. N. N. Fedyakin, B. V. Deryagin, A. V. Novikova, and M. V. Talaev, Dokl. AN SSSR, 165(4):878 (1965).
5. B. Derjaguin (Deryagin), Discussion of the Faraday Society on the Stability of Colloids in Aqueous and Nonaqueous Media, London, 1966.

MOLECULAR MODELS FOR THE STRUCTURE OF WATER

Yu. V. Gurikov

Khlopin Radium Institute, Leningrad

Models for the structure of water are reviewed, with particular attention to the model. which takes into account the formation of hydrogen bonds from the cavity molecules to the molecules of the ice-type framework. It is shown that the cavity molecules join into groups because they interact with the framework as a whole, the cause of this interaction being the cooperative formation of a three-dimensional network of hydrogen bonds. The theory implies that it is possible for there to exist a structural modification of liquid water having an elevated density and a reduced vapor pressure.

Structure of the Framework of Water

Liquid water has a structure resembling that of ice, as is clear from x-ray scattering [1-4], cold neutron scattering [5-7], fluctuations in the coordination number [8, 9], and long-range correlation in the density fluctuations in small volumes [10]. Samoilov [12, 13] used Frenkel's [11] concepts to devise a model for water in which the molecules are placed as in the framework of ice, the difference being that water does not have long-range order. Several two-structure models [14-17] are based on the similarity of the structures of water and ice. It is supposed that water is an ideal mixture of regions in which the molecules are disposed as in ice together with regions in which the hydrogen bonds are broken and the order approximates to close-packed.

The ice structure thus appears on two types of model, which differ in the distribution of the framework defects.

1. Two-State Models. Models of this type are characterized by a uniform distribution of the molecules over the holes in the framework. Frameworks of clathrate type [18, 19] have also been considered (Pauling's model [20, 21]), in which it is assumed that the molecules may lie either at the nodes of the framework and participate in hydrogen bonds or are unbonded and can rotate freely in the holes. However, the electrostatic energy has been calculated for several of these ice-type and clathrate structures, and it has been found that the ice lattice is the stronger when allowance is made for the interactions in the first and second coordination spheres [22].

2. Mosaic Models. The system is divided into small regions having either an ordered framework structure or a denser structure with broken hydrogen bonds. The Frank-Nemethy-Scheraga model [23-25] is the most detailed in this class. Nemethy and Scheraga

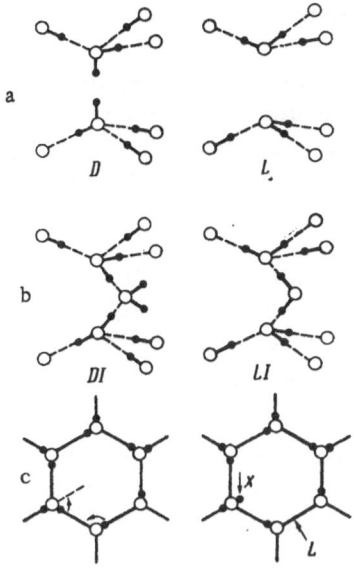

Fig. 1. Defects in a hydro-
gen-type framework: a)
Bjerrum [34] DL defects;
b) Haas [38] DI and LI defects; c) Dunitz [37] X defect.

have used such models to relate the concentration of molecules with a given number of bonds to the size of the ice-
type groups. It was found that a group contains 90 molecules
at 0°C and 20 at 100°C.

Similar results have been obtained [26] from electronic
absorption spectra, but the results [26] are open to some
doubt, as is the Nemethy-Scheraga model [27-30], which has
difficulty in explaining IR and Raman results for the stretching vibrations of water molecules. Wall and Hornig [31] estimated only 5% of the molecules are in ice-type groups, on
the basis of the O−H stretching bands in HDO, and not more
than 5% of molecules are free from bonds. Stevenson [32]
has shown that the concentration of monomers with no hydrogen bonds may not exceed 1% even at 100°C. It is also
unclear how the Nemethy-Scheraga model can explain the
large fluctuations in the coordination number in water [8]
or the large increase (by a factor 10^5) in the self-diffusion
coefficient on melting [28, 33]. These difficulties may be
overcome in models in which the structure is treated as that
of ice but distorted by defects.

State of the Molecules in Holes

The populations of states of higher energy increase
with the temperature. In water, breakage of hydrogen bonds
(which is equivalent to production of Bjerrum D and L defects [34], Fig. 1a) is accompanied
by filling of the holes, with general disorganization of the framework, which has been quantitatively explained in Pople's [35] model of curved hydrogen bonds and in a model [36] based on
diffuse energy bands each corresponding to several discrete states.

The free molecules perform activated jumps from one orientation to another; the framework molecules have their functional groups (protons or unpaired electrons) turned to adjacent
holes for part of the time. Dunitz [37] has shown that energy favors rotation of a framework
molecule through 60° to direct a proton into an adjacent hole and simultaneous formation of an
L defect (Fig. 1c) rather than generation of a pair (D and L) of defects. Also, formation of a
hydrogen bond to the molecule in this cavity will further stabilize the X defect. Haas [38] assumed participation in the hydrogen bonds of the cavity molecules. The times of dielectric
and mechanical relaxation agree with the characteristic time for self-diffusion in ice, and
Haas explained this via DI and LI defects (Fig. 1b), such a defect being a combination of framework D and L defects with a molecule in a cavity, which is linked to the participants in the D
or L defect by a pair of hydrogen bonds. Of course, production of DI and LI defects is facilitated
in the liquid state [33].

Linking of cavity molecules to framework molecules is in accordance with the spectro-
scopic data, which indicate not only that water contains no monomeric molecules [27, 29, 31,
32] but also that a certain band (3620 cm^{-1} [39-41]) in the stretching modes of H_2O can be ascribed to defective (curved) hydrogen bonds. Unfortunately, it is not presently practicable to
obtain complete experimental information on the state of the molecules in the cavities. Some
conclusions can be drawn by relating filling of holes in the framework to the difference between
the thermodynamic parameters of the ice-type and denser disordered structures [42, 43]; it
appears [44] that about 20% of the bonds are broken in the disordered structure, and this is
supposed by the persistence of appreciable isotopic differences between the enthalpies for the
disordered structures of H_2O and D_2O [45] (Fig. 2).

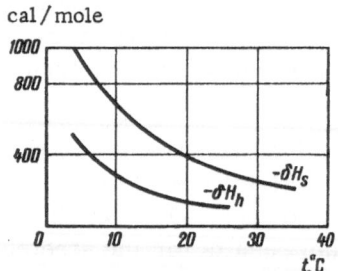

cal/mole

Fig. 2. The isotope effect in the enthalpies δH_s (ice-type structure) and δH_h (disordered structure) as calculated from the equations of the two-structure model [45].

Here we may note some attempts at quantitative interpretation of the thermodynamic parameters of water via models involving various assumptions on the nature and magnitude of the interaction between a cavity molecule and its environment. In particular, Frank and Quist [18] have calculated the thermodynamic parameters of water on the basis of a model of clathrate type, in which it was assumed that the molecules in the cavities were free to rotate. Krestov [46] used x-ray data [3] to calculate the the thermodynamic parameters of the cavity molecules and found that these cannot be considered as completely free from bonds. His results imply that a framework molecule is bound about 3 times as strongly as a cavity one.

Samoilov's model has been modified [29, 47] by taking into account the possibility of formation of hydrogen bonds between molecules in cavities and those in the framework. The results are described briefly below.

The Extended Samoilov Model

This model is based on an ice-type framework with defects of four types: a) curvature of framework hydrogen bonds to the point of rupture to give a DL defect pair, b) vacant sites in the framework (Frenkel's defects [11]), c) filled cavities in the framework, d) formation and dissociation of Haas DI and LI defects. The free energy of such a molecule is put as

$$\gamma F_w = \gamma F_0(\rho_0) + RT\gamma - \frac{1}{2}\ln 2 \cdot RT x_h - RT\gamma\{n_h\} + RT\{x_s\}$$
$$+ \frac{1}{2}RT\{x_h\} + 2x_s^2 f_s + 3x_s x_h f_h + 2x_s^2 RT[x] + 2x_s x_h(1-x_s x)(RT[t] + x_s^2 xt f_r), \qquad (1)$$

in which

$$\{x\} = x\ln x + (1-x)\ln(1-x),$$
$$[x] = \ln 2 + \{x\} + x\frac{f_H}{RT}.$$

Also, $F_0(\rho_0)$ is the free energy of an ideal gas of density ρ_0, $1-x_s$ is the concentration of Frenkel defects, $1-x$ is the concentration of D and L defects, t is the proportion of undissociated DI and LI defects, x_h is the proportion of filled holes in the framework, and $\gamma = \rho_m/\rho_0 = x_s + x_h/2$ is the ratio of the density of water to the density of a framework free from inclusions. Further, x_s and x_h are related to n_h (the fraction of molecules entering the holes) and to the density by

$$x_s = (1-n_h)\gamma, \quad x_h = 2n_h\gamma.$$

Also, f_s and f_h are the contributions to the free energy from various interactions (dispersion, dipole—dipole, repulsion, etc.) for each pair of neighbors in the framework and for each pair of adjacent molecules of which one is in a hole and the other at a framework site, and f_H is the change in free energy on formation of a hydrogen bond in the framework (i.e., on recombination on a DL pair).

The physical significance of parameter f_r needs to be considered in somewhat more detail. This quantity is not dependent on the structural state of the framework and is introduced via

$$\psi_r = x_s^2 x f_r,$$

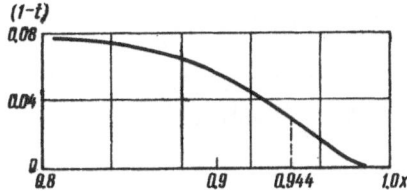

Fig. 3. Relation of framework disruption x to degree of dissociation of DI and LI defects via DI → D + I, in which I denotes an unbonded molecule in a hole. There is no stable turning point in the free energy for x < 0.81.

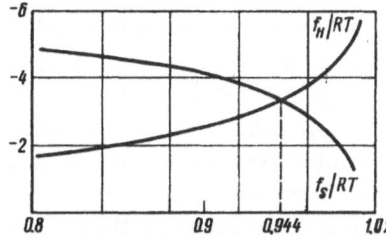

Fig. 4. Relation of f_H/RT (free energy of a hydrogen bond) and of f_s/RT (contribution to the free energy from dispersion and other interactions) to framework disruption x.

in which $x_s^2 x$ is the proportion of persisting bonds in the framework (a measure of the strength of the latter) and ψ_r is the change in the free energy due to distortion of the ice-type framework when a hydrogen bond is formed from a molecule in a hole to the environment. This change is governed by two factors: 1) the curvature of the defective bonds, which are less favored by energy than bonds in the framework; 2) the cooperative character of the formation and rupture of hydrogen bonds in water [23]. Frank showed from the donor−acceptor theory of the hydrogen bond [48] that transfer of unpaired electrons from a molecule to a proton on an adjacent molecule was accompanied by division of the charge, one of the interacting molecules acquiring an acid character and the other a basic character, which in turn leads to strengthening of the hydrogen bonds from these molecules to their neighbors. The effect is the more pronounced the greater the charge separation, i.e., the stronger the original bond.

When a framework molecule forms a hydrogen bond to a hole molecule, the former must lose some fraction of its bonds to its neighbors in the framework. The process is cooperative, so it leads to weakening of the hydrogen-bond system over a substantial region in the framework, which increases the rotational mobility of the molecules and thus greatly increases the entropy of the framework. The effect will obviously increase with the strength of the framework as a whole. If many molecules are involved in the cooperative rupture of hydrogen bonds, effect 2) above will make the main contribution to f_r.

Equation (1) has been used [47] to determine the numerical values for f_H, f_s, and f_h for several values of f_r in this general model, the initial data being the density, vapor pressure, and compressibility of water. The calculations were performed for the triple point of water, where two additional conditions for three-phase equilibrium are obeyed. It was assumed that the lattice constant was not changed on melting, and that only the concentration of vacancies in the framework is affected by change in temperature and pressure.

From f_H, f_s, and f_h we can calculate the parameters x, t, x_s, x_h, and n_h via equations for a minimum in the free energy, which is considered as a function of the three free parameters x, t, and n_h for γ = constant. It is found (Fig. 3) that t does not deviate from unity by more than 8% for any value of f_r. Also, $1 - t$ cannot exceed 3% (Fig. 4) if we restrict consideration to cases where $f_H < f_s$ (only in such cases can we expect a three-dimensional hydrogen-bond network to appear). The concentration of Frenkel defects is $1 - x_s \approx 10^{-5}$ (negligible). The most interesting and important result is that the observed compressibility can be obtained only for $f_r/RT < -0.183$. A negative f_r implies that hydrogen bonds from the hole molecules to the environment are thermodynamically more favored, which agrees with the above interpretation of f_r as a characteristic of the cooperative disruption of the ice-type framework around a filled hole.

Linking of Hole Molecules

Occurrence of a second solution is one of the unexpected results from the above calculation. In cases where $f_H < f_s$, the free energy has a further minimum, where the proportion of

TABLE 1. Properties of Water in the State Corresponding to the
Second Minimum in the Free Energy

f_r/RT	x_h	γ	x	$\dfrac{F_w^{(2)}}{RT}$	$\dfrac{F_w^{(2)}-F_w^{(1)}}{RT}$	v, cm³/mole
—0.2	0.959	1.48	0.678	—12.227	—0.077	13.3
—0.5	0.995	1.50	0.640	—12.387	—0.237	13.1

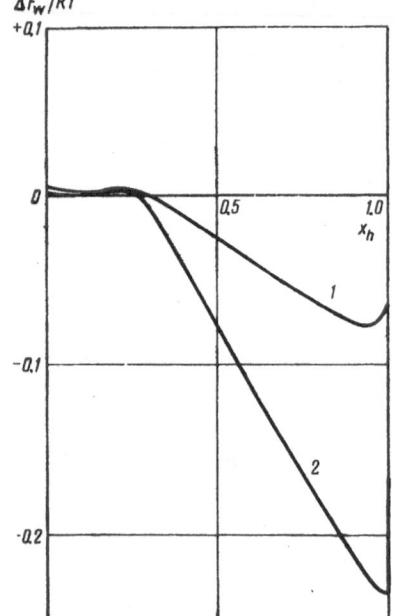

Fig. 5. Free energy $\Delta F_w/RT$ of water as a function of density x_h for P and T constant and f_r/RT of: 1) —0.2; 2) —0.5.

filled holes is much larger, and hence the density is greater; the concentrations of the various types of defect are also much increased (Table 1). Figure 5 shows that the second minimum is deeper than the one corresponding to ordinary water. This state will subsequently be called the disordered structure.

This result casts a new light on the structure of water and on the character of the thermal motion of its molecules. The existence of regions having a higher density and a lower thermodynamic potential indicates that it is favorable for the hole molecules to link up, the framework then acting as an internal three-dimensional reinforcement. These regions are not permanent; they arise and vanish in 10^{-10} to 10^{-11} sec [23], and so their occurrence does not conflict with the high mobility of water molecules. We may speak of density fluctuations in the gas of hole molecules. The geometry of the framework persists but is much less regular in detail (Table 1). The microheterogeneity is confirmed by direct calculation [10] of the correlation of the density fluctuations in small volumes.

This linking of hole molecules is the physical basis for the two-structure model. Direct calculations [44, 49] have been made of the thermodynamic parameters of the structures for this model via comparison of the properties of H_2O and D_2O, on the assumption that the molar volume and enthalpy for the ice-type structure at 0°C are as for ice itself. These calculations showed that the denser structure has the lower free energy (Fig. 6), so passage of a molecule into a framework hole is thermodynamically possible.

Of course, ordinary water cannot be considered as a mixture of two substances corresponding to the two minima in the free energy. The existence of a second minimum corresponding to a higher density indicates only that hole molecules can link up on account of a special interaction via the framework. The theory of heterophase fluctuations [11] indicates that nuclei of a new phase having a lower free energy will, if they exceed a certain critical size, increase in size without limit. The two-structure model is meaningful only if we understand the disordered structure as the set of all (even spatially separated) hole molecules, i.e., if this state is dispersed at the molecular level. Interaction between the hole molecules indicates that they must be considered as a real gas in the external field produced by the framework.

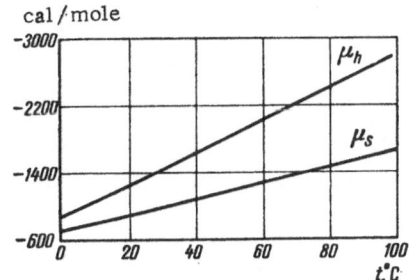

Fig. 6. Chemical potentials μ_s and μ_h of the ice-type and disordered structures in the two-structure model.

As a state with lower free energy should be able to exist independently, we naturally seek to establish whether it is observed, and if so, why not all water is in this thermodynamically more stable state. It is clear that the state does exist, since Deryagin et al. [50-53] have observed unusual properties in a daughter water column growing in a capillary not far from a parent drop having the properties of ordinary water. The daughter column has a higher density, lower free energy, and higher viscosity and thermal expansion. A more surprising feature is that the vapor pressure is not dependent on the radius of the capillary. Deryagin has concluded that this liquid phase is a modification of ordinary water and that the special properties of anomalous water cannot be ascribed to the direct action of the capillary walls. These properties also agree well with those that should be observed in the disordered structure in isolated form.

Frenkel's theory of heterophase fluctuations [11] explains why not all water exists in this form. Nucleation of a new phase (superscript 2) is unfavorable if the nuclei have a radius less than

$$R = \frac{2\sigma v}{F_w^{(1)} - F_w^{(2)}},\tag{2}$$

in which σ is the surface tension at the interface and v is the molar volume of the new phase.

It has been shown [53] that in fact $(F_w^{(1)} - F_w^{(2)})/RT = \ln(p_a/p_s) = -0.0726$ at 20°C, where p_a and p_s are the vapor pressures of anomalous and standard water.

Now $F_w^{(1)} - F_w^{(2)} \approx 42$ cal/mole, which is close to the free energy difference for the two minima for $f_r/RT = 0.2$ (Table 1). We may estimate R by taking $F_w^{(1)} - F_w^{(2)} \approx 40$ cal/mole and $v \approx 13$ cm^3/mole. As σ is unknown, we use $\sigma \approx 70$ dyne/cm (the value for the interface between ordinary water and the vapor) as a very crude approximation. Then (2) gives $R \approx 150$ Å. On the other hand, Fisher's results [10] give the radius of inhomogeneity in water as ~15 Å, i.e., much less, and so there is only a negligible probability of formation of nuclei capable of growth, and the actual probability is zero, because we have cooperative formation and disruption of the three-dimensional hydrogen bond network. This cooperative feature means that the groups of hole molecules break up simultaneously, not by loss of single molecules. It might thus appear that water is supersaturated relative to the disordered structure, but separation of the latter as a distinct phase is hindered by the rapid molecular exchange involved in the formation and disruption of short-lived molecular associations.

To conclude, I consider the causes of the especial stability of groups of hole molecules. Marchi and Eyring [19] consider that the ice-type structure (with fourfold coordination) must be accompanied by monomeric molecules, since there is a marked increase in entropy of bond rupture; but there are [32] hardly any such molecules in water. The hole molecules link up in the above model on account of feedback between hole filling and generation of D and L defects.

The equation defining the concentration of D and L defects is [47]

$$\ln \frac{x}{1-x} = -(1 - x_h) \frac{f_H}{RT} - x_h (1 - 2x) \frac{f_r}{RT} + x_h \ln 2.$$

Here we have put t = 1 and x_s = 1 for simplicity. It is readily seen that this equation has a solution x > 1/2 for f_H < 0 and f_r < 0, the concentration increasing with f_H, and so, for any f_r,

$$\frac{dx}{df_H} < 0. \tag{3}$$

The following is the equation for the concentration of hole molecules on the basis of the above assumptions:

$$RT \ln(1 - x_h) = 4f_h - \frac{4}{3}f_s - \frac{4}{3}RT \ln(1 - x) - \frac{8}{3}x_h x^2 f_r.$$

Then

$$\frac{dx_h}{df_H} = -\frac{4(1 - x_h)}{3(1 - x)} \cdot \frac{1 - 4x_h x (1 - x)\frac{f_r}{RT}}{1 - \frac{8}{3}(1 - x_h) x^2 \frac{f_r}{RT}} \cdot \frac{dx}{df_H}.$$

For f_r < 0, we have from (3) that

$$\frac{dx_h}{df_H} > 0. \tag{4}$$

Combination of (3) and (4) gives

$$\frac{dx}{dx_h} < 0. \tag{5}$$

There are two consequences here: 1) accumulation of molecules in the holes disrupts the framework of hydrogen bonds; 2) disruption of the framework favors passage of molecules into the holes.

We can now envisage the formation of groups of hole molecules as follows. The proportion of D and L defects is increased in some fairly large region around a filled hole as a result of cooperative formation and rupture of hydrogen bonds. From (5), the probability of a molecule entering a hole is increased for a disrupted framework. Moreover, the molecules in the holes are stabilized by van der Waals interaction with the environment. The advantage in free energy arises because the mean coordination number in a framework with filled holes is about 10, as against only four for the lattice of ice.

The conclusion about linking of hole molecules is a consequence of the original basic assumption that the strength of the framework as a whole governs the formation of hydrogen bonds from the hole molecules to the framework, which is an expression of the cooperative production and disruption of the hydrogen-bond system.

I am indebted to B. V. Deryagin and V. M. Vdovenko for valuable advice and comments.

References

1.   J. D. Bernal and R. H. Fowler, J. Chem. Phys., 1:515 (1933).
2.   J. Morgan and B. E. Warren, J. Chem. Phys., 6:666 (1938).
3.   M. D. Danford and H. A. Levy, J. Am. Chem. Soc., 84:3965 (1962).
4.   G. W. Brady and W. J. Romanow, J. Chem. Phys., 32(1):306 (1960).
5.   K. E. Larsson and V. Dahlborg, J. Nucl. Energy, 16:81 (1962).
6.   K. S. Singwi and A. Sjolander, Phys. Rev., 119:863 (1960).

7. Yu. V. Gurikov, Zh. Strukt. Khim., 4:824 (1963).
8. I. Z. Fisher and V. K. Prokhorenko, Dokl. AN SSSR, 123:131 (1958).
9. V. K. Prokhorenko, O. Ya. Samoilov, and I. Z. Fisher, Dokl. AN SSSR, 125:356 (1959).
10. I. Z. Fisher and V. I. Adamovich, Zh. Strukt. Khim., 4:819 (1963).
11. Ya. I. Frenkel', Kinetic Theory of Liquids, Izd. AN SSSR, Moscow—Leningrad (1959).
12. O. Ya. Samoilov, in: Structure of Aqueous Electrolyte Solutions and Ion Hydration, Izd. AN SSSR, Moscow (1957).
13. O. Ya. Samoilov, Zh. Fiz. Khim., 20:1411 (1946).
14. L. Hall, Phys. Rev., 73:775 (1948).
15. K. Grjotheim and J. Krogh-Mor, Acta Chem. Scand., 8:1193 (1954).
16. G. Wada, Bull. Chem. Soc. Japan, 34:955 (1961).
17. C. M. Davis and T. A. Litovitz, J. Chem. Phys., 42:2563 (1965).
18. H. S. Frank and A. S. Quist, J. Chem. Phys., 34:604 (1961).
19. R. P. Marchi and H. Eyring, J. Phys. Chem., 68:221 (1964).
20. L. Pauling, The Hydrogen Bond, London (1959).
21. G. G. Malenkov, Dokl. AN SSSR, 137:1354 (1961).
22. G. G. Malenkov and O. Ya. Samoilov, Zh. Strukt. Khim., 6:9 (1965).
23. H. S. Frank, Proc. Roy. Soc., A247:481 (1958).
24. G. Nemethy and H. A. Scheraga, J. Chem. Phys., 36:3382 (1962).
25. G. Nemethy and H. A. Scheraga, J. Chem. Phys., 41:680 (1964).
26. K. Buijs and G. P. Choppin, J. Chem. Phys., 39:2035 (1963).
27. D. F. Hornig, J. Chem. Phys., 40:3119 (1964).
28. O. Ya. Samoilov and T. A. Nosova, Zh. Strukt. Khim., 6:798 (1965).
29. Yu. V. Gurikov, Zh. Strukt. Khim., 6:817 (1965).
30. V. Vand and W. A. Senior, J. Chem. Phys., 43:1869 (1965).
31. T. T. Wall and D. F. Hornig, J. Chem. Phys., 43:2079 (1965).
32. D. P. Stevenson, J. Phys. Chem., 69:2145 (1965).
33. Yu. V. Gurikov, Zh. Strukt. Khim., 5:188 (1964).
34. N. Bjerrum, Dan. Mat. Fys. Medd., 27(1):7 (1951).
35. J. A. Pople, Proc. Roy. Soc., A205:163 (1951).
36. V. Vand and W. A. Senior, J. Chem. Phys., 43:1878 (1965).
37. J. D. Dunitz, Nature, 197:860 (1963).
38. C. Haas, Phys. Letters, 3:128 (1962).
39. Z. A. Gabrichidze, Optika i Spektr., 19:575 (1965).
40. Z. A. Gabrichidze, in: The State of Water in Living Organisms, Izd. LGU (1966), p. 94.
41. Yu. V. Gurikov, in: The State of Water in Living Organisms, Izd. LGU (1966), p. 103.
42. O. Ya. Samoilov, Zh. Strukt. Khim., 4:499 (1963).
43. V. M. Vdovenko, Yu. V. Gurikov, and E. K. Legin, in: The State of Water in Living Organisms, Izd. LGU (1966), p. 64.
44. V. M. Vdovenko, Yu. V. Gurikov, and E. K. Legin, Zh. Strukt. Khim., 8:403 (1967).
45. V. M. Vdovenko, Yu. V. Gurikov, and E. K. Legin, Zh. Strukt. Khim., 8:524 (1967).
46. G. A. Krestov, Zh. Strukt. Khim., 5:909 (1964).
47. Yu. V. Gurikov, Zh. Strukt. Khim., 7:8 (1966).
48. N. D. Sokolov, Usp. Fiz. Nauk, 57:205 (1955).
49. V. M. Vdovenko, Yu. V. Gurikov, and E. K. Legin, Zh. Strukt. Khim., 7:819 (1966).
50. N. N. Fedyakin, Kolloidn. Zh., 24:497 (1962).
51. B. V. Deryagin and N. N. Fedyakin, Dokl. AN SSSR, 147(2):402 (1962).
52. N. N. Fedyakin, B. V. Deryagin, A. V. Novikova, and M. V. Talaev, Dokl. AN SSSR, 165(4):878 (1965).
53. B. V. Deryagin, M. V. Talaev, and N. N. Fedyakin, Dokl. AN SSSR, 165(3):597 (1965).

VISCOSITIES OF AQUEOUS SOLUTIONS
IN THE CAPILLARIES OF SILICA GEL

Z. M. Tovbina

Institute of Physical Chemistry,
Academy of Sciences of the Ukrainian SSR, Kiev

Measurements of effective solute diffusion coefficients are used to deduce the viscosities of aqueous solutions in hydrophilic porous media. It is found that, for capillary radii < 100 Å, the viscosity rises rapidly and that the structuring revealed by NMR is retained throughout the temperature range in which the liquid phase exists.

Deryagin et al. [1-3] have shown that the liquid near the surface of a solid has anomalous properties; in particular, the viscosity in the boundary layer differs substantially from that in the bulk. Structuring in small capillaries must have a marked effect on infiltration and diffusion in porous media. However, no allowance has previously been made in diffusion studies for the anomalous properties of aqueous solutions in hydrophilic capillaries.

I have previously [4, 5] derived an equation for the effective diffusion coefficient in a porous medium of globular structure:

$$D^{\bullet} = \frac{D \varepsilon \eta_0 / \eta}{[1 + 0.274 (1 - \varepsilon)]^2 (1 + 2.4 r / r_c)} \, , \tag{1}$$

in which D is the coefficient of free diffusion, r is the hydrodynamic radius of the molecule of the diffusing substance, ε is the porosity of the material (silica gel), η_0 is the bulk viscosity, and η is the viscosity in capillaries of radius r_c.

The following factors have been taken into account in (1) as reducing solute diffusion rates in porous media:

1) Mechanical blocking of the diffusion flux by the solid skeleton;

2) Extension of the diffusion path because the capillaries bend;

3) Retardation of the molecules by the fixed walls; and

4) Anomalous viscosity of the solution in capillaries.

20

It has been shown [5] that (1) does allow D* to be calculated. Measurements have been made of D* for silica gels differing in porous structure. The diffusion rates for one of the compounds were used to find η_0/η, which was used to deduce D* for the other compounds, the values agreeing well with those found by experiment. It is easy to measure the diffusion rates of solutes accurately, so such measurements can be used to study the viscosities of aqueous solutions in the capillaries of hydrophilic substances having globular structures.

The objects of the present paper are: 1) to show that η/η_0 calculated from (1) do characterize the relative viscosities of solutions in porous media; 2) to examine the effects of various factors on the viscosities of aqueous solutions in the capillaries of silica gels differing in porous structure.

Experimental

The D* for iodine, methanol, and sucrose were determined by the diaphragm method [5], whose principle is as follows. A vessel is divided into two parts by a silica gel diaphragm, with pure solvent on one side and a solution of known concentration on the other. The amount of solute passing through the membrane is measured over a given interval. Fick's law then gives D* from the area and thickness of the membrane. The deviations between D* from parallel experiments usually do not exceed 10%.

The objection might be made that the resulting diffusion rates in capillaries may be affected by the special properties of the water near the hydrophilic surface of the porous material. Deryagin and Milekhina [6, 7] have shown that capillary osmosis is produced by the special properties of water in a boundary layer; the pure water differs in chemical potential from the solution, which causes the water to migrate along the surfaces of the capillary walls, which leads to a pressure difference between the two sides, the result being a compensating flow within the volume of the capillaries, which can cause an apparent increase in the diffusion rate and hence D*. This makes it necessary to elucidate the effects of capillary osmosis on the results.

Capillary osmosis was detected in my experiments. Sometimes stresses produced microcracks in the membranes. The diffusion resistance of these is much greater than that of the membrane, because the latter has at least 100 times the free area of the former, but these cracks substantially increased the diffusion rate[†]; this must be due to capillary osmosis. The hydrodynamic resistance of the membrane in the absence of cracks is so large that capillary osmosis can hardly have affected the results, but tests were made for a possible effect of osmotic pressure here, for which purpose D* was measured as a function of sucrose concentration.

The osmotic pressure increases with the concentration, so it would be expected that the η/η_0 given by (1) should decrease regularly as the concentration increases. However, Table 1 shows that η/η_0 hardly alters when the concentration is increased by a factor of 20.

Capillary osmosis should also produce a difference between the liquid levels (increase on the solution side), but none was observed, even in runs lasting 48 h. Here the rate of convective transport for liquid in capillaries of radius 10-100 Å is vanishingly small, and a simple calculation shows that the amount of material transported in this way is less than that transported by diffusion by a factor of 100. A further test was to measure D* in the presence of differences Δh at least 10 times greater than those produced by capillary osmosis (Table 2).

These results show that osmotic pressure is virtually without effect on the diffusion rates of solutes in these silica gels.

[†] Of course, these results were not used in the subsequent calculations.

TABLE 1. Relative Viscosities from (1) for Sucrose
Solutions in Silica Gel Capillaries

Spec.	r_c, Å	$C_{12}H_{22}O_{11}$, M	$D \times 10^4$, cm²/min	$D^* \times 10^4$, cm²/min	η/η_0
2	72	0.05	3.0	0.56	3.0
2	72	0.5	2.46	0.48	2.9
5	24	0.05	3.0	0.17	4.8
5	24	0.5	2.46	0.13	5.2
5	24	1.0	2.28	0.12	5.1

TABLE 2. D^* for Sucrose in Silica Gel
(r_c = 72 Å, 20°C) in Cells with Differences of Level

Δh, mm	$D^* \cdot 10^4$, cm²/min	Δh, mm	$D^* \cdot 10^4$, cm²/min
+10.84	0.67	−6.66	0.60
+ 5.46	0.70	−8.92	0.72
0.0	0.63		

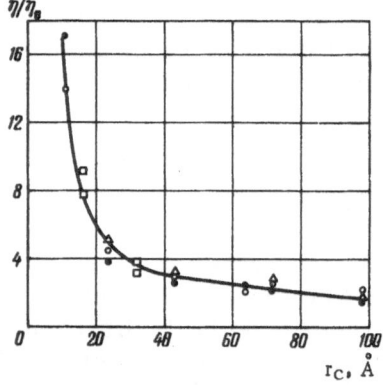

Fig. 1. Relation of η/η_0 for
aqueous solutions to capillary
radius r_c for silica gel at 20°C
as deduced from diffusion of:
•) I_3^-; ○) CH_3OH; △) $C_{12}H_{22}O_{11}$;
also □) as deduced from wetting
kinetics.

A further test was made on how far the η/η_0 given by (1) actually characterize the relative viscosities of aqueous solutions in the pores of silica gel by deriving η/η_0 from the diffusion rates of three substances differing in nature and particle size (Fig. 1). This figure also shows η/η_0 calculated from Musienko's [8] measured wetting rates for silica gel.[†] The values in Fig. 1 all agree well, so the η/η_0 actually do characterize the viscosity in the capillaries.

The relation of η/η_0 to r_c was tested by NMR for the water (not a diffusion method). The line widths in NMR spectra increase considerably as the mobility of the water molecules is reduced, so the linewidth ΔH provides a direct indication of the mobility. Figure 2 shows results obtained in collaboration with V. V. Strelko on a high-resolution spectrometer; ΔH shows a clear inverse relation to r_c, and the curve is similar to that for η/η_0 as a function of r_c (Fig. 1). The NMR results thus confirm the latter relation.

The effects of other factors were also examined. Zakhavaeva and Lopatina [10] have shown that electrolytes, which disrupt the structure of water near hydrophilic surfaces, reduce the viscosity in the boundary layer. It follows from (1) that D^* should be raised by addition of electrolytes. The effects of KI on the diffusion rate for iodine were examined for two specimens of silica gel; D^* for iodine increased rapidly as the KI concentration increased,

_____

† Washburn's equation [9] was used in calculating the viscosity from the wetting rate.

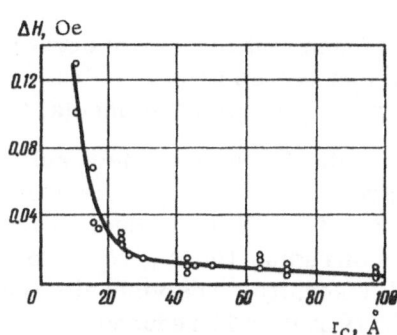

Fig. 2. NMR signal width for water in silica gel as a function of r_c.

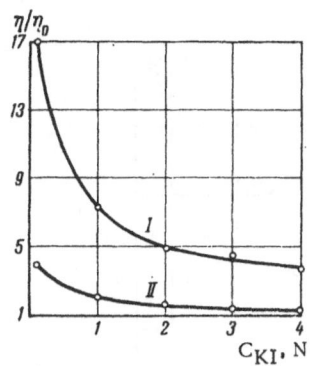

Fig. 3. Relation of η/η_0 to KI concentration for silica gels with mean pore radii (Å) of: I) 11; II) 24.

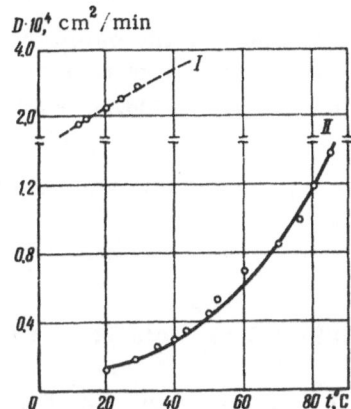

Fig. 4. Coefficients of: I) free; II) bound diffusion for sucrose as functions of temperature (silica gel, $r_c = 24$ Å).

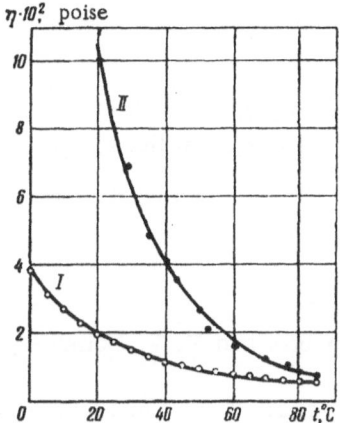

Fig. 5. Viscosity of sucrose solution as a function of temperature: I) in bulk; II) in capillaries ($r_c = 24$ Å).

while η/η_0 (Fig. 3) fell regularly. This confirms the conclusion [10] that electrolytes disrupt the water structure produced in the capillaries of hydrophilic substances.

Disruption in the boundary layer should also increase with temperature, so D* should increase more rapidly with temperature than does the coefficient of free diffusion. Measurements were made (Fig. 4) of D* as a function of temperature for sucrose in silica gel of mean pore radius 24 Å; the expected effect is observed.

These D* were used to determine η as a function of temperature (Fig. 5); as would be expected, the capillary viscosity falls much more rapidly than the bulk viscosity. The respective heats of activation were found to be 8.4 and 3.8 kcal/mole.

The straight lines for log $\eta = f(1/T)$ intersect (Fig. 6) only near 100°C, which shows that the structured state in the capillaries persists throughout the temperature range in which water is liquid at normal pressure.

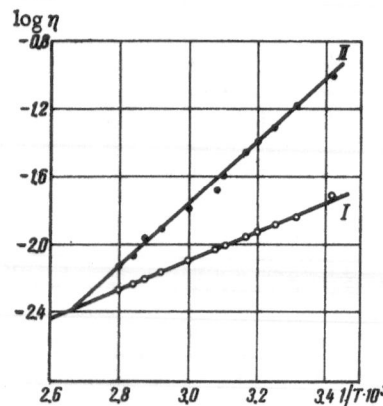

Fig. 6. Relation of log η for
sucrose solution to temperature:
I) in bulk; II) in capillaries (r =
24 Å).

Conclusions

1. It has been found that measurements of the effective diffusion coefficient can be used to calculate the viscosities of aqueous solutions in porous media.

2. Diffusion results for iodine, methanol, and sucrose in silica gel have been used to calculate the relative viscosity as a function of capillary radius. It is found that even capillaries of radius 100 Å give viscosities differing from the bulk viscosity. The viscosity increases rapidly as the capillary radius is reduced.

3. NMR measurements confirm that the liquid in a hydrophilic porous medium has a definite structure.

4. Electrolytes are found to disrupt the boundary layer structure of water; the effective diffusion coefficient increases rapidly with the electrolyte concentration.

5. The viscosity of water in the capillaries of silica gel has been measured as a function of temperature; the heat of activation for viscous flow is E = 8.4 kcal/mole. The structure arising near hydrophilic surfaces is found to exist in the range 0-100°C.

I am indebted to Professor D. N. Strazhesko for general guidance and valuable advice in this work.

References

1. B. V. Deryagin and E. V. Obukhov, Zh. Fiz. Khim., 7:297 (1936); Acta Phys. Chim. URSS, 5:1 (1936).
2. B. V. Deryagin, V. V. Karasev, N. N. Zakhavaeva, and V. P. Lazarev, Zh. Tekh. Fiz., 27(5):1076 (1957).
3. V. V. Karasev and B. V. Deryagin, Zh. Fiz. Khim., 33(1):100 (1959).
4. Z. M. Tovbina, Ukr. Khim. Zh. (in press).
5. D. N. Strazhesko and Z. M. Tovbina, Ukr. Khim. Zh. (in press).
6. B. V. Deryagin, G. P. Sidorenkov, E. A. Zubashchenko, and E. V. Kiseleva, Kolloidn. Zh., 9(5):335 (1947).
7. M. M. Milekhina, Kolloidn. Zh., 23:173 (1961).
8. V. P. Musienko, Dissertation, Kiev (1963).
9. Z. W. Wolkowa, Kolloid.-Z., 69:280 (1934).
10. N. N. Zakhavaeva and A. I. Lopatina, Kolloidn. Zh., 24:455 (1962).

THERMAL EXPANSION OF H_2O AND D_2O IN SMALL PORES

V. V. Karasev, B. V. Deryagin, and E. N. Khromova

*Surface Phenomena Division, Institute of Physical Chemistry,
Academy of Sciences of the USSR*

Water in bodies with small pores is found not to have a density maximum at $+4°C$. Heavy water under the same conditions gives a less pronounced maximum, which is shifted towards lower temperatures. Water released from the pores is found to have the properties of ordinary bulk water.

There is much evidence for distinctive structure-sensitive properties in water in layers attached to hydrophilic surfaces. One such property is the thermal expansion, with a minimum in the volume at $+4°C$. It has been shown [1, 2] that the expansion of water on a silicate surface is anomalous and does not have a minimum volume at $+4°C$. This result is confirmed here and is shown not to have a trivial explanation. Similar results are also reported for heavy water.

Three principal methods have previously been used in research on the expansion of water in pores: dilatometric, thermodynamic, and gravimetric.

In the gravimetric method [3], the measurement was of the apparent weight of carbon powder previously impregnated with water and immersed in water. Deviations from Archimedes's law were found in the range 0 to $+4°C$, which was considered to be due to anomalous change in the density of the water in the pores. It was concluded that this water did not have a density maximum between 0 and $5°C$. It is not stated in the original paper at what pressure the powder was degassed, and it would appear that the powder was not heated during outgassing, so there is no certainty that the pores were highly evacuated and hence completely filled with water. Moreover, it is not known how well this carbon specimen was wetted by water, although it is clear that poor wetting would lead to incomplete filling, on account of the convex menisci in the capillaries. An effect analogous to anomalous expansion could be observed if the filling were incomplete, on account of the expansion of the air.

The thermodynamic method [4] was applied via the expansion of water adsorbed in the pores of zeolites; thermodynamic relationships were used to deduce the coefficient of expansion. It is clear from the results that there is only a small difference from the expansion of bulk

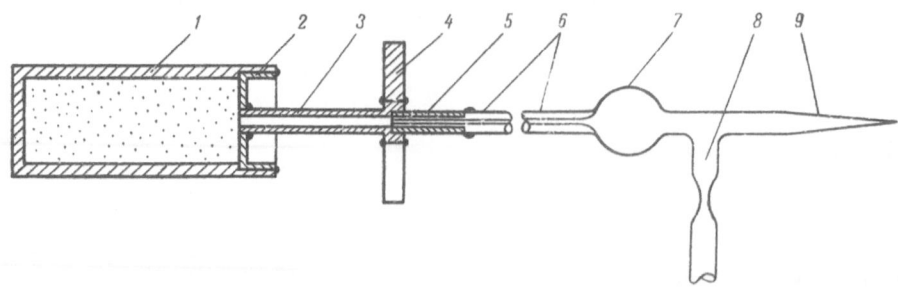

Fig. 1. Apparatus for the thermal expansion of water in small pores.

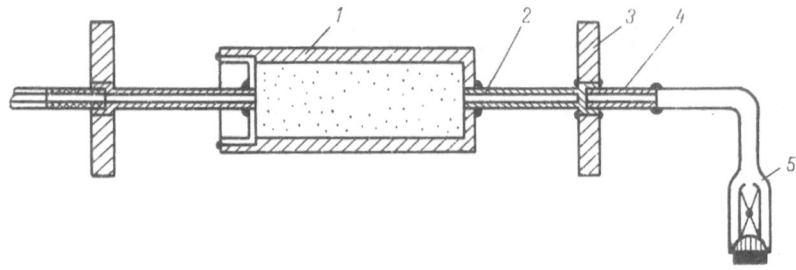

Fig. 2. Apparatus for verifying the degree of evacuation.

ordinary water. The method was usable only between 30 and 80°C, and the most interesting range (0 to +4°C) could not be covered.

The dilatometric method allows direct measurement of the volume as a function of temperature. Fedyakin [1] applied this method to water in glass capillaries of diameters down to 400 Å and found that there was no minimum at +4°C. We have also used the dilatometer method; preliminary results have already been reported [2].

We used U-333 white silica gel, which is very insoluble in water [5] and has very narrow pores both within the particles and between them (when compacted), on account of the small particle size. Taking the density of this filter as 2.1 g/cm^3 (that of fused quartz), the mean pore diameter was found by Deryagin's method [6] as ~100 Å. The powder must be compressed in order to reduce the size of the pores between the particles, for which purpose we tested copper and steel cylinders. However, it was at once found that the thermal expansion of the cylinder completely masked the effect. It was subsequently found that the expansion of water in the pores between 0 and 4°C was comparatively small, the water column sinking on account of expansion of the cylinder. For this reason, an invar cylinder was made, whose coefficient of expansion is only 10^{-6}. Calculation showed that volume increase in the invar would lower the column by less than 1 cm, while anomalous expansion of the water would exceed this.

The silica gel was repeatedly washed with water and was dried at about 150°, as for use in chromatography. It was then pressed into the invar cylinder 1 (Fig. 1) at 1000 atm, the amount being determined by weighing the vessel before and after the pressing. The cylinder was then closed with the thin stainless-steel lid 2, which bore a silver-soldered tube 3 bearing a copper cooling fin 4. The lid was attached to the cylinder with tin solder, and the covar tube 5 was soldered inside the fin, this being sealed to the glass capillary 6 (diameter 0.3 mm). It was not possible to silver-solder the covar tube to the other tube, since this caused cracking in the covar-glass joint. The cooling fin kept the tin-soldered joint cool, and so this joint did not loosen even when the cylinder was heated to 400°C. At the end of the glass capillary lies a sphere 7, which bears the tapering tube 9 and the tube 8 with a construction for sealing.

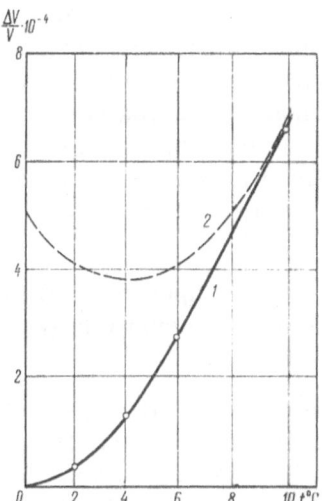

Fig. 3. Thermal expansion of ordinary water: 1) in thin pores; 2) in bulk (from tables).

Fig. 4. Apparatus for testing the anomalous activity of water in small pores.

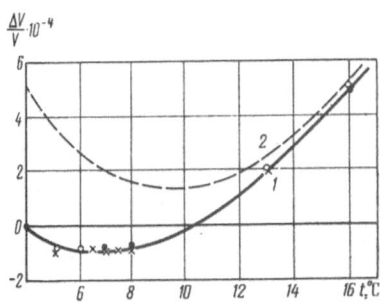

Fig. 5. Thermal expansion of heavy water: 1) in small pores; 2) in bulk (from tables).

The degree of evacuation was verified with a special apparatus (Fig. 2), in which the bottom of vessel 1 bore an additional tube 2 with cooling fin 3 also attached to a cover tube 4 leading to an LT-2 thermocouple gauge 5. The cylinder was heated to 400°C. It was found on connecting the system to a vacuum train that the LT-2 gauge registered a pressure of 10^{-4} mm Hg, so the silica gel had been relatively outgassed even at the bottom of the cylinder.

The water was outgassed in the glass bulb by repeated freezing and thawing under vacuum. At the end of this process, the bulb was sealed off. At the end of pumping, the dilatometer was sealed off, the device was tilted, and the tip was broken. Atmospheric pressure forced the water into the dilatometer. When the filling was complete, the level of the water was set in the capillary so that it was at the lowest point on the scale at 0°C. The amount of water in the pores was determined by weighing the dilatometer before and after filling. The expansion was followed from the height of the water column.

Figure 3 shows that the expansion in the pores is monotonic and shows no minimum volume (maximum density) at 4°C, which confirms the preliminary results [2]. There are two possible sources of error in these experiments which must now be considered.

1. Air Bubbles in Pores

If the pores contain air as well as liquid, the expansion of the air could mask the anomalous expansion at 0 to +4°C. A simple calculation shows that 0.2 cm^3 of air would be sufficient, which would correspond to a residual pressure after evacuation of about 7 mm Hg. Consequently, evacuation to 10^{-4} mm Hg provided a pressure less by a factor 10^5 than the one required to explain the effect. A direct test for the absence of air in the pores after filling with water was by applying a pressure of 300 mm Hg to the water column, which would compress any air bubbles. Calculation shows that an air volume of 0.2 cm^3 (sufficient to balance the volume reduction in the water) would cause a displacement of 30 cm in the column in response to an excess pressure of 0.1 atm. However, no comparable depression was found even with a pressure of over 300 mm Hg.

2. Effects of Soluble Impurities in the Silica Gel on the Expansion

A test for this was to measure the expansion of the water after recovery from the pores. A small dilatometer (Fig. 4) was made, which consisted of a pyrex glass bulb 1, tube 2, and capillary 3 (diameter 0.1 mm). Part of the metal tube, together with the capillary, was cut off from the dilatometer cylinder. On heating the cylinder, the water from the pores passed into the rest of the tube by expansion, from which it was drawn into the bulb and capillary of the small dilatometer, tube 2 being closed by the rubber tube 4 and bung 5. The meniscus in the capillary sank on heating the bulb from 0 to 4°C but rose again at higher temperatures, which showed that the water displaced from the pores had its maximum density at +4°C, i.e., had the normal properties of bulk water. It is therefore impossible to explain the altered expansion in terms of partial solution of the silica gel. It proved impossible to do analogous experiments with carbon, because the pores could not be filled with water (carbon is not wetted by water).

It is reasonable to suppose that the density of water in pores at high temperatures differs very little from that of bulk water [4], so the present results indicate that the density in pores at 0 to +4°C is greater than for the bulk phase.

Other liquids with anomalies are similarly of interest, e.g., heavy water. Here all the conditions were as for ordinary water. Figure 5 shows the expansion of heavy water: 1) in pores; 2) in bulk. Here curve 2 has been displaced vertically to coincide with curve 1 at 20°C. It is clear that there is a minimum for heavy water in pores, but that the depth is substantially less than that for the bulk phase, and the position lies at lower temperatures. One difficulty here was that it was not possible to remove the ordinary water completely from the pores, so isotope exchange occurred. Complete removal of water from silica gel requires heating to over 1000°C, which is impracticable, since invar alters the properties at such temperatures.

References

1. N. N. Fedyakin, Dokl. AN SSSR, 138:1389 (1961).
2. V. V. Karasev, B. V. Deryagin, and E. N. Efremova, Kolloidn. Zh., 24(4):471 (1962).
3. F. Goldmann and M. Polanyi, Z. Phys. Chem., 132:321 (1928).
4. R. M. Barrer and B. E. Fender, J. Phys. Chem. Solids, 21:1 (1961).
5. W. A. Bloor and H. W. Webb, Trans. Brit. Ceram. Soc., 38:66 (1939).
6. B. V. Deryagin, Dokl. AN SSSR, 53(7):627 (1946).

THICKNESS OF THE EQUILIBRIUM FILM IN A CAPILLARY PARTLY FILLED WITH LIQUID

V. D. Sobolev and Z. M. Zorin

Two different modifications of the adsorption film are detected from the pressure dependence of the condensation rate for water vapor in glass and quartz capillaries.

The adsorption film inside a capillary is important in relation to capillary condensation in porous sorbents and to anomalies in the behavior of liquids in capillaries.

Methods

A capillary sealed at one end (Fig. 1) is placed in a lucite thermostat set up on a microscope stage. The open end is brought into contact with water, which enters under capillary forces, usually 3-4 cm into a capillary 9-10 cm long. Then the open end is closed with freshly distilled mercury or epoxide resin. The length of the liquid column as a function of time is then observed at ×100 with illumination from the side. If the liquid lost from the column is uniformly distributed over the surface, the equivalent film thickness is

$$h = \frac{rl'}{2l} \, ,$$

in which r is the radius of the capillary, l' is the reduction in the column length, and l is the free part of the capillary.

Especial attention was given to two points. Firstly, we wished to establish whether an equilibrium film in contact with the liquid in a cylindrical capillary can produce a partially filled capillary. The system was therefore left for several days at a carefully controlled temperature, the TS-24 thermostat being modified for the purpose by a switching system (Fig. 2). A timer pulse passes to the program unit, which turns on the motor and thermostat relay. After 50 sec (the time needed for the vacuum-tube relay to warm up), the heater is switched on (if the temperature is below the set value). When the maximum temperature has been reached, the heater is switched off and the cooling water is turned on, and the thermostat works in the usual way for 3.5 min. Then the thermostat and cooling are turned off for 7.5 min. This system maintained a temperature close to room temperature to 0.05°C for several days.

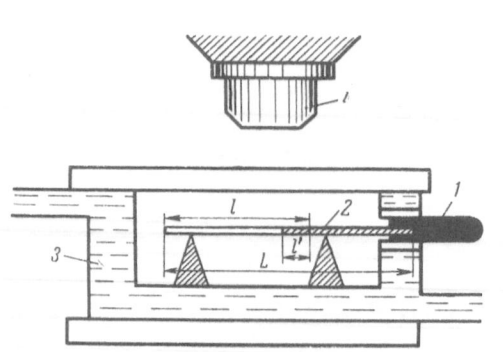

Fig. 1. Apparatus for observing the evaporation of a liquid column in a capillary: 1) mercury seal; 2) liquid column; 3) thermostat; 4) microscope.

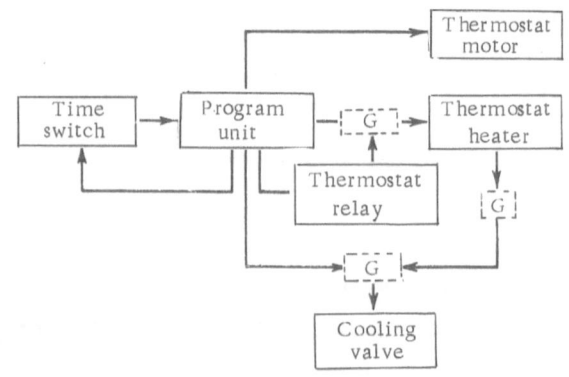

Fig. 2. Block diagram of the thermostat control (G = gate).

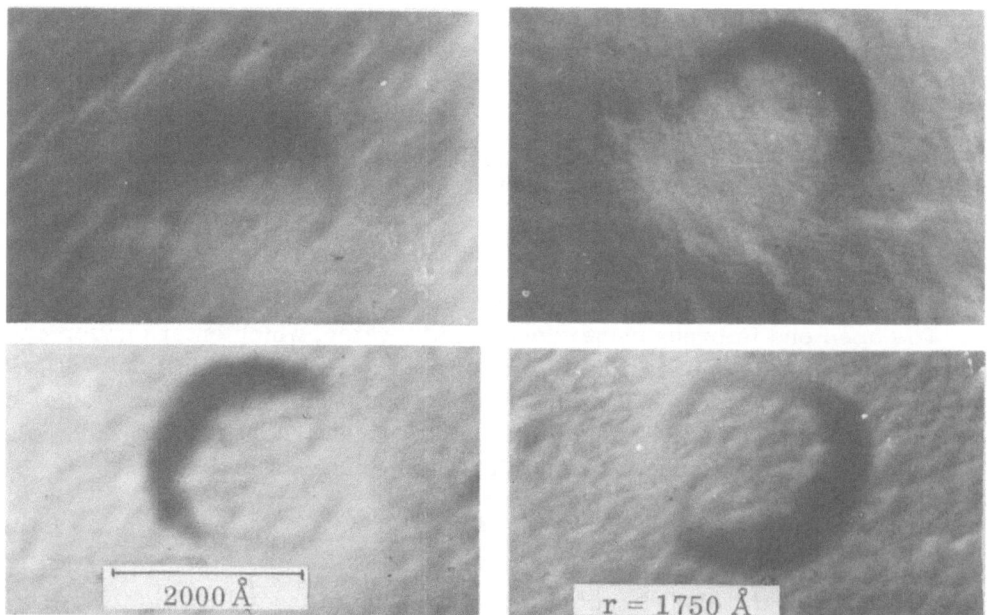

Fig. 3. Electron micrograph of end of capillary.

The formula for the film thickness contains r, which was measured by Fedyakin's method [1], which has been criticized, but (in our view) without justification. We compared the radius as measured thermometrically with that measured by electron micrography.

A difficulty arose in electron micrography that the replica contained not only the capillary but also other inhomogeneities that could be taken as the capillary. Errors were eliminated by making the replica as follows. The cut end of the capillary was pressed onto a tin foil that had been carefully polished mechanically or electrolytically. The capillary was broken 10–15 times and 7 or 8 indents were made on the foil, from which a carbon replica was prepared. This replica was shadowed by a molecular beam of chromium at 10° (30–40 Å film). Photographs showing indents identical in size and shape were selected. There were five such (Fig. 3) in our experiment, which gave a reliable determination of r as 1750 ± 200 Å.

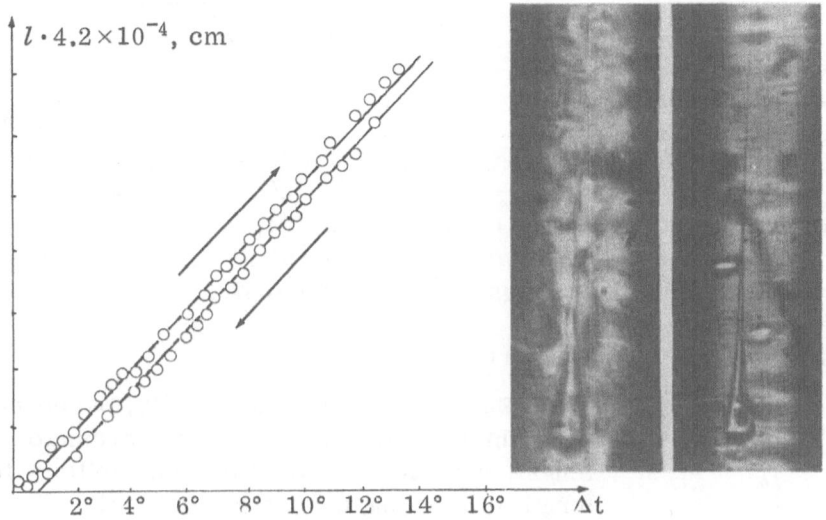

Fig. 4. Thermal expansion of water in capillary.

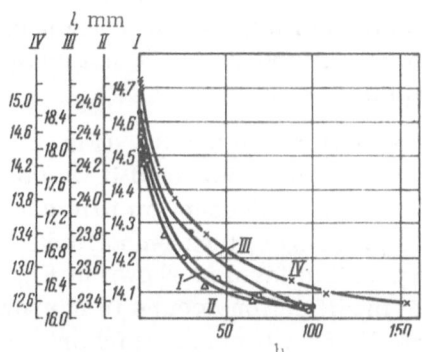

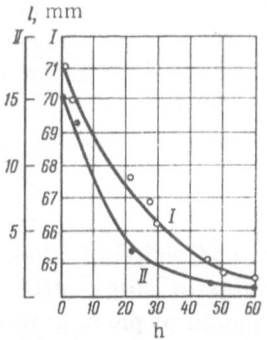

Fig. 5. Evaporation of water columns in glass capillaries: I) r = 20,000 Å, L = 101 mm, l = 86.9 mm, l' = 0.5 mm, h = (58 + 15) Å; II) r = 13,500 Å, L = 119 mm, l = 96 mm, l' = 0.9 mm, h = (63 ± 15) Å; III) r = 5500 Å, L = 125.9 mm, l = 109.8 mm, l' = 2.3 mm, h = (58 ± 15) Å; IV) r = 4400 Å, L = 113.6 mm, l = 101.0 mm, l' = 2.6 mm, h = (57 ± 15) Å.

Fig. 6. Evaporation of water columns in glass capillaries: I) r = 3650 Å, L = 95.7 mm, l = 30.7 mm, l' = 6.2 mm, h = (370 ± 40) Å; II) r = 2750 Å, L = 68.9 mm, l = 67.9 mm, l' = 14.2 mm, h = (285 ± 30) Å.

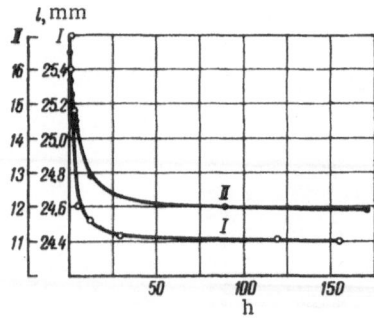

Fig. 7. Evaporation of water columns in quartz capillaries: I) r = 13,000 Å, L = 90.8 mm, l = 66.4 mm, l' = 1.16 mm, h = (113 ± 20) Å; II) r = 3900 Å, L = 98.1 mm, l = 86.2 mm, l' = 4.5 mm, h = (113 ± 15) Å.

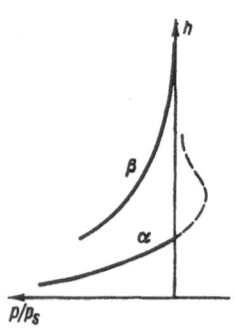

Fig. 8. Possible form for the adsorption isotherms of water on glass.

Figure 4 shows the thermal expansion of water in the expanded end of the capillary (shown in two mutually perpendicular projections by the side). The radius is

$$r = \frac{1}{1.51} \sqrt{\frac{\beta v}{\pi \, \text{tg} \, \alpha}}$$

in which β = 2.1 × 10^{-4} deg^{-1} (coefficient of bulk expansion of water), v = 2.55 × 10^{-8} cm^3 (volume of wide end), tan α = $\Delta l / \Delta t$ = 2.06 × 10^{-3} cm/deg was deduced from the curve, and the 1.51 arises because the capillary forms a cylindrical lens. This gave r = 1920 ± 200 Å. The results from the two methods thus agree satisfactorily.

Results

Figures 5-7 show the results, the abscissa being the time (h) from the start, while the ordinate is the column length. Values are given for each capillary for r, the total length L, the length l containing the film, h, and l'.

Discussion

The curves show that h is dependent on r; for glass capillaries with r > 0.4 μ (Fig. 5), h is 50-60 Å, which corresponds to the thickness of the water film formed on a flat glass surface near saturation [2], so it is reasonable to suppose that this film is a boundary phase not completely wetted by the bulk liquid.

Glass capillaries with r of 0.2-0.3 μ (Fig. 6) gave h of 300-400 Å. This is difficult to understand if we do not assume [3] that different types of adsorption film can be produced at high relative pressures, these modifications differing in structure and shape of adsorption isotherm (a possible form is shown in Fig. 8). The α modification occurs in wide capillaries whereas the β modification occurs in narrow ones, the latter being capable of passing into the bulk liquid as the film becomes thicker. Deryagin [4] has considered rigorously the equilibrium between the meniscus of a liquid partly filling a pore and a polymolecular adsorption layer; he has also shown [5, 6] that the rate of equilibrium is governed by diffusion of vapor through the gas and by the flow of liquid in the film. If the negative pressure in the film increases as the film becomes thinner, the film should flow towards the thinner part, i.e., the two processes should act in the same direction. The isotherm for the negative pressure must be known in order to calculate the equilibration time. The problem is made more difficult by the fact that the vapor pressure at the sealed end of the capillary increases as time passes. The equilibrium thicknesses of 60 and 400 Å are attained in roughly equal times (60 h), although the capillary is thinner in the second case and so diffusion is slower. This feature indicates that motion in the film plays a substantial part in a thick wetting film.

Figure 7 shows the results for quartz capillaries; here h was 100-130 Å for capillaries of radius 1.3 and 0.4 μ, the equilibration time being about 25 h, which is much less than the time that might be expected on the assumption that the main transport is by diffusion. It is less than the time for glass capillaries. The flow of a liquid in a film is proportional to grad Π, in which Π is pressure, so the adsorption isotherm for water on quartz is presumably steeper than that for water on glass.

Conclusions

1. Capillary radii have been measured with the electron microscope and have been found to agree well with values derived by Fedyakin's thermometric method.

2. An equilibrium polymolecular film can be formed in a closed glass or quartz capillary partly filled with water.

3. The film thickness in a glass capillary is substantially dependent on the radius. It is considered that different modifications of the adsorption layer correspond to different film thicknesses.

4. Transport in the film is considered to play a substantial part in the equilibration.

We are indebted to N. N. Fedyakin for assistance in this work.

References

1.   N. N. Fedyakin, Zh. Fiz. Khim., 36:1450 (1962).
2.   B. V. Deryagin and Z. M. Zorin, Zh. Fiz. Khim., 29(10):1755 (1955); B. V. Derjaguin (Deryagin) and Z. M. Zorin, in: Proceedings of the Second International Congress on Surface Activity, Vol. 2, Butterworths, London (1957), p. 145.
3.   Z. M. Zorin, in: Research in Surface Forces, Vol. 2, Consultants Bureau, New York (1966), p. 134.
4.   B. V. Deryagin, Dokl. AN SSSR, 113:842 (1957).
5.   B. V. Deryagin, Kolloidn. Zh., 17:207 (1955).
6.   B. V. Deryagin, S. V. Nerpin, and N. V. Churaev, Kolloidn. Zh., 26(3):301 (1964).

NEW DATA ON THE THERMAL CONDUCTIVITY
OF THIN FILMS OF WATER

M. S. Metsik and G. T. Timoshchenko

Zhdanov University, Irkutsk

An improved method is used to show that structurally ordered films of water between sheets of mica have an anomalously high thermal conductivity.

Preliminary results have been given [1] on the thermal conductivity of thin films of water on the surfaces of mica crystals. These films are structured by the surface and have anomalously high thermal conduction. Here we report new experimental data on this.

The measurements were made on stacks of 1000 mica sheets of equal thickness (20-40 μ), the gaps between the sheets being filled with water. The thickness of the water layers was adjusted via the pressure on the stack. At the center of the stack was a copper-constantan thermocouple made of wires 0.12 mm thick. The rate of heating or cooling of the stack was measured over a range of 3-5° in a thermostat whose temperature was kept constant to 0.01-0.02°. The thermal conductivity of the film is [2-4] given by

$$\lambda'' = \frac{cpZ_{av}}{\pi^2\left(\frac{1}{X^2}+\frac{1}{Y^2}\right)}\frac{m_2-m_1}{Z_2-Z_1} - \frac{\lambda_m^{\perp}Z_{av}}{\left(\frac{1}{X^2}+\frac{1}{Y^2}\right)}\frac{\left(\frac{1}{Z_2^2}-\frac{1}{Z_1^2}\right)}{(Z_2-Z_1)} + \lambda_m^{\|},$$

in which $\lambda_m^{\|}$ and $\lambda_m^{\perp}$ are the components of the thermal conductivity of mica respectively along the plates and normal to them; X, Y, and Z are the dimensions of the plates along the x, y, and z axes; m_1 and m_2 are the rates of cooling under regular conditions with $m \to \infty$ for the corresponding Z_1 and Z_2 set by the compression of the stack; and c and ρ are the specific heat and density of the stack with allowance for the absorbed water ($c_m = 0.21$ cal/g, $\rho = 2.78$ g/cm^3, $\lambda_m^{\|} = 1.4 \times 10^{-3}$ and $\lambda_m^{\perp} = 9 \times 10^{-3}$ cal/cm · sec · deg.

First we repeated experiments with stacks made of ordinary mica plates obtained from a mica factory, which had not been treated in any way. Curve 1 shows the variation in λ with film thickness h for this case. Anomalous conduction clearly sets in as h is reduced, and λ'' rises to 9×10^{-2} cal/cm · sec · deg when h = 0.15 μ. Curve 2 shows λ after the crystals had been washed in boiling water, which leaches the surface and weakens the surface field, λ being

34

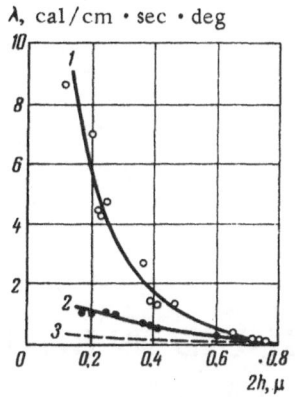

λ, cal/cm · sec · deg

Fig. 1. Relation of film conductivity λ between mica crystals to film thickness 2h for films: 1) of water without treatment of the crystals after cleavage; 2) of water with crystals washed in boiling water; 3) of glycerol.

greatly reduced. Curve 3 is the result when the water is replaced by glycerol, which crystallizes only with difficulty; no anomaly is observed.

These results give rise to the following conclusions:

1. Water films of thickness ∼0.1 μ on freshly cleaved mica crystals have an anomalously high thermal conductivity, which may be considered as a consequence of the structural perfection produced by the molecular field of the surface.

2. The thermal conductivity of the films is reduced by washing the crystals in boiling water.

3. The anomalous conduction is virtually absent for glycerol, which crystallizes only with difficulty.

We are indebted to B. V. Deryagin for a discussion of the results.

References

1.   M. S. Metsik and O. S. Aidanova, in: Research in Surface Forces, Vol. 2, Consultants Bureau, New York (1966), p. 169.
2.   M. S. Metsik, Izv. Vyssh. Ucheb. Zaved., Fizika, No. 4, 29 (1958).
3.   M. S. Metsik, Izv. Vyssh. Ucheb. Zaved., Fizika, No. 5, 131 (1960).
4.   A. D. Dmitrovich, Determination of the Thermophysical Properties of Building Materials, Gosstroiizdat, Moscow (1963), p. 58.

SHEAR ELASTICITY OF A LIQUID IN BULK
AND IN BOUNDARY LAYERS

U. B. Bazaron, B. V. Deryagin, and A. V. Bulgadaev

Buryat Joint Research Institute, Siberian Division, Academy of Sciences of the USSR,
and Surface Phenomena Division, Institute of Physical Chemistry,
Academy of Sciences of the USSR

Resonance measurements at 7.5×10^5 Hz reveal a measurable bulk shear elasticity for various liquids (water, acetone, benzene, alcohols, acetic acid, CCl_4, oil). The shear modulus of a nonpolar liquid is independent of the layer thickness. A polar liquid shows a sharp increase in the shear elasticity at distances less than 10^{-5} cm from the surface of the quartz.

It has been found that thin boundary layers of a liquid acquire properties different from those of the bulk liquid under action of the surfaces of solids. For instance, viscosity measurements [1, 2] reveal a special boundary viscosity for polar and also for nonpolar liquids containing a little polar material, there being a sharp boundary between the bulk liquid and the layer with elevated viscosity. Polymolecular adsorption [3] also reveals special properties of thin layers of polar liquids. These results demonstrate the existence of special boundary phases, and hence of a phase boundary between the bulk liquid and the boundary layer.

Other measurements have recently been made on boundary layers. For example, it has been shown [4, 5] that water in pores and capillaries of diameter < 0.1 μ does not have a maximum density at $+4°C$, and also [6] that a film of water on mica has a thermal conductivity increased by more than an order of magnitude. Altered thermodynamic parameters have also been demonstrated [7, 8] for boundary layers of liquids.

The above evidence indicates that boundary layers may have form elasticity, i.e., a measurable shear modulus. This has been shown to be so [9] for a layer of water between glass surfaces, by reference to the marked increase in the vibrational frequency of a convex lens supported on a thin strip as the bottom of a vessel filled with water was approached. The damping was slight at small amplitudes, so the damping correction could be neglected, and the effective shear modulus (deduced from the period of oscillation) could be equated to the true value. This effective modulus was found to decrease rapidly as the layer thickness increased and could not be observed for layers thicker than 0.15 μ.

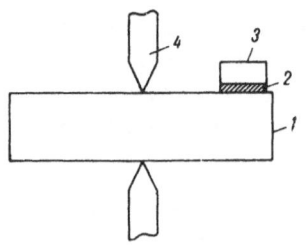

Fig. 1. Quartz crystal
with additional coupling:
1) crystal; 2) liquid film;
3) plate; 4) holder.

The special properties of layers between solid surfaces are of interest because they have a bearing on various aspects of surface and colloid phenomena, in particular the mechanical properties of colloids and the mechanism of boundary friction.

We have examined the shear elasticity of boundary layers via a more convenient method that provides measurement of the shear modulus at very small amplitudes, where stress relaxation is probably less important.

Mason [10] has reported measurement of the bulk shear modulus for liquids of high viscosity via the reflection of transverse ultrasonic waves from a boundary with fused silica. The amplitude and phase of the reflected waves were measured with and without liquid at the surface. Of the few recent studies, we may mention one [11] concerned with application of Mason's method to very viscous liquids at various frequencies and over a wide temperature range.

The method is similar to one proposed by Mandel'shtam and Khaikin for examining the interaction between two bodies in contact. A rectangular quartz crystal rod is set in vibration along its length at its fundamental resonant frequency, the horizontal face then performing tangential movements. This face is in contact with a second solid. In particular, in this way it has been shown [12] that the frictional forces for small amplitudes of vibration (A = 10^{-7} cm) are of elastic type.

In our case, the second body was a prism (plate) of fused silica, and there was a film of liquid between the two bodies (Fig. 1). The crystal was of X-5° section, 36 mm long, 12 mm wide, and 5 mm thick. The working horizontal face was perpendicular to the optic axis. The plate was 8 × 5 × 4.5 mm and was virtually at rest during the vibrations, since the coupling provided by the liquid film was known to be weak and could not transmit the very large accelerations of the crystal, whose natural frequency was 74.4 kHz.

We measured the shift in the resonant frequency of the crystal as a function of the film thickness. Any increase in resonant frequency would indicate a shear elasticity in the liquid film, since the resonant frequency could only be reduced by dissipative forces, e.g., internal friction. If the effective shear modulus is deduced from the frequency shift without allowance for dissipation and relaxation, the result can only be an underestimate of the true value. The X-5° section has a zero value for Poisson's ratio, but interaction with the low-frequency bending modes occurs with the above dimensions [13]. There is therefore a small normal component in the motion which produces a negative frequency shift, which is virtually independent of the film thickness and of the nature of the liquid, and so does not need to be considered.

The crystal was fixed in a holder made of two steel needles located on the nodal line at the middle of the crystal. The electrodes, which had an air gap of 0.5 mm, were employed because the crystal had to be removed from the holder for cleaning after each measurement. The crystal was excited by a crystal-stabilized 75 kHz source modulated by a low-frequency oscillator whose frequency was known to ±1 Hz. Balanced modulation was used to suppress the carrier frequency, while one sideband was selected by a tuned circuit. The output amplitude was adjustable up to about 50 V, and the voltage was applied to the first pair of electrodes, which were placed on the two side faces at one end of the crystal. The forced motion produced an emf recorded by a second pair of electrodes similarly placed at the other end of the crystal, which was used in monitoring the vibration and in measuring the resonant frequency of ±1 Hz. The modulation frequency was adjusted to bring one of the sidebands to the resonant frequency, which caused the amplitude of the vibration to increase. An oscilloscope recorded the voltage produced by the second pair of electrodes (< 1 V) and also provided rapid and accurate measurement of the resonant frequency, which was essential with volatile liquids of low viscosity, since the film thickness might alter during the measurement. Meter instruments are very inconvenient on account of their slow response, which means that the resonance might not be noticed, especially at small oscillation amplitudes.

We can readily deduce a general equation relating the frequency shift Δf for a crystal of mass M due to interaction with the plate of mass m via the liquid film; but a simpler approximate formula may be used, since the plate is virtually at rest, on the basis of the conditions

$$\frac{m}{M} \ll 1, \tag{1}$$

$$\frac{H}{\lambda} \ll 1, \tag{2}$$

$$\frac{f_1}{f_0} \ll 1, \tag{3}$$

in which λ is the wavelength of the shear wave produced in the film of thickness H by the motion of the crystal, whose natural frequency is f_0, while f_1 is the natural frequency of the plate due to the elastic coupling with the crystal via the liquid film.

These conditions were obeyed in our experiments, since M = 7.04 g and m = 0.4 g. Condition (2) was found to be obeyed in every case, because λ was about 100 μ, while H was usually only a few μ. Condition (3) was also found to be obeyed in all the experiments.

Condition (2) allows us to neglect the inertial forces on the film elements and to assume that the shear stress at any point in the film is a function of time alone and does not depend on the distance to the quartz crystal. In other words, we can employ a quasistatic approach to the film deformation in response to the shear stress, although the latter is actually a function of time. The plate can then be considered as elastically coupled to the crystal, and the rigidity K of the coupling may be expressed as follows, on the assumption that the film is uniform:

$$K = \frac{SG}{H}, \tag{4}$$

in which G is shear modulus and S is the area of the base of the plate.

We may use Rayleigh's method to calculate the Δf produced by the coupling to the plate, which is based on determination of the natural frequency by equating the maximal values of U_P and U_K, the potential and kinetic energies of the system. We equate the changes in U_P and U_K on applying the plate, on the assumption that the amplitude remains unchanged, and get

$$\Delta U_K = \pi^2 M A^2 \Delta (f_0)^2 = \Delta U_P = \frac{k}{2} A^2. \tag{5}$$

Here we have assumed that the effect of the coupled mass on U_K can be neglected, as the plate is virtually immobile and H is small. Then

$$K = 4\pi^2 M f_0 \Delta f, \tag{6}$$

and from (4) we get that

$$G = \frac{4\pi^2 M H f_0 \Delta f}{S}. \tag{7}$$

Formula (7) still applies if conditions (1)-(3) are obeyed even though G becomes complex on account of dissipative force, if Δf is then understood as a complex frequency shift. However, the concept of G for a liquid is novel, so we judge plausible to restrict consideration here to measurement of the effective shear modulus (equal to the real part of the complex shear modulus) which is sufficiently interesting. Further, the loss angle would be of real interest only if it could be measured over a wide range, which presents technical difficulties. No sound phenomenological treatment can be based on measurements at only one frequency.

If the liquid is equated to a Kelvin body by adding up the elastic and viscous stresses (an ungrounded assumption), the effective shear modulus G_e deduced from (7) via the real part of $\triangle f$ is equal to the shear modulus of a Kelvin body. If we equate the liquid to a Maxwell body with a relaxation time τ (also an ungrounded assumption), the Maxwell shear modulus G is related to G_e by

$$G_e = \frac{G}{1 + \left(\frac{1}{\omega\tau}\right)^2} = \frac{G}{1 + \left(\frac{G}{\omega\eta}\right)^2},$$ (8)

which shows that $G > G_e$. However, there is no basis for assuming that the resulting G is more "true" than G_e, since use of the Maxwell model is usually not justified for frequencies such that $\omega\tau < 1$. Boltzmann's elastic aftereffect equations give the best results for solids, and these are equivalent to a broad spectrum of relaxation frequencies [15].

It is to be expected that the films of polar liquids will be inhomogeneous in properties in the transverse direction, so the result from (7) is some mean $\overline{G(H)}$. To determine the true G as a function of distance from the crystal, we split the film up into elementary layers parallel to the crystal surfaces and at distances z. The shear in a layer dx is

$$dx = \frac{T dz}{G(z) \cdot S},$$ (9)

in which T is the tangential shear stress.

Integration of (9) gives us for K that

$$\frac{X}{T} \equiv \frac{1}{K} = \frac{1}{S} \int_0^H \frac{dz}{G(z)},$$ (10)

in which X is the shear of the entire film.

As the two surfaces are similar in nature, we assume that the film is symmetrical in properties, with G(z) = G(H − z), so (10) becomes

$$\frac{1}{K} = \frac{2}{S} \int_0^h \frac{dz}{G(z)},$$ (11)

in which h = H/2.

Differentiation of (11) and use of (6) gives

$$G(h) = \frac{8\pi^2 M f_0}{S} \cdot \frac{1}{\dfrac{d(\triangle f)^{-1}}{dh}},$$ (12)

for the shear modulus at the median plane of the film.

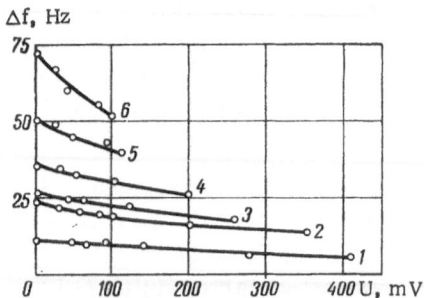

Fig. 2. Δf as a function of voltage U produced by the crystal (proportional to vibration amplitude) for oil films of thickness (μ): 1) 7; 2) 3.25; 3) 2.8; 4) 2.05; 5) 1.4; 6) 1.04.

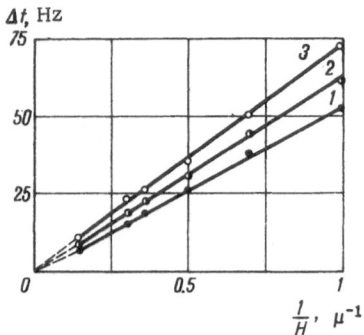

Fig. 3. Δf as a function of 1/H for oil at constant shear angles for amplitudes of U/H (mV/μ) of: 1) 100; 2) 50; 3) 0.

Equation (7) shows that Δf and H appear as a product, so if G is independent of thickness, the dependence of Δf on H should be hyperbolic. Any deviation from this relation indicates the occurrence of a special boundary elasticity. However, (12) shows that it is more convenient to examine $1/\Delta f$ as a function of h when there are anomalies, since this then indicates variation in G also. We shall use both relationships in considering the experimental results.

The liquids were carefully purified by standard techniques. The crystal and plate were treated briefly with chromic acid and then were cleaned in a glow discharge [14]. The crystal was at once placed in the holder and coated with the liquid on a clean glass rod, on which the plate was placed. The film thickness was then determined as follows.

The light from a monochromator fell almost at right angles on the plate, and an interference pattern was produced in reflection. Slight pressure on the plate produced a uniform thickness. The wavelengths were then measured for two successive cases of destructive interference (path difference in the film an odd number of half-wavelengths), which gave the film thickness to 0.01 μ. The resonance frequency was measured almost simultaneously. The film thickness was adjusted by squeezing liquid from the gap, so the measurements were begun at the large thicknesses.

It is necessary to allow for the dissipative forces in the case of a viscous liquid, as these produce negative shifts; moreover, the shear elasticity may be nonlinear. For this reason, we plotted Δf as a function of the amplitude for constant H, the result being extrapolated to zero amplitude.

Figure 2 shows such curves for oil. It is clear that Δf is rather dependent on the amplitude, especially for H small.

Figure 3 shows Δf as a function of 1/H, in which curve 3 was obtained by extrapolation to zero amplitude (curves 1 and 2 are for finite amplitudes). The three straight lines converge to the origin, which indicates that G is independent of H, and also that the highest G relates to zero amplitude. The linearity and convergence show that the loss angle due to hysteresis is independent of H and is the same at all points, i.e., an oil film is homogeneous in dissipative forces as well as in elasticity.

Figure 4 shows Δf (as extrapolated to zero amplitude) as a function of 1/H for water, acetone, and benzene. The first two show large deviations from linearity at small H. All three liquids clearly have finite G, but the mean G increases greatly towards small H for the polar liquids.

When boundary elasticity occurs, it is more convenient to consider $1/\Delta f$ as a function of H/2; Fig. 5 shows such curves for benzene and CCl_4. The lines are straight and converge to

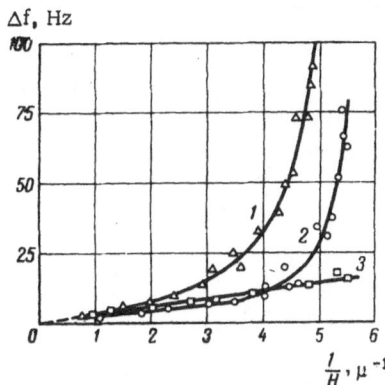

Fig. 4. Δf as a function of $1/H$ for: 1) water; 2) acetone; 3) benzene.

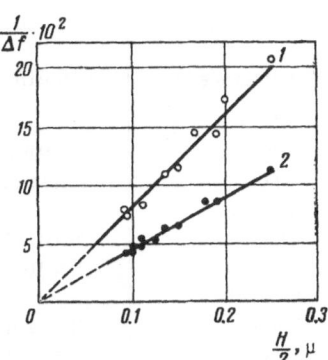

Fig. 5. $1/\Delta f$ as a function of $H/2$ for: 1) benzene; 2) CCl_4.

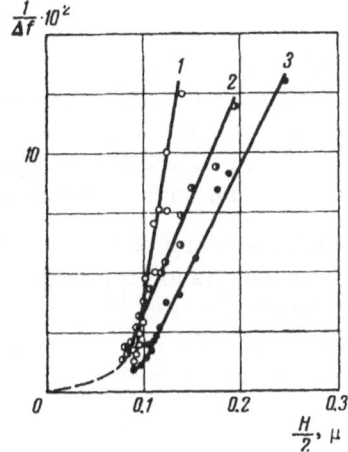

Fig. 6. $1/\Delta f$ as a function of $H/2$ for: 1) butanol; 2) hexanol; 3) acetic acid.

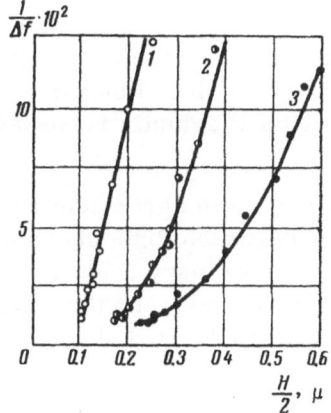

Fig. 7. $1/\Delta f$ as a function of $H/2$ for: 1) triple-distilled water; 2) double-distilled water; 3) distilled water.

the origin, so (12) shows that these two liquids have no special boundary elasticity, at least within the experimental errors.

Figure 6 shows similar results for acetic acid, hexanol, and butanol (all polar). The lines are straight within the range of H used, but they do not converge to the origin, which means that G is constant for $h > 10^{-5}$ cm, and the value can be deduced from (12). The expected behavior for hexanol at smaller h is indicated by the broken line, with $\Delta f \to \infty$ as $H \to 0$, since the line cannot meet the axis for $H > 0$, with Δf going from $+\infty$ to $-\infty$. These liquids therefore have special mechanical properties at thicknesses less than 10^{-5} cm. Similar results were obtained for other liquids.

Line 1 in Fig. 7 gives results for triple-distilled water (in the last instance, without boiling in a quartz vessel). As for other polar liquids, the result is a straight line not passing through the origin. Curves 2 (double-distilled water) and 3 (single-distilled) are not straight

TABLE 1. Shear Elasticities of Liquids

Substance	Purification	t, °C	Bulk G, dyn/cm^2	Boundary layer, μ
Acetone	Distilled	25	0.42×10^4	0.08
Water	Triple distilled	25	1.1×10^4	0.09
Benzene	Absolute	14	1.3×10^4	0
CCl$_4$	Distilled	16	2.3×10^4	0
Acetic acid	Distilled	17	1.9×10^4	0.06
Ethanol	Absolute	16	1.0×10^4	0.06
Butanol	The same	18	0.7×10^4	0.07
Hexanol	The same	26	0.9×10^4	0.08
Octanol	The same	25	1.2×10^4	0.08
Oil	Filtered	14	3.7×10^5	0

lines, and they show that the region of elevated G extends substantially further from the surface. It has previously been found [9] that the thickness of the boundary layer is dependent on the purity for water, but further study is needed in order to elucidate the causes.

We get the minimum possible G for the boundary layer by joining the bottom ends of the straight lines to the origin. The value is about 10 times the bulk value. The curves of Fig. 6 also give the thickness of the layer with special properties. The least possible thickness for the anomalous layer is found by extending the lines to meet the abscissa, while an upper limit is the point of least H actually recorded by experiment. The two values are close together in magnitude, so we may take the mean of the two as the thickness of the anomalous boundary layer.

The results are in agreement with the existence of special boundary phases [1-3]. All the liquids have a measurable bulk shear modulus, including ones of low viscosity, at the frequencies used. On ordinary concepts, G would be expected to be observable at about 10^{10} Hz for liquids of low viscosity, so the results are somewhat unexpected. Table 1 gives the G calculated from (7) and (12), and also the thicknesses of the boundary layers with special properties.

References

1. B. V. Deryagin, V. V. Karasev, and Z. M. Zorin, Transactions of the Conference on the Nature of the Liquid State, Kiev (1953); B. V. Derjaguin (Deryagin) and Z. M. Zorin, Proceedings of the Second International Congress on Surface Activity, Vol. 2, Butterworths, London (1957), p. 145; B. V. Derjaguin (Deryagin) and V. V. Karasev, Ibid., Vol. 3, p. 531.
2. V. V. Karasev and B. V. Deryagin, Zh. Fiz. Khim., 33:1 (1959).
3. B. V. Deryagin and Z. M. Zorin, Dokl. AN SSSR, 98(1):93 (1954).
4. N. N. Fedyakin, Dokl. AN SSSR, 138:1389 (1961).
5. V. V. Karasev, B. V. Deryagin, and E. N. Efremova, Kolloidn. Zh., 24(4):471 (1962).
6. M. S. Metsik and O. S. Aidanova, in: Research in Surface Forces, Vol. 2, Consultants Bureau, New York (1966), p. 169; M. S. Metsik and G. T. Timoshchenko, this volume, p. 34.
7. Yu. M. Popovskii and B. V. Deryagin, Dokl. AN SSSR, 159:897 (1964).
8. B. V. Deryagin, S. V. Nerpin, and M. A. Arutyunyan, Dokl. AN SSSR, 160(2):337 (1965).
9. B. V. Deryagin, Zh. Fiz. Khim., 3:29 (1932); Z. Phys., 84:657 (1933).
10. W. Mason, Piezoelectric Crystals and Their Application to Ultrasonics, D. Van Nostrand, New York (1950).
11. R. Meister, C. Marhoeffer, R. Sciamanda, and T. Litovitz, J. Appl. Phys., 5:854 (1960).

12. S. É. Khaikin, L. P. Lisovskii, and A. E. Salomonovich, Dokl. AN SSSR, 24:134 (1939).
13. W. Cady, Piezoelectricity; an Introduction to the Theory and Applications of Electro-mechanical Phenomena in Crystals, McGraw-Hill, New York (1946).
14. V. V. Karasev and G. I. Izmailova, Zh. Tekh. Fiz., 24:871 (1954).
15. B. V. Derjaguin (Deryagin), Beitr. angew. Geophys., 4:452 (1934).

TEMPERATURE DEPENDENCE OF THE SHEAR
ELASTICITY OF A VISCOUS LIQUID

U. B. Bazaron, A. V. Bulgadaev,
and B. V. Deryagin

Buryat Joint Research Institute, Siberian Division, Academy of Sciences of the USSR;
Institute of Physical Chemistry, Academy of Sciences of the USSR

A resonance method is used to show that the shear modulus of a 10^{-4} cm film is inversely related to temperature for wax, rosin, and various oils.

The first measurements of shear elasticity for liquids were made on very viscous or viscoplastic liquids; usually the temperature dependence was studied, for instance for rosin, guaiacol resin, and other substances near their softening points [1]. The method was to oscillate a cylinder placed in the liquid, which was contained in a cylinder of larger diameter. The shear modulus was found to be substantially temperature-dependent.

A composite rod has also been used [2] with many substances. For instance, for rosin the shear modulus fell by more than an order of magnitude over a narrow temperature range. However, these methods are applicable only to substances whose viscosities are 10^6 poise or so.

Mason [3, 4] described two methods that would accept lower viscosities. The first employed torsional oscillations, while the second employed the amplitude and phase of a wave reflected from a solid–liquid interface. The lower limit of viscosity in these was about 10 poise.

A recent study [5] employed Mason's method at several frequencies. The shear modulus is given as a function of temperature for butadiol, methylpentadiol, and hexatriol, the results being roughly as given by Kornfeld, except that they extended to lower viscosities, i.e., to higher temperatures.

We have used the Mandel'shtam-Khaikin method, which employs the alteration of the resonance frequency of a quartz plate due to the additional coupling produced by a film of liquid [6].

The film is placed between the quartz plate and a small plate (Fig. 1). The shift in the resonance frequency may be measured to ±1 Hz with a suitable circuit. The system of Fig. 1 is placed in a thermostat to record the temperature dependence.

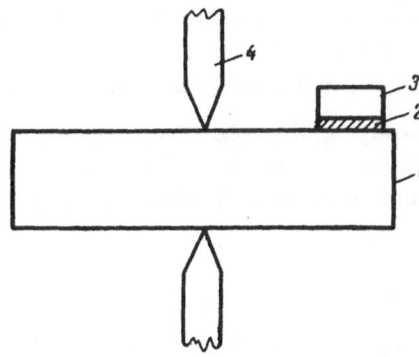

Fig. 1. 1) Quartz plate; 2) liquid; 3) small plate; 4) holder.

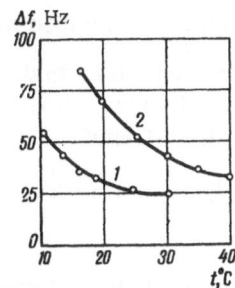

Fig. 2. Change Δf in resonance frequency as a function of t for 1) vaseline; 2) castor oil.

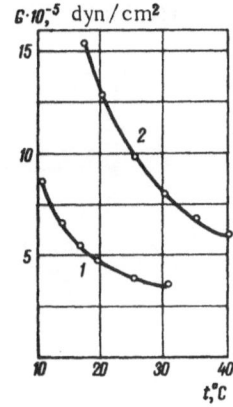

Fig. 3. G as a function of t for 1) vaseline; 2) castor oil.

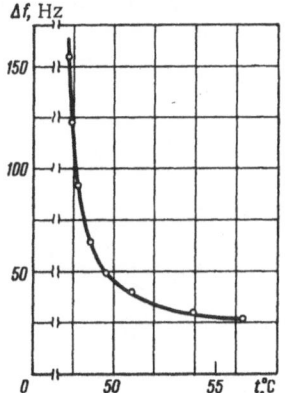

Fig. 4. Δf as a function of t for paraffin wax near the solid−liquid transition point.

Theory [6] gives the shear modulus as

$$G = \frac{4\pi^2 M f_0 \Delta f H}{S},$$ (1)

in which M is the mass of the quartz plate, whose resonance frequency is f_0, Δf is the frequency shift, H is the thickness of the liquid film, and S is the area of contact.

The quartz plate is cleaned in a glow discharge and is set up in the holder. A drop of liquid is placed on it with a glass rod, followed by the small plate. The thickness of the film is determined by an optical method, and then the system is placed in a thermostat (temperature controlled to 0.1°C) for the resonance measurements.

Figure 2 shows results for vaseline and castor oil in film thicknesses of 2.05 and 2.5 μ respectively. Figure 3 shows the result from (1). Figure 4 shows Δf near the solid−liquid transition for paraffin wax; it is clear that G is largely independent of t away from the transition point, but that there is a rapid variation within 3-4° of that point. The curve is taken up to a Δf of 150 Hz, but values up to 1000 Hz were measured, which cover the complete solid—liquid transition region. This is possible because G is small for solid wax near the transition point.

The present results show that this method is widely applicable, especially to the dynamic characteristics of lubricants, solutions, gels, and structured colloids. The results are of significance to the physics of surface phenomena, the mechanics of colloids, and physical mechanics generally.

References

1.   M. P. Volarovich, B. V. Deryagin, and A. A. Leont'eva, Zh. Fiz. Khim., 8:479 (1936); Acta Physicochim. URSS, 5:617 (1936).
2.   M. Kornfeld, Elasticity and Strength of Liquids [Russian translation], Gostekhizdat, Moscow (1951).
3.   W. P. Mason, Piezoelectric Crystals and Their Application to Ultrasonics, D. Van Nostrand, New York (1950).
4.   W. P. Mason, Trans. Am. Soc. Mech. Eng., 69:359 (1947).
5.   R. Meister, C. Marhoeffer, R. Scimanda, and T. Litovitz, J. Appl. Phys., 5:854 (1960).
6.   U. B. Bazaron, B. V. Deryagin, and A. V. Bulgadaev, Dokl. AN SSSR, 160:799 (1965); this volume, p. 36.

SECTION II

SURFACE FORCES IN THIN FILMS

PERMANENT ELECTRIC DIPOLE MOMENTS
OF COLLOIDAL PARTICLES

N. A. Tolstoi, A. A. Spartakov, and A. A. Trusov

Zhdanov University, Leningrad

Methods and apparatus have been developed for research on the electrooptic effect in colloidal solutions.
It has been found that the particles in a polar medium possess permanent electric dipole moments μ,
whose origin is related to the spontaneous orientation of the polar molecules adsorbed on the particles.

Introduction

A colloidal solution subject to an external field may become optically anisotropic. The field may be electric, magnetic, or acoustic, or it may be due to laminar flow. Whatever the mode of production, the quantity usually examined is the birefringence.

It has been shown [1-3] that the anisotropy of the extinction, i.e., the dichroism [1], is more expedient to study than the birefringence for colloidal solutions, at least when the particles are large. It has also been shown [2-4] that colloidal particles often have permanent dipole moments of considerable magnitude in water and other polar liquids, and this moment is ascribed [2, 3] to spontaneous orientation of the polar solvent molecules (or additional solute molecules) adsorbed on the particles. These permanent moments can be demonstrated from three kinds of response to electric fields.

Response to Rectangular Voltage Pulses

1. A parallel beam from the source S (Fig. 1a) passes through the parallel-plate condenser K containing the colloidal solution. The light is polarized either perpendicular to the electric field E or parallel to E. The voltage applied to the electrodes of K varies as shown in Fig. 1b (Π_1 field), with E^2 = constant but varying sign. The alternating frequency ν is varied from 1 to 3 kHz. In some cases, the field varies as in Fig. 1e (Π_2 field). The light is modulated by passage through K and falls on the photocell PC, whose output is recorded by oscilloscope O. The modulation is produced by orientation reversal and takes the form shown schematically in Fig. 1, c and d. Each sign reversal produces a new modulation wave, and the curve for the transmitted intensity I shows a kink at each sign reversal.

The field may orient the particles in two ways:

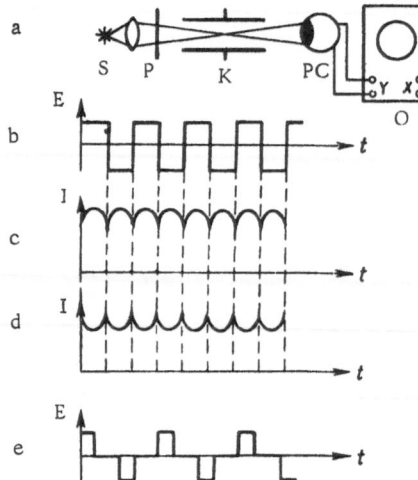

Fig. 1. a) Optical system of the apparatus for examining the optical response of a colloid to rectangular field pulses: S) light source, P) polaroid, K) cell, PC) photoelectric cell, O) oscilloscope; b) rectangular voltage pulses (Π_1 field); c) and d) curves for light intensity transmitted with light polarized: c) along field; d) perpendicular to field; e) rectangular pulses of Π_2 field.

a) The field polarizes the particles (or the surrounding double layer); then the induced dipole experiences a torque, which is proportional to E^2, i.e., does not change sign when the field is reversed, since E^2 = constant, and this produces a constant orientation, if a particle has only an induced moment, i.e., no modulation can occur.

b) The field acts on a permanent dipole moment. Then sign reversal in the field reverses the sign of the torque, and the orientation will change at each field reversal, i.e., the light will be modulated (if the particles are optically anisotropic, as is usually the case). Modulation of I in a Π_1 field is thus direct evidence for a permanent dipole moment. This is an advantage of a Π_1 field over the sinusoidal fields often used in previous work.

2. To determine whether this response occurs in any colloidal system, we examined over 50 colloids with various dispersed phases and dispersion media. These colloids were made as follows: 1) replacement of a solvent (usually alcohol) by water (e.g., for aromatic organic compounds), 2) direct solution in water (dyes), 3) mechanical production in the medium (water, alcohol, benzene, etc.).

Modulation occurred always when the polar liquids (water and alcohol) were used. Nonpolar liquids (CCl_4, C_6H_6) gave no modulation in the Π_1 field. This means that a permanent dipole moment occurs only when the particles are immersed in a medium whose molecules themselves have permanent dipole moments, and so the permanent moment of a particle is of surface origin, being due to spontaneous unipolar orientation of polar molecules (e.g., of water) adsorbed on the particle. We have to assume that such a molecule is adsorbed in such a way as to have a component of its permanent dipole tangential to the surface. A set of conforming dipoles can resemble a surface ferroelectric domain. Evidence of this comes from the following observations.

a. A colloid in a nonpolar medium might give no modulation in a Π_1 field perhaps because the suspension is unstable; but this objection is removed by the observation that such suspensions produce modulation in a Π_2 field, i.e., when orientation and disorientation can occur for particles having only induced moments.

b. Addition of a highly polar compound (e.g., nitrobenzene) to a nonpolar solvent (e.g., benzene) produces modulation in a Π_1 field, i.e., when modulation can be due only to reorientation of particles (in our case, talc or bentonite) having permanent dipole moments.

From this we conclude that permanent dipole moments and modulation by a Π_1 field are not specific to particular colloidal particles but occur for all colloidal particles (unless, perhaps, if these are spherical) placed in a polar medium.

3. The unspecific modulation in a Π_1 field does not mean that the modulation curves are identical; each colloidal solution gives its own curve, and these curves may be classified qualitatively into the four types shown in Fig. 2, where C denotes maximum clarity of the solution in the intervals between successive sign changes, while T denotes minimum transmission.

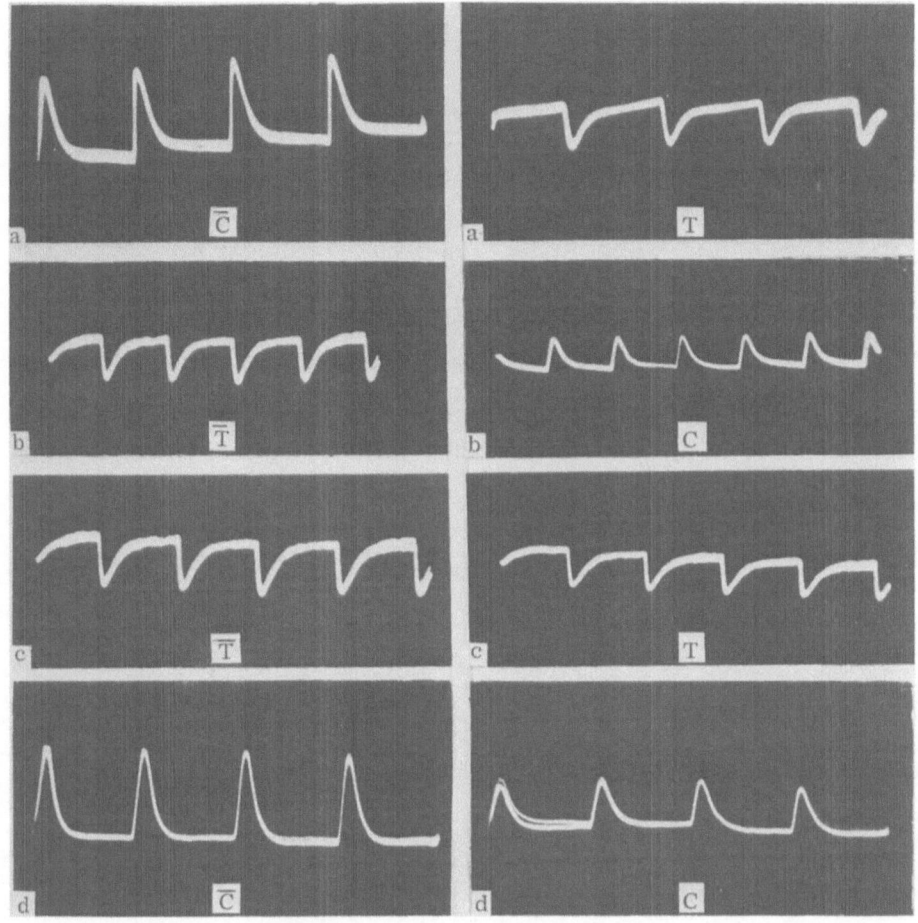

Fig. 2. Types of simple modulation curve: a) $\overline{C}T$; b) $\overline{T}C$; c) $\overline{T}T$; d) $\overline{C}C$.

A bar above a letter indicates that this applies for light polarized along E. The four basic types of modulation are therefore $\overline{C}T$, $\overline{T}C$, $\overline{T}T$, and $\overline{C}C$.

It is not difficult to classify the observed curves. For instance, the scale of the polarizer P is set with the light vector along E (horizontal in the apparatus as described), and it is seen that the humps are directed upwards, while the sharp peaks (at the instants of switching) are directed downwards (Π_1 field). If the oscilloscope is connected so that intensity increase corresponds to upward deflection, the humps correspond to maximum intensity, and so we have $\overline{C}$ modulation. If with P turned through 90° the humps are directed downwards, the colloid gives type $\overline{C}T$ modulation.

The shape of the modulation curve is not determined solely by the particles, since a given colloid can give curves of different types in accordance with the conditions of preparation. This is obviously an effect from the size and shape of the particles, other things being equal.

A colloidal solution is usually polydisperse and polymorphic. Such a colloid contains several groups of particles differing in geometrical characteristics. Each group gives its own modulation curve in a Π_1 field, and the resultant curve may have more than one turning point in a half-cycle (Fig. 3). The first (fast) one in such a case corresponds to rotation of the small particles, while the second (slow) one corresponds to larger particles which have less rotational mobility.

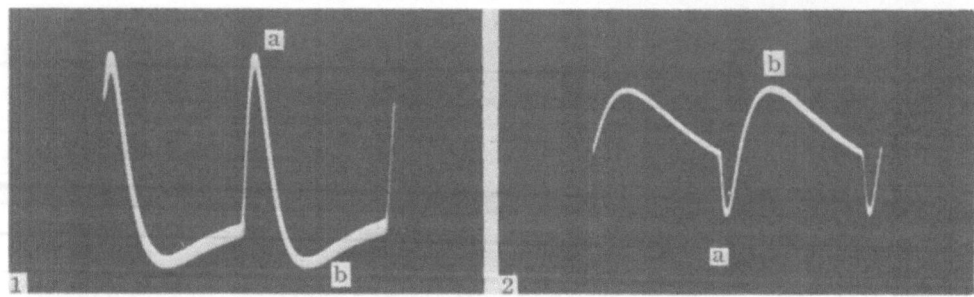

Fig. 3. Complex modulation curves with the light polarized: 1) along E; 2) perpendicular to E; a) first (fast) turning point; b) second (slow) turning point.

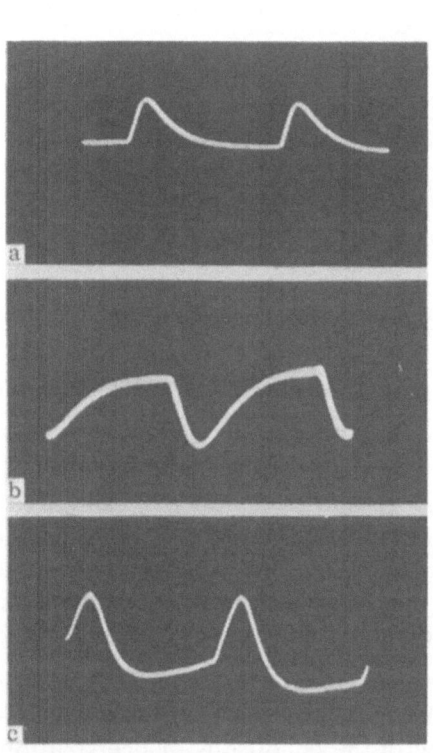

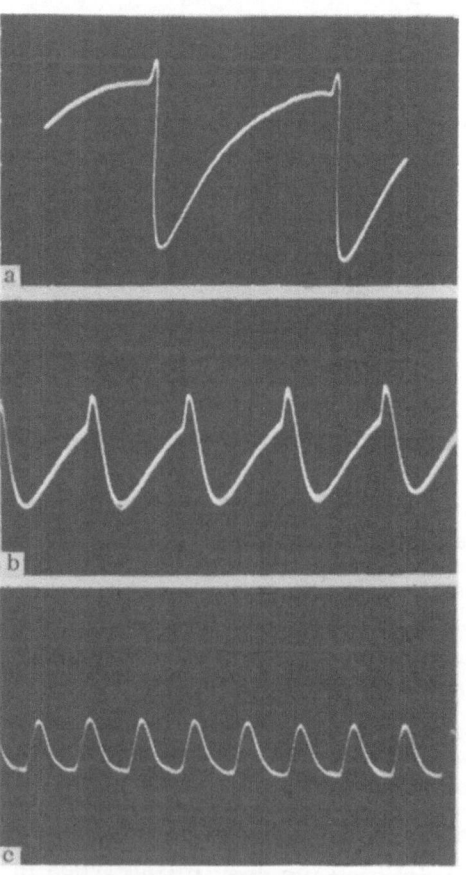

Fig. 4. a) and b) Simple modulation curves of two different colloids; c) composite curve produced by mining these colloids.

Fig. 5. Effects of frequency Π_1 field on the shape of a composite modulation curve. Frequency (Hz): a) 5; b) 50; c) 200.

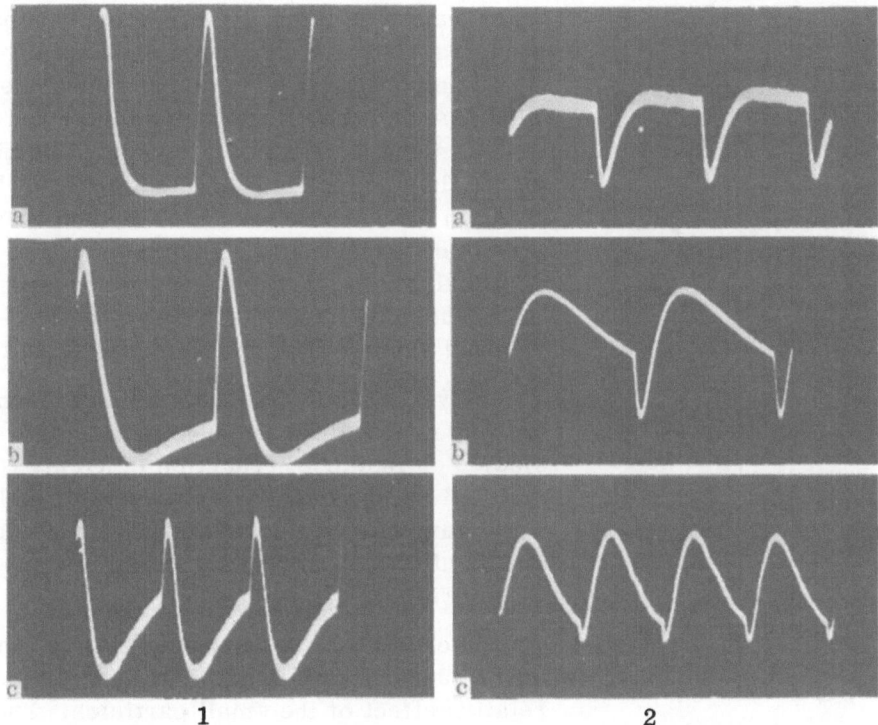

Fig. 6. Effects of λ on the modulation curve. a) Blue; b) yellow; c)
red light polarized: 1) along the field; 2) across the field.

Various experiments confirm these views. If the large particles are removed by centrifugation, the modulation curve usually becomes simple, with only the fast peak persisting. Again,
it is possible to prepare two chemically identical solutions differing only in particle size, each
of which gives a simple modulation curve (Figs. 4a and 4b), in one of which it is due to rotation
of small particles (fast turning point) and in the other to rotation of large particles (slow turning point). A mixture of these gives a composite containing slow and fast parts (Fig. 4c).

A composite curve may become a simple one with increase of the frequency of the Π_1 field.
At the higher frequency, the large particles do not have time to rotate through any considerable
angle during a half-cycle and so produce very little modulation. The slow turning point therefore decreases in amplitude more rapidly than does the fast one as the frequency is increased,
and at some frequency it vanishes entirely. This frequency dependence again indicates the surface nature of the permanent dipole moment. If the permanent dipole were of bulk origin, the
orienting torque and the frictional torque would both be proportional to the particle volume, and
the ratio would not be dependent on the latter, so the rotational mobility would be independent
of the particle size. The kinetics of particle motion would then be the same for small and large
particles, and increase in frequency of the Π_1 field would not cause the small particles to have
a dominant effect. The higher mobility of small particles thus arises because the rotational
frictional torque is proportional to d^3 (particle size d), while the dipole moment is proportional
to d^2. Quantitative evidence for this is given below.

4. In relation to composite modulation curves, the possibility arises of distinguishing
effects from groups of particles by varying the wavelength of the light. Here we have to consider the optical mechanism of the effect. The beam interacts with the particles and is scattered, which weakens the beam, which is a result of diffraction by bodies whose size d is comparable with the wavelength λ. Diffraction by small spherical particles leads to a complicated
(not monotonic) dependence on d when d and λ are comparable, the intensity passing through

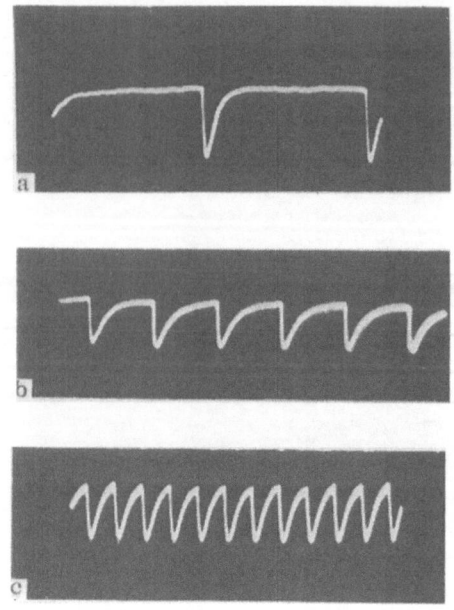

Fig. 7. Curves produced by tobacco smoke with Π_1 fields of frequencies (Hz): a) 2; b) 50; c) 2000.

maxima and minima, which correspond roughly to integer ratios of d and λ (diffraction resonance). This picture is fairly complicated even for spherical particles and becomes even more complicated for others. All the same, the effects can be interpreted qualitatively as due to rotation of elongated particles, which alters the size of the equivalent sphere. Whether the rotation produces an increase or decrease in effective size is dependent on the shape, on λ, and on the state of polarization. Particularly pronounced modulation is produced by particles whose effective spheres correspond closely to diffraction resonance for the given λ.

Even for polydisperse and polymorphic particles (as in a real colloid), change in λ affects the equivalent size. For instance, reduction in λ may transfer the main effect to particles of somewhat smaller size, which may have different optical parameters; this should affect the maxima and minima, as is actually found (Fig. 6).

Reduction in λ causes a relative increase in the initial fast maxima or minima, i.e., an increase in the relative effect of the small particles. These effects from varying λ are apparent even with fairly crude monochromatization.

5. So far we have considered colloids (liquids such as water and benzene). We have also done some very interesting tests with aerosols [5]. A lucite cell was filled with tobacco smoke, which gave a pronounced response to a Π_1 field (Fig. 7). The magnitude of the effect under fixed conditions was dependent on the time of exposure to water vapor; no effect was observed if the smoke had not been exposed to the vapor, which shows that here again we have surface permanent dipoles due to orientation of water molecules. The effect persists up to frequencies of about 3 kHz, i.e., much larger than for liquids.

The Electrooptic Effect in a Rotating Electric Field

(R Field)

1. The beam from the source S (Fig. 8) passes through the polarizer P and the cell C to the photocell PC. The cell contains three electrodes placed at the vertices of an equilateral triangle with their axes parallel. These are connected to a three-phase sine-wave source (angular frequency ω), and the field E at the center of the equilateral triangle rotates with angular velocity ω. The particles oriented by E rotate with angular velocity ω with a phase lag α. This oriented system is dichroic and so acts as an imperfect polaroid.

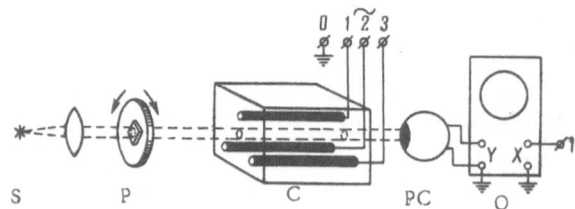

Fig. 8. Apparatus for examining the behavior of a colloid in a rotating electric field: S) light source, P) polarizer, C) cell fitted with electrodes, PC) photocell, O) oscilloscope.

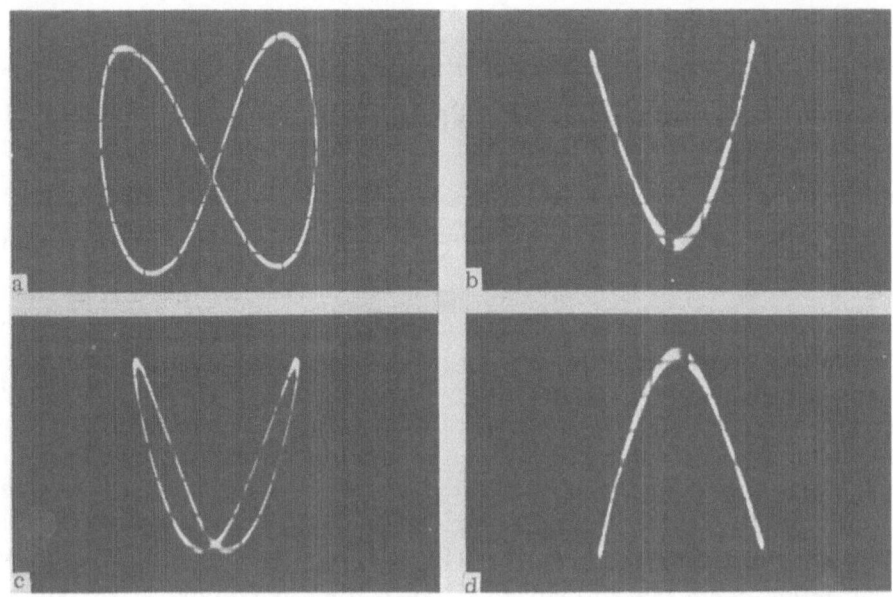

Fig. 9. Examples of second-order Lissajous figures obtained in
measuring α.

The rotation relative to P produces modulation at a frequency 2ω whose depth is dependent on the quality of the second polarizer. The output from PC passes to the Y plates of the oscilloscope. Rotation of P merely affects the phase of the sine wave, and in this way it is possible to deduce α in relation to E. One of the objects of the method is to measure α as a function of the various parameters.

2. The X plates are fed from one of the three phases, e.g., lead 1, at frequency ω. This interacts with the 2ω sine wave to produce a second-order Lissajous figure, e.g., Fig. 9a; the shape of the figure is dependent on the position of P, and this may be set to give the shape of Fig. 9b. This means that the axis of strongest absorption in the cell is parallel to the transmission axis of P when the voltage on lead 1 passes through zero, and so E is horizontal. (The oscilloscope is connected so that the upward deflection increases with the intensity.)

Measurement of the inclination of the transmission axis of P to the horizontal would give α, but a much more convenient method is to interchange leads 2 and 3, which causes the Lissajous figure to take the form of Fig. 9c. We now turn P through an angle γ to restore the figure of Fig. 9b; then $\alpha = \gamma/2$, provided that P is turned in the sense of the initial rotation of the field. If $\gamma > 180°$, there is a lag $\alpha < 90°$ between E and the axis of transmission formed by the particles; in that case, the entire process can be repeated for the figure as shown in Fig. 9d. Under favorable conditions, it is possible to measure α to about 0.5°.

3. Let a polydisperse colloid contain i types of particle, for each of which we have for an R field that

$$\mu_i E \sin \alpha_i + \Delta\chi_i E^2 \sin 2\alpha_i = \eta V_i [p_i] \cdot \omega. \tag{1}$$

The left side of (1) represents the torque exerted by the field, while the right side represents the frictional torque. Here μ_i is the permanent electric dipole moment of a particle, $\Delta\chi_i$ is the anisotropy in the polarizability, $[p_i]$ is the form factor of a particle, V_i is the particle volume, η is the viscosity of the medium, and α_i is the α for class i.

The assumptions are as follows:

1) The particles are symmetrical bodies of rotation, e.g., uniaxial ellipsoids;

2) The directions of the axis of the rigid dipole and of the axes of optical dichroism coincide with the geometrical axes of the ellipsoids;

3) Brownian motion is neglected, i.e., the particles are large or E is high;

4) The particle concentration is small enough for the particles not to interact one with another and not to alter the viscosity.

Then E and ω for given parameters of species i define α_i via (1), which of course has physically meaningful solutions only for $\omega < \omega_{max}^i$ (since $\sin \alpha_i$ and $\sin 2\alpha_i$ cannot exceed 1). Physically this means that the particles can follow the field only when the rotational friction (proportional to ω) does not exceed the external torque, so ω_{max}^i is dependent on E and on i. Particles for which $\omega > \omega_{max}^i$ do not participate in producing the regular rotating dichroic pattern. For each i with $\omega < \omega_{max}^i$ there is a distance α_i for given E and ω. The resultant pattern is a superposition of the patterns for the various i, and the observed α is dependent on all the α_i, each corresponding to a number m_i of particles, with $\sum_i m_i = n$, in which n is the number of particles. It might seem that interpretation of α for a polydisperse or polymorphic system would be impracticable, but we shall see that very valuable conclusions can be drawn even in this case.

We first assume that $\mu = 0$ or is so small that it makes no appreciable contribution to the torque under the conditions used. Then (1) is replaced by

$$\Delta\chi_i E^2 \sin 2\alpha_i = [p_i] \cdot V_i \cdot \eta \cdot \omega \tag{2}$$

or

$$\sin 2\alpha_i = \frac{\omega}{E^2} \cdot \eta \cdot \frac{[p_i] \cdot V_i}{\Delta\chi_i}. \tag{3}$$

The resultant α for particular E and ω is dependent on all the α_i:

$$\alpha = f_1(m_1 \sin 2\alpha_1 \ldots m_n \sin 2\alpha_n) = f_1\left(m_1 \frac{\omega}{E^2} \eta \frac{[p_1] V_1}{\Delta\chi_1} \ldots m_n \frac{\omega}{E^2} \eta \frac{[p_n] V_n}{\Delta\chi_n}\right). \tag{4}$$

Whatever the form of f_i, measurements at other E and ω such that ω/E^2 = constant should give the same α, since α_i is unaltered if ω changes in proportion to E^2, and hence α also does not alter. Then

$$\alpha = \alpha\left(\frac{\omega}{E^2}\right). \tag{5}$$

If (5) is obeyed, this shows that $\mu = 0$, or that the first term in (1) is negligible.

We now assume that μ is so substantial that under the given conditions (e.g., given range in E) the second term in (1) is negligible relative to the first, i.e., the rotation is due to the tendency of the permanent dipole to set along E. Then

$$\mu_i E \sin \alpha_i = [p_i] V_i \cdot \eta \cdot \omega \tag{6}$$

or

$$\sin \alpha_i = \frac{\omega}{E} \eta \frac{[p_i] V_i}{\mu_i}. \tag{7}$$

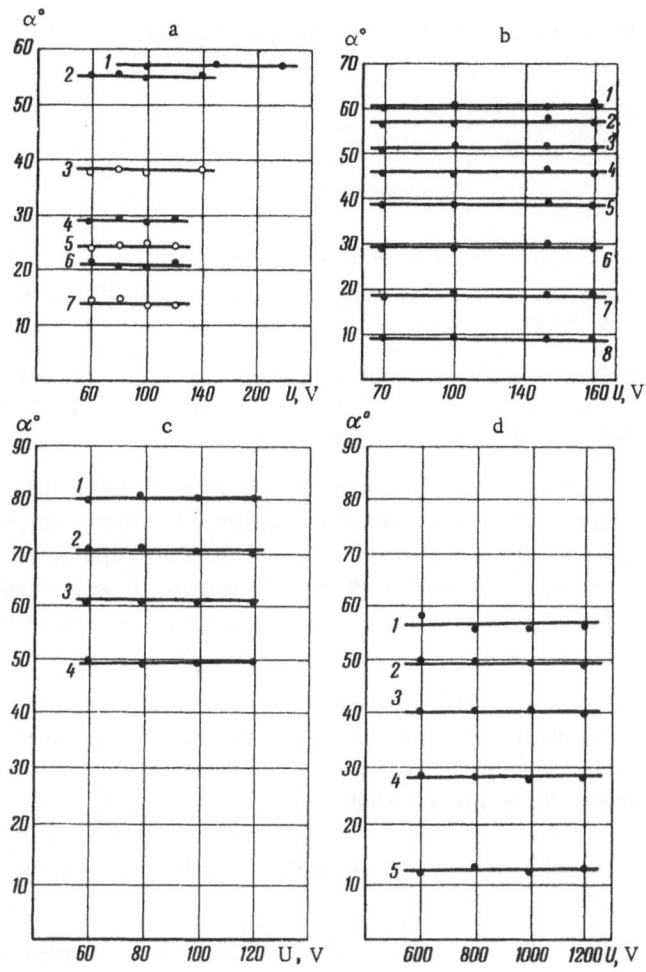

Fig. 10. Dependence of α on field U for given ν/U or ω/E, where $\omega = 2\pi\nu$ and $E = 2U$ is the field strength (V/cm): a) bentonite in water: $\bullet$) suspension of low dispersion, $\bigcirc$) suspension of high dispersion for γ/U of: 1) 1; 2, 3) 0.5; 4, 5) 0.2; 6, 7) 0.1; b) suspension produced by replacing alcohol by water in an anisaldazine solution in alcohol for ν/U of: 1) 1.4; 2) 1.2; 3) 1; 4) 0.8; 5) 0.6; 6) 0.4; 7) 0.2; 8) 0.1; c) colloidal electrolyte benzopurpurine in water at ν/U of: 1) 0.3; 2) 0.2; 3) 0.15; 4) 0.1; d) aerosol (tobacco smoke treated with water vapor) for ν/U of: 1) 0.5; 2) 0.4; 3) 0.3; 4) 0.2; 5) 0.1.

The result for given E and ω is

$$\alpha = f_2(m_1 \sin \alpha_1 \ldots m_n \sin \alpha_n) = f_2\left(m_1 \frac{\omega}{E} \eta \frac{[p_1]V_1}{\mu_1} \ldots m_n \frac{\omega}{E} \eta \frac{[p_n]V_n}{\mu_n}\right) \tag{8}$$

and

$$\alpha = \alpha\left(\frac{\omega}{E}\right). \tag{9}$$

If (9) is obeyed, the permanent dipole moment dominates the rotation; again (9) is not dependent on the dispersion.

If neither (9) nor (5) is obeyed, i.e., if

$$\alpha = \alpha \left(\frac{\omega}{F(E)} \right), \tag{10}$$

with F(E) increasing more rapidly than as E but less rapidly than as E^2, this indicates that μ is finite but only such as to be comparable with the induced moment. Of course, this again applies for a definite range in E.

These are only preliminary theoretical considerations, but they are sufficient to provide a new quantitative approach. The method is quite independent of the qualitative method of demonstrating permanent dipoles with Π fields (Section 2).

4. The effect in an R field is as common as that in a π_1 field [6], as may be illustrated for aqueous colloidal solutions of various origins: mineral (bentonite), organic (anisaldazine), and biological (B. coli), as well as a colloidal electrolyte (benzopurpurine) and an aerosol consisting of tobacco smoke treated with water vapor. Figures 10 and 11 show that, in every case, a given α corresponds to the same ω / E for α not too large, i.e., (9) is obeyed, so all of these particles have μ dominant.

5. It follows from (3) that parameters such as to make the right side of (3) greater than 1 will not allow the particles to follow the rotation. The largest possible α for $\mu = 0$ is 45°, since fluctuation increase in α with $\alpha > 45°$ will reduce the rotational torque, with the result that regular rotation ceases. This means that $\mu \neq 0$ if $\alpha > 45°$, which gives a third means of demonstrating a permanent dipole moment. Figures 10 and 11 give results for various E and ω, and they show that α can substantially exceed 45°, which again shows that these particles have $\mu \neq 0$.

6. Equation (1) involves the assumption that rotational Brownian motion can be neglected. The validity of this can be examined via the modulation depth A, which is proportional to the amplitude of the Lissajous figure. If Brownian motion plays an appreciable part, the axes will not rotate always in the plane of E. The spread due to the Brownian motion will be inversely related to E, and so A will tend to increase. If the Brownian motion is slight, A will not tend to increase with E. For this reason, A as a function of E must be examined at fixed ω / E, on account of possible polydispersion.

Figure 12 shows several examples of A as a function of E for ω / E = constant. It is clear that A is independent of E for α small, so we have conditions such that the motion can be treated from the viewpoint of classical mechanics. Of course, this itself demonstrates that μ dominates the process. If the Brownian motion is appreciable (e.g., for α large), it is found that this approach is still applicable, since (9) often is obeyed for α large, and the $\alpha = \alpha(\omega / E)$ relation begins to be violated only for α approaching 90°. This in no way affects the conclusions drawn for the case of absence of Brownian motion.

Magnitude of μ

1. The R-field method allows one to deduce μ if the shape and size of the particle are known [7]. Anisodimensional particles are usually also polydisperse and polymorphic, most of the exceptions being particles of biological origin: bacteria, free cells, bacteriophages, viruses, protein molecules, etc. In particular, suspensions of bacteria can serve as model colloid systems. We have used bacterial suspensions, and we found that α is constant for a fixed ω / E, so (6) allows us to deduce μ from the volume and shape given by microscopy (Table 1).

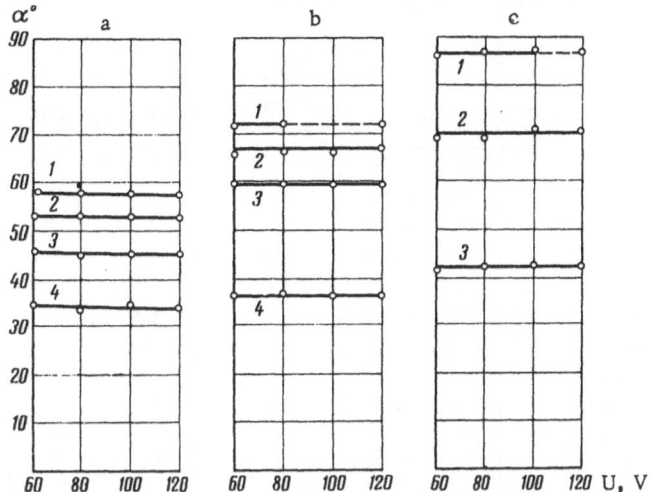

Fig. 11. Dependence of α on U for given ν/U for aqueous suspensions of: a) B. prodigosum for ν/U of: 1) 0.5; 2) 0.3; 3) 0.2; 4) 0.1; b) B. coli for ν/U of: 1) 0.4; 2) 0.3; 3) 0.2; 4) 0.1; c) B. fluorescens for ν/U of: 1) 0.3; 2) 0.2; 3) 0.1.

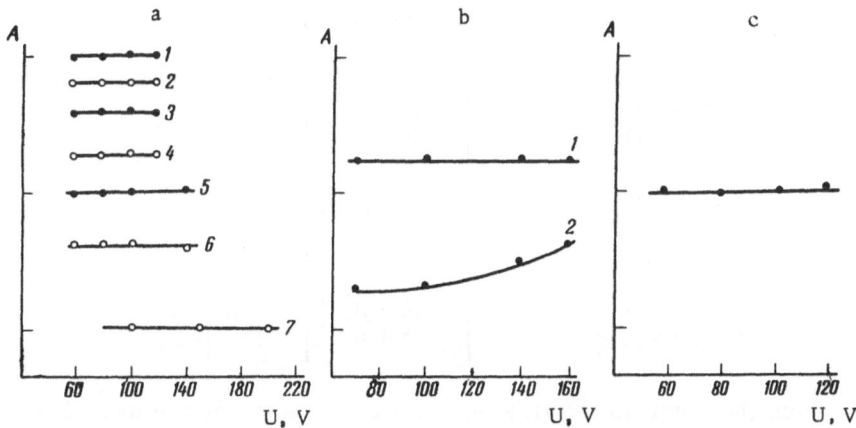

Fig. 12. Dependence of A (relative units) on U for given ν/U for aqueous suspensions of: a) bentonite [O) low dispersion, ●) high dispersion] for ν/U of: 1, 2) 0.1; 3, 4) 0.2; 5, 6) 0.5; 7) 1.0; b) anisaldazine at ν/U of: 1) 0.1; 2) 0.5; c) B. coli at $\nu/U = 0.1$.

2. We have developed the use of Π pulses of the second kind in order to deduce μ for particles whose shape and size are not known in advance.

The pulse shape is shown in Fig. 1e. The particles tend to set with their long axes parallel to the field during the pulses, while exponential relaxation to a random orientation occurs between pulses. The orientation buildup and relaxation are observed with an oscilloscope.

The relaxation curve is described by

$$J = J_0 e^{-6D_r t}. \tag{11}$$

TABLE 1. Permanent Dipole Moments of Biological Particles

Particle	Volume V, cm^3	Ratio p = a/b of semiaxes	μ, electro-static cgs units	φ
Bacteria:				
prodigosum	2.5×10^{-13}	2.2	4.1×10^{-12}	83°
coli	3.1×10^{-13}	2.0	4.5×10^{-12}	83°
fluorescens	3.2×10^{-13}	2.6	5.7×10^{-12}	82°
Proteins:				
myoglobin	2.2×10^{-20}	2.6	1.7×10^{-16}	77°
horse serum hemoglobin	18.2×10^{-20}	9	11×10^{-16}	77°
horse serum globulin	8.6×10^{-20}	1.6	4.8×10^{-16}	73°
horse serum albumin	9.0×10^{-20}	6	3.8×10^{-16}	82°
egg albumin	5.8×10^{-20}	5	2.5×10^{-16}	82°
Tobacco mosaic virus	6×10^{-17}	23	4.1×10^{-14}	82°

TABLE 2. Mean Dipole Moments
and D_r for Colloidal Particles

Colloid	μ, electro-static cgs units	D_r, cgs	φ
Anisaldazine	1.7×10^{-13}	170	85°
Anisylidene-benzidene	1.6×10^{-13}	180	85°
Azoxyanisol	2.1×10^{-12}	3.8	85°
B. coli	4.3×10^{-12}	2.1	83.5°
Anil brown	4.9×10^{-13}	68	83°
Direct orange	1.3×10^{-12}	17	83°
Primuline	1.6×10^{-12}	11	83°
Congo red	1.6×10^{-12}	11	83°
Benzopurpurine	2.4×10^{-12}	5.7	83°
Bentonite	9×10^{-13}	17	85°

This curve gives the rotational diffusion coefficient, e.g., by the use of a τ meter such as we have developed, in which the Π_2 pulses are fed to an RC integrator (R and C variable) at the same time as to the cell. The integrated output goes to the X plates of the oscilloscope to provide an exponential sweep of scale 1/RC. If the time constant is the same as that from the photocell, the trace on the screen is a straight line. Then the known R and C give

$$\frac{1}{\tau} \equiv \frac{1}{RC} = 6D_r, \tag{12}$$

in which

$$D_r = \frac{kT}{\eta \cdot V \cdot [p]}. \tag{13}$$

From α and the D_r of (6) and (13) we get

$$\mu = \frac{kT\omega}{D_r \cdot E \sin \alpha}. \tag{14}$$

In this way the mean μ for a polydisperse and polymorphic solution can be determined. Table 2 gives μ and D_r determined in this way.

Direction of the Permanent Dipole Moment

in the Particle

This direction relative to the long axis of the particle (an elongated body of rotation) can be deduced by combining the Π_1 and R methods.

The amplitude decreases as the frequency of the Π_1 pulses is increased (Fig. 1, c and d), and the effects of the permanent dipole are lost at very high frequencies. However, the effects on the induced dipole persist, and these have a sign independent of the sign of the field, since the induced dipole reverses in sign along with the field, and the particle experiences always a torque tending to set it along the field. This will occur for any elongated body whose dielectric constant ε differs from the dielectric constant ε_0 of the medium, no matter what the sign of $\varepsilon - \varepsilon_0$. This effect due to form anisotropy may be complicated by an effect due to dielectric anisotropy. We have found that, at high frequencies, colloidal particles tend to set their long axes along the Π_1 field, and this produces an imperfect polarizer, whose axes in this case are fixed. It is not difficult to determine whether the transmission or absorption axis lies along the field. This establishes whether the long axis is the axis of greatest or least absorption for light polarized along this axis.

The R field gives an angle $0° < \alpha < 90°$ for the permanent dipole relative to the field (motion in the range $90° < \alpha < 180°$ is unstable). The Lissajous figure indicates at once which axis of the polarizer follows the field.

In all the cases we have so far examined, μ has been directed along the long axis of the particle.

Possible Structure of a Surface Domain

1. The results on μ agree well with the view that this moment is related to unipolar orientation of the polar adsorbed molecules (here H_2O). In fact, μ is less by about a factor ten than the arithmetic sum of the dipole moments of the water molecules capable of forming a monolayer on the surface of a particle. If an adsorbed monomolecular film is considered as a single domain (Fig. 14c), we can estimate the angle φ between the surface and the axes of the water dipoles.

We introduce $\mu_1 = \mu_0/S_0$, in which μ_0 is the dipole moment of an adsorbed molecule and S_0 is the area taken up on the surface. Then the resultant μ for an ellipsoid of rotation (semiaxes a and b) will be dependent on φ and on the dipole moment per cm^2:

$$\mu = \pi^2 \mu_1 \cdot a \cdot b \cdot \cos \varphi \tag{5}$$

or, if the ellipsoid is characterized by its volume V and ratio $p = a/b$,

$$\mu = \pi^{1/3} \cdot \left(\frac{3}{4}\right)^{2/3} \cdot \mu_1 \cdot \sqrt[3]{V^2 \cdot p} \cdot \cos \varphi . \tag{6}$$

We take $\mu_0 = 1.84 \times 10^{-18}$ cgs units and $S_0 = (30 \times 10^{-24})^{2/3} = 9.6 \times 10^{-16}$ cm^2 to get from (6) and (16) that φ is 82-83° for these bacteria (Table 2).

2. It is of interest to apply these arguments to particles similar in nature but very different in size. Values have been given [8, 9] for the permanent dipole moments of protein molecules, whose volumes are less than those of bacteria by six orders of magnitude; these

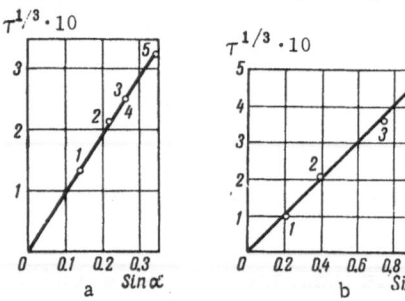

Fig. 13. Relation of $\tau^{1/3}$ to $\sin \alpha$ for $\nu/U = 8 \times 10^{-2}$ sec^{-1}/V for: a) dyes in water: 1) anil brown; 2) direct orange; 3) Congo red; 4) primuline; 5) benzopurpurine; b) water suspensions of: 1) anisaldazine; 2) bentonite; 3) azoxyanisole.

μ were deduced from the dielectric constants of aqueous protein solutions. If we assume that these μ arise by orientation of the dipoles in a water film and not from the molecules themselves, we get from (15) or (16) via [8] the values of φ given in Table 1, which also gives the φ calculated from measurements [10] of μ for tobacco mosaic virus.

The size difference between a bacterium and a protein molecule is very great, so the similarity in the φ is very good evidence that the permanent dipoles of protein molecules are of surface origin.

3. R and Π_2 fields together allow one to demonstrate the surface nature of the rigid dipoles in colloidal particles and also to estimate the inclination of the polar molecules adsorbed on the surface. We represent a colloidal particle as an ellipsoid of rotation via (13) and (16) to get

$$\mu = \pi^{4/3} \cdot \left(\frac{3}{4}\right)^{2/3} \cdot \mu_1 \left(\frac{kT\tau}{\eta}\right)^{2/3} \cdot \sqrt[3]{\frac{36p}{[p]^2}} \cdot \cos \varphi. \tag{17}$$

Now $Z(p) \equiv (36p/[p]^2)^{1/3}$ is close to one for p not very large, i.e., not very elongated particles, so we take $Z(p)$ as 1 to get from (12), (14), and (17) that

$$\cos \varphi = \frac{4\omega (6kT\tau\eta^2)^{1/3}}{\pi^{4/3} \cdot \mu_1 E \sin \alpha}. \tag{18}$$

It follows from (11) that the dependence of α on τ will be as follows for colloidal particles of different size but the same shape (perhaps not exactly the same shape) if the molecules have surface μ and equal φ:

$$\sin \alpha = \frac{4\omega (6kT\eta^2)^{1/3}}{\pi^{4/3} \cdot \mu_1 E \cdot \cos \varphi} \tau^{1/3}. \tag{19}$$

Figure 13 shows $\sin \alpha$ as a function of $\tau^{1/3}$ for fixed E and ω for colloidal solutions similar in nature; φ can be deduced from the slope of the straight lines, and the values are given in Table 2. These φ for particles differing in chemical nature are almost the same as the φ found for biological colloidal particles, so we may say that the nature and mode of formation of the dipole moments are the same in the two cases.

4. We need to demonstrate the unipolar orientation of the dipoles in the monomolecular film (Fig. 14b). The main argument here is that this disposition is more favorable as regards energy than the disposition with $\varphi = 90°$, which gives no resultant dipole (Fig. 14a).

Such a monomolecular film constitutes a double electrical layer, which should not be confused with an ionic double layer, and whose parts may be considered roughly as consisting of continuously distributed surface charge (density σ). There is a force of attraction between the two parts, which tends to reduce the thickness. If $\varphi = 90°$, this force is everywhere normal to the surface and is balanced by the elasticity of the molecules along the dipole axis. The resulting equilibrium is, however, unstable.

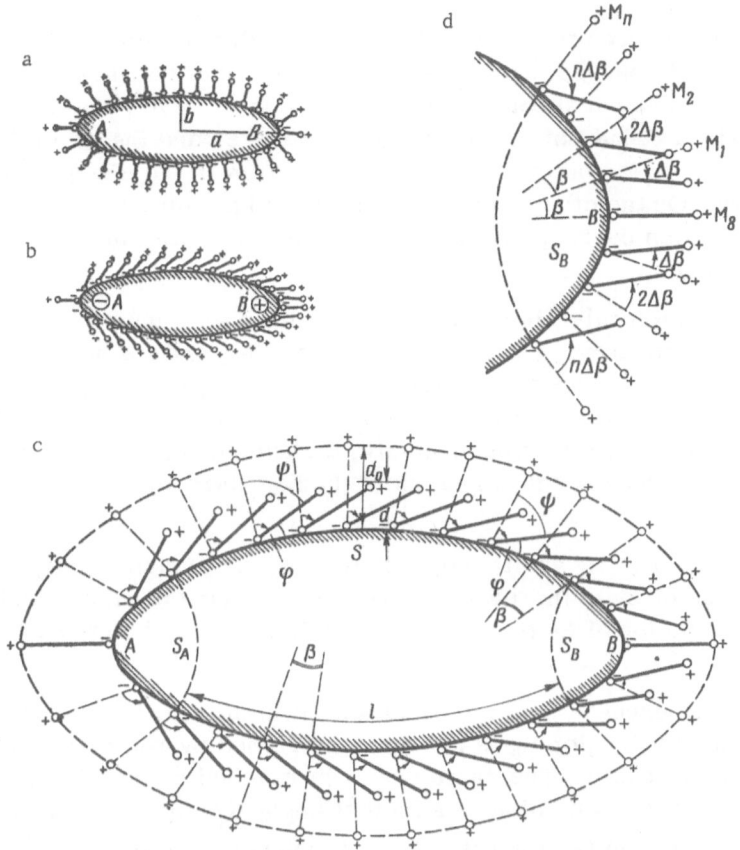

Fig. 14. Colloidal particle with a monomolecular film of water (or other polar compound). The drawing is to be considered as rotated around the axis AB: a) molecules normal to surface ($\mu = 0$); b) molecules inclined to surface ($\mu \neq 0$); c) spontaneous polarization accompanied by reduction in the thickness of the double layer from d_0 to d on area S and by shift in the center of gravity of the outer charges towards B; d) detailed structure of the irregular part S_B.

Consider the case $\varphi \neq 90°$. On the main part S (Fig. 14c) of the surface, the inclination $\varphi < 90°$ causes the two parts of the double layer to come together, the separation being reduced from d_0 to $d = d_0 \sin \varphi$ or, introducing $\psi = 90° - \varphi$, to $d = d_0 \cos \psi$.

The electrostatic energy per unit area of the double layer is $2\sigma^2 d$, so

$$F = 2\sigma^2 \operatorname{tg} \psi$$

or for ψ small

$$F \approx 2\sigma^2 \psi \tag{20}$$

is the force per cm² tending to displace the outer part of the double layer from the position $\varphi = 90°$ or $\psi = 0°$. This force increases with the deviation from $\varphi = 90°$, which represents instability and a tendency for the spontaneous polarization of the double layer to increase.

Here we must consider the behavior on parts S_A and S_B (Fig. 14c), where the regular disposition ($\varphi = $ constant) of the axes does not occur because the axes have to fit together consistently over the surface as a whole.

Let β be the angle between the axes of two adjacent dipoles for $\varphi = 90°$; β differs from zero because the surface is curved, and β varies with the local curvature. (Of course, β is very small, because the area per dipole is very small.) If $\varphi < 90°$, the β can retain their original values over all of S, but not in the irregular parts S_A and S_B; β in S_A must increase, which leads to separation of the outer charges and reduction in the energy, and so the behavior in S_A favors instability. On the other hand, the β on S_B (Fig. 14d) should decrease and even may become negative, which corresponds to energy increase. We must now consider how this affects the general pattern.

Consider the polar molecule M_B at the point B, whose axis must be normal to the surface from considerations of symmetry. Let the angle between M_B and its nearest neighbor M_1 be reduced from β to $\beta - \Delta\beta$. The axes of molecules such as M_1 now lie at $90 - \Delta\beta$ to the surface. Molecules M_2 next to M_1 may also lie at an angle $\Delta\beta$, so the axes of M_2 are inclined to M_B at $2\Delta\beta$ and to the surface at $90 - 2\Delta\beta$. Consequently, the axes of M_n (zone n) lie at $90 - n\Delta\beta$ to the surface, and so for n large enough, the axes at the boundary of S_B will conform to those in area S, and $\varphi = 90 - n\Delta\beta$ or $\psi = n\Delta\beta$.

This can be attained for $\Delta\beta$ arbitrarily small, i.e., for an arbitrarily small pressure on S_B, provided that the size of S_B is increased appropriately. On the other hand, the force exerted on unit length of the perimeter of S_B from S is proportional to the length l of region S in the direction AB.

Thus, if the particle exceeds some critical size (which cannot be determined from these semiquantitative arguments), region S_B is always constrained by region S, and the double layer becomes polarized spontaneously. The same result is reached by considering the energy. The force needed to bring the axes of dipoles to a mutual angle $\Delta\beta$ may reasonably be taken as elastic for $\Delta\beta$ small, so the energy change ΔW_{S_B} in region S_B is proportional to $(\Delta\beta)^2$ and to the number of dipoles, i.e., to the area of S_B. As $\psi = n\Delta\beta$, $\Delta W_{S_B} = C_{S_B} \cdot \psi^2$. The energy change ΔW_S on S is also proportional to ψ^2: $\Delta W_S = -C_S(l) \cdot \psi^2$, and so the change ΔW in the total energy as a function of ψ is

$$\Delta W(\psi) = \Delta W_{S_B}(\psi) - \Delta W_S(\psi) = [C_{S_B} - C_S(l)]\psi^2. \tag{21}$$

Now $C_S(l)$ increases with the particle size, whereas C_{S_B} is independent of particle size, which means that $\Delta W(\psi) < 0$ if the particle exceeds the critical size, so the $\varphi = 90°$ array is unstable and the double layer becomes polarized spontaneously.

Area S_A and S_B have been taken as sharply defined in order to simplify the discussion, whereas the boundaries are actually not sharp; but this has no effect on the general argument. If (15) is applied for φ for bacteria or proteins, or (18) for φ for colloidal particles, this involves the assumption that S_A, $S_B \ll S$. It would be very difficult to obtain an exact solution to the problem, since φ must be dependent on the geometry of the adsorbed molecules, among other features.

These permanent dipoles in colloidal particles raise the question of the dipole moment of the ionic double layer (Stern layer). Here we merely note that our observed μ and φ should be considered only as effective quantities, since the dipole will produce an electrical image in the material of the particle and also rearrangement of the ions in the Stern layer. The observed (effective) μ must therefore be less than the true value, while the true φ must be less than the value calculated.

Conclusions

1. Three methods have been developed for examining the electrooptic in colloidal solutions, which employ Π_1, Π_2, and R fields.

2. Dichroism of conservative type (dityndalism or double scattering is the main electro-optic response of a colloid.

3. Periodic modulation in a Π_1 field shows that the colloid particles have permanent dipole moments μ in water or other polar media.

4. Further evidence for μ is that the lag angle α for given E and ω can exceed 45°.

5. Constancy of α for fixed ω/E is evidence not only for the presence of μ but also that it plays the dominant part.

6. A finite μ is always present for a colloid in water, at least.

7. Permanent dipole moments have been measured in water for colloids of various types (anisaldazine, azoxyanisole, anisylidenebenzidine, benzopurpurine, anil brown, Congo red, bentonite suspension, and suspensions of bacteria).

8. In every case, the axis of μ lies along the longest axis of the particle.

9. The permanent dipole has a surface origin and is proportional to the surface area of the particle, other things being equal.

10. This μ is due to spontaneous unipolar orientation of adsorbed water molecules on the surface of the particle. Such a film of polar molecules may be compared with a ferroelectric domain.

11. If the adsorbed film is taken as monomolecular, and if it is assumed that μ is due solely to this, then the angle φ between the dipole in a molecule and the surface is 83 ± 2° in most cases. These values have been obtained with monomorphic monodisperse particles (bacteria) and with polydisperse suspensions made in various ways.

References

1. N. A. Tolstoi and P. P. Feofilov, Dokl. AN SSSR, 66:617 (1949).
2. N. A. Tolstoi, Dokl. AN SSSR, 110:893 (1955).
3. N. A. Tolstoi, A. A. Spartakov, and G. I. Khil'ko, Kolloidn. Zh., 22:705 (1960).
4. N. A. Tolstoi and A. A. Spartakov, Kolloidn. Zh., 28:580 (1966).
5. A. A. Spartakov and N. A. Tolstoi, Zh. Éksp. Teor. Fiz., 29:385 (1955).
6. N. A. Tolstoi, A. A. Spartakov, and A. A. Trusov, Opt. i Spektroskopiya, 19:826 (1965).
7. A. A. Spartakov, N. A. Tolstoi, and A. A. Trusov, Opt. i Spektroskopiya, 20:535 (1966).
8. G. T. Edsall, Fortschr. Chem. Forsch., 1:119 (1949).
9. N. R. Oncley, Chem. Rev., 30:433 (1942).
10. C. T. O'Konski, J. Am. Chem. Soc., 79:4815 (1957).

THE ELECTRICAL POLARIZABILITY OF ANISODIAMETRIC COLLOIDAL PARTICLES IN AQUEOUS SOLUTIONS

S. P. Stoilov

Institute of Physical Chemistry, Bulgarian Academy of Sciences

Light scattering by colloids in ac fields is discussed, and it is concluded that the electrical polarizability of colloid particles larger than 200 Å in aqueous solutions is due to the ionic double layers.

Introduction

The electrokinetic potential ζ is used [1-3] to characterize the electrical properties of colloidal particles, and the value of this is deduced from electrokinetic studies. This has given valuable information on the compositions of colloidal solutions and also the basis of the current theory of lyophobic systems. On the other hand, there is considerable difficulty in deriving reliable values for the particle charges from the ζ potential, partly because the theory is inadequately developed and also because there is uncertainty as to the dielectric constant and viscosity in the diffuse electrical layer. Highly conflicting results have been obtained [4] from attempts to deduce the distance to the slip plane from the ζ potential.

Another way of characterizing the electrical properties of colloidal solutions has been proposed [5-9], which appears to be free from these disadvantages, namely derivation of quantities analogous to those used in characterizing the electrical properties of molecules (the electrical polarizability and the dipole moment). It is usually assumed that these are purely bulk properties of the particles [7-12].

Here I show that the electrical polarizability of a colloidal particle is not always due to the bulk properties and that the surface can make a substantial contribution. The mechanism of the surface electrical polarizability is discussed.

Surface Electrical Polarizability

It has been shown [13] from experiments with potassium polysulfates that the polarizability of the colloidal particles is of surface origin. A later study [14] indicated the place of shift of the changes that determines the polarizability (a layer directly adjoining the surface). In these measurements, the electrical conductivity of a colloidal solution was determined as a function of field. The result was confirmed [15, 16] via the dispersion of the dielectric constant for polystyrene suspensions.

In 1955 O'Konski [17] discussed polarizability due to surface charge displacement without details of the mechanism. In 1957 O'Konski and Haltner [18] measured the birefringence of colloidal particles in an electric field and showed that the effect for suspensions of tobacco mosaic virus was entirely due to surface electrical polarizability. They demonstrated dispersion in the effect at several kHz, as well as pH and concentration dependence. In 1960 O'Konski [19] gave a detailed theory of surface electrical polarizability and calculated the dielectric constants of colloidal solutions for particles of various shapes.

Mandel [20] considered theoretically the surface electrical polarizability of cylindrical particles; he considered that the charge displacement occurs at the surface. Dukhin [21] has recently given theoretical relations between the polarizability, charge, and electrolyte concentration. When these are complete, it should be possible to make a comparison with experiment.

Polarizability of Cylindrical Particles and the Change in Light Scattering in an External Electrical Field

This method has given reliable evidence on the dominant part played by surface polarizability. Here I give results for two objects differing greatly in length: the monomer and dimer of tobacco mosaic virus. The measurements gave preliminary estimates of the particle length and indicated at what length the surface polarizability may be expected to predominate over the bulk effect. The apparatus has previously been described [22, 23], and details of the objects were given there.

The quantity α characterizes [24] the change in scattered intensity in an electric field

$$\alpha = \frac{J_E}{J_0} - 1, \tag{1}$$

in which J_E is the scattered intensity in the field and J_0 is the same for zero field.

Theory [25, 26] indicates that the slope of $\alpha(E^2)$ at low degrees of orientation (orientation energy $< kT$) gives the polarizability:

$$\alpha = \frac{3}{4J_0} GE^2 F(K), \tag{2}$$

in which

$$G = \frac{p^2}{3k^2T^2} - \frac{\gamma}{3kT},$$

$$F(K) = \frac{J_0}{3} - \frac{\sin 2K}{4K^3} - \frac{1}{2K^2},$$

$$K = K'l \sin\frac{\theta}{2}.$$

Here p is the permanent dipole moment of a particle along its long axis, k is Boltzmann's constant, T is absolute temperature, γ is particle polarizability, $K' = 2\pi/\lambda$, λ is the wavelength of the light in the medium, l is particle length, θ is the angle between the directions of illumination and observation, E is field, and J_0 is given [25-27] by

$$J_0 = \frac{1}{K} \int_0^{2K} \frac{\sin x}{x} dx - \left(\frac{\sin K}{K}\right)^2.$$

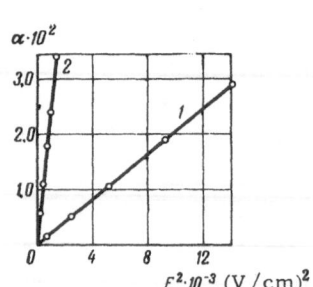

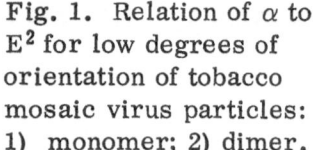

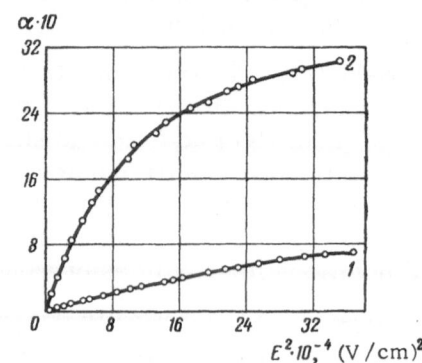

Fig. 1. Relation of α to E^2 for low degrees of orientation of tobacco mosaic virus particles: 1) monomer; 2) dimer.

Fig. 2. Relation of α to E^2 for arbitrary degrees of orientation of tobacco mosaic virus particles: 1) monomer; 2) dimer.

Formula (2) follows also from Wippler's formula [25]. In our case the field was perpendicular to the plane of observation and illumination.

Figure 1 shows $\alpha(E^2)$ for low degrees of orientation for frequencies large relative to the relaxation frequencies of the particles but small relative to the relaxation frequencies of the charges whose displacement determines γ, i.e., for frequencies in the plateau region on the dispersion curves [26, 28].

There are two ways of deducing l, which is needed in the calculation of γ: from electron micrographs and from the saturation value α_∞, as found by extrapolation of $\alpha(E^2)$ (Fig. 2). The two values are almost the same [23] and are $l_1 = 0.29 \mu$ for the monomer and $l_2 = 0.79 \mu$ for the dimer. The values given by (2) are then

$$\gamma_1 = 1.8 \cdot 10^{-13} \text{ cgs} \quad \text{and} \quad \gamma_2 = 9.7 \cdot 10^{-13} \text{ cgs.}$$

These l are mean values, as the objects are very much polydisperse, so the γ are also mean values. A rough estimate indicates that the results for both objects are close to the weighted-mean values, but a little higher.

The γ for these objects can be calculated for two cases: 1) when they are due to the bulk properties; 2) when they are due to displacement of surface charges.

In the first case the formula is [29, 30]

$$\gamma_b = \frac{4\pi(\varepsilon_1 - \varepsilon_2) - \dfrac{(\varepsilon_1 - \varepsilon_0)(\varepsilon_2 - \varepsilon_0)}{\varepsilon_0}(L_1 - L_2)}{\left(4\pi - \dfrac{\varepsilon_1 - \varepsilon_0}{\varepsilon_0}L_1\right)\left(4\pi - \dfrac{\varepsilon_2 - \varepsilon_0}{\varepsilon_0}L_2\right)} V, \quad (3)$$

in which ε_1 and ε_2 are the dielectric constants along the long and short axes of the particle respectively, ε_0 is the dielectric constant of the media, L_1 and L_2 are steric factors dependent on the ratio of the lengths of the axes [29], and V is particle volume.

The values $\varepsilon_1 = \varepsilon_2 = 3$, $\varepsilon_0 = 80$, $L_1 = 0$, and $L_2 = 2\pi$ [18] give $\gamma_{b_1} = 2.9 \times 10^{-16}$ cgs and $\gamma_{b_2} = 7.9 \times 10^{-16}$ cgs.

TABLE 1. Relation of Polarizability to Particle Length

Object	l, μ	α_∞	γ_{obs} (cgs)	γ_b (cgs)	γ_s (cgs)
Virus-1 ..	0.29	1.25	$1.8 \cdot 10^{-13}$	$2.9 \cdot 10^{-16}$	$1.6 \cdot 10^{-14}$
Virus-2 ..	0.79	4.67	$9.7 \cdot 10^{-13}$	$7.9 \cdot 10^{-16}$	$4.1 \cdot 10^{-14}$

Maxwell's formula [31] gives the γ due to displacement of surface charges:

$$\gamma_s = \frac{\varepsilon_0}{24 \ln q} l^3, \tag{4}$$

in which q is the ratio of the major and minor semiaxes of the ellipsoid of rotation representing the particle.

These values are much closer to the observed ones, namely $\gamma_{s_1} = 1.6 \times 10^{-14}$ cgs and $\gamma_{s_2} = 4.1 \times 10^{-14}$ cgs; the observed γ exceed γ_b by more than two orders of magnitude, whereas they exceed γ_s by less than one order of magnitude.

There is a relation of the observed γ to l, though only preliminary conclusions can be drawn, as we have data for only two lengths.

We see that γ_1/γ_2 falls between the square and the cube of the ratio of particle lengths. All the theories of surface polarizability [14-20] give γ as proportional to l^3, which indicates when the surface mechanism predominates. The bulk polarizability is proportional to l (at constant cross section), so the two polarizabilities would be of the same order for l of 200-300 Å, or values less than those used above by a factor ten. Particles with $l < 200$ Å are thus polarized by the bulk mechanism. However, the particles cease to be rods if the length is greatly reduced. This conclusion serves to explain the results of Krause and O'Konski [32], where the surface mechanism was not detected in the electrical birefringence of solutions of small protein molecules.

It follows from recent systematic measurements [33] of γ as a function of frequency and ionic strength that it is possible to distinguish parts of this surface polarizability as due to charge displacement at the surface and in the diffuse part of the double electrical layer. In most cases, the two are comparable, so it is very likely that the ions at the phase interface have considerable mobility along the surface.

References

1. A. Sheludko, Colloid Chemistry, Sofia (1963).
2. H. R. Kruyt, Colloid Science, Elsevier, Barking (England) (1949-1952).
3. S. S. Voyutskii, Textbook of Colloid Chemistry, Goskhimizdat, Moscow (1961).
4. W. G. Eversole and W. W. Boardmen, J. Chem. Phys., 9:789 (1941).
5. N. A. Tolstoi, Dokl. AN SSSR, 710:893 (1955).
6. C. T. O'Konski and B. H. Zimm, Science, 111:113 (1950).
7. G. Schwarz, Z. Physik, 145:563 (1955).
8. A. Peterlin and H. Stuart, Z. Physik, 112:139 (1938).
9. M. V. Vol'kenshtein, Molecular Optics, GITTL, Moscow (1951).
10. H. Benoit, Ann. Phys., 6:561 (1951).
11. I. Jr. Tinoco, J. Am. Chem. Soc., 77:4486 (1955).
12. Sh. Ikeda, J. Chem. Phys., 38:2839 (1963).

13.   M. Eigen and G. Schwarz, Z. Phys. Chem., 4:516 (1955).

14.   G. Schwarz, Z. Phys. Chem., 19:286 (1959).

15.   G. Schwarz, J. Phys. Chem., 66:2636 (1962).

16.   H. P. Schwan, G. Schwarz, J. Maczuk, and H. Pauly, J. Phys. Chem., 68:2626 (1962).

17.   C. T. O'Konski, J. Chem. Phys., 23:1559 (1955).

18.   C. T. O'Konski and A. Haltner, J. Am. Chem. Soc., 79:5634 (1957).

19.   C. T. O'Konski, J. Phys. Chem., 64:605 (1960).

20.   M. Mandel, Mol. Phys., 4:489 (1961).

21.   S. S. Dukhin, this volume, p. 287.

22.   A. Sheludko and St. Stoilov, Izv. Inst. Fiz. Khim. Bolg. Akad. Nauk, 2:191 (1962).

23.   St. Stoilov, A. Sheludko, and R. Chernev, Godishnik Sofiskiya Univ. (in press).

24.   A. Sheludko, Godishnik Sofiskaya Univ., 54:229 (1959-1960).

25.   C. Wippler, J. Chim. Phys., 51:122 (1954).

26.   St. Stoilov, Izv. Inst. Fizikokhim. Bulgar. Akad. Nauk, 4:155 (1964).

27.   H. C. van de Hulst, Light Scattering by Small Particles, Wiley, New York (1957).

28.   St. Stoilov, Izv. Inst. Fizikokhim. Bulgar. Akad. Nauk, 4:167 (1964).

29.   A. Peterlin and H. Stuart, Hand- u. Jahrbuch Chem. Phys., Vol. 8, Part 1B, Leipzig (1943).

30.   R. Gans, Ann. Phys., 62:331 (1920).

31.   J. C. Maxwell, A Treatise on Electricity and Magnetism (3rd ed.), Vol. 2, Oxford (1892).

32.   S. Krause and C. T. O'Konski, Biopolymers, 1:503 (1963).

33.   St. Stoilov and I. Petkanchin, Godishnik Sofiskiya Univ. (in press).

ELECTRON MICROSCOPY OF SURFACE EFFECTS

G. I. Distler

Institute of Crystallography, Academy of Sciences of the USSR

Decoration methods have been developed for examining structural defects that form active centers in insulators and semiconductors.

Surface phenomena are very important in physics, chemistry, and biology. Adsorption, catalysis, adhesion, friction, coagulation, peptization, nucleation, enzyme processes, and surface processes in semiconductors are all examples of surface phenomena.

Many surface phenomena occur on crystalline surfaces, but these are not the boundaries of ideal crystal lattices, since they have various defects, vacancies, impurity centers, block boundaries, etc. Physicochemical methods have been the main ones used in research on solid surfaces; these give average values for the surface as a whole. Direct structural methods are required in order to determine the nature, structure, number, and distribution of inhomogeneities. Transmission electron microscopy is one such method. The main method has been that of replicas, which reveals only surface geometry, not surface activity.

Decoration methods have recently come into general use [1-3], in which a chemically very inert material (gold or platinum) is caused to crystallize selectively at certain sites during vacuum deposition. This method has revealed monatomic steps and points of emergence of dislocations in alkali halides. The method has been extended to vitreous materials [4]. However, the method reveals principally the surface relief, and other methods are required, which employ specific interactions, i.e., reaction at active sites.

New electron-microscope methods for this purpose have been developed at this institute, which are based on selective crystallization [5, 6] at surfaces with special properties or on selective adsorption [7]. The crystallization is produced by chemical reaction, the active centers being nucleation initiators. Selective adsorption of colloidal particles occurs at electrically active centers. These methods of chemical and electrical decoration block the active centers, so less active centers can then be detected by altering the decoration conditions, e.g., by increasing the supersaturation. The particle size distribution is an indication of the activity distribution of the centers, since the crystallization begins selectively at the most active centers, and hence these centers produce crystals larger than those arising later at less active centers. Decorating crystals identical in size represent centers similar in activity.

71

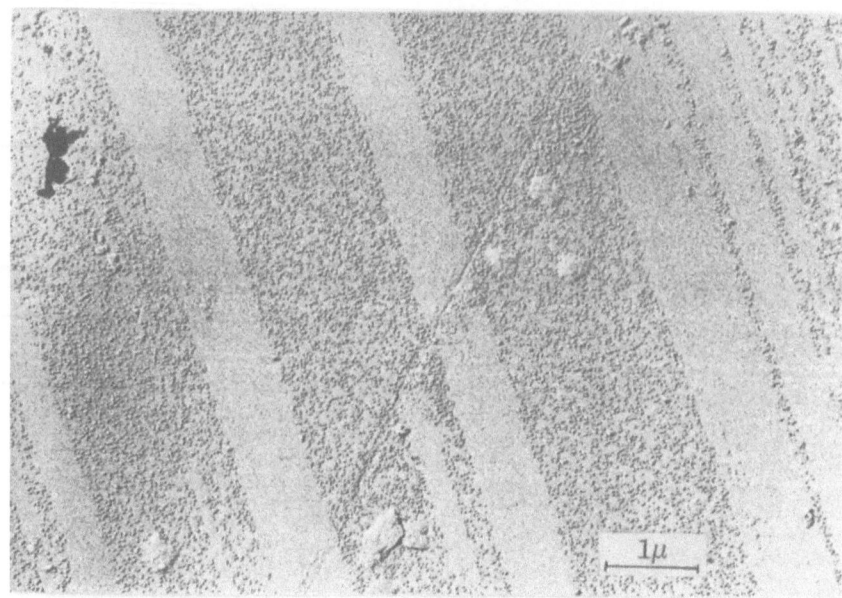

Fig. 1. Chemical decoration of a silicon surface parallel to the
growth axis.

Various methods of chemical and electrical decoration have been used, but the most experiments have been done via the crystallization of PbS, which is produced in aqueous solution
by a very slow reaction of lead acetate with thiourea, whose rate is dependent on the temperature, the concentrations, and the balance between the reagents. The very earliest stages (most
active sites) can therefore be observed. Electrical decoration is produced with colloidal PbSe,
which is formed in aqueous solution by reaction of lead acetate with selenourea. The deposited
particles in both methods are detached from the surface with gelatin; they may subsequently be
shadowed for viewing in the Hitachi-11 electron microscope. These methods have been applied
to silicon, mica, quartz, and other materials.

Very interesting results have been obtained with silicon single crystals grown by Czochralski's method, which have a stratified distribution of structural defects on account of periodic variation in the growth rate. These crystals are excellent test objects, because the decoration patterns are closely related to the crystal growth.

Figure 1 shows the PbS pattern on a boron-doped silicon crystal cut parallel to the growth
axis. The decoration is also stratified, the mean size of the PbS crystals being about 200 Å.
If the reaction is stopped rapidly, no crystals appear between the most active bands, which shows
that the reaction is highly selective. The active bands vary in width from 0.5 to 10 μ, while the
density of active sites within a band is $4-5 \times 10^{10}$ cm^{-2}. Increase in reaction time (which implies increase in supersaturation [8]) causes crystals to appear in the less active areas (Fig. 2)
with the same density. These crystals correspond to a different, less active, type of center.
Sections perpendicular to the growth axis (Fig. 3) show marked inhomogeneity but no stratifica-
tion. The banding in certain orientations shows that the active centers are structural defects.

The PbS forms three-dimensional crystals at sites which differ greatly in activity from
adjacent areas and which appear to act independently; they are the most reactive parts of the
surface. The patterns show that the activity differences between faces are due mainly to the
defects, not to the atomic structure of the ideal faces.

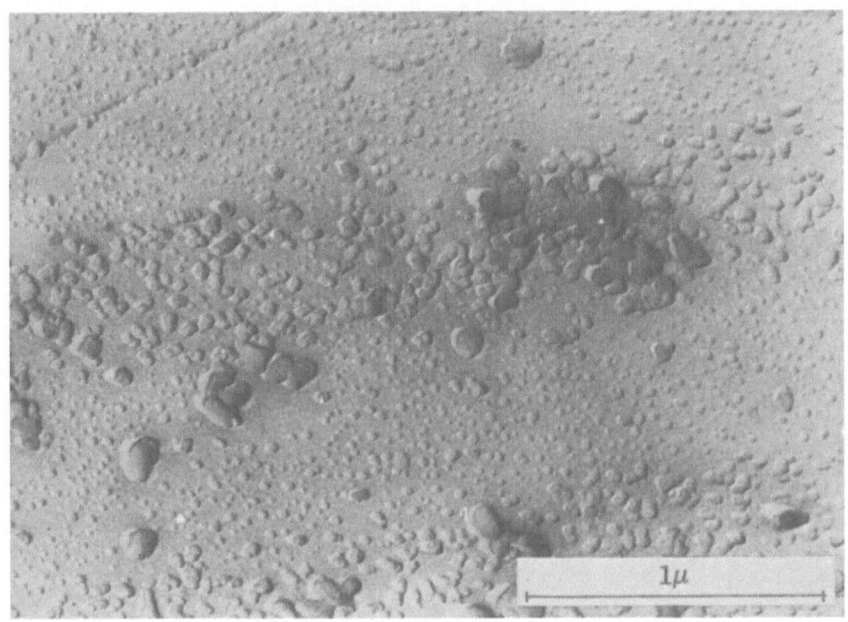

Fig. 2. Chemical decoration of a silicon surface at elevated supersaturation.

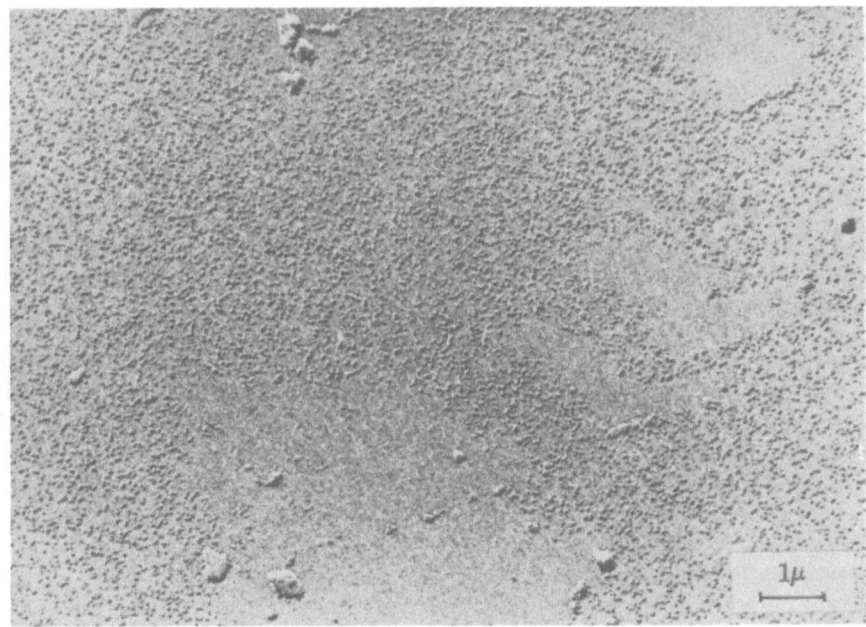

Fig. 3. Chemical decoration of a silicon surface perpendicular to
the growth axis.

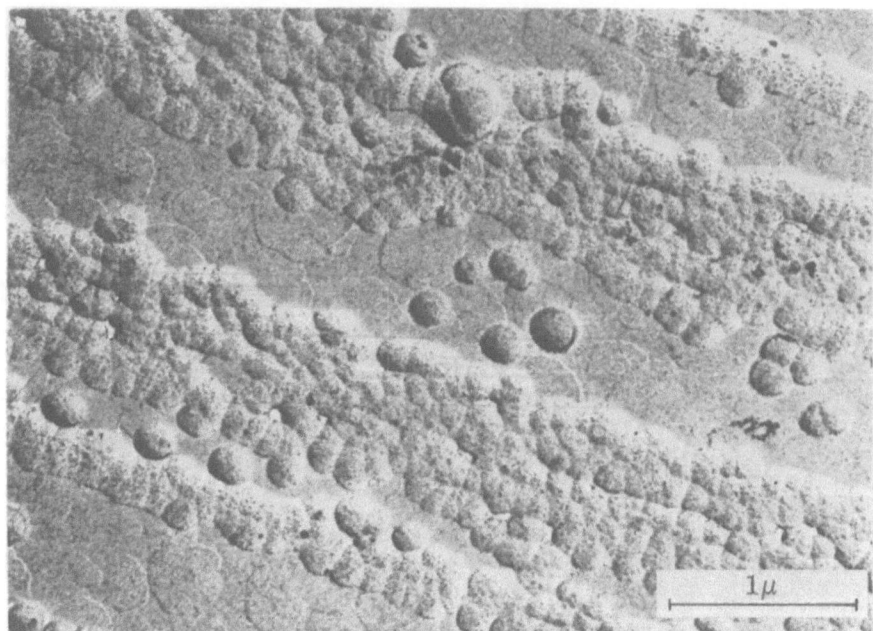

Fig. 4. Secondary decoration of mica by PbS crystals.

These active sites in silicon differ in structure. Bands with randomly disposed centers are accompanied by ones with selectively disposed centers. For instance, there are clumps of centers at the edges of the bands, which indicate segregation of defects at boundaries between parts differing in structure.

Cleaved surfaces of mica crystals give very interesting patterns. First, separate thin crystals are formed, which gradually spread to cover the entire surface; then secondary decoration occurs [9-11] (Fig. 4). These patterns correspond to cleavage steps, which are not decorated directly. The secondary decoration occurs because the steps are not revealed until other more active sites have been blocked.

Decoration with colloids is the second method. Many structural defects (vacancies, dislocations, etc.) are electrically charged and so have effects distinct from those of adjacent ideal areas. We have used PbSe particles 50-500 Å in size, which provide adequate resolution and also indicate the electrical microrelief.

Figure 5 shows that a silicon crystal grown by Czochralski's method adsorbs the particles in layers when it is cut parallel to the growth axis. The width and disposition of the bands are as in the crystallization of PbS, the adsorption density being $2-5 \times 10^{10}$ cm^{-2}. The patterns also have the morphology found with PbS. The electrical activity of impurity centers is seen with stratified p-n junctions, in which the regions have charged impurities of different signs (Fig. 5).

The structure of the boundaries in these p-n junctions can be seen [12], and these boundaries are [13] ones between phases differing in real structure. Figure 6 shows a p-n junction decorated with PbSe at two different magnifications. The crystals within a band are randomly disposed, but there are clumps at the boundaries, and these clumps repeat at intervals of roughly 0.2-0.4 μ, which indicates a complicated mutual penetration of the phases.

Epitaxial silicon films have been produced on (111) silicon wafers by crystallization from the vapor, and these give interesting PbSe patterns (Figs. 7 and 8). There are randomly placed hexagonal (occasionally square) groups of particles, which vary in size and degree of order,

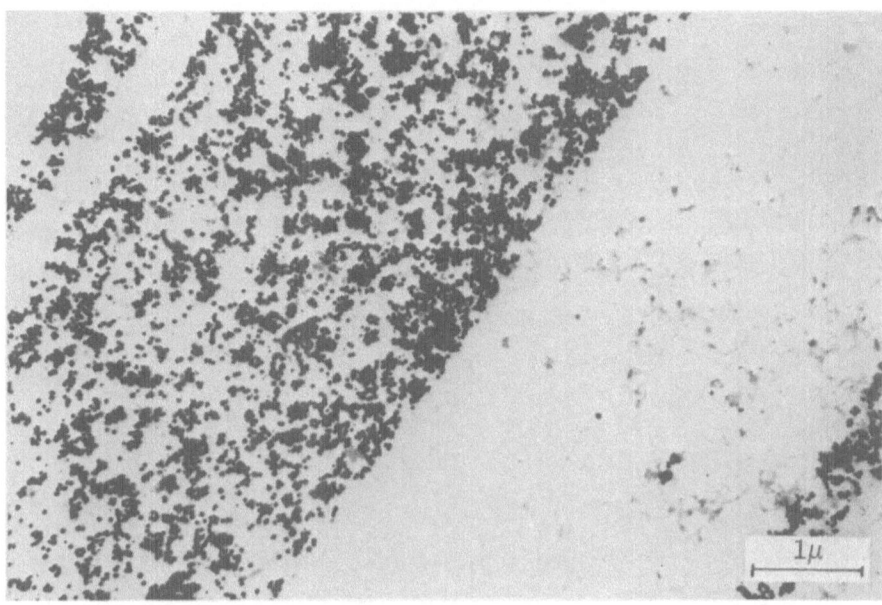

Fig. 5. Electrical decoration by PbSe particles of a silicon surface
parallel to the growth axis.

especially at the margins. These groups extend up to several μ in size, while the density is 10^8 cm^{-2}. There are also isolated particles giving a density (with the groups) of $\sim10^{10}$ cm^{-2}. The ordering in the groups would appear to reflect the detailed structure of the active areas; areas where the order is disturbed may represent the presence of impurities of reverse sign or other types of defect. Such groups have been produced only on epitaxial silicon films, which indicates the importance of a surface unaffected by polishing or etching. Each group represents a clump of charged defects of one sign and essentially reveals a local p-n junction.

The area actually covered by the charged active centers is substantially less than the area influenced by these centers, so a surface revealed as inhomogeneous by these direct methods may behave as more or less homogeneous in respect of certain surface processes.

The active areas are structural defects, which have excess free energy and which allow surface reactions to occur with the least activation energies. If the defect concentration is high, the defects tend to form clumps, and these act as nuclei of a new phase if they become large.

The theory of dilute solutions of weak electrolytes has been applied [14] to impurities that occur in semiconductors as isolated atoms or ions, or pairs of these. Local clumps of defects can be compared with colloidal particles, so the principles of colloid chemistry can be applied, the dispersion medium being the ideal lattice. The defects interact via this lattice in the bulk, while at the surface they interact with the external medium as fairly independent active particles.

These surface centers can exert forces of various types: electrostatic, dipole, and van der Waals (as for colloidal particles). A difference from a colloidal solution (where there are particles of one sign only) is that the surface centers can have different signs. The various types of surface defect produce various types of surface forces, and the mechanisms are similar to those for colloidal systems.

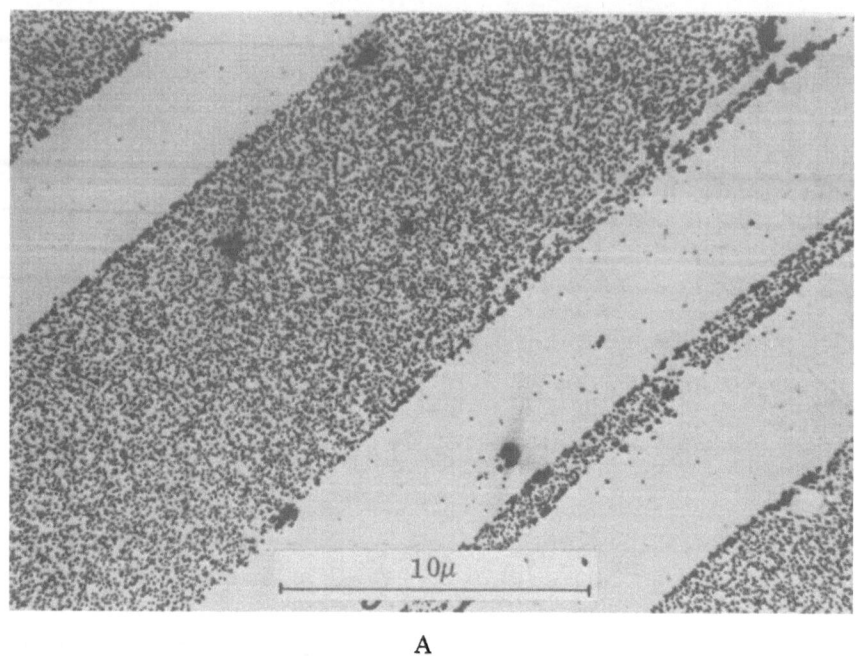

A

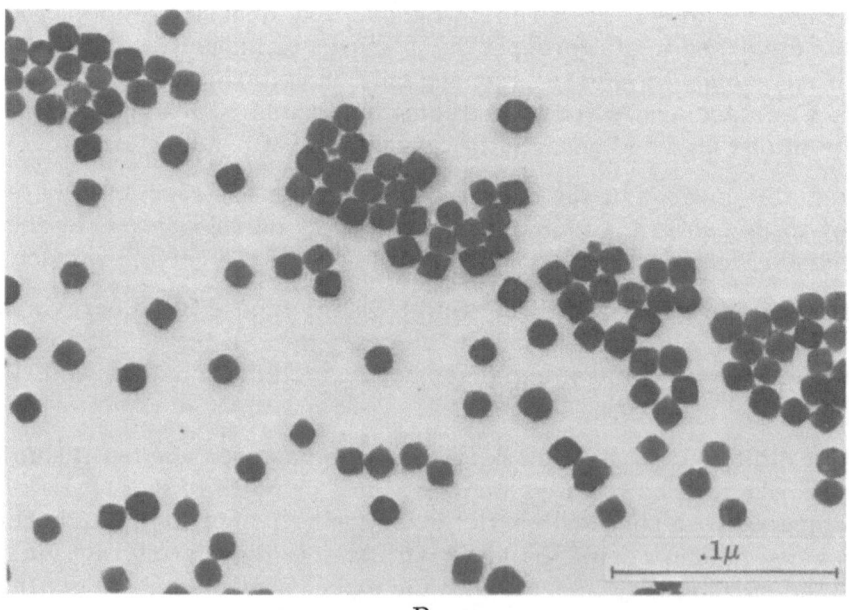

B

Fig. 6. Electrical decoration of the fine structure of a stratified
p-n junction at different magnifications.

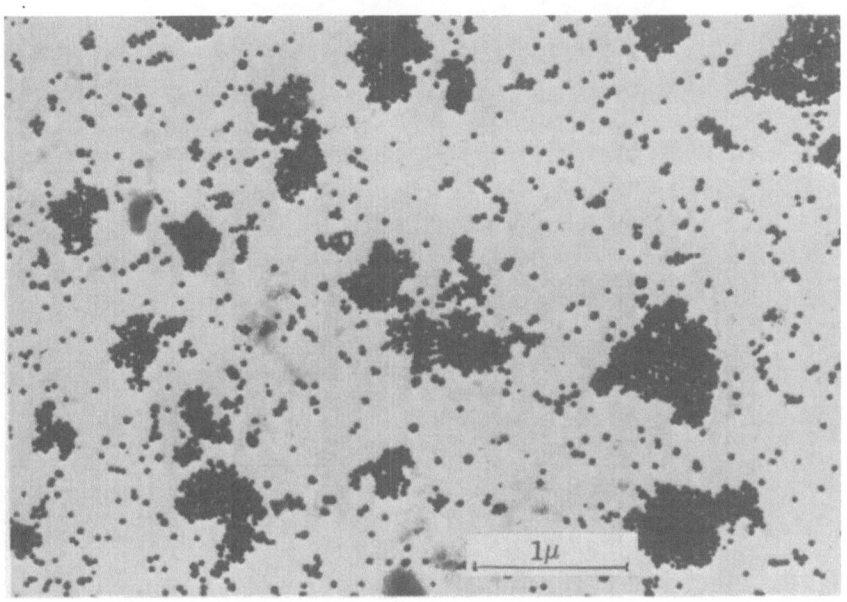

Fig. 7. Electrical decoration of epitaxial films of silicon.

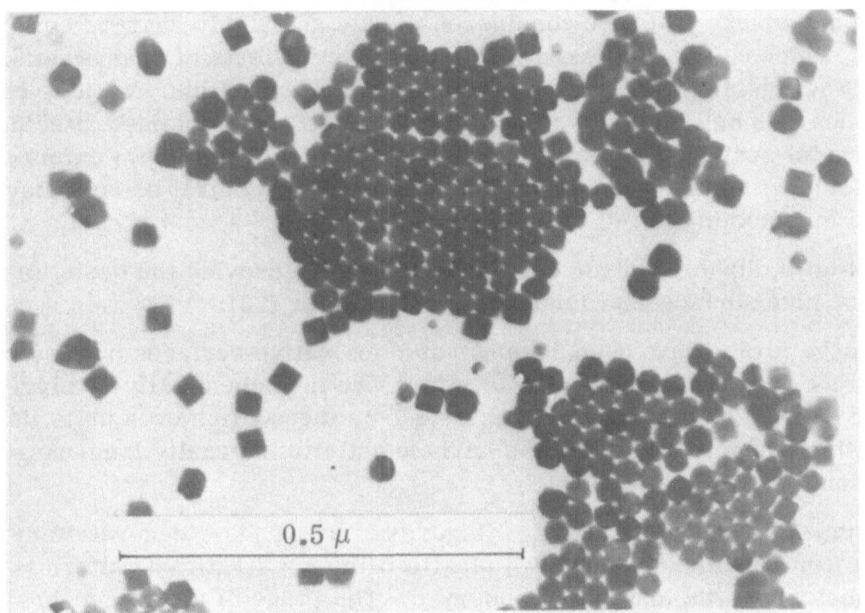

Fig. 8. Electrical decoration of epitaxial films of silicon.

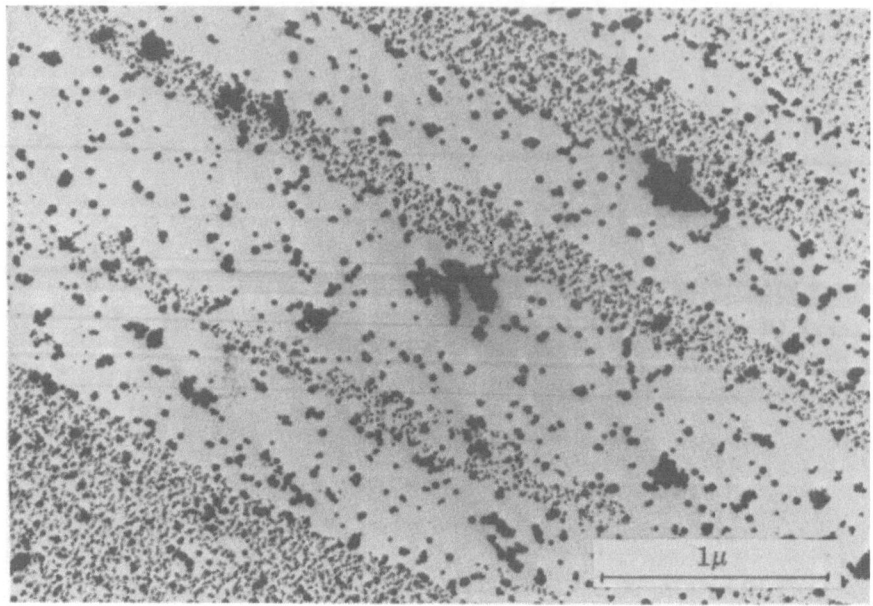

Fig. 9. Electrical decoration of a basal face of a quartz single crystal via an intermediate plastic film 300 Å thick.

Deryagin has considered the interaction and coagulation of colloidal particles and has developed a theory incorporating the electrostatic repulsion of the diffuse double ionic layers. Deryagin [15] and others [16] have considered weakly and highly charged colloidal particles of various sizes, which can be compared with the charged defects of real crystals. Deryagin's theory of heterocoagulation [17] can be applied to defect activities. Various observed surface effects (adhesion of small particles to solid surfaces [18, 19], oriented attachment of microcrystals to crystal surfaces [20]) are probably related to a substantial extent to interaction of charged particles with surface defects. The activity of structural defects may also make itself felt via dipole interaction [21].

Electrostatic, dipole, and van der Waals forces all provide the basis for long-range effects and hence give surface phenomena a bulk character [22].

Long-range forces have been demonstrated for active surfaces by electron microscopy† after deposition of neutral films of plastic, which are then chemically or electrically decorated. The stratified decoration of silicon is reproduced on these films even up to thicknesses over 1000 Å [23], which demonstrates that the surface activity is usually long-range and has a large sphere of action.

Quartz crystals show this clearly. Figures 9 and 10 show decoration by PbSe particles on basal faces coated with 300 and 750 Å plastic films. A stratified pattern is expected on the basis of the sector growth, and this is observed. The range of action appears to extend well beyond 750 Å. A very interesting feature is that there are two systems of bands, which take up particles of different sizes: one system has particles of 250-300 Å, while the other has ones of 50-150 Å.

The theory of the interaction of colloidal particles indicates that the radius of action is dependent on the size of the particle, other things being equal. In other words, the least active centers can be revealed only by interaction with particles exceeding a certain critical size.

_____

† This topic is the subject of another paper in this volume [25].

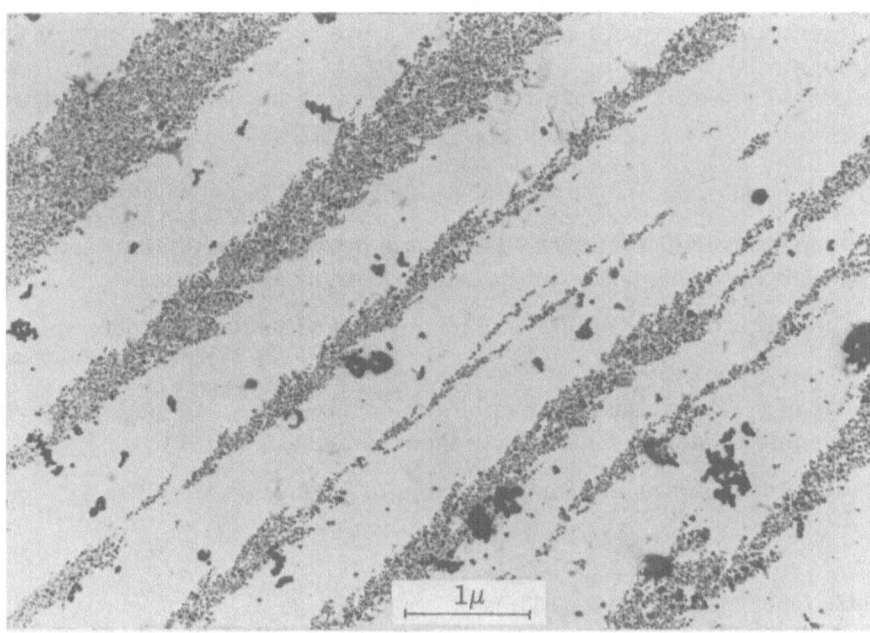

Fig. 10. Electrical decoration of a basal face of a quartz single
crystal via an intermediate plastic film 750 Å thick.

The illustrations show that this approach is justified; for instance, the quartz crystals have at
least two types of center, which interact with PbSe particles of different sizes. A plastic film
750 Å thick takes the particles outside the radius of action of centers of one type, and these
are not seen. Figure 10 shows that the action of the other centers persists to greater distances.

Decoration shows that the active defects on silicon single crystals have a density of
10^{10}-10^{11} cm^{-2}, which corresponds very closely to the density of fast surface states revealed
by electrophysical measurements [13], which are also discrete and have a fairly constant density but which may be donor or acceptor in nature. The electron micrographs show that the
active centers correspond in number and properties to the fast surface states, and also that
these centers are groups of impurities and vacancies. We may therefore conclude that these
centers are mainly responsible for those states and lie at the interface between the semiconductor and the oxide film. The slow surface states lie at the interface between the oxide film
and the gas [13], and these have a density of 10^{13}-10^{15} cm^{-2}. Rzhanov [24] has shown that ad-
sorption−desorption processes convert one type of state to the other, but no structural models
for these states have previously been available.

These long-range effects of surface defects would appear to explain the origin of the slow
states and their relation to the fast states. Electrically active surface defects act via the oxide
film by the long-range mechanism (or directly) on the gas adsorption; a single defect may interact with many adsorbed molecules, and the latter may affect the electrophysical properties
of the defects, perhaps so strongly as to convert fast states to slow ones in some cases. There
is thus an interconnected system of fast and slow states, the two types interacting via the long-
range mechanism. The fast states must always arise, on account of the defect structure of the
surface, whereas the production and properties of the slow states are dependent mainly on the
external conditions.

Groups of electrically active defects (especially impurities and vacancies) are the common important factor in the activity of crystal surfaces. Electron microscopy enables one to correlate many apparently unrelated results by reference to the actual defect structures of solids.

Conclusions

1. New electron-microscope methods have been developed (chemical and electrical decoration), which provide high resolution in the study of the real structure of solid surfaces.

2. Structural defects (mainly groups of impurities and vacancies) determine the activities of crystal surfaces.

3. The activity of a crystal surface can be interpreted via concepts of structural defects as colloidal particles in an inactive medium (an ideal lattice).

I am indebted to L. D. Kislovskii for a valuable discussion.

References

1. G. A. Bassett, Phil. Mag., 3:1042 (1958).
2. C. Sella, P. Conjeaud, and J. J. Trillat, Compt. Rend., 249:1987 (1959).
3. H. Bethge, Phys. Status Solidi, 2:1 (1962); 2:775 (1962).
4. W. Skatulla, Silikattechnik, 13:19 (1962); Struktur und Kristallisation der Glaser, Leipzig (1965), p. 211.
5. G. I. Distler, S. A. Daryusina, and Yu. M. Gerasimov, Dokl. AN SSSR, 154:1328 (1964).
6. G. I. Distler, S. A. Daryusina, and J. M. Guerassimov (Yu. M. Gerasimov), Third European Conference on Electron Microscopy, Prague (1964), p. 307.
7. G. I. Distler, Yu. M. Gerasimov, and N. M. Borisova, Dokl. AN SSSR, 165:329 (1965).
8. G. I. Distler and S. A. Kobzareva, Fiz. Tverd. Tela, 7:2450 (1965).
9. G. I. Distler and S. A. Daryusina, Kristallografiya, 7:266 (1962).
10. G. I. Distler and S. A. Daryusina, Kristallografiya, 9:119 (1964).
11. G. I. Distler, in: Growth of Crystals, Vol. 5B, Consultants Bureau, New York (1968), p. 51.
12. G. I. Distler and L. D. Kislovskii, Fiz. Tverd. Tela, 8:600 (1966).
13. A. Many, S. Goldstein, and N. B. Grover, Semiconductor Surfaces, Amsterdam (1965).
14. F. A. Kroger and H. I. Vink, Physica, 20:950 (1954); F. A. Kröger, H. J. Vink, and J. Van den Boomgaard, Z. Phys. Chem., 203:1 (1954); H. Reiss, C. S. Fuller, and F. I. Morin, Bell. Syst. Tech. J., 35:535 (1956); C. S. Fuller and H. Reiss, J. Chem. Phys., 27:318 (1957).
15. B. V. Deryagin, Izv. AN SSSR, OMEN, seriya khim., No. 5, 1153 (1937); Kolloidn. Zh., 7:285 (1941); B. V. Deryagin and L. D. Landau, Zh. Éksp. Teor. Fiz., 15:662 (1945).
16. E. Verwey and J. Overbeek, Theory of the Stability of Lyophobic Colloids, New York–Amsterdam (1948).
17. B. V. Deryagin, Kolloidn. Zh., 16(6):425 (1954).
18. A. Buzagh, Kolloid-Z., 47:370 (1929); 51:105, 230 (1930).
19. N. A. Fuks, Dokl. AN SSSR, 103:635 (1957).
20. M. P. Schaskolskaja and A. W. Schubnikow, Z. Kristallogr., A85:1 (1933); A. V. Shubnikov, How Crystals Grow, Izd. AN SSSR, Moscow–Leningrad (1935).
21. N. A. Tolstoi, Dokl. AN SSSR, 100:893 (1955); Opt. i Spektroskopiya, 19:826 (1965).
22. B. V. Deryagin, in: Research in Surface Forces, Vol. 2, Consultants Bureau, New York (1966), p. 3.
23. G. I. Distler and S. A. Kobzareva, Dokl. AN SSSR, 172:1069 (1967).

24.   A. V. Rzhanov, in: Surface Properties of Semiconductors, Izd. AN SSSR, Moscow (1962), p. 101.
25.   G. I. Distler and S. A. Kobzareva, this volume, p. 82.

LONG-RANGE ACTION OF THE SURFACE FORCES OF SOLIDS

G. I. Distler and S. A. Kobzareva

Institute of Crystallography, Academy of Sciences of the USSR

Decoration on the scale of electron microscopy shows that there are long-range surface forces that affect topo-
chemical reactions at relatively large distances from the surface. It is found that epitaxial growth can
occur at distances out to 1500 Å from the active surface of mica, while the active centers on silicon
have effects extending out to 1 μ.

Long-range forces have long attracted attention, because many effects in various areas
are difficult to explain without such forces. For instance, the surface of a solid affects the
structure of the adjacent layer of liquid [1], ordered colloidal systems are formed in solutions
[2], adsorption occurs [3], there are long-range actions in catalysis [4], globular proteins form
fibers [5], sols and suspensions show gelatinization [6], and crystals show mosaic growth [7].

The long-range forces of solids are of bulk character [8]. Until recently, solid surfaces
producing heterogeneous processes were considered as ideal. Mechanisms of surface reactions were discussed mainly in relation to lattice parameters and symmetry, bond type, lattice
energy, atomic and ionic radii, etc. Structural defects were assigned only a secondary role,
since there were no direct methods of examining the long-range forces of solids at the microscopic level.

Chemical decoration has been used at this institute [9] to examine the defect structure
of real surfaces with very high resolution. The method is based on selective crystallization of
the decorating agent at the active sites; it allows one to determine the character and distribution of the surface forces, including the long-range ones.

Long-range surface processes are very varied in type, but the mechanisms have a certain
common basis in the real structure of the surface. We have examined the long-range forces
at the surfaces of mica single crystals and have related them to the real surface structure.

Experimental

Freshly cleaved plates of mica (muscovite) were coated on the cleaved sides with neutral
plastic films and were placed in the reaction mixture, which consisted of solutions of 0.4%
lead acetate, 0.2% thiourea, and 0.28% caustic potash taken in the ratio 1:3:3. The reaction

TABLE 1. Thickness of Parlodion Film as a Function
of Solution Volume

Vol., ml	p, g	s, cm^2	h, Å
0.018	3.0×10^{-5}	4.5	400
0.036	6.0×10^{-5}	4.5	800
0.045	7.5×10^{-5}	4.5	1000
0.072	12.0×10^{-5}	4.8	1500
0.090	15.0×10^{-5}	4.5	2000
0.135	22.5×10^{-5}	4.5	3000

was run at 20°C for 1.5-2 h, which corresponded to the start of formation of a mirror layer of PbS on the side not bearing the plastic film. The supersaturation in this reaction increased with time, which allowed decoration of centers differing in activity [10].

Plastic films of the required thickness were produced by uniform deposition of a 0.16% solution of Parlodion in amyl acetate. The volume of solution was measured with a burette giving 0.009 ml per drop, which contained 15 μg of plastic. The film thickness h (cm) was

$$h = \frac{p}{sd},$$

in which p is film weight (g), s is area (cm^2) covered by the film, d (g/cm^3) is film density (1.66 g/cm^3) (Table 1). The thicknesses were measured to ±30%.

The specimen was then washed in distilled water and dried in air. A carbon film (~50 Å) was deposited on top. The lead sulfide, the carbon film, and the plastic layer were removed with gelatin, which was then dissolved in water; the plastic was partially dissolved in amyl acetate to reduce its thickness, and then the specimen was examined in transmission in a Hitachi-11 microscope. Some of the specimens were examined by means of replicas, which meant that the specimen was unaffected by the electron beam. Diffraction measurements were used to test for epitaxy.

Results

In a previous study of direct deposition of PbS on mica [11] we found that initially there was epitaxy of flat crystals ~1500 Å in diameter and of irregular shape (Fig. 1). The electron diffraction pattern shows that these particles have an oriented disposition. Increase in supersaturation causes the production of isometric crystals only 300 Å in size, with completely random orientations, which gave a powder pattern. There were about 10^9 epitaxial particles per cm^2.

Essentially the same results were obtained with plastic films up to several 1000 Å in thickness (Fig. 2), but epitaxy occurred only up to about 1500 Å, as is clear from the change in the diffraction pattern (Fig. 2). The density of the planar crystals was almost independent of the thickness of the plastic film.

Discussion

We have shown [10] that PbS crystallizes very selectively at active centers (structural defects), mainly impurities and vacancies, often electrically charged. The reaction responds to the type of center in accordance with the supersaturation, as the centers differ in activation energy for nucleation.

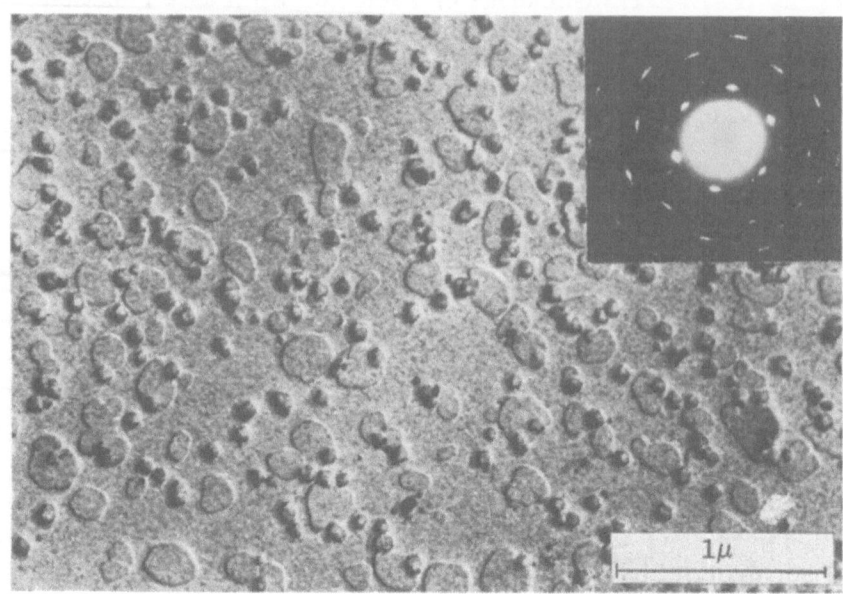

Fig. 1. Oriented crystallization of PbS on mica directly.

Fig. 2. Oriented crystallization of PbS on mica via a parlodion film
~1500 Å thick.

   In the early stages, two morphologically distinct types of crystal (planar and isometric) are formed. The constant occurrence of these indicates that the surface has two essentially different types of center. The first type has an orienting action and produces epitaxy, whereas the second does not. The disposition of the centers determines whether the crystals are uniformly distributed (Fig. 2) or form bands (Fig. 3).

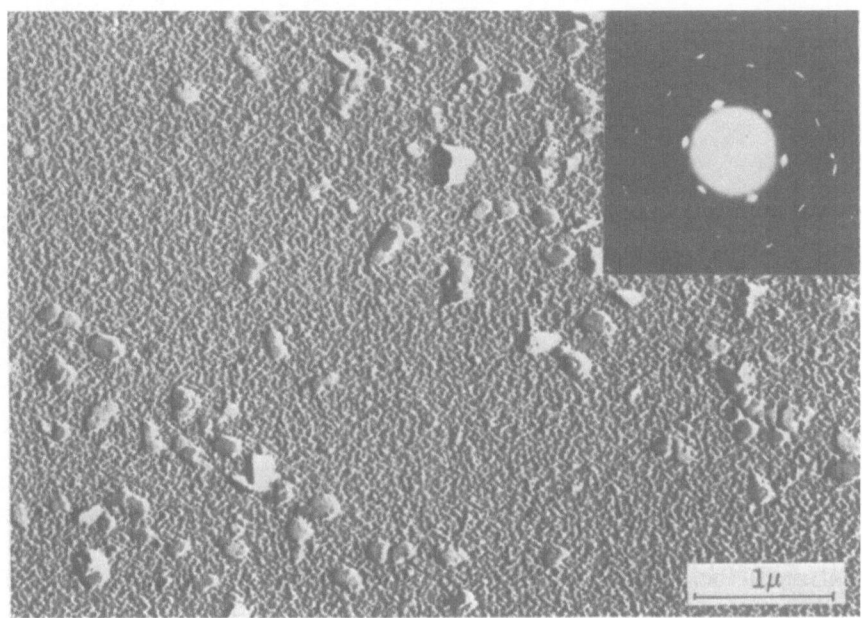

Fig. 3. Oriented crystallization of PbS on mica corresponding to a
definite pattern of centers.

The epitaxial centers have an effect extending out to about 1500 Å. This was the thickness found [12] for plastic films used in epitaxy of ammonium iodide on mica. As the ranges of action are the same, we may conclude that the same epitaxial centers are involved. The centers continue to act as nucleation initiators out to several 1000 Å, but the orienting action is lost beyond 1500 Å.

The other centers have no orienting action but a long range of action; they very closely resemble the long-range centers of silicon,[†] whose radius of action is over 1μ, but which also do not produce epitaxy.

The 1000-1500 Å radius of action for the epitaxial type indicates that these centers are not point defects but aggregates of defects. The active centers of solids may thus be considered as for the interaction of colloidal particles in solution [14], which involves forces of electrostatic, van der Waals, and dipole types. The range and type of the long-range interaction are determined by the electrical structure and size of the centers. The epitaxial centers seem to have a regular structure, which is determined firstly by the symmetry and strength of the bonds in the ideal lattice, and secondly by the physicochemical conditions of formation. The orienting action is dependent on this regularity and symmetry. It is not essential for the active centers to have a continuous structure; a center can consist of groups of defects interspersed with patches of ideal lattice. However, the electric fields of these groups in an active center are mainly parallel.

The second type of center has an irregular (amorphous) structure, which accounts for the lack of orienting action. The very large radius of action may be due to a large effective size, since the formation of a center of this type is not restricted by the conditions for formation of a regular structure. A center of this type may consist of many defects differing in elec-

---

[†] See our previous paper [13] for active centers of the second type that do not have an orienting action.

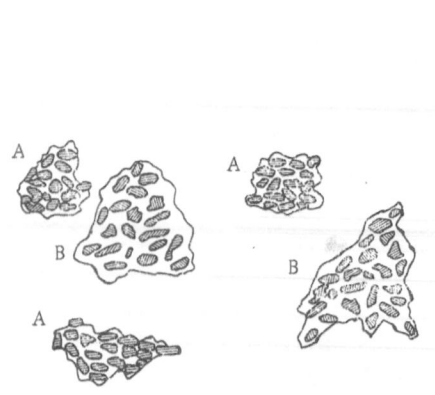

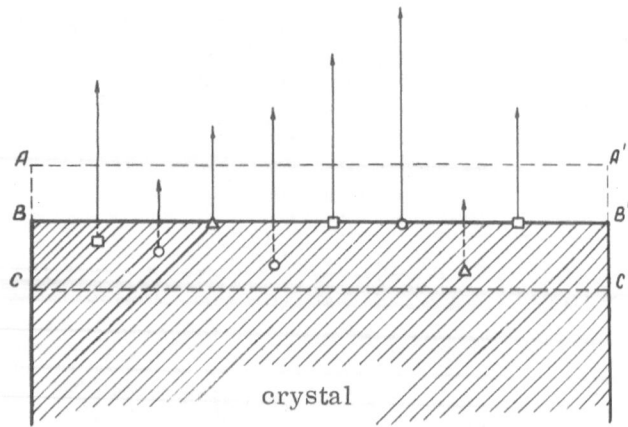

Fig. 4. Models of: A) regular (epit-axial) centers, B) irregular ones. The direction of the hatching in-dicates the orientation of the individu-al parts.

Fig. 5. Schematic diagram of the surface forces of a real crystal: Δ) regular (epitaxial) centers; ○) and □) irregular centers, BB') geometric surface of the crystal, AA') effective active surface, BB'CC') active layer adjacent to surface.

trical structure; moreover, the forces may differ in nature from those of epitaxial centers. Figure 4 shows models for both types of center.

Our experiments indicate that the long-range action has a complicated character. The long-range effects are made up of the effects from a whole range of local centers as well as from the influence of the surface as a whole. Microscopic examination reveals localized centers with a large radius of action, which appear specifically in relation to the nature of the surface defects.

Macroscopic examination reveals the integral long-range effect of the whole surface, which is then to be considered as quasi-homogeneous with a certain average defect structure.

Long-range effects arise not only from the bounding surface in the geometrical sense but also from a layer adjacent to the surface, which contains various active centers. The thickness of this layer is governed by macroscopic parameters such as the dielectric constant and by the nature of the defects; it may exceed 1000 Å. The overall long-range action is then due to a finite surface layer.

Figure 5 shows a scheme for the surface forces of a real solid, which reflects the variety of type and position in the active centers. This model indicates that the effective surface of a solid may be displaced some way from the geometrical surface.

The mode of action of these long-range forces is of considerable interest; electrostatic and molecular forces may be involved [15]. Often the medium in which the action is exerted (vacuum, air, nonpolar liquids, etc.) has no specific effect; but a polar liquid as a medium may have a specific effect. For instance, water may produce effects by a messenger mechanism. The surface of a solid affects the structure and orientation of the adjacent layer of liquid, and this may affect the next layer via the altered array of dipole moments and hydrogen bonds, and so on. This transfer process can extend to a considerable distance, e.g., to 10 μ or so [16].

In principle, the long-range effects may be produced by both mechanisms acting togeth-er, with one or other predominating in accordance with the conditions. However, it may be that the messenger mechanism predominates when water is the medium, as in our experiments, because the plastic films contain micropores of molecular size, which are not revealed by the electron microscope, and these may serve for messenger transfer, as in the structuring of water in glass capillaries [17].

Messenger transfer in the case of these films has the specific feature that there are numerous molecular pores above any single active center. The individual areas in a center produce various types of water structure in the pores, and this transmits the structural information out to large distances. If the areas in a center are identical in orientation (Fig. 4), each pore transmits the same information, and so we get epitaxy. Oriented adsorption may have a similar mechanism. Of course, oriented adsorption of water molecules may occur at surface defects, which may result in rigid dipoles with very large dipole moments [18]. In that case, the oriented long-range action is produced via dipole forces, not via the messenger mechanism.

The present results explain many long-range processes and confirm the need to study the real structures of solid surfaces, as these largely determine the mechanisms of many surface processes.

Conclusions

1. It has been shown that mica has long-range surface forces due to local centers representing structural defects. It has been found that there are two types of active center, one of which has an orienting action.

2. Models are considered for the surface as a whole and for the active surface centers, as well as possible mechanisms for long-range surface forces. The most likely one is that parts of a center act via microcapillaries containing structured water when plastic films are used.

3. There is considerable general significance in the results from correlation of long-range surface forces from mica with the surface defect structure.

We are indebted to L. D. Kislovskii for valuable advice and to I. F. Efremov for a fruitful discussion.

References

1. B. V. Deryagin and V. V. Karasev, Kolloidn. Zh., 15(5):365 (1953); Dokl. AN SSSR, 101(2):289 (1955).
2. J. D. Bernal and J. Fankuchen, J. Gen. Physiol., 25:111 (1941).
3. A. Rothen, Science, 112:331 (1950).
4. G. I. Distler and S. A. Kobzareva, in: Methods of Research on Catalysts and Catalytic Reactions, Vol. 2, Novosibirsk (1965), p. 17.
5. D. F. Waugh, Macromolecular Complexes, ed. M. V. Edds, Ronald Press, New York (1961), p. 3-18.
6. I. F. Efremov, Kolloidn. Zh., 18:276 (1956).
7. N. N. Sheftal', in: Growth of Crystals, Vol. 1, Consultants Bureau, New York (1958), p. 5.
8. B. V. Deryagin, in: Research in Surface Forces, Vol. 2, Consultants Bureau, New York (1966), p. 3.
9. G. I. Distler, S. A. Daryusina, and Yu. M. Gerasimov, Dokl. AN SSSR, 154:1328 (1964).
10. G. I. Distler and S. A. Kobzareva, Fiz. Tverd. Tela, 7:2450 (1965).
11. G. I. Distler and S. A. Daryusina, Kristallografiya, 7:107 (1962).
12. R. S. Bradley, Z. Kristallographie, 96:499 (1937).
13. G. I. Distler, S. A. Kobzareva, and V. S. Chudakov, Fiz. Tverd. Tela, 9:269 (1967).

14. G. I. Distler, this volume, p. 71.

15. B. V. Deryagin, Izv. AN SSSR, OMEN, Ser. Khim., No. 5, 1153 (1937); Kolloidn. Zh., 6:291 (1940); B. V. Deryagin and L. D. Landau, Zh. Éksp. Teor. Fiz., 15:662 (1945); B. V. Deryagin, I. I. Abrikosova, and E. M. Lifshits, Usp. Fiz. Nauk, 64(3):493 (1958); A. Many, J. Goldstein, and N. B. Grover, Semiconductor Surfaces, North-Holland Publishing Company, Amsterdam (1965).

16. G. L. Mikhnevich and V. G. Zaremba, Kolloidn. Zh., 24:491 (1962).

17. N. N. Fedyakin, Dokl. AN SSSR, 138:1389 (1961).

18. N. A. Tolstoi, Dokl. AN SSSR, 100:893 (1955); Opt. i Spektroskopiya, 19:826 (1965).

POLYMOLECULAR ADSORPTION OF WATER VAPOR ON FRESHLY CLEAVED MICA AND ELECTRICAL SURFACE CONDUCTIVITY

V. D. Perevertaev and M. S. Metsik

Zhdanov University, Irkutsk

Isotherms and heat of adsorption are given for water on mica crystals. The adsorption film thickness is shown to be related to the surface conductivity and to the bulk conductivity.

Adsorption Film Thickness and Surface and Bulk Conductivities

A method previously described [1] has been applied to the relation of film thickness to electrical surface conductivity. The film thickness increases rapidly immediately after the mica has been split; then follows a maximum, and afterwards a relatively slow decay to some more or less equilibrium value. The change may be less regular if the humidity is close to saturation, and the equilibrium state is reached 30-60 min after cleavage. Similar kinetic curves have been observed from the surface conductivities of quartz and glass [2, 3], which have been ascribed to a phase transition in the surface film (from an amorphous state formed initially to a quasi-crystalline one). This transition is naturally accompanied by a reduction in the surface conductivity, so it is of interest to measure the conductivity and film thickness simultaneously. Figure 1 shows typical curves for: 1) surface conductivity, 2) film thickness on freshly cleaved mica, and these curves are clearly similar.

Such parallel measurements have given the first direct evidence on the film bulk conducivity (curve 3 of Fig. 1), which is defined by

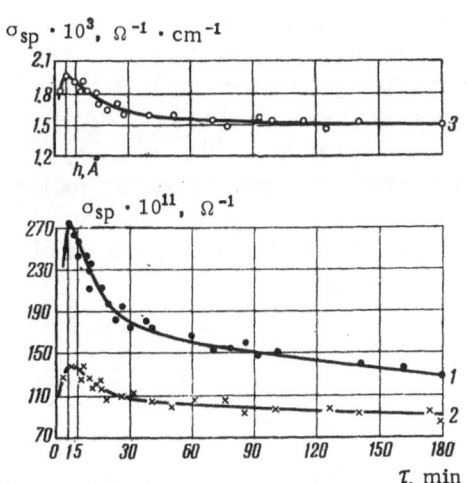

Fig. 1. Curves for: 1) surface conductivity; 2) film thickness; 3) film bulk conductivity for freshly cleaved mica crystals at $p/p_s \approx 1$ and 21°C.

$$\sigma_{sp} = \frac{\sigma_s}{h}. \qquad (1)$$

89

TABLE 1. Film Thickness h, Surface Conductivity σ_s, and Film Bulk
Conductivity σ_{sp} as Functions of p/p_s at 21°C.

p/p_s	Maximal value			Equilibrium value		
	h, Å	$\sigma_s \cdot 10^{11}$, Ω^{-1}	$\sigma_{sp} \cdot 10^3$, $\Omega^{-1} \cdot cm^{-1}$	h, Å	$\sigma_s \cdot 10^{11}$, Ω^{-1}	$\sigma_{sp} \cdot 10^3$, $\Omega^{-1} \cdot cm^{-1}$
0,1	15	9	0.6	5	1.2	0,24
0,2	20	14	0.7	10	1,9	0,19
0,3	30	21	0.7	15	4,5	0.3
0.5	40	31	0.77	15	7	0.47
0,6	55	43	0.78	20	9.8	0.49
0,8	80	100	1.25	40	32	0,8
0,9	100	156	1.56	60	60	1
0.95	115	195	1.7	65	84	1.3
0,98	120	220	1.83	70	98	1.4
~1	145	280	1.93	80	120	1.5

Note: Means from several measurements are given in the table.

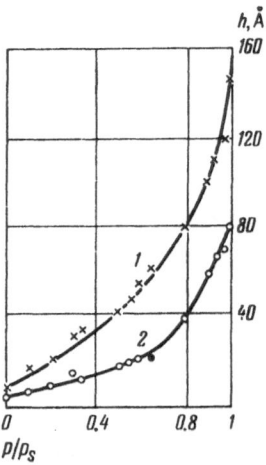

Fig. 2. Adsorption isotherms
for 21°C for water on mica
crystals for: 1) maximal thickness; 2) equilibrium thickness.

Humidity Dependence of Film Thickness and Conductivity

Table 1 shows that the equilibrium film thickness
h is dependent on the vapor pressure and varies from a
monolayer at p/p_s near zero to ~80 Å near saturation
(at 21°C), the maximal thickness at $p/p_s \approx 1$ being ~140 Å.
A few tests gave thicknesses up to 300 Å near saturation,
which is evidently to be ascribed to a state of supersaturation.

Increase in h is accompanied by increase in surface
conductivity σ_s and bulk conductivity σ_{sp}, whose σ_s increases rather more rapidly than h, whereas σ_{sp} increases
rather less rapidly (as p/p_s goes from 0.1 to ~1,
the equilibrium h increases by a factor 14, σ_s by
a factor 100, and σ_{sp} by a factor 7). The h at p/p_s
close to saturation were mostly measured visually (polarizing goniometer). The correction for potassium ions in
the film did not exceed the error in the measurement of
the relative humidity ($\Delta p/p_s \approx 0.01$).

Adsorption Isotherm

This isotherm may be derived from h as a function of p/p_s for the maximal and steady-
state values of h (Fig. 2).

Figure 2 shows that the isotherm intersects the saturation ordinate. A similar isotherm has
been reported [4, 5] for polar liquids on glass and has been explained as occurring because the
transition to the bulk phase occurs not by gradual thickening of the adsorption film but sud-
denly by occurrence of nuclei of the bulk liquid in a layer of finite thickness.

Relation of Conductivity to Film Thickness

Figure 3 shows plots of log σ_s and log σ_{sp} against h. The relation of σ_s to h is

$$\sigma_s = \sigma_{0n} e^{\beta h}. \tag{2}$$

TABLE 2. Heat of Adsorption of Water on Mica

h, Å	$t_1 = 21°C$	$t_2 = 50°C$	$q_x = 14.9 \ln p_2/p_1$	
	p_1, mm Hg	p_2, mm Hg	$\ln p_2/p_1$	q_x, kcal/mole
6	2.3	5.6	0.91	14
6	2.6	9.4	1.27	19
8	3.9	11.2	1.05	16
11	6.0	16.8	1.03	15
15	8.2	18.7	0.82	12

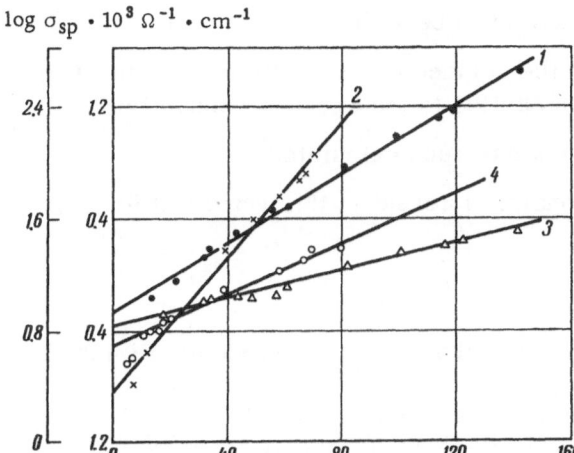

$\log \sigma_{sp} \cdot 10^3 \, \Omega^{-1} \cdot cm^{-1}$

Fig. 3. Dependence of h for: 1, 3) maximal; 2, 4) equilibrium values of: 1, 2) σ_s; 3, 4) σ_{sp}.

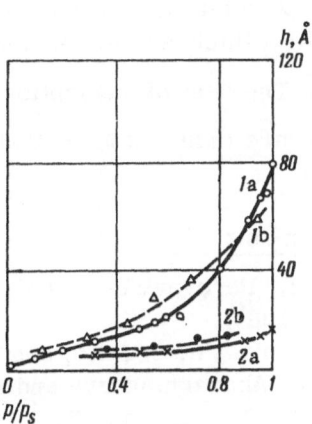

Fig. 4. a) Adsorption isotherms for: 1) 21°C; 2) 50°C; b) desorption isotherms for: 1) 21°C; 2) 50°C for equilibrium film thickness.

Gokhbert [6] has reported a similar law for old mica surfaces at high p/p$_s$. An analogous relation applies for σ_{sp} as a function of h.

Heat of Adsorption

Clausius's equation [7]

$$q_x = RT^2 \left(\frac{\partial \ln p}{\partial T} \right)_x \tag{3}$$

is used to calculate the differential heat of adsorption q$_x$, in which x is the adsorption.

We integrate (3) on the assumption that q$_x$ is constant in a narrow temperature range. Then

$$q_x = \frac{2.3 R T_1 T_2 \ln \frac{p_2}{p_1}}{T_2 - T_1}, \tag{4}$$

in which p$_1$ and p$_2$ are the equilibrium pressures above the adsorbent at T$_1$ and T$_2$ for a fixed amount of adsorbed material.

Isotherms for at least two temperatures (21 and 50°C in Fig. 4) allow one to derive q_x from (4). Table 2 shows that the difference between the calculated q_x [8] and the measured values lie within the limits of error of experiment.

The mean value of q_x is 15 kcal/mole. Energies given elsewhere [8] are 16.3 kcal/mole near aluminum atoms and 12.1 kcal/mole in oxygen depressions.

Conclusions

1. Kinetic curves are reported for the film thickness and conductivity of water on freshly cleaved mica; the two run parallel.

2. Adsorption isotherms for water on mica have been drawn up.

3. A relation $\sigma_s = \sigma_{0s} \exp(\beta h)$ applies to the surface conductivity σ_s as a function of adsorbed film thickness h. An analogous relation applies to the specific conductivity.

4. The heat of adsorption of water on mica has been calculated.

We are indebted to B. V. Deryagin for constant interest in this work and for critical discussions.

References

1.   V. D. Perevertaev and M. S. Metsik, Izv. Vysshikh. Uchebn. Zavedenii, Fizika, No. 2, 77 (1964).
2.   S. Kyrosaki, S. Saito, and G. Sato, J. Chem. Phys., 23:10 (1955).
3.   I. F. Abdrakhmanova and B. V. Deryagin, Dokl. AN SSSR, 120(1):94 (1958).
4.   B. V. Deryagin and Z. M. Zorin, Zh. Fiz. Khim., 29(6):1010 (1955).
5.   B. V. Deryagin and Z. M. Zorin, Zh. Fiz. Khim., 29(10):1755 (1955).
6.   B. M. Gokhberg, Zh. Éksp. Teor. Fiz., 1:275 (1930).
7.   B. V. Il'in, Nature of Adsorption Forces, Gos. Izd. Tekh.-Teor. Lit., Moscow−Leningrad (1952).
8.   M. S. Metsik, in: Research in Surface Forces, Vol. 1, Consultants Bureau, New York (1963), p. 50.

ELASTIC DEFORMATION OF SEDIMENTS
IN A STEADY ELECTRIC FIELD

O. G. Us'yarov and I. F. Efremov

Lensovet Technological Institute, Leningrad

Elastic deformation accompanies the plastic irreversible deformation when a dc field acts on a polytrifluorochloroethylene sediment in a polar organic liquid. This elastic deformation is due to a boundary phase at the surface of the particles, which has anomalous properties and which prevents direct contact between the particles.

There are many papers on the properties of boundary liquid layers. Deryagin et al. [1] showed that these layers have singular mechanical and thermodynamic parameters; for instance, it was shown [1] that the layer near the solid (first 30-40 layers of molecules) has a viscosity greater than that of the bulk liquid sometimes by one order of magnitude. The observed discontinuity in the viscosity–wall distance function allows the boundary layer to be considered as a special phase. Anomaly of viscosity has also been observed [2] in the flow of liquid in narrow gaps. Deryagin and Kusakov [3] have recorded isotherms for the disjoining pressure exerted by water films, for which it is claimed [4] that the molecular component is responsible for the special mechanical properties at $0.2-0.3\mu$ from the surface. It has been deduced that the molecular state and structure are altered near a wall from the heat of wetting [5], the dielectric constant [6], the nondissolving volume [7], the thermal conductivity [8], and various other properties of colloids.

One method of research on boundary phases is deformation testing of sediments, pastes, and other such highly concentrated dispersions. For instance, the various forms of bound fluid can be deduced from the rate of release of water in response to centrifuging or external pressure for clays, peat, etc. [9]. Rebinder et al. [10] have shown that there are effects from solvate adsorption layers, which have a finite shear strength, on the strength of such structures.

We have examined the consolidation of a polytrifluorochloroethylene (PTFCE) sediment in normal aliphatic alcohols in an external electric field and have found that the mode of deformation is very much determined by the uneven particle distribution, which in turn is dependent on the balance of the forces between particles [11].

Predominance of molecular attraction over repulsion causes the particles to adhere to form a loose sediment with fairly large pores free from the action of disjoining pressure, whose volume is reduced irreversibly during consolidation. The adhesion forces may be characterized via the external load needed to start flow, and the rate of plastic deformation falls

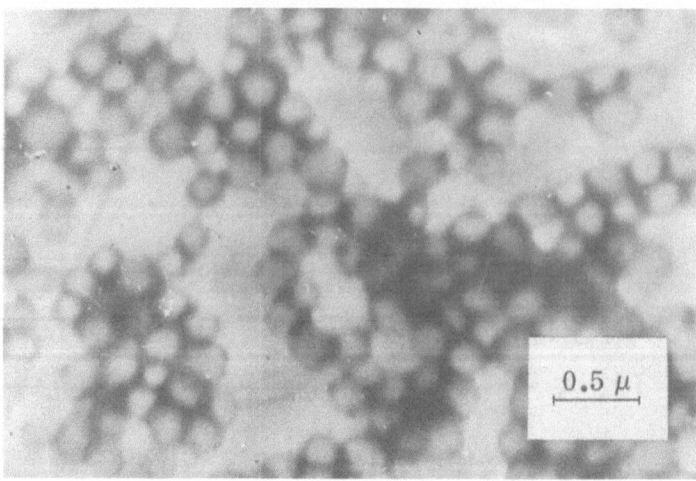

Fig. 1. Electron micrograph of PTFCE particles.

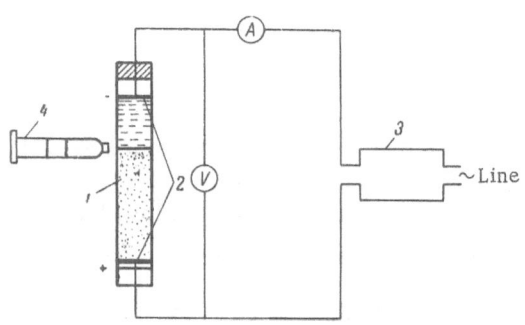

Fig. 2. Apparatus for measuring the deformation of a sediment in an electric field: 1) cylindrical cell containing sediment; 2) electrodes; 3) rectifier; 4) microscope.

as the pores are filled. The irreversible consolidation ceases when the external load is balanced by the increasing adhesion forces due to increase in the number and closeness of the contacts. The plastic deformation is accompanied by reversible (elastic) deformation, because the liquid films exert a disjoining pressure whose origin may be molecular or ionic (electrostatic). Here we report elastic consolidation of sediments in an electric field.

Method

We used a PTFCE suspension whose spherical particles were 0.2–0.3 μ in size (Fig. 1), the liquids being acetone, hexane, and n-alcohols. The polymer was first washed with acetone and then with distilled water until the wash water had a constant conductivity (2×10^{-6} ohm$^{-1} \cdot$ cm^{-1}); it was then dried at 80°C and was kept in a desiccator over a freshly roasted calcium oxide. The alcohols were dried and redistilled, the purity criteria being the electrical conductivity and refractive index. The suspensions were prepared by mixing the polymer with a liquid and exposing the mixture to ultrasound; the stability was characterized via the sedimentation rate and the limiting volume of the sediment. The electrokinetic potential of the particles was deduced from the rate of electroosmosis. The compression was produced in a system consisting of a cylindrical cell ($S = 1$ cm^2) fitted with two parallel nickel electrodes having a separation of 50 mm (Fig. 2), which were supplied by a rectifier with an output up to 3 kV. The height of the sediment layer was measured to ±0.01 mm with a traveling microscope. The cell was loaded with 5 ml of a suspension of 31.5% volume concentration and a potential difference of 2 kV was applied to the electrodes, which caused rapid coagulation, deposition, and consolidation [12]. The completion of deformation was established from the cessation of movement in the upper boundary of the sediment; the height corresponding to limiting compression was 4 ± 0.1 cm. The power supply was then switched off and the specimen was left for 7 days for stress relaxation to occur. Then elastic consolidation was produced in a field of strength up to 200 V/cm, the equilibrium deformation being measured as well as the rates of compression and decompression.

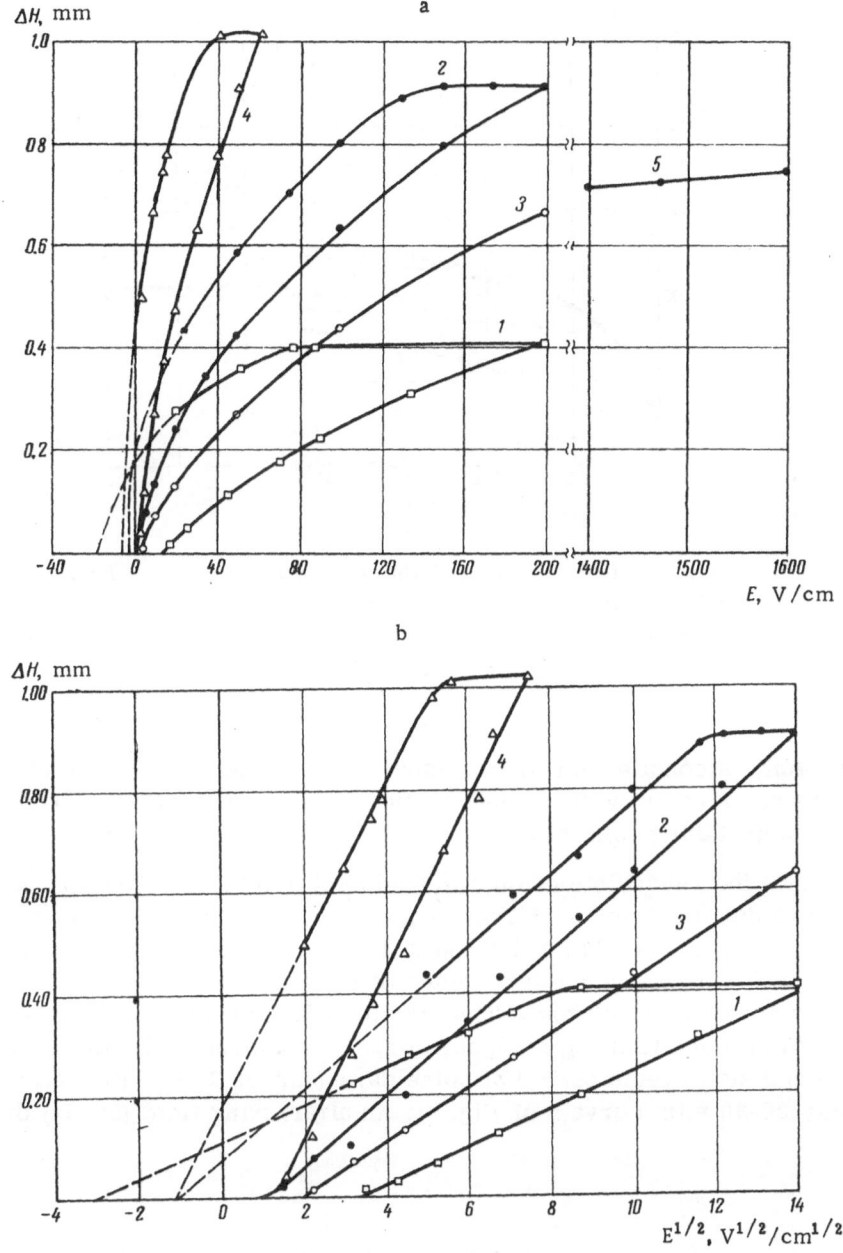

Fig. 3. Relation of elastic deformation ΔH for PTFCE as a function of: a) E; b) $E^{1/2}$ for various media: 1) butanol; 2) hexanol; 3) octanol; 4) acetone; 5) hexanol. Distance between electrodes 1.4 cm, height of sediment layer on completion of plastic deformation 1.2 cm.

Results

The stability of a PTFCE suspension decreases in the following sequence of alcohols: butyl, amyl, hexyl, octyl. This is accompanied by a tendency for the electrokinetic potential to fall: -1.2, -1.2, -1.0, -1.0 mV. In acetone (ζ potential -10 mV) and butyl alcohol, the sediment was formed from the bottom, which is usually considered [13] as a sign of stability; but the sedimentation was accompanied by coagulation, which led to relatively loose packing

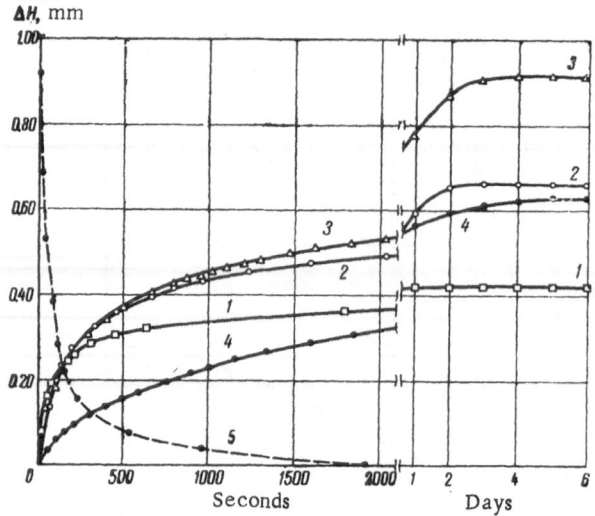

Fig. 4. Relation of ΔH (height increase) of PTFCE sediment previously consolidated at 200 V/cm (curve 5) to relaxation time at zero field in various alcohols: 1) butanol; 2) amyl alcohol; 3, 5) hexanol; 4) octanol.

of the sediment. Amyl alcohol gave two interfaces, while in hexane (ζ potential 0), hexanol, and octanol there was more pronounced aggregation, the sedimentation curve having the form characteristic of an unstable suspension.

Figure 3 shows the reversible change in sediment height as a function of field strength for compression and decompression. Hexane showed no consolidation. In no case did the circuit current exceed 50 μA. There is a threshold field E_0, below which there is no elastic deformation, while for $E > E_0$ there are two values of E corresponding to a given ΔH for the compression and decompression conditions. The elastic deformation increases with E, but even 1600 V/cm failed to produce limiting compression. On removal of the field, the sediment expanded to its original size over a period of some days (curves 1–4 of Fig. 4), whereas compression took only 20–30 min (curve 5 of Fig. 4). An alternating field (50 Hz) did not produce compression.

Discussion

The elastic deformation appears to be due to boundary phases with anomalous properties. For instance, we might suppose that the field compression is due to particle deformation. The particles at the top press on the underlying ones, so the contact forces will vary with height. The force is maximal at the lower (positive) electrode and is given [14] roughly by

$$F_{max} = n_0 \varepsilon r \zeta E, \tag{1}$$

in which n_0 is the number of layers of particles in the complete sediment layer, r is the radius of a particle, ε is the dielectric constant of the medium, and ζ is the electrokinetic potential.

Cubic packing and E = 200 V/cm (a field seldom exceeded in the tests) give $F_{max} = 5 \times 10^{-5}$ dyne for hexanol. Taking the area of contact between two particles as 10^{-12} cm², we get a pressure that is less by an order of magnitude than the pressure of 1.4×10^8 dyne/cm² needed for 0.1% deformation [15]. The forces compressing the higher layers are much less than F_{max} and are obviously insufficient to deform the polymer.

It is also unexplained (if we assume mechanical compression of the particles) why E_0 for butanol and acetone are high, while the ζ potentials and ε for these are largest, since (1) shows that these should increase F_{max}. The value calculated for F_{max} is an overestimate, since we have neglected polarization effects produced by the field. Moreover, PTFCE is crystalline [15], and elastic deformation of this cannot produce hysteresis in compression–decompression cycles. Further, no effect was found from varying the vessel diameter (an effect would be expected if there were wall friction). It is also considered [16] that direct contact does not occur between colloidal particles bearing surface charges at pressures of the order of 100 atm.

It is also impossible to explain the expansion on removing the field by the particle diffusion at the surface of sediment; observation under the microscope reveals no relative displacement of the particles. We have found similar volume changes in a sediment in studying the swelling in alcohols for PTFCE powder previously consolidated by vibration.

The disjoining pressure due only to electrostatic effects in this case cannot explain the behavior in compression and decompression, since suspensions in amyl and hexyl alcohols are unstable and the potential curves have maxima not exceeding 0.1 kT; and the depth of the potential well corresponding to the long distances is inadequate to hold the particles, on account of the large thickness of the diffuse ion atmospheres.[†] These considerations indicate that the polymer particles are surrounded by liquid boundary layers that exert repulsion.

Consider the field-induced deformation of the boundary phases of two spherical particles in a sediment. For the equilibrium state we have

$$F + R - R_{ip} \pm P_\tau = 0, \tag{2}$$

in which F is the force due to interaction of the particles with the electric field, which is dependent on the position of the particles in the sediment, R is the resultant repulsion developed in the gap between the particles (the sum of van der Waals forces, electrostatic ion repulsion, and elastic interaction of the boundary phases), R_{ip} is the force from polarization interaction of the particles, and P_τ is the force that the layer will withstand on account of plasticity and slow stress relaxation, compression corresponding to the – sign and decompression to +.

As a low-frequency ac field produces no visible elastic deformation, R_{ip} cannot be important, so (2) becomes

$$F + R \pm P_\tau = 0. \tag{3}$$

For any deformation of the boundary phases

$$F - F' = 2P_\tau,$$

in which F and F' are the forces causing approach of the particles in the compression and decompression states for an identical degree of deformation, and

$$P_\tau = \frac{ner\zeta(E - E')}{2}, \tag{4}$$

in which n is the number of overlying layers of particles.

As the particles do not come together for $E = E_0$,

$$P_\tau = n_0 er\zeta E_0. \tag{5}$$

[†] It was assumed that $\zeta = \psi_0$ [17] in the calculations.

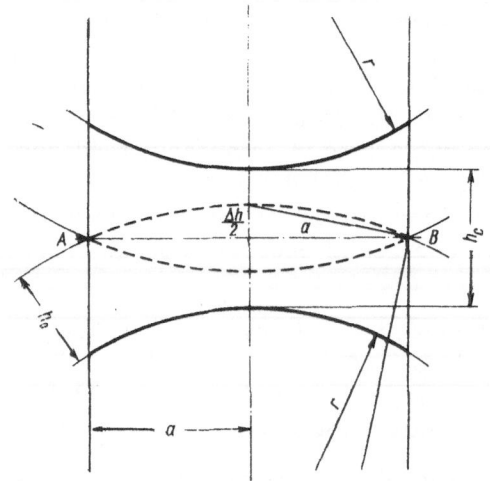

Fig. 5. Overlap of boundary layers between approaching particles.

For spherical surfaces separated by a layer of viscoplastic liquid of thickness h_c, the yield value τ_0 is [18]

$$P_\tau = 2\pi\tau_0 \left(ar - h_c^{1/2} r^{3/2} \text{ arc tg } \frac{a}{\sqrt{h_c r}} \right), \qquad (6)$$

in which a is the radius of the area of contact, which equals half the chord AB (Fig. 5).

The P_τ given by (5) are integral values, since the τ_0 for boundary phases are generally functions of the distance from the solid surface. From geometry,

$$a = \sqrt{\Delta h \left(r + h_0 - \frac{\Delta h}{4} \right)},$$

in which h_0 is the thickness of the boundary phases and $\Delta h = 2h_0 - h_c$.

This h_c corresponds to the distance between particles for $E \leq E_0$ and can be estimated on the assumption of a uniform particle distribution in a sediment maximally consolidated by a field:

$$h_c = \sqrt{\frac{4\pi}{3} \frac{v d_0 r^3}{m}} - 2r,$$

in which v is the volume of the sediment in that state and m is the mass of polymer, whose density is d_0. The h_c for PTFCE in hexanol is ~350 Å, or 40 layers of alcohol molecules.

The ΔH for maximal compression (E = 1600 V cm) gives the reduction Δh in the gap between particles as ~210 Å. Assuming cubic close packing, we can calculate n and $h_c \approx \Delta h = \Delta H/n$. Although various assumptions are made in this calculation, the resulting h_c is acceptable; but the gaps between particles are never reduced to zero, and the lower layers produce much of the compression, so the real h_c must be somewhat more than 210 Å. Therefore, coagulated PTFCE particles in hexanol retain between them layers about 30 alcohol molecules thick, which prevent direct contact. This state corresponds to partial overlap of the boundary phases, whose thickness may [3, 4] substantially exceed $h_c/2$.

From (5) and (6), we get $\tau_0 \approx 600$ dyne/cm^2 for hexanol with $E_0 = 1$ V/cm, $\varepsilon = 12$, and $r = 1.5 \times 10^{-5}$ cm, on the assumption that $h_c = 300$ Å and $h_0 = 500$ Å (i.e., 50 layers of alcohol molecules). This τ_0 is probably somewhat of an overestimate because we have neglected polarization effects, which reduce the field strength in a dielectric; but it is of the right order of magnitude and agrees with the yield value found for boundary phases by filtration [19].

R = 0 if the sediment is kept at zero field for a time greater than the relaxation time for the plastic stresses. In this thermodynamic-equilibrium state, which corresponds to partial overlap of the boundary layers, we have

$$R_m - R_i - R_c = 0,$$

in which R_m is the molecular attraction of the particles, R_i is the ionic electrostatic repulsion, and R_c is the repulsion arising from overlap of the boundary phases.

Deformation in response to a small external load is very slow and cannot be measured within the times used, which explains why E_0 occurs.

The compression and swelling of the sediment can be explained only if allowance is made for the viscosity and other properties of the medium. The rates of these processes are dependent on the filtration resistance in passage through pores of all sizes [20]. However, it has been shown for the deformation of clay soils [21] that the role of the filtration resistance is not large when the entire external load is taken up by the boundary-layer forces, and its effects on elastic consolidation can be neglected. In that case, dh/dt for a viscoplastic medium of yield value τ_0 and viscosity η is determined [22] only by the shape and separation of the bodies:

$$F + R = f_1(\tau_0,\ h) + f_2(dh/dt,\ h).$$

A previously field-consolidated sediment expands because the molecular attraction is exceeded by the repulsive forces from overlapping of boundary layers and ionic atmospheres. The main part is played by relaxation-type creep when $R \to f_1(\tau_0, h)$, and this basically determines the duration of the process. Analogous effects occur in the swelling of dispersions. For instance, it has been found for the swelling of clay that most of the water ($\sim 90\%$) is taken up in the first 3-5 h. The production of hydrate layers (or double ionic layers) on the particles gives rise to strong repulsion, but the equilibrium state is attained only after several days [23]. The yield value for boundary phases may cause [24] a slow increase in adhesive forces when solid surfaces are in contact, as is often observed in the formation of gels [25] and which can affect the migration of water in porous systems [26]. We have also observed elastic deformation in response to electrical and centrifugal fields in other sediments, e.g., polyacrylonitrile and copolymers based on this.

Conclusions

1. A study has been made of the elastic deformation of polytrifluorochloroethylene sediments exposed to electric fields in various organic liquids. The special boundary phases on the particles have thicknesses >150 Å and produce a repulsion that prevents the particles from coming together.

2. The effects of plasticity in these phases are discussed in relation to the rates of approach of particles, and an estimate is made of the yield value of the boundary phases.

References

1.   B. V. Dergyagin, G. M. Strakhovskii, and D. S. Malysheva, Zh. Éksp. Teor. Fiz., 16:171 (1946); B. V. Deryagin and V. V. Karasev, Dokl. AN SSSR, 101:289 (1955); B. V. Deryagin, N. N. Zakhavaeva, S. V. Andreev, and A. A. Milovidov, Research in Surface Forces, Vol. 1, Consultants Bureau, New York (1963), p. 110; B. V. Deryagin, Dokl. AN SSSR, 109(5):967 (1956).
2.   B. V. Deryagin and M. A. Krylov, in: Viscosities of Liquids and Colloidal Solutions, Vol. 1, Izd. AN SSSR, Moscow—Leningrad (1944), p. 52; B. V. Deryagin, N. N. Zakhavaeva, and A. M. Lopatina, Research in Surface Forces, Vol. 1, Consultants Bureau, New York (1963); p. 141; N. N. Zakhavaeva and A. M. Lopatina, Kolloidn. Zh., 24:455 (1962); M. P. Volarovich and N. V. Churaev, Research in Surface Forces, Vol. 2, Consultants Bureau, New York (1966), p. 212; S. V. Nerpin and N. F. Bondarenko, Dokl. AN SSSR, 114:833 (1937).
3.   B. V. Deryagin and M. M. Kusakov, Izv. AN SSSR, Ser. Khim., No. 5, 1119 (1937).
4.   G. A. Korchinskii, Kolloidn. Zh., 19:307 (1957).

5. A. V. Dumanskii, Izv. AN SSSR, Ser. Khim., No. 5, 1165 (1937); Kolloidn. Zh., 12:319 (1950); A. V. Dumanskii and E. F. Nekryach, Kolloidn. Zh., 13:20 (1951); 15:91 (1953); 17:171 (1955).

6. A. V. Dumanskii and O. D. Kurilenko, Kolloidn. Zh., 12:326 (1950); D. A. Frank-Kamenet-skii and B. I. Sedunov, Usp. Fiz. Nauk, 79:617 (1963); N. V. Afanas'ev and M. S. Metsik, Research in Surface Forces, Vol. 2, Consultants Bureau, New York (1966), p. 181; B. Jaconcon, J. Am. Chem. Soc., 77:2919 (1955).

7. A. V. Dumanskii, Lyophilicity in Disperse Systems, Izd. Voronezh. Univ. (1940), p. 56.

8. Yu. M. Popovskii and B. V. Deryagin, Dokl. AN SSSR, 159(4):897 (1965).

9. Z. B. Olmsted, Dep. Agr. Techn., Bull, No. 9, 526 (1937); A. I. Kotov, Dissertation, Leningrad Institute of Water Transport Engineers (1955); A. I. Kotov and S. V. Nerpin, Summaries of Papers at the Fourth All-Union Conference on Colloid Chemistry, Izd. AN SSSR, Moscow (1958), p. 209.

10. P. A. Rebinder, Transactions of the Third All-Union Conference on Colloid Chemistry, Izd. AN SSSR, Moscow (1956), p. 7.

11. O. G. Us'yarov, I. S. Lavrov, and I. F. Efremov, Kolloidn. Zh., 27:787 (1965).

12. O. G. Us'yarov, I. S. Lavrov, and I. F. Efremov, Kolloidn. Zh., 28:596 (1966).

13. V. I. Baranova and I. S. Lavrov, Kolloidn. Zh., 24:252 (1962); F. Szanto and B. Varkonyi, Koll. Z., 191:123 (1963).

14. H. G. Hamaker and K. J. W. Verwey, Trans. Faraday Soc., 36:180 (1940); L. Milička, Acta F. R. N. Univ. Comen. IV, 11-12, Chimia, 4:621 (1960).

15. D. D. Chegodaev, Z. K. Naumova, and Ts. S. Dunaevskaya, Fluoroplastics, Goskhimizdat, Moscow (1960); Foreign Commercial Polymer Materials and Their Components, Izd. AN SSSR, Moscow (1963), p. 258.

16. G. H. Bolt and R. D. Miller, Soil Sci. Soc. America Proc., 19:285 (1955); G. H. Bolt, Soil Sci. Soc. America Proc., 17:210 (1953).

17. J. H. Shchenkel and J. A. Kitchener, Trans. Faraday Soc., 56:161 (1960).

18. S. V. Nerpin, Tr. Leningr. Inst. Vod. Inzh. Transp., 25:37 (1958).

19. S. V. Nerpin and N. F. Bondarenko, Summaries of Papers at the Twentieth International Congress of Theoretical and Applied Chemistry, Izd. Nauka, Moscow (1965), p. 73.

20. N. V. Kozubov, Tr. Leningr. Inst. Vod. Inzh. Transp., 26:126 (1959).

21. A. I. Kotov and S. V. Nerpin, Summaries of papers at the Fourth All-Union Conference on Colloid Chemistry, Izd. AN SSSR, Moscow (1958), p. 209.

22. S. V. Nerpin, A. I. Kotov, and V. A. Raev, Tr. Leningr. Inst. Vod. Inzh. Transp., 26:105 (1959).

23. F. D. Ovcharenko, S. P. Nichiporenko, N. N. Kruglitskii, and V. F. Tretinin, Research in Physicochemical Mechanics of Disperse Clay Materials, Izd. Naukova Dumka, Kiev (1965), p. 69.

24. B. V. Deryagin and A. D. Malkina, Kolloidn. Zh., 12:431 (1950); B. V. Deryagin, Trans-actions of the All-Union Conference on Colloid Chemistry, Izd. AN SSSR, Kiev (1952), p. 26.

25. I. F. Efremov, Tr. Leningr. Tekhnol. Inst. im. Lensoveta, 37(1):107, 120 (1957); Kolloidn. Zh., 18:276 (1956); I. F. Efremov and S. V. Nerpin, Dokl. AN SSSR, 113:846 (1957).

26. B. V. Deryagin and S. V. Nerpin, Dokl. AN SSSR, 99(6):1029 (1954); 100(1):17 (1955).

SECTION III

THERMODYNAMICS AND STABILITY OF THIN FILMS

THERMODYNAMICS OF FILMS

A. I. Rusanov

Zhdanov University, Leningrad

The four main aspects of film thermodynamics are considered: methods of considering thick and thin films, film stability and elasticity, curved films, and participation of films in phase equilibria.

There are not many papers on the thermodynamics of films, but certain results appear clear [10]. Here I give a general survey of the thermodynamics of films.

Methods of Thermodynamic Consideration

of Films

a. Thick Films. First we must consider what is meant by a film. Consider the equilibrium of the three phases α, β, and γ in Fig. 1; we assume that phase γ, which lies between phases α and β, is much smaller in extent in the direction perpendicular to the interfaces than in the other two directions. We also suppose that the $\alpha\gamma$ and $\beta\gamma$ interfaces are parallel. The system therefore consists of these three phases and the interfaces $\alpha\gamma$ and $\beta\gamma$. A thick film is phase γ together with interfaces $\alpha\gamma$ and $\beta\gamma$; the thickness of such a film is determined either by the sum τ' of the thicknesses of phase γ and of the $\alpha\gamma$ and $\beta\gamma$ surface layers (if the latter are considered as having a finite thickness) or by the distance τ between the bounding surfaces, if these are considered by Gibbs's method.

We shall term the film the inhomogeneous region between phases α and β. This definition formally coincides with the Gibbs definition of a surface of discontinuity and may be supplemented by stating the specific features of the film: the inhomogeneous region between phases α and β may have a thickness much greater than an ordinary dividing surface, may include a homogeneous part, and may contain components virtually absent from phases α and β.

Unlike an interface, a film can exist when phases α and β are identical [10].

This definition implies that the basic equation of the Gibbs theory of capillarity is completely applicable to films [10]:

$$d\varepsilon^f = T d\eta^f + \sum_i \mu_i dm_i^f + \gamma dS, \qquad (1)$$

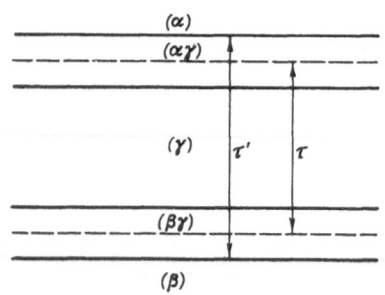

Fig. 1. Schematic representation of a film: τ' is film thickness and τ is the distance between the Gibbs dividing surfaces.

in which ε^f, η^f, and m_i^f are the excess energy, entropy, and mass of component i, T is temperature, μ_i is the chemical potential of component i, γ is film tension, and S is film area.

If the film is considered as having a finite thickness τ', we can use instead of (1) an equation containing the actual quantities (instead of the excess ones), which is analogous to the equation for surface layers of finite thickness:

$$de^{(f)} = Td\eta^{(f)} - PdV^{(f)} + \sum_i \mu_i dm_i^{(f)} + \gamma dS, \qquad (2)$$

in which $\varepsilon^{(f)}$, $\eta^{(f)}$, $V^{(f)}$, and $m_i^{(f)}$ are respectively the energy, entropy, volume, and mass of component i of the film, while P is pressure.

Equations (1) and (2) play a fundamental part in the thermodynamics of surface phenomena; they have the same form for interfaces and films, so all the methods for surface layers [1] are completely applicable to films [10].

On the other hand, the fundamental equation for the energy of a thick film can be obtained by simple summation of the corresponding equations for phases γ and for interfaces $\alpha\gamma$ and $\beta\gamma$, which can be considered either by Gibbs's method or as having a finite thickness. We get

$$de^{(f)} = Td\eta^{(f)} - PdV^{(f)} + \sum_i \mu_i dm_i^{(f)} + \sigma^{(\alpha\gamma)} dS^{(\alpha\gamma)} + \sigma^{(\beta\gamma)} dS^{(\beta\gamma)}, \qquad (3)$$

in which $\sigma^{(\alpha\gamma)}$ and $\sigma^{(\beta\gamma)}$ are surface tensions and $S^{(\alpha\gamma)}$ and $S^{(\beta\gamma)}$ are the corresponding surface areas.

Equations (2) and (3) are equivalent, but they reflect different approaches; the first describes a film without reference to the internal structure, while the second is more detailed and is based on a very simple structural model, namely a homogeneous part enclosed between two inhomogeneous parts.

Comparison of (2) and (3) gives us for planar films that

$$\gamma = \sigma^{(\alpha\gamma)} + \sigma^{(\beta\gamma)}. \qquad (4)$$

b. Thin Films. Both of the above approaches may be extended to thin films. This is obvious for the first, since (1) and (2) involve no assumptions about the film structure and thickness. We cannot transfer (3) automatically to thin films, since the energies of the surface layers cease to be additive (this feature may serve as the definition of a thin film), and the energy becomes an explicit function of the film thickness. However, we can still remain within the framework of the second approach for phase γ (at equilibrium) by introducing into (3) a correction for the explicit dependence of the energy on the thickness. If the Gibbs approach is used for the interfaces, this correction is

$$\frac{\partial \varepsilon^{(\alpha\gamma,\,\beta\gamma)}}{\partial \tau} d\tau,$$

in which $\varepsilon^{(\alpha\gamma,\,\beta\gamma)}$ is the excess energy of the two interfaces.

Then the equilibrium between the film and phases α and β gives us the condition

$$P^{(\alpha)} = P^{(\beta)} = P^{(\gamma)} - \frac{1}{S} \frac{\partial \varepsilon^{(\alpha\gamma,\ \beta\gamma)}}{\partial \tau}, \tag{5}$$

in which the P are the pressures in the corresponding phases.

The difference $P^{(\alpha)} - P^{(\gamma)}$ is Deryagin's [2] disjoining pressure Π, whose sign is always the opposite of that of the derivative $\partial \varepsilon^{(\alpha\gamma,\beta\gamma)}/\partial \tau$:

$$\Pi \equiv P^{(\alpha)} - P^{(\gamma)} = P^{(\beta)} - P^{(\gamma)} = - \frac{1}{S} \cdot \frac{\partial \varepsilon^{(\alpha\gamma,\ \beta\gamma)}}{\partial \tau}. \tag{6}$$

It is clear from (5) that the distinctive feature of this case is that the standards for calculating the excess quantities σ for the $\alpha\gamma$ and $\beta\gamma$ interfaces are taken at different pressures, which means that the surface tensions $\sigma^{(\alpha\gamma)}$ and $\sigma^{(\beta\gamma)}$ found in this way should be dependent on the positions of the dividing surfaces even if these are planar. Kondo's method [3] may be used in order to show that the relation is linear and is defined by

$$\left(\frac{d\sigma^{(\alpha\gamma)}}{d\tau} \right)^* = -\Pi, \tag{7}$$

in which the asterisk indicates that the derivative relates only to an imaginary (nonphysical) displacement of the dividing surface $\alpha\gamma$ at fixed position of the dividing surface $\beta\gamma$.

This argument shows that (3) becomes as follows on going to thin films:

$$d\varepsilon^{(f)} = T d\eta^{(f)} - P dV^{(f)} + \sum_i \mu_i dm_i^{(f)} + (\sigma^{(\alpha\gamma)} + \sigma^{(\beta\gamma)} + \tau\Pi)\, dS, \tag{8}$$

in which $P \equiv P^{(\alpha)} = P^{(\beta)}$.

Then comparison with (2) gives

$$\gamma = \sigma^{(\alpha\gamma)} + \sigma^{(\beta\gamma)} + \tau\Pi. \tag{9}$$

Relation (8) connects the two approaches to thin films: in the first, the main role is played by the tension of the film as a whole, while in the second the film is characterized via Π and the tensions in the interfaces. This Π is the principal characteristic of the film in the second approach since (7) shows that we can always find a position for the dividing surfaces such that the surface tensions of these become zero.

Film Elasticity

By this is meant the tendency for the tension to increase on stretching. It is usual to distinguish the equilibrium elasticity from the nonequilibrium one. The first corresponds solely to thermodynamic equilibrium in the entire system, while the second may represent either disequilibrium in the film or disequilibrium between the medium and an equilibrium film. In the second case, the nonequilibrium elasticity may be discussed to a first approximation via equilibrium thermodynamics if we consider as a closed system the film and the small part of the medium in equilibrium with it. The degree of deviation from equilibrium will here be characterized by the volume of the closed system, which increases as equilibrium is approached. The nonequilibrium state in the film itself will not be considered in this paper.

Consider the causes of film elasticity. There is a relation between the elasticity and the thermodynamic condition for stability of the film and of the heterogeneous system containing the film:

$$d\gamma dS + dT d\eta - dP dV + \sum_i d\mu_i dm_i > 0. \tag{10}$$

From (10) we can derive many particular inequalities for elasticity under various conditions. The derivatives must be taken with allowance for the rule implied by the variance of a heterogeneous system involving surface phenomena [4]: the number of fixed extensive parameters should not be less than the total number of bulk and surface phases. If this condition is not obeyed, the tension becomes a constant and elasticity cannot occur. For instance, the constancy of T, P, and μ in a completely open system means that increase in the area of the film will imply production of this at the expense of the bulk phases without change in tension (a nonvariant bulk-surface phase process).

It is difficult to assume constancy of the entropy and volume of film during stretching in practice, and one usually employs the masses of the components as fixed extensive parameters. Then the above rule may be formulated as follows: film elasticity can occur only in closed and partially open systems, and the number of immobile components (ones whose masses are constant [4]) should not be less than the total number of bulk and surface phases. In particular, elasticity may occur for a thick film with identical surfaces if the film contains not less than two immobile components. This number is reduced to one for a thin film (one not containing a homogeneous internal region).

Film elasticity should change in a regular fashion on going from a closed system to a partly open one and further by increase in the number of mobile components. This relation may be derived qualitatively via the abbreviated Le Chatelier—Brown principle applied to heterogeneous systems [5], which implies that film elasticity should fall in the sequence

$$\left(\frac{d\gamma}{dS}\right)_{\eta, V, m_1, m_2, \ldots} \geqslant \left(\frac{d\gamma}{dS}\right)_{T, V, m_1, m_2, \ldots} \geqslant \left(\frac{d\gamma}{dS}\right)_{T, P, m_1, m_2, \ldots}$$
$$\geqslant \left(\frac{d\gamma}{dS}\right)_{T, P, \mu_1, m_2, \ldots} \geqslant \cdots \geqslant \left(\frac{d\gamma}{dS}\right)_{T, P, \mu_1, \ldots, \mu_i, m_{i+1}, \ldots} = 0, \tag{11}$$

in which i is such that the number of component masses remaining fixed is equal to the total number of bulk and surface phases. The length of the chain in (11) is dependent on the number of components and on the total number of phases.

The quantitative theory of equilibrium elasticity is based on consideration of the elasticity mechanism; thin films differ from thick ones in this respect. Gibbs's adsorption mechanism for thick films involves alteration in the component distribution between bulk and surface phases, and it can occur also for thin films either by alteration of the distribution between parts of the film (if this is fairly thick) or between the film and the medium (when an open film is in a closed or partly open equilibrium system). However, at all stages in the thinning of a thin film there is a specific elasticity mechanism related to the occurrence of Π and of interaction between the surface layers.

The adsorption mechanism is simple and readily accessible to thermodynamic consideration; here theory gives rigorous relations between the elasticity, the surface tension, and the composition of the surface layers. The following is the expression for a two-component closed thick film [6]:

$$E \equiv S\frac{d\gamma}{dS} = \frac{2\frac{dm^{(S)}}{dS}\left(\frac{S}{m^{(S)}}\right)^2\left(\frac{d\sigma}{dx^{(\gamma)}}\right)^2}{\left(\frac{\partial^2\zeta}{\partial x^2}\right)^{(\gamma)}\left(\frac{m^{(\gamma)}}{2m^{(S)}} + \frac{dx^{(S)}}{dx^{(\gamma)}}\right)}, \tag{12}$$

in which E is elastic modulus, m is the mass of all components, σ is surface tension, x is the molar fraction of one of the components, ζ is the Gibbs molar thermodynamic potential, and the superscripts (S) and (γ) relate to the surface layer and bulk phase respectively. If allowance is made also for the external bulk phases, (γ) relates to all together, and all volume quantities become overall ones.

The following are the implications of (12): 1) E is positive for a stable film, 2) E is the greater the higher the surface activity of one component with respect to the other, 3) E at constant composition is the greater the smaller the film thickness and the more rapid the stretching ($m^{(\gamma)}/2m^{(S)}$ is reduced in both cases), 4) E increases with the concentration at low concentrations and becomes zero when one of the components vanishes.

Curved Films

Both of the above approaches can be applied to curved films if the dividing surfaces are chosen suitably. We assume for simplicity that all the surfaces are spherical and concentric, so that

$$r^{(\beta\gamma)} - r^{(\alpha\gamma)} = \tau, \tag{13}$$

in which the r are the radii of curvature of the corresponding surfaces $(r^{(\beta\gamma)} > r^{(\alpha\gamma)})$. This means that curvature produces terms of the same form in (3), no matter whether the film is thick or thin:

$$C^{(\alpha\gamma)}dr^{(\alpha\gamma)} + C^{(\beta\gamma)}dr^{(\beta\gamma)}.$$

The difference is that the C for a thin film will be dependent not only on the curvature but also on the interaction between the interfaces; if there is attraction ($\partial\varepsilon^{(\alpha\beta,\,\beta\gamma)}/\partial\tau > 0$), $C^{(\alpha\gamma)}$ will be less for a thin film, while $C^{(\beta\gamma)}$ will be larger, the converse being the case if there is repulsion ($\partial\varepsilon^{(\alpha\beta,\,\beta\gamma)}/\partial\tau < 0$).

The conditions for mechanical equilibrium will formally take the same form as in the theory of capillarity:

$$P^{(\alpha)} - P^{(\gamma)} = \frac{2\sigma^{(\alpha\gamma)}}{r^{(\alpha\gamma)}} + \frac{C^{(\alpha\gamma)}}{S^{(\alpha\gamma)}}, \tag{14}$$

$$P^{(\gamma)} - P^{(\beta)} = \frac{2\sigma^{(\beta\gamma)}}{r^{(\beta\gamma)}} + \frac{C^{(\beta\gamma)}}{S^{(\beta\gamma)}}, \tag{15}$$

but the above features of the C will mean that the pressure differences will have altered values for a thin film. It is clearly possible for the pressure difference at one boundary of a thin film to be zero or to have a sign the converse of that given by the ordinary theory of capillarity; at such a boundary we have competition between capillary effects due to the curvature and to thinning of the film.

In the theory of curved films, as for curved surfaces generally, we have to consider how to choose the positions of the dividing surfaces. Of course, for the film as a whole, which is characterized by (1), we can always find a surface of tension, which allows us to extend to curved films not only this equation but also the entire thermodynamic treatment of curved surface layers [1]. In particular cases the problem is to be solved on the balance between the effects of curvature and thinning. A surface of tension can be defined for both boundaries if the film is sufficiently thick and sufficiently curved, but for a slightly curved and very thin film we can find such a surface only for the boundary at which the pressure difference retains the sign given by the ordinary theory of capillarity.

If a surface of tension exists for each boundary of a thin film, the position will not be that for a thick film, the two surfaces coming together if there is attraction and vice versa. Moreover, the σ and positions of these surfaces will determine the tension γ and position r of the surface of tension for the film as a whole via [6]:

$$\gamma = \alpha^{2/3}\beta^{1/3}, \tag{16}$$

$$r = \alpha^{-1/3}\beta^{1/3}, \tag{17}$$

in which

$$\alpha \equiv \frac{\sigma^{(\alpha\gamma)}}{r^{(\alpha\gamma)}} + \frac{\sigma^{(\beta\gamma)}}{r^{(\beta\gamma)}}, \quad \beta \equiv \sigma^{(\alpha\gamma)}(r^{(\alpha\gamma)})^2 + \sigma^{(\beta\gamma)}(r^{(\beta\gamma)})^2$$

(all quantities relate to the corresponding surface of tension). Relations (16) and (17) reveal the connection between the two thermodynamic approaches as applied to curved films.

Participation of Films in Heterogeneous Equilibria

A full discussion must include polyphase equilibria involving films. If the film as a whole is considered via (1) and (2) for this purpose, we can use all the methods in the thermodynamics of heterogeneous equilibria in conjunction with the above treatment of surface phenomena. We merely have to replace the surface tension by the corresponding tension for the film in all relationships involving surface phenomena. A special feature is introduced by the fact that a film differs from an interface in that on both sides it may be in contact with the same phase; also, in certain cases (e.g., free films), it is convenient to have relationships in terms of the composition variables of the film. However, both of these features are easily incorporated into the equations.

For thick films, the argument is merely that of the usual theory of heterogeneous systems with surface phenomena. For thin films, we have to distinguish whether the γ phase is or is not present as a separate bulk phase. Participation of phase γ in the equilibrium is of special interest, since it enables us to introduce Π directly into the thermodynamic equations, this being defined as the difference between the pressures of the two bulk phases and being directly measurable.

As an example we consider the system of equations for equilibrium of a planar thin film of finite thickness (taken as a whole) with phases α, β, and γ in terms of the composition variables of phase γ [6]:

$$\left.\begin{aligned}
v^{(f)}d\Pi - sd\gamma &= \eta_{\gamma f}dT - v_{\gamma f}dP^{(\gamma)} + \sum_{i=1}^{n-1}\sum_{k=1}^{n-1}(x_i^{(f)} - x_i^{(\gamma)})\zeta_{ik}^{(\gamma)}dx_k^{(\gamma)}, \\
v^{(\alpha)}d\Pi &= \eta_{\gamma\alpha}dT - v_{\gamma\alpha}dP^{(\gamma)} + \sum_{i=1}^{n-1}\sum_{k=1}^{n-1}(x_i^{(\alpha)} - x_i^{(\gamma)})\zeta_{ik}^{(\gamma)}dx_k^{(\gamma)}, \\
v^{(\beta)}d\Pi &= \eta_{\gamma\beta}dT - v_{\gamma\beta}dP^{(\gamma)} + \sum_{i=1}^{n-1}\sum_{k=1}^{n-1}(x_i^{(\beta)} - x_i^{(\gamma)})\zeta_{ik}^{(\gamma)}dx_k^{(\gamma)},
\end{aligned}\right\} \tag{18}$$

in which s is the molar surface of the film as a whole, v is molar volume, η is molar entropy, x_i is the molar fraction of component i, $\zeta_{ik} \equiv \partial^2\zeta/\partial x_i\partial x_k$, and n is the number of components.

This system relates Π and the film tension to T, external pressure, and the composition of phase γ. It also allows us to derive direct relations between Π and the film tension.

The following are some important examples and some relationships implied by (18). We assume that phases α and β are identical and are liquid or gaseous, T and the external pressure being constant, which corresponds, in particular, to formation of a film between drops or bubbles of identical size within the phase γ [7]. We assume for simplicity that we have a two-component system. Then from (18) we get

$$\left(\frac{d\Pi}{dx_1^{(\gamma)}}\right)_{T,\,P^{(\gamma)}} = \frac{(x_1^{(\alpha)} - x_1^{(\gamma)})\,\zeta_{11}^{(\gamma)}}{v^{(\alpha)}}, \tag{19}$$

$$\frac{1}{\tau'}\left(\frac{d\gamma}{dx_1^{(\gamma)}}\right)_{T,\,P^{(\gamma)}} = -\left[\frac{x_1^{(f)} - x_1^{(\gamma)}}{v^{(f)}} - \frac{x_1^{(\alpha)} - x_1^{(\gamma)}}{v^{(\alpha)}}\right]\zeta_{11}^{(\gamma)}, \tag{20}$$

$$\frac{1}{\tau'}\left(\frac{d\gamma}{d\Pi}\right)_{T,\,P^{(\gamma)}} = 1 - \frac{v^{(\alpha)}(x_1^{(f)} - x_1^{(\gamma)})}{v^{(f)}\,(x_1^{(\alpha)} - x_1^{(\gamma)})}, \tag{21}$$

in which τ' is the thickness of the film.

The following rules are implied by these relationships for a liquid film in contact with a vapor: a) addition of the more volatile component raises Π and conversely; b) if the film is enriched in the least volatile component relative to phase γ, addition of this will reduce the film tension, and conversely. It also follows from (21) that the tension and Π may change in opposite sense.

A liquid film in contact with its own vapor has $v^{(\alpha)} \gg v^{(f)}$ far from the critical state, and (21) becomes

$$\frac{1}{\tau'}\left(\frac{d\gamma}{d\Pi}\right)_{T,\,P^{(\gamma)}} \approx -\frac{v^{(\alpha)}\,(x_1^{(f)} - x_1^{(\gamma)})}{v^{(f)}\,(x^{(\alpha)} - x_1^{(\gamma)})}. \tag{22}$$

If the film is surrounded by a sparingly soluble gas (e.g., a film between two air bubbles in a liquid), (21) becomes

$$\left(\frac{d\gamma}{d\Pi}\right)_{T,\,P^{(\gamma)}} \approx \tau'. \tag{23}$$

Consider now the other case of equilibrium, which relates especially to attachment of drops and bubbles. Let phase α be the drop of bubble, while phase β is a body with a plane surface insoluble in phases α and γ. If the additional deformation of phase β in the surface layer can be neglected, it is readily shown that the above isobaric-isothermal relations still apply provided that by $v^{(f)}$ we understand the molar volume not of the whole film but of the part less the surface layer of phase β (the adsorption film). Also, τ' is replaced by the thickness τ_a of this part. Then (23), for example, becomes

$$\left(\frac{d\gamma}{d\Pi}\right)_{T,\,P^{(\gamma)}} \approx \tau_a. \tag{24}$$

For a bubble adhering to a solid, τ_a is the thickness of the liquid layer between the two. Frumkin [8] has derived an analogous formula for this case.

Relation (24) still applies when phases α and β are insoluble bodies, as in Deryagin's experiments [9] on Π. For this case, τ_a is the thickness of the liquid layer between the bodies. Here the condition of constancy of T and P in (24) may be supplemented with the condition of constant composition in phase γ, which can always be obeyed if the mass of phase γ is large enough relative to the masses of the surface layers of the film.

References

1.   A. I. Rusanov, Thermodynamics of Surface Phenomena, Izd. LGU (1960).
2.   B. V. Deryagin, Kolloidn. Zh., 17(3):207 (1955).
3.   S. Kondo, J. Chem. Phys., 25:662 (1956).

4.   A. I. Rusanov, Kolloidn. Zh., 27:428 (1965).

5.   A. I. Rusanov and M. M. Shul'ts, Vestnik LGU, Seriya Fiz. i Khim., No. 4, p. 60 (1960);
      A. I. Rusanov, Zh. Fiz. Khim., 40:3134 (1966).

6.   A. I. Rusanov, Kolloidn. Zh., 28:551 (1966); 29:141, 148, 237 (1967).

7.   B. V. Deryagin and A. S. Titievskaya, Dokl. AN SSSR, 89(6):1041 (1953).

8.   A. N. Frumkin, Zh. Fiz. Khim., 12:337 (1938).

9.   B. V. Derjaguin (Deryagin) and E. W. Obuchov, Acta Physicochim. URSS, 5:1 (1936);
      B. V. Deryagin and M. M. Kusakov, Izv. AN SSSR, Ser. Khim., No. 5, p. 741 (1936);
      No. 5, p. 1119 (1937).

10.   B. V. Deryagin et al., Research in Surface Forces, Vol. 2, Consultants Bureau,
      New York (1966), pp. 3, 9, 17.

DISTRIBUTION FUNCTIONS AND PRESSURE TENSOR
FOR A FILM OF A SIMPLE LIQUID

A. I. Rusanov and F. M. Kuni

Zhdanov University, Leningrad

Asymptotic formulas are derived for the one- and two-particle distribution functions within a thin planar film of a simple liquid on the assumption that the forces between molecules at large distances are of van der Waals type. The formulas are used to find expressions for the pressure tensor and for the film tension.

Here we consider asymptotic formulas for the behavior of a thin film during transformation to a thick one. We envisage a planar film of pure liquid with van der Waals interaction between the molecules separated by large distances in two other phases or in vacuum. The method is based on strict functional expansions previously proposed [1] and is not restricted to the approximations usually employed in statistical mechanics (superposition, hyperchain, etc.). The basic assumption is the physical condition for the decay of correlation between particles towards large separations, which must be the case for a system lacking long-range order, e.g., a liquid or gas. In particular, a recently derived asymptotic formula [2] will be used for the complete two-particle correlation function of a homogeneous liquid:

$$G_0(r) = \frac{\gamma}{r^6} \quad (r \to \infty), \tag{1}$$

in which r is the distance between the two particles and γ is a constant.

We assume that the liquid interlayer and adjacent phases are mutually insoluble; the interfaces are termed bounding faces, whose separation H is the thickness of the liquid layer. The region filled by the liquid film is denoted by V. The distances h and s from the bounding surfaces obey h + s = H, and the distribution function within the film varies with these distances. It is possible to derive this relationship rigorously for h, s, and H large via functional expansions if the convergence accelerates as h and s increase. The second functional series previously proposed [1] meets this condition when the asymptotic formula (1) is used, and this will be the one to be employed.

We compare the state of the film in the field of the two adjacent phases with the state of a homogeneous liquid filling the same volume V. An important point is the choice of $u_0(r)$, the one-particle accessory external field, to provide homogeneity of the liquid in this region.

111

We assume[†] that this field is given by the following formula for van der Waals forces and for h and s sufficiently large:

$$u_0(r) = -a_0 \left(\frac{1}{h^3} + \frac{1}{s^3}\right), \tag{2}$$

in which $a_0 = \frac{1}{6}\pi\rho_0 A_0$, ρ_0 is the density of the homogeneous liquid, and A_0 is the constant for the attractive part of the molecular interaction in the liquid.

In other words, we assume that $u_0(\vec{r})$ for h and s sufficiently large coincides with the field produced in the middle of V by the homogeneous liquid outside this volume.

An external potential u(r) is produced in the internal part of the film by the surrounding phases and is described (as for van der Waals forces) by

$$u(r) = -\frac{a_1}{h^3} - \frac{a_2}{s^3}, \tag{3}$$

in which $a_1 = \frac{1}{6}\pi\rho_1 A_{01}$; $a_2 = \frac{1}{6}\pi\rho_2 A_{02}$; and ρ_1 and ρ_2 are the densities of these surrounding phases, while A_{01} and A_{02} are the constants for the attractive parts of the interactions between the liquid molecules and the molecules of the two phases respectively.

The statistical calculations will be performed for a grand canonical ensemble. The one- and two-particle distribution functions are of the greatest interest, so only these are considered here, though they are also used to find the pressure tensor, while the film tension is considered at the end of the paper.

One-Particle Distribution Function

Symmetry considerations imply that this function $\rho^{(1)}(\vec{r})$ within the liquid film is a function of h and s only. The following functional expansion relates the one-particle distribution functions of the film and the homogeneous liquid:

$$\rho^{(1)}(\vec{r}) = e^{-\beta\Delta u(\vec{r_1})}\left\{\rho_0 + \sum_{m=1}^{\infty}\frac{1}{m!}\int_V g_0^{(m+1)}(\vec{r_1},\ldots,\vec{r_{m+1}})\prod_{i=2}^{m+1}[e^{-\beta\Delta u(\vec{r_i})} - 1]\,d\vec{r_i}\right\}, \tag{4}$$

in which $\rho^{(1)}(\vec{r_1})$ is that function (the local density) at point $\vec{r_1}$ within the film, $g_0^{(m+1)}(\vec{r_1},\ldots,\vec{r_{m+1}})$ is the Mayer-Ursell distribution function of order m + 1 for the homogeneous liquid, $\beta = 1/kT$, k is Boltzmann's constant, T is temperature, $\Delta u(\vec{r_i}) = u(\vec{r_i}) - u_0(\vec{r_i})$; and $u(\vec{r_i})$ and $u_0(\vec{r_i})$ are respectively the potential of the external field and the accessory potential, which are given for h and s large by (3) and (2). The integration is carried over the region V between the bounding surfaces, which have a separation H.

We begin our examination of the right-hand side in (4) with the first term in the sum, which is denoted by $I_1(\vec{r_1})$. With $g_0^{(2)}(\vec{r_1},\vec{r_2}) = \rho_0^2 G_0(r_{12})$, in which $r_{12} = |\vec{r_1} - \vec{r_2}|$, we put this as

$$I_1(\vec{r_1}) = \rho_0^2\int_V G_0(r_{12})[e^{-\beta\Delta u(\vec{r_i})} - 1]\,d\vec{r_2}. \tag{5}$$

[†] It can be shown rigorously by the diagram technique that (2) is obeyed to quantities of higher order in 1/h and 1/s, and also that field $u_0(r)$ produces negligibly small perturbations in the one- and two-particle distribution functions relative to the values for a truly homogeneous liquid.

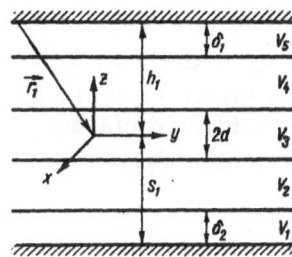

Fig. 1.

We transfer the origin to point $\vec{r_1}$ (Fig. 1), which is assumed to lie at large distances h_1 and s_1 from the bounding surfaces; the z axis is perpendicular to those surfaces. Planes parallel to those surfaces are used to divide V into regions V_1 $(-s_1 \leq z \leq -s_1 + \delta_1)$, V_2 $(-s_1 + \delta_1 \leq z \leq -d)$, V_3 $(-d \leq z \leq d)$, V_4 $(d \leq z \leq h_1 - \delta_2)$, and V_5 $(h_1 - \delta_2 \leq z \leq h_1)$, in which, d, δ_1, and δ_2 are so large that (1) applies outside V_3, while (3) and (2) apply for $(\vec{r})$ and $(\vec{r})$ respectively outside V_1 and V_5:

$$u(\vec{r}) = -\frac{a_1}{(h_1 - z)^3} - \frac{a_2}{(s_1 + z)^3}, \tag{6}$$

$$u_0(\vec{r}) = -a_0 \left[\frac{1}{(h_1 - z)^3} + \frac{1}{(s_1 + z)^3} \right]. \tag{7}$$

At the same time, the film is assumed to be sufficiently thick for us to have h_1, $s_1 \gg d$, δ_1, δ_2.

Then the integral in (5) splits up into five integrals, each of which is calculated separately. Formulas (6) and (7) may be used in integration over V_3. Taking only terms[†] of order $1/h_1^3$ and $1/s_1^3$, we have

$$\int_{V_3} G_0(r_{12}) [e^{-\beta \Delta u(\vec{r_2})} - 1] d\vec{r_2} = \beta \left(\frac{a_1 - a_0}{h_1^3} + \frac{a_2 - a_0}{s_1^3} \right) \int_{V_3} G_0(r_{12}) d\vec{r_2} + O\left(\frac{1}{h_1^4}, \frac{1}{s_1^4} \right). \tag{8}$$

The following identity applies to a grand canonical ensemble:

$$\int_{V_\infty} G_0(r_{12}) d\vec{r_2} = \frac{\chi_0}{\beta} - \frac{1}{\rho_0}, \tag{9}$$

in which χ_0 is the isothermal compressibility of the homogeneous liquid $[\chi_0 \equiv (1/\rho_0)(\partial \rho_0 / \partial P_0)_T]$ and P_0 is pressure; then

$$\int_{V_3} G_0(r_{12}) d\vec{r_2} = \frac{\chi_0}{\beta} - \frac{1}{\rho_0} - \int_{V_\infty - V_3} G_0(r_{12}) d\vec{r_2}. \tag{10}$$

The integral remaining on the right in (10) is easily calculated via (1).

The integrals over V_2 and V_4 are calculated via (1), (6), and (7). The integrals over V_1 and V_5 produce contributions of order $1/h_1^4$ and $1/s_1^4$, and they may be neglected in the asymptotic formulas. The results for the various integrals in (5) are combined to give

$$I_1(\vec{r_1}) = \beta \rho_0 \left(\frac{a_1 - a_0}{h_1^3} + \frac{a_2 - a_0}{s_1^3} \right) \left(\frac{\rho_0 \chi_0}{\beta} - 1 \right) + O\left(\frac{1}{h_1^4}, \frac{1}{s_1^4} \right). \tag{11}$$

The next term in (4) is denoted by $I_2(\vec{r_1})$:

$$I_2(\vec{r_1}) \equiv \frac{1}{2} \int_V g_0^{(3)}(\vec{r_1}, \vec{r_2}, \vec{r_3}) [e^{-\beta \Delta u(\vec{r_2})} - 1][e^{-\beta \Delta u(\vec{r_3})} - 1] d\vec{r_2} d\vec{r_3}. \tag{12}$$

[†] The symbol $O(1/h_1^n, 1/s_1^n)$ denotes a quantity of order $h_1^{-k} s_1^{-n+k}$, for $h_1 \to \infty$ and $s_1 \to \infty$, in which k may take the values 0, 1, 2,..., n, in which n is the order of smallness of $O(1/h_1^n, 1/s_1^n)$.

It can be shown that this is of a higher order of smallness than $I_1(\vec{r_1})$.   The contribution from the integral in (12) is of the sixth order of smallness when points $\vec{r_2}$ and $\vec{r_3}$ both lie in V_3, on account of the factors in square brackets; it can therefore be neglected. If either r_{12} or r_{13} is greater than d, we can use the condition for the decline in the correlation at large distances, which may, for example, be expressed via the formula for the superposition approximation:

$$g_0^{(3)}(\vec{r_1}, \vec{r_2}, \vec{r_3}) = \rho_0^3 [G_0(r_{12})G_0(r_{13}) + G_0(r_{12})G_0(r_{23})$$
$$+ G_0(r_{13})G_0(r_{23}) + G_0(r_{12})G_0(r_{13})G_0(r_{23})]. \tag{13}$$

The first term on the right in (13) is substituted into $I_2(\vec{r_1})$ to give the product of two integrals which is proportional to $[I_1(\vec{r_1})]^2$. A contribution of the same order comes from the last term on the right in (13), because $|G_0(r_{23})| \leqslant M$, in which M is a constant.

The other two terms make equal contributions, so we have

$$I_2(\vec{r_1}) = \rho_0^3 \int\limits_V G_0(r_{12})G_0(r_{23}) [e^{-\beta\Delta u(\vec{r_2})} - 1][e^{-\beta\Delta u(\vec{r_3})} - 1]\, d\vec{r_2}\, d\vec{r_3} + O\left(\frac{1}{h_1^6}, \frac{1}{s_1^6}\right) \tag{14}$$

or, from (5)

$$I_2(\vec{r_1}) = \rho_0 \int\limits_V G_0(r_{12}) [e^{-\beta\Delta u(\vec{r_2})} - 1] I_1(\vec{r_2})\, d\vec{r_2} + O\left(\frac{1}{h_1^6}, \frac{1}{s_1^6}\right). \tag{15}$$

The integral appearing here may be calculated in the same way as for (5), namely by dividing V and choosing δ_1 and δ_2 in such a way that (11) applies within the film outside regions V_1 and V_5. The result is a quantity of the fourth order of smallness in $1/h_1$ and $1/s_1$; the same order is to be ascribed to the entire expression in (12).

It can be shown that all the subsequent terms in (4) are of order of smallness in $1/h_1$ and $1/s_1$ not less than the fourth, so the main contribution to the asymptote for the one-particle distribution function within the film comes from the first term in the sum in (4), and for h_1 and s_1 large we may put the relation as

$$\rho^{(1)}(\vec{r_1}) = e^{-\beta\Delta u(\vec{r_1})} [\rho_0 + I_1(\vec{r_1})] + O\left(\frac{1}{h_1^4}, \frac{1}{s_1^4}\right). \tag{16}$$

We substitute expressions (2), (3), and (11) into this and expand the exponential in series form to get (omitting the subscript 1 to $\vec{r}$, h and s) that

$$\rho^{(1)}(\vec{r}) = \rho_0 + \rho_0^2\chi_0\left(\frac{a_1 - a_0}{h^3} + \frac{a_2 - a_0}{s^3}\right) + O\left(\frac{1}{h^4}, \frac{1}{s^4}\right). \tag{17}$$

Formula (17) is the desired asymptotic formula for the local density in the internal part of a film of sufficiently great thickness. As h + s = H, the formula can be applied for h and s variable with H constant, and also when H varies with h or s constant. The coefficients are readily calculated from experimental data for the homogeneous liquid.

If the adjacent phases interact strongly with the molecules of the liquid film ($a_1 > a_0$, $a_2 > a_0$), we see from (17) that the density within the film will be greater than the density of the homogeneous liquid and will increase as the bounding surfaces are approached. Formula (17) becomes as follows for a free liquid film:

$$\rho^{(1)}(\vec{r}) = \rho_0 - \rho_0^2\chi_0 a_0 \left(\frac{1}{h^3} + \frac{1}{s^3}\right) + O\left(\frac{1}{h^4}, \frac{1}{s^4}\right), \tag{18}$$

which shows that the density is then less than that for the homogeneous liquid and will fall as the surfaces of the film are approached.

Two-Particle Distribution Function

The same approach is used to find this function within a thin film. We use the functional expansion

$$g^{(2)}(\vec{r}_1, \vec{r}_2) = e^{-\beta[\Delta u(\vec{r}_1) + \Delta u(\vec{r}_2)]}\left\{g_0^{(2)}(\vec{r}_1, \vec{r}_2) + \sum_{m=1}^{\infty}\frac{1}{m!}\int_V g_0^{(m+2)}(\vec{r}_1, \ldots, \vec{r}_{m+2})\prod_{i=3}^{m+2}[e^{-\beta\Delta u(\vec{r}_i)} - 1]\,d\vec{r}_i\right\}, \quad (19)$$

in which $g^{(2)}(\vec{r}_1, \vec{r}_2)$ is the two-particle Mayer-Ursell function within the film.

Functions $g^{(2)}(\vec{r}_1, \vec{r}_2)$ and $g_0^{(2)}(\vec{r}_1, \vec{r}_2)$ have simple relations to the two-particle distribution functions:

$$g^{(2)}(\vec{r}_1, \vec{r}_2) = \rho^{(2)}(\vec{r}, \vec{r}_2) - \rho^{(1)}(\vec{r}_1)\rho^{(1)}(\vec{r}_2), \tag{20}$$

$$g_0^{(2)}(\vec{r}_1, \vec{r}_2) = \rho_0^{(2)}(\vec{r}_1, \vec{r}_2) - \rho_0^2, \tag{21}$$

which allows us to establish a relation between $\rho^{(2)}(\vec{r}_1, \vec{r}_2)$ and $\rho_0^{(2)}(\vec{r}_1, \vec{r}_2)$ by examination of (19).

It can be shown that the first term on the right in (19) makes the main contribution to the asymptote for the two-particle distribution function for a thin film for h and s large (as r_{12} is assumed finite, distances $\vec{r}_1$ and $\vec{r}_2$ to the bounding surfaces are of the same order: $h_1 \approx h_2 \approx$ h; $s_1 \approx s_2 \approx$ s). The problem thus reduces to that of finding the integral

$$\int_V g_0^{(3)}(\vec{r}_1, \vec{r}_2, \vec{r}_3)[e^{-\beta\Delta u(\vec{r}_3)} - 1]\,d\vec{r}_3, \tag{22}$$

which is calculated by the method used for the integral in (5).

We divide V into V_1-V_5, taking the width of V_3 such that (13) applies outside V_3. Formulas (1) and (13) allow us to calculate the integrals over V_2 and V_4, and also to estimate the integrals over V_1 and V_5 as having a higher order of smallness. We take the linear part of the exponential in the integral over V_3, and the residual integral is calculated via a relation derived[†] for a large canonical ensemble:

$$\int_{V_\infty} g_0^{(3)}(\vec{r}_1, \vec{r}_2, \vec{r}_3)\,d\vec{r}_3 = 1/\beta\,(\partial g_0^{(2)}(\vec{r}_1, \vec{r}_2)/\partial\mu)_T - 2g_0^{(2)}(\vec{r}_1, \vec{r}_2), \tag{23}$$

in which μ is the chemical potential of the liquid. The integration is carried over the infinite space of the homogeneous liquid.

The final result from (19) for h and s large is

$$g^{(2)}(\vec{r}_1, \vec{r}_2) = g_0^{(2)}(\vec{r}_1, \vec{r}_2) + \left(\frac{a_1 - a_0}{h^3} + \frac{a_2 - a_0}{s^3}\right)(\partial g_0^{(2)}(\vec{r}_1, \vec{r}_2)/\partial\mu)_T + O\left(\frac{1}{h^4}, \frac{1}{s^4}\right) \tag{24}$$

or

$$\rho^{(2)}(\vec{r}_1, \vec{r}_2) = \rho_0^{(2)}(\vec{r}_1, \vec{r}_2) + \left(\frac{a_1 - a_0}{h^3} + \frac{a_2 - a_0}{s^3}\right)(\partial\rho_0^{(2)}(\vec{r}_1, \vec{r}_2)/\partial\mu)_T + O\left(\frac{1}{h^4}, \frac{1}{s^4}\right). \tag{25}$$

Formula (25) describes the behavior of the two-particle distribution function in the internal part of a thin film of sufficiently great thickness. As $(\partial\rho_0^{(2)}/\vec{r}_1, \vec{r}_2)/\partial\mu)_T$ is positive, the general behavior of this function is similar to that of the local density.

[†] The general recurrence relation between $\int_{V_\infty} g_0^{(n)}(\vec{r}_1, \ldots, \vec{r}_n)\,d\vec{r}_n$ and $g_0^{(n-1)}(\vec{r}_1, \ldots, \vec{r}_{n-1})$ is of similar form.

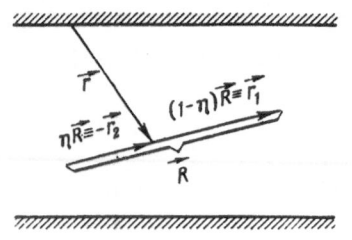

Pressure Tensor

    The distribution functions and intermolecular potential together allow one to calculate the pressure tensor for an inhomogeneous system. The concept of a pressure tensor is not unambiguous [3]; here we use the definition of [4], which is mathematically the most complicated but which best reflects the physical essence. The formula is

Fig. 2.

$$\hat{P}\,(\vec{r}) = kT\rho^{(1)}\,(\vec{r})\,\hat{1} - \frac{1}{2}\int_{V_\infty} \frac{\vec{R}\times\vec{R}}{R}\,\Phi'(R)\,d\vec{R}\int_0^1 \rho^{(2)}\,(\vec{r} - \eta\vec{R}, \vec{r} - \eta\vec{R} + \vec{R})\,d\eta,$$

$$(26)$$

in which $\hat{P}\,(\vec{r})$ is the pressure tensor at point $\vec{r}$; $\hat{1}$ is unit tensor, $\vec{R}$ is the vector joining two particles and passing through the point $\vec{r}$ (Fig. 2), and $\Phi(R)$ is the intermolecular potential.

    In our case, there are four parts $\hat{P}_1$–$\hat{P}_4$ to the complete pressure tensor when $\vec{r}$ is within the film at large distances h and s from the first and second phases respectively. Each of these parts corresponds to an intermolecular potential. $\hat{P}_1$ arises from interaction of molecules within the liquid film, $\hat{P}_2$ and $\hat{P}_3$ arise from interaction of the film molecules with the first and second phases respectively and $\hat{P}_4$ arises from interaction of the phases surrounding the film, e.g., the solids enclosing the film.

    We transfer the origin of coordinates to $\vec{r}$ and set the z axis perpendicular to the bounding surfaces towards the first phase. We then perform the transformation R_x, R_y, R_z, $\eta \rightarrow x_1$, y_1, z_1, z_2, where x_1, y_1, and z_1 are the components of the vector $\vec{r}_1 \equiv (1-\eta)\vec{R}$, z_2 is a component of the vector $\vec{r}_2 = -\eta\vec{R}$, and we integrate only with respect to positive values of z_1, discarding simultaneously the factor 1/2 before the integral in (26). Then we get the following expressions for the tensors:

$$\hat{P}_1 = kT\rho^{(1)}\,(0)\,\hat{1} - \int_{-\infty}^{+\infty} dx_1 \int_{-\infty}^{+\infty} dy_1 \int_0^h dz_1 \frac{\vec{r}_1\times\vec{r}_1}{r_1 z_1^3}\int_{-s}^0 dz_2\,(z_1-z_2)^2 \Phi_0'\left[\frac{r_1(z_1-z_2)}{z_1}\right]\rho^{(2)}\left(\frac{z_2}{z_1}\,\vec{r}_1, \vec{r}_1\right),$$

$$(27)$$

$$\hat{P}_2 = -\int_{-\infty}^{+\infty} dx_1 \int_{-\infty}^{+\infty} dy_1 \int_h^{+\infty} dz_1 \frac{\vec{r}_1\times\vec{r}_1}{r_1 z_1^3}\int_{-s}^0 dz_2\,(z_1-z_2)^2 \Phi_{01}'\left[\frac{r_1(z_1-z_2)}{z_1}\right]\rho^{(2)}\left(\frac{z_2}{z_1}\,\vec{r}_1, \vec{r}_1\right),$$

$$(28)$$

$$\hat{P}_3 = -\int_{-\infty}^{+\infty} dx_1 \int_{-\infty}^{+\infty} dy_1 \int_0^h dz_1 \frac{r_1\times r_1}{r_1 z_1^3}\int_{-\infty}^{-s} dz_2\,(z_1-z_2)^2 \Phi_{02}'\left[\frac{r_1(z_1-z_2)}{z_1}\right]\rho^{(2)}\left(\frac{z_2}{z_1}\,\vec{r}_1, \vec{r}_1\right),$$

$$(29)$$

$$\hat{P}_4 = -\int_{-\infty}^{+\infty} dx_1 \int_{-\infty}^{+\infty} dy_1 \int_h^{\infty} dz_1 \frac{\vec{r}_1\times\vec{r}_1}{r_1 z_1^3}\int_{-\infty}^{-s} dz_2\,(z_1-z_2)^2 \Phi_{12}'\left[\frac{r_1(z_1-z_2)}{z_1}\right]\rho^{(2)}\left(\frac{z_2}{z_1}\,\vec{r}_1, \vec{r}_1\right),$$

$$(30)$$

in which Φ_0, Φ_{01}, Φ_{02}, Φ_{12} are the potentials for the following pair interactions: between molecules of the liquid, between these molecules and the first phase, the same for the second phase, and between the two phases.

    The corresponding tensor for a homogeneous liquid is

$$\hat{P}_0 = kT\rho_0\hat{1} - \int_{-\infty}^{+\infty} dx_1 \int_{-\infty}^{+\infty} dy_1 \int_0^{+\infty} dz_1 \frac{\vec{r}_1\times\vec{r}_1}{r_1 z_1^3}\int_{-\infty}^0 dz_2\,(z_1-z_2)^2\,\Phi_0'\left[\frac{r_1(z_1-z_2)}{z_1}\right]\rho_0^{(2)}\left[\frac{r_1(z_1-z_2)}{z_1}\right].$$

$$(31)$$

Subtraction of (31) from (27) gives

$$\hat{\Phi}_1 = \hat{p_0} + kT \left[\rho^{(1)}(0) - \rho_0\right] \hat{1} - \int\limits_{-\infty}^{+\infty} dx_1 \int\limits_{-\infty}^{+\infty} dy_1 \int\limits_{0}^{h} dz_1 \frac{\vec{r_1} \times \vec{r_1}}{r_1 z_1^3}$$

$$\times \int\limits_{-s}^{0} dz_2 (z_1 - z_2)^2 \Phi_0' \left[\frac{r_1(z_1 - z_2)}{z_1}\right] \left\{\rho^{(2)}\left(\frac{z_2}{z_1}\vec{r_1}, \vec{r_1}\right) - \rho_0^{(2)}\left[\frac{r_1(z_1 - z_2)}{z_1}\right]\right\}$$

$$+ \int\limits_{-\infty}^{+\infty} dx_1 \int\limits_{-\infty}^{+\infty} dy_1 \int\limits_{0}^{h} dz_1 \frac{\vec{r_1} \times \vec{r_1}}{r_1 z_1^3} \int\limits_{-\infty}^{-s} dz_2 (z_1 - z_2)^2 \Phi_0'\left[\frac{r_1(z_1 - z_2)}{z_1}\right] \rho_0^{(2)}\left[\frac{r_1(z_1 - z_2)}{z_1}\right]$$

$$+ \int\limits_{-\infty}^{+\infty} dx_1 \int\limits_{-\infty}^{+\infty} dy_1 \int\limits_{h}^{\infty} dz_1 \frac{\vec{r_1} \times \vec{r_1}}{r_1 z_1^3} \int\limits_{-\infty}^{0} dz_2 (z_1 - z_2)^2 \Phi_0'\left[\frac{r_1(z_1 - z_2)}{z_1}\right] \rho_0^{(2)}\left[\frac{r_1(z_1 - z_2)}{z_1}\right]. \tag{32}$$

The expression may be used to find $\hat{P}_1$. The main difficulty here arises over the first integral on the right in (32). To derive this, we again divide the region V into five parts (Fig. 1), δ_1 and δ_2 being taken such that (18) and (25) apply for $\vec{r_1}$ and $\vec{r_2}$ outside regions V_1 and V_5 within the film (they apply also, of course, for the points where we seek to determine the pressure tensor). The width of V_3 is chosen such that the intermolecular potential may be approximated by its attractive part if either $\vec{r_1}$ or $\vec{r_2}$ lies outside this layer:

$$\Phi_0\left[\frac{r_1(z_1 - z_2)}{z_1}\right] = -\frac{A_0 z_1^6}{r_1^6 (z_1 - z_2)^6}, \qquad \Phi_0'\left[\frac{r_1(z_1 - z_2)}{z_1}\right] = \frac{6A_0 z_1^7}{r_1^7 (z_1 - z_2)^7} \tag{33}$$

and also such that the condition for weak correlation is obeyed:

$$\rho^{(2)}\left(\frac{z_2}{z_1}\vec{r_1}, \vec{r_1}\right) = \rho^{(1)}\left(\frac{z_2}{z_1}\vec{r_1}\right)\rho^{(1)}(\vec{r_1}). \tag{34}$$

The difficulty then reduces to that of finding the integral over region V_3 ($0 \le z_1 \le d$, $-d \le z_2 \le 0$). To calculate this, we use (25) and represent the integral as a combination of three integrals:

$$\left(\frac{a_1 - a_0}{h^3} + \frac{a_2 - a_0}{s^3}\right)\frac{\partial}{\partial\mu}\left\{\left(\int\limits_{-\infty}^{+\infty} dx_1 \int\limits_{-\infty}^{+\infty} dy_1 \int\limits_{0}^{+\infty} dz_1 \int\limits_{-\infty}^{0} dz_2\right.\right.$$

$$- \int\limits_{-\infty}^{+\infty} dx_1 \int\limits_{-\infty}^{+\infty} dy_1 \int\limits_{0}^{d} dz_1 \int\limits_{-\infty}^{-d} dz_2 - \int\limits_{-\infty}^{+\infty} dx_1 \int\limits_{-\infty}^{+\infty} dy_1 \int\limits_{d}^{\infty} dz_1 \int\limits_{-\infty}^{0} dz_2\right)$$

$$\left.\times \frac{\vec{r_1} \times \vec{r_1}}{r_1 z_1^3}(z_1 - z_2)^2 \Phi_0'\left[\frac{r_1(z_1 - z_2)}{z_1}\right]\rho_0^{(2)}\left[\frac{r_1(z_1 - z_2)}{z_1}\right]\right\}_T. \tag{35}$$

The first integral in (35) is shown by (31) to be simply $kT\rho_0 - P_0$, in which ρ_0 and P_0 are the density and pressure in the homogeneous liquid. Differentiation with respect to μ transforms this integral to $\rho_0(kT\rho_0\chi_0 - 1)$. The other integrals in (35) resemble all the other integrals in (32) in being calculable via (18), (33), and (34).

The final result is as follows. From the symmetry conditions, $\hat{P}_1$ can be characterized by only two components: the normal one $P_{1N} = P_{1zz}$ and the tangential one $P_{1T} = P_{1xx} = P_{1yy}$. The following is the asymptotic formula for the components of $\hat{P}_1$:

$$P_{1N(T)} = P_0 + \rho_0\left(\frac{a_1 - a_0}{h^3} + \frac{a_2 - a_0}{s^3}\right) + \rho_0 a_0 \alpha_{N(T)}\left(\frac{1}{h^3} + \frac{1}{s^3}\right) - \rho_0 a_0 \alpha_{N(T)}\frac{1}{H^3} + O\left(\frac{1}{h^4}, \frac{1}{s^4}\right), \tag{36}$$

in which $\alpha_N = 1$, $\alpha_T = 1/4$.

Tensors $\hat{P}_2$-$\hat{P}_4$ are calculated solely via the asymptotic formulas of (18) and (34) for the distribution functions and of (33) for the intermolecular potentials, since all integrals in (28)-(30) have the ranges of variation in $\vec{r}_1$ and $\vec{r}_2$ separated by layers of thickness h, s, or H. The results are

$$P_{2N\,(T)} = \rho_0 a_1 \alpha_{N\,(T)} \left(\frac{1}{H^3} - \frac{1}{h^3} \right) + O\left(\frac{1}{H^4} \right), \tag{37}$$

$$P_{3N\,(T)} = \rho_0 a_2 \alpha_{N(T)} \left(\frac{1}{H^3} - \frac{1}{s^3} \right) + O\left(\frac{1}{H^4} \right), \tag{38}$$

$$P_{4N\,(T)} = \frac{\pi \rho_1 \rho_2 A_{12} \alpha_{N\,(T)}}{6H^3} + O\left(\frac{1}{H^4} \right), \tag{39}$$

in which A_{12} is the constant in the attractive part of the potential for the interaction between the molecules of the first phase and those of the second phase.

We add (36)-(39) to get an asymptotic formula for the complete pressure tensor within the film. The components are as follows:

$$P_N = P_0 + \frac{B}{H^3} + O\left(\frac{1}{H^4} \right), \tag{40}$$

$$P_T = P_0 + \frac{3}{4} \rho_0 \left(\frac{a_1 - a_0}{h^3} + \frac{a_2 - a_0}{s^3} \right) + \frac{B}{4H^3} + O\left(\frac{1}{h^4}, \frac{1}{s^4} \right), \tag{41}$$

in which

$$B \equiv \rho_0 (a_1 + a_2 - a_0) - \frac{1}{6} \pi \rho_1 \rho_2 A_{12} = -\frac{1}{6} \pi (\rho_0 \rho_1 A_{01} + \rho_0 \rho_2 A_{02} - \rho_0^2 A_0 - \rho_1 \rho_2 A_{12}). \tag{42}$$

As would be expected, P_N is independent of the position within the film; hence, although our calculations relate to a point somewhere in the internal part of the thin film, formula (40) is strictly applicable to any part of the film and characterizes the deviation of the normal pressure P_N from P_0 on going from thick films to thin ones. This formula also allows us to find the disjoining pressure Π, which is equal to the difference between P_N and P_0 [5]:

$$\Pi \equiv P_N - P_0 = \frac{B}{H^3} + O\left(\frac{1}{H^4} \right). \tag{43}$$

For a free liquid film under vacuum

$$B = -\frac{1}{6} \pi \rho_0^2 A_0 < 0. \tag{44}$$

If a vacuum takes the place of the liquid layer between the phases (e.g., when two solids come together under vacuum)

$$B = -\frac{1}{6} \pi \rho_1 \rho_2 A_{12} < 0. \tag{45}$$

In both cases, (43) gives this pressure as negative.

The relationship described by (43) begins to become operative only when the properties of the thin film become apparent, and so it inevitably appears on passing from thick films to thin ones. Deryagin and Abrikosova [6] found on bringing together solids that the cleaving pressure was inversely proportional to the cube of the distance.

There have been various studies of the cause of this pressure; for instance, the case has been discussed [7, 8] where the gap between the bodies is filled with a liquid whose dielectric constant differs from one. These papers employed a macroscopic approach, but they involved the very important assumptions that the liquid film was homogeneous and had the properties of the bulk phase. Deryagin and Martynov [9] examined the features due to loss of homogeneity for the simple case of a film of ideal gas whose particles are attracted by molecular forces to the surfaces of the interacting bodies. The present formula (43) makes proper allowance for film inhomogeneity, while not involving any assumption as to the smallness of ρ_0. In the limit $\rho_0 \to 0$ (ideal gas), our result agrees with that of [9].

Film Tension

From (40) and (41) we have

$$P_N - P_T = \frac{3}{4}\left[\frac{B}{H^3} - \rho_0\left(\frac{a_1 - a_0}{h^3} + \frac{a_2 - a_0}{s^3}\right)\right] + O\left(\frac{1}{h^4}, \frac{1}{s^4}\right),$$ (46)

which characterizes the tension in an elementary layer of unit thickness lying within the film parallel to the bounding surfaces at large distances s and h from the latter. This is termed the local tension. Formula (46) becomes as follows for a free film:

$$P_N - P_T = \frac{1}{8}\pi\rho_0^2 A_0\left(\frac{1}{h^3} + \frac{1}{s^3} - \frac{1}{H^3}\right) + O\left(\frac{1}{h^4}, \frac{1}{s^4}\right),$$ (47)

which shows that the local tension is minimal for h = s = H/2 (middle of the film). A thin film between the strong adsorbents is seen from (46) to show the converse effect (a maximum in the local tension somewhere in the middle of the film). Formula (46) gives for an elementary layer precisely at the center that

$$P_N - P_T = \frac{C}{H^3} + O\left(\frac{1}{H^4}\right),$$ (48)

in which

$$C = \frac{1}{8}\pi(15\rho_0^2 A_0 - 7\rho_0\rho_1 A_{01} - 7\rho_0\rho_2 A_{02} - \rho_1\rho_2 A_{12}).$$ (49)

Formula (48) describes the dependence of the local tension at the center on the film thickness for H sufficiently large. The magnitude of this tension is proportional to $1/H^3$, while the sign is positive for a free film but negative for a film between strong adsorbents (the latter agrees with views on adsorption films [10]).

Formula (46) applies only to the local tension within the film and cannot be used to find the tension γ of the film as a whole (with consideration not only of the liquid film between the phases but also of the surface layers of these phases). For this purpose we may use (43) in conjunction with thermodynamic relations between γ and the disjoining pressure [11, 12]. From

$$\frac{d\gamma}{d\Pi} = H$$ (50)

we get directly that

$$\gamma = \sigma_1 + \sigma_2 + \frac{3B}{2H^2} + O\left(\frac{1}{H^3}\right),$$ (51)

where the physical meaning of the constant of integration $\sigma_1 + \sigma_2$ is the sum of the surface tensions of the thick film derived from the thin film for H → ∞. From (44) and (51) it follows for a free film of liquid that the tension increases with H. This fact indicates instability in one-component films.

The $\hat{P}$ defined by (40)-(42) relates to the system as a whole, which consists of the liquid film and the surrounding phases, since it includes the forces of interaction between the first and second phases. The $\hat{P}$ for the film alone, but of course incorporating the interaction of the film with the two phases, is also given by (40)-(42) if we put $A_{12} = 0$:

$$\widetilde{P}_N = P_0 + \frac{\widetilde{B}}{H^3} + O\left(\frac{1}{H^4}\right), \tag{52}$$

$$\widetilde{P}_T = P_0 + \frac{3}{4}\rho_0\left(\frac{a_1 - a_0}{h^3} + \frac{a_2 - a_0}{s^3}\right) + \frac{\widetilde{B}}{4H^3} + O\left(\frac{1}{h^4}, \frac{1}{s^4}\right), \tag{53}$$

in which

$$\widetilde{B} = \frac{1}{6}\pi(\rho_0\rho_1 A_{01} + \rho_0\rho_2 A_{02} - \rho_0^2 A_0). \tag{54}$$

The local tension of the liquid film alone is now

$$\widetilde{P}_N - \widetilde{P}_T = \frac{3}{4}\left[\frac{\widetilde{B}}{H^3} - \rho_0\left(\frac{a_1 - a_0}{h^3} + \frac{a_2 - a_0}{s^3}\right)\right] + O\left(\frac{1}{h^4}, \frac{1}{s^4}\right). \tag{55}$$

Also, the Π for the film alone, $\widetilde{\Pi}$, equals Π less the direct interaction between the two phases and is given by

$$\widetilde{\Pi} \equiv \widetilde{P}_N - P_0 = \frac{\widetilde{B}}{H^3} + O\left(\frac{1}{H^4}\right). \tag{56}$$

We can also consider the surface tension $\widetilde{\gamma}$ of the film alone, which is

$$\widetilde{\gamma} = \widetilde{\sigma}_1 + \widetilde{\sigma}_2 + \frac{3\widetilde{B}}{2H^2} + O\left(\frac{1}{H^3}\right), \tag{57}$$

in which $\widetilde{\sigma}_1$ and $\widetilde{\sigma}_2$ are analogous to σ_1 and σ_2 but refer to the film.

Conclusions

Asymptotic formulas for $H \rightarrow \infty$ have been derived for the one-particle distribution function, the two-particle distribution function, the pressure tensor, and the tension in the liquid film. These have been derived on the assumption that van der Waals forces act between the molecules at large distances throughout the system.

References

1. F. M. Kuni, Vestnik LGU, No. 4, p. 11 (1965).
2. J. Enderby, T. Gaskell, and N. H. March, Proc. Phys. Soc., 85:217 (1965).
3. S. Ono and S. Kando, The Molecular Theory of Surface Tension in a Liquid [Russian translation], IL, Moscow (1963).
4. J. H. Irving and J. G. Kirkwood, J. Chem. Phys., 18:817 (1950).
5. B. V. Deryagin, Kolloidn. Zh., 17(3):207 (1955).
6. B. V. Deryagin and I. I. Abrikosova, Zh. Éksp. Teor. Fiz., 30:993; 31:3 (1956); Dokl. AN SSSR, 108:214 (1956); J. Phys. Chem. Solids, 5(1-2):1 (1958).
7. E. M. Lifshits, Zh. Éksp. Teor. Fiz., 29:94 (1955).
8. I. E. Dzyaloshinskii, E. M. Lifshits, and L. P. Pitaevskii, Zh. Éksp. Teor. Fiz., 37:229 (1959); Usp. Fiz. Nauk, 73:381 (1961).

9. B. V. Deryagin and G. A. Martynov, Dokl. AN SSSR, 144(4):825 (1962).
10. A. I. Frumkin, Zh. Fiz. Khim., 12:33 (1938).
11. A. I. Rusanov, this volume, p. 103.
12. B. V. Derjaguin (Deryagin) and Yu. V. Gutop, in: Research in Surface Forces, Vol. 2, Consultants Bureau, New York (1966), p. 17; Kolloidn. Zh., 27(5):674 (1965).

PRESSURE-TENSOR DISTRIBUTION AND EQUILIBRIUM OF A FREE FILM OF AN ELECTROLYTE SOLUTION

B. V. Deryagin and Yu. V. Gutop

Surface Phenomena Division, Institute of Physical Chemistry,
Academy of Sciences of the USSR

The Poisson-Boltzmann equation is used to derive the Maxwellian pressure tensor for a free film of an electrolyte solution, which provides the electrostatic components of the tension and pressure.

A thermodynamic treatment has been given [1-3] for the equilibrium of a free film consisting of three components, of which only two are simultaneously present in the bulk phase A enclosing the film. Here we consider by the methods of molecular statistics a free film in which a component is a strong electrolyte. First we consider the case where the film communicates at the ends with the bulk phase of which it is part. The distribution of the pressure tensor is discussed in order to clarify in detail the mechanical equilibrium. The method of calculating the electrical component of the film tension is simultaneously verified.

Poisson's equation is as follows for points in the film sufficiently remote from the ends:

$$\frac{d^2\varphi}{dz^2} = -\frac{4\pi}{\varepsilon}\rho, \tag{1}$$

in which $\varphi(z)$ is the potential at a distance z from the plane of symmetry of the film, ε is the dielectric constant (assumed constant), and $\rho(z)$ is the density of the electric charge.

We assume that ρ satisfies Boltzmann's equation:

$$\rho = z_1 n_1 e \exp\left(-\frac{z_1 e\varphi + \Phi^+}{\theta}\right) - z_2 n_2 e \exp\left(\frac{z_2 e\varphi - \Phi^-}{\theta}\right), \tag{2}$$

in which e is the electronic charge, z_1 and z_2 are the valencies of cation and anion respectively, n_1 and n_2 are the concentrations of the cations and anions in the volume communicating with the film, $\theta = kT$, T is absolute temperature, k is Boltzmann's constant, and Φ^+ and Φ^- are the specific adsorption potentials for cation and anion.

122

We assume φ such that $z_i e\varphi/\theta \ll 1$, and so the exponentials in (2) may be expanded as series in φ with only the linear terms retained. Then, putting $z_1 n_1 = z_2 n_2 = n_0$, we have

$$\rho = e n_0 \left[e^{-\frac{\Phi+}{\theta}} - e^{-\frac{\Phi-}{\theta}} - \frac{e\varphi}{\theta} \left(z_1 e^{-\frac{\Phi+}{\theta}} + z_2 e^{-\frac{\Phi-}{\theta}} \right) \right]. \tag{2'}$$

We assume that the adsorption potentials can be neglected at a certain distance d from the interface, which can be taken as small, so that to a first approximation we can put

$$\varphi(h) \simeq \varphi(h - d) = \varphi_0,$$

in which h is half the film thickness.

Then Poisson's equation for the film in the region $0 \leq z \leq h - d$ becomes

$$\frac{d^2\varphi}{dz^2} = \frac{4\pi e^2 n_0 (z_1 + z_2)}{\varepsilon\theta} \varphi \tag{3}$$

or

$$\frac{d^2\varphi}{dz^2} - \varkappa^2\varphi = 0, \tag{3'}$$

in which

$$\varkappa^2 = \frac{4\pi e^2 n_0 (z_1 + z_2)}{\varepsilon\theta}. \tag{4}$$

The following expression is [4] the solution to this equation that satisfies the condition $\frac{d\varphi}{dz}\big|_{z=0} = 0$, which follows from the symmetrical distribution of the potential:

$$\varphi = \varphi_1 \operatorname{ch} \varkappa z, \tag{5}$$

in which φ_1 is the potential in the plane z = 0.

As the film as a whole is electrically neutral, we have

$$\int_0^{h-d} \rho \, dz = - \int_{h-d}^{h} \rho \, dz = - Q, \tag{6}$$

in which Q is the charge on the surface layer.

Using (2') for $\rho(z)$ and the above behavior of φ in the range h − d to h, we get from (6) that

$$\frac{(z + z_2) e^2 n_0}{\theta} \int_0^{h} \varphi \, dz = e n_0 \left[l_1 - l_2 - \frac{e\varphi_0}{\theta} (z_1 l_1 + z_2 l_2) \right], \tag{7}$$

in which the constants

$$l_1 = \int_{h-d}^{h} e^{-\frac{\Phi+}{\theta}} dz \quad \text{and} \quad l_2 = \int_{h-d}^{h} e^{-\frac{\Phi-}{\theta}} dz \tag{8}$$

are dependent on the specific adsorbability of the ions.

The integration is performed with use of (5), and also of the fact that the second term in the square brackets in (7) can be neglected for $e\varphi_0/\theta \ll 1$ which gives the potential at the center of the film as

$$\varphi_1 = \frac{\theta}{e} \frac{\varkappa l}{z_1 + z_2} \frac{1}{\operatorname{sh} \varkappa h}, \tag{9}$$

in which

$$l = l_1 - l_2 = \int_{h-d}^{h} \left(e^{-\frac{\Phi^+}{\theta}} - e^{-\frac{\Phi^-}{\theta}} \right) dz \cong \int_{0}^{\infty} \left(e^{-\frac{\Phi^+}{\theta}} - e^{-\frac{\Phi^-}{\theta}} \right) dy, \tag{10}$$

with y = h − z and l an adsorption constant independent of the choice of d.

Then the potential at the interface is found as

$$\varphi_0 = \frac{\theta}{e} \frac{\varkappa l}{z_1 + z_2} \coth \varkappa h. \tag{11}$$

The total numbers $N_\pm$ of ions per unit area of film are

$$N_\pm = 2n_{1,2} \int_{0}^{h} e^{-\frac{\Phi^\pm}{\theta}} \left(1 \mp \frac{z_{1,2} e \varphi}{\theta} \right) dz.$$

Performing the integration and neglecting the second term in the bracket for the surface layer d, we have

$$N_\pm = 2n_{1,2} \left[h + l_{1,2} - d \mp \frac{z_{1,2}}{z_1 + z_2} l \right]. \tag{12}$$

The quantities

$$l_{1,2} - d = \int_{0}^{\infty} \left(e^{-\frac{\Phi^\pm}{\theta}} - 1 \right) dy \tag{13}$$

are also independent of the choice of d. The condition of electrical neutrality gives us a relation between N_- (number of anions) and N_+ (number of cations):

$$z_1 N_+ = z_2 N_-. \tag{14}$$

Consider now the tension σ in the film, which deviates from twice σ_0 (the surface tension of the bulk solution) by an amount that increases as h decreases.

We use Shcherbakov's method [5, 6] to deduce σ by expressing it as an integral:

$$\sigma = \int_{0}^{\infty} (P_N - P_T) dy, \tag{15}$$

in which y is distance measured from the surface of the liquid inwards, while P_N and P_T are respectively the normal and tangential components of the pressure tensor for the film.

We consider an elementary parallelepiped and calculate the resultant forces from interaction of volume elements (macroscopic approach) or molecules (molecular-statistical approach) separated by the faces of the parallelepiped, which allows us to determine the components of the pressure tensor and the derivatives of these with respect to the coordinates, which are needed for the conditions for hydrostatic equilibrium. In the very simple case of an isotropic hydrostatic tensor, we calculate the sum of the components of the interaction forces normal to the faces.

Necessary conditions in order to obtain single-valued tensor components are that the dimensions of the parallelepiped must be large relative to the radius of action of the forces but small relative to the distances within which the external bulk forces (e.g., gravity) can substantially alter the pressure or density of the liquid. The pressure concept is based on the division of forces into short-range (molecular) and long-range (e.g., gravity). The choice of the arbitrary boundary between these classes can affect the pressure tensor, so the derivation is not absolute, which explains the occurrence of the concepts of internal (cohesion) pressure and thermal pressure in the theory of liquids and real gases. It is possible to exclude the long-range forces by appropriate choice of the pressure tensor, these forces being transformed to equivalent short-range ones, as Faraday and Maxwell did for electrostatic forces. An analogous substitution is possible for molecular forces, since these are of electromagnetic fluctuation nature [7].

In the theory of capillarity, molecular attraction is usually considered as a long-range force whose radius of action is of the order of the thickness of the surface transition layer; but unambiguous derivation of P_T requires us to consider an area whose linear dimension along the y axis is extremely small relative to the thickness of the transition layer, and hence small relative to the radius of the molecular forces. In that respect, completely unambiguous derivation of P_T is impossible [8]; we can incorporate unambiguously in P_T only the repulsive forces, whose radius of action is smaller. If P_T and P_N are determined in this way, we arrive at the concept of thermal pressure.

The thermal pressure is balanced by the internal (cohesion) pressure at distances from the surface greater than the thickness of the transition layer, and the latter pressure is isotropic and clear in significance within the bulk liquid. If, however, the system of molecular forces is replaced by the equivalent distribution of electromagnetic fluctuation forces, the short range of the latter allows one to include these (after time averaging) in the pressure tensor, and also the ion interaction forces, if these are replaced by the Maxwell electrostatic stress tensor. Then we can derive rigorously the right side in (15) by considering P_N and P_T as quantities dependent on the electrostatic and molecular electromagnetic forces in the surface layer. Here, however, the lower limit must be taken as $-\infty$, since the forces extend into the gas phase.

Shcherbakov's method [5] gives us for the tension of a free film

$$\sigma = \int\limits_{-\infty}^{+\infty} (P_N - P_T)\, dz,$$

in which z is reckoned from the symmetry plane.

We calculate σ by representing it as

$$\sigma = \sigma_E + \sigma_M = -2 \int\limits_{0}^{h-d} \frac{\varepsilon E^2}{4\pi}\, dz + \sigma_M, \tag{16}$$

in which $E = -\partial\varphi/\partial z$ is the electric field strength and σ_M is the part of σ dependent on the forces acting in the surface layers of thickness d, which equals the radius of action of the adsorption and van der Waals surface forces, while $-\varepsilon E^2/4\pi$ is the difference between the normal and tangential components of the Maxwell tensor.

We use (5) and (9) to integrate and get for σ_E that

$$\sigma_E = -\frac{\varepsilon\theta^2}{8\pi e^2} \frac{(\varkappa l)^1 \cdot \varkappa}{(z_1 + z_2)^2} \frac{[\operatorname{sh}(2\varkappa h) - 2\varkappa h]}{\operatorname{sh}^2 \varkappa h}. \tag{17}$$

It is extremely difficult to calculate σ_E with allowance for the molecular and electrostatic interactions between adsorbed ions, since one has to consider the discontinuous distribution of these in the surface layer (see p. 188 of [8]). However, it would be illogical to substitute for σ_M in our formulas with allowance for these interactions, since the ion adsorption has been calculated on the basis of Boltzmann's formula, i.e., without these interactions. We therefore neglect them, on the assumption that the adsorption layer is sufficiently rarefied. Then we assume [8] that

$$\sigma_M = \sigma_0 - 2\theta\,(N_S^+ + N_S^-),\tag{18}$$

in which $N_S^\pm$ is the number of ions per unit area in each surface layer of thickness d, or

$$\sigma_M = \sigma_{M\infty} - 2\theta\,(\Delta N_S^+ + \Delta N_S^-),\tag{19}$$

in which $\sigma_M\infty$ is σ_M for $h \to \infty$, in which ΔN_S^+ and ΔN_S^- are the differences of the actual N_S^+ and N_S^- for a given h from those for $h = \infty$.

From Boltzmann's theorem

$$N_S^+ + N_S^- = \int_{h=d}^{h} \left[\left(n_1 \exp\left(-\frac{\Phi^+ + z_1 e\varphi}{\theta}\right) + n_2 \exp\left(-\frac{\Phi^- - z_2 e\varphi}{\theta}\right)\right)\right] dz$$

$$= \int_{h-d}^{h} \left[n_1 e^{-\frac{\Phi^+}{\theta}} + n_2 e^{-\frac{\Phi^-}{\theta}}\right] dz - \frac{e\varphi n_0}{\theta} \int_{h-d}^{h} \left(e^{-\frac{\Phi^+}{\theta}} - e^{-\frac{\Phi^-}{\theta}}\right) dz.\tag{20}$$

Then

$$\Delta N_S^+ + \Delta N_S^- = -\frac{n_0 e l\,(\varphi_0 - \varphi_\infty)}{\theta} = -\frac{Q\,(\varphi_0 - \varphi_\infty)}{\theta},\tag{21}$$

in which

$$Q = e n_0 \int_{h-d}^{h} \left(e^{-\frac{\Phi^+}{\theta}} - e^{-\frac{\Phi^-}{\theta}}\right) dz = e n_0 l.\tag{22}$$

Here Q is the charge per unit area on each surface layer of thickness d and φ_∞ is the surface potential for $h \to \infty$.

Substitution for φ_0 from (11) then gives

$$\Delta N_S^+ + \Delta N_S^- = -\frac{n_0\,(\varkappa l)^2}{\varkappa\,(z_1 + z_2)}\,(\text{cth}\,\varkappa h - 1).\tag{23}$$

Then (19) gives

$$\sigma_M = \sigma_{M\infty} + \frac{2\theta n_0\,(\varkappa l)^2}{\varkappa\,(z_1 + z_2)}\,(\text{cth}\,\varkappa h - 1).\tag{24}$$

We add σ_E to (24) to get

$$\sigma = \sigma_M + \sigma_E = \sigma_\infty + \frac{\theta n_0\,(\varkappa l)^2}{\varkappa\,(z_1 + z_2)}\left(\frac{\text{sh}\,\varkappa h \cdot \text{ch}\,\varkappa h + \varkappa h}{\text{sh}^2\,\varkappa h} - 1\right),\tag{25}$$

in which σ_∞ is σ for $h \to \infty$.

It follows from (25) that

$$\frac{d\sigma}{dh} = -\frac{2\theta n_0\,(\varkappa l)^2}{(z_1 + z_2)} \cdot \frac{\varkappa h\,\text{ch}\,\varkappa h}{\text{sh}^3\,\varkappa h}.\tag{26}$$

The outward pressure Π [2] is defined as the difference between the pressure in the phase adjacent to the film and the pressure in the liquid phase communicating with the film. On the assumption that the film has three components (water, gas, electrolyte), while the adjacent phase has two (water, gas), and assuming that the chemical potentials of the adjacent phase are constant, we have

$$d\Pi = \frac{n_1}{N_+} d\sigma. \tag{27}$$

We suppose that the electrolyte is completely dissociated. Integration of (27) over the thickness of the film gives

$$\Pi = -\int_h^\infty \frac{n_1}{N_+} \frac{d\sigma}{dh} dh. \tag{28}$$

Substitution for $d\sigma/dh$ from (26) and for N_+ from (22) gives

$$\Pi = \frac{\theta n_0 (\varkappa l)^2}{2(z_1 + z_2)} \cdot \frac{1}{sh^2 \varkappa h}. \tag{29}$$

Here we have neglected all terms independent of h in (12) as a first approximation.

The Π given by this new method agrees with the value previously derived by one of us [14], which indicates that it is correct to calculate σ for a free film of an electrolyte solution by dividing it into two components that are calculated essentially independently. The agreement also shows (which is not by itself obvious) that we can use these methods to calculate the electrostatic component of Π for free films. Similar calculations for the van der Waals component of Π are more difficult, because there is the unsolved problem (see, however, [9]) of calculating the tangential components of the pressure tensor for the electromagnetic fluctuation field, which is [7] equivalent to the field of the molecular forces.

From (19) and (21) we get an equation analogous to the electrocapillarity equation:

$$d\sigma_M = -2Q d(\varphi_0 - \varphi_\infty). \tag{30}$$

This is evidently true more generally.

From (27) and (30) we have

$$d\sigma_E - 2Q d(\varphi_0 - \varphi_\infty) = \frac{N_+}{n_1} d\Pi,$$

in which σ_E is the energy of the electrostatic field in the film, while the differentials correspond to change in film thickness with the chemical potentials μ_2 and μ_3, and also the pressure P_1, constant in the film.

This study has involved various simplifying assumptions, which restrict the range of direct application of the results. These have been made in order to simplify the mathematical operations, thereby revealing more clearly the physical essence of this new approach to the mechanical equilibrium of a thin film.

References

1.   B. V. Deryagin, G. A. Martynov, and Yu. V. Gutop, Kolloidn. Zh., 27(3):357 (1965); Research in Surface Forces, Vol. 2, Consultants Bureau, New York (1966), p. 9.
2.   B. V. Deryagin and Yu. V. Gutop, Kolloidn. Zh., 27(5):674 (1965); Research in Surface Forces, Vol. 2, Consultants Bureau, New York (1966), p. 17.

3.   B. V. Deryagin, in: Research in Surface Forces, Vol. 2, Consultants Bureau, New York (1966), p. 3.

4.   B. V. Deryagin, Izv. AN SSSR, OMEN, Ser. Khim., No. 5, p. 1119 (1937); Acta Physico-chim. URSS, 10:333 (1939).

5.   L. M. Shcherbakov, in: Research in Surface Forces, Vol. 2, Consultants Bureau, New York (1966), p. 26.

6.   G. Bakher, Kapillarität und Uberflächenspannung, Handbuch der Experimentalphysik, Vol. 6, Leipzig (1928).

7.   B. V. Deryagin, I. I. Abrikosova, and E. M. Lifshits, Usp. Fiz. Nauk, 64:493 (1958); J. Phys. Chem. Solids, 5(1-2):1 (1958).

8.   S. Ono and S. Kondo, The Molecular Theory of Surface Tension in a Liquid [Russian translation], IL, Moscow (1963).

9.   A. I. Rusanov and F. M. Kuni, this volume, p. 111.

BOUNDARY PHASE—BULK LIQUID TRANSITION

Yu. M. Popovskii

Higher Marine Engineering College, Odessa

It is shown that the specific heat of the boundary phase in nitrobenzene (thickness <150 mµ) on glass is less than that of the bulk phase. On heating the boundary phase is converted continuously into the bulk phase, which absorbs heat.

It has been shown [1, 2] that much of the liquid in a mixture of nitrobenzene with glass powder exists as a boundary phase, whose specific heat is less than that of the bulk phase. Here I report the specific heat of nitrobenzene in this system as a function of T and layer thickness. This gives the conditions of stable existence of the boundary phase and also allows one to estimate the thermodynamic parameters characterizing the transition to the bulk liquid.

The measurements were made with an adiabatic calorimeter with continuous heat input. The inner adiabatic casing was made of 0.08 mm brass and was heated by the passage of a current whose magnitude was controlled by a thermocouple. This casing had low thermal inertia and could maintain a temperature difference (relative to the calorimetric vessel) of 0.005-0.01°C. The heating rate and amount of material in the calorimeter during use were the same as during calibration, so that a correction could be applied for the thermal inhomogeneity of the calorimetric vessel.

The glass powder was produced by grinding up clean glass in a porcelain mortar. The surface was activated by exposure to a glow discharge for 30 min with continuous stirring. The liquids (benzene and nitrobenzene) were purified by distillation (twice) at atmospheric pressure and were stored in sealed tubes. The specific surface S of the powder used in the main series of experiments was determined by passage of a rarefied gas [3] as 0.84 ± 0.02 m^2/g.

Small amounts of nitrobenzene were evenly distributed over the powder as follows. A measured amount of nitrobenzene was introduced into benzene sufficient to wet all the powder, which was placed in the bottom of the layer of glass powder in the calorimeter vessel. The powder was stirred to mix and consolidate it, and then was kept at 20°C and 10-30 mm Hg for several hours until the benzene had been removed completely. The test for complete removal of the benzene was a constant low rate (0.5-1.3 mg/min) of weight loss at 5-10 mm, which cor-

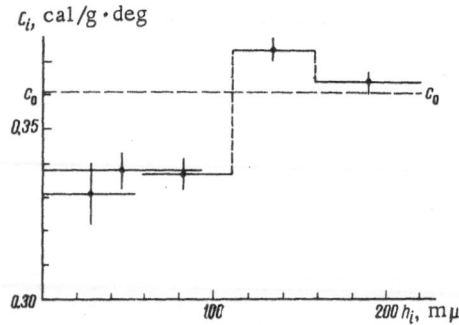

Fig. 1. Specific heat c_i of nitrobenzene mixed with glass powder as a function of distance h_i from the surface of the glass at 29.2°C.

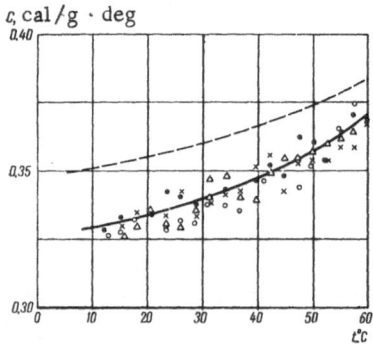

Fig. 2. Mean specific heat of the boundary phase of nitrobenzene as a function of temperature for film thickness (mμ) of: ○) 56; ●) 93; ×) 100; △) 153. The broken line is the curve for the bulk liquid.

responds to the rate of evaporation of nitrobenzene. After complete removal of the benzene, the system was consolidated to a porosity δ of 0.45-0.47. The vessel was weighed and sealed, and the thermal capacity was measured. The film thickness was then increased by opening the vessel and introducing more nitrobenzene as above.

The thermal capacity of the nitrobenzene was determined as the difference between the values for the mixture and the pure glass powder. The value for the mixture was measured to 0.2-0.3%, so the result for c_i was accurate to 2-5%. Even the largest amounts of nitrobenzene were such that the mean thickness of the nitrobenzene film[†] was less than the mean radius of the pore channels as calculated from Slichter's formula [4], i.e., the liquid formed menisci at contacts between particles as well as films on the particles.

As c_i is a function of the distance h_i from the substrate, we may measure the thermal capacity Q_i as a function of thickness h_i and calculate c_i by formula

$$c_i = \frac{Q_{i+1} - Q_i}{\rho S m_1 (h_{i+1} - h_i)},$$

in which ρ is the density of the liquid and m_1 is the mass of the glass powder.

Figure 1 shows the c_i as a function of h_i from a series of measurements at 29.2°C. The horizontal lines indicate the range in h_i for which the corresponding c_i were calculated, while the vertical lines indicate the possible error of measurement. The broken line represents the bulk c_i for this temperature. We see that c_i is independent of h_i within the error of measurement for $h_i < 110$ mμ and is less than the bulk c_i. We may say that, in this range of h_i, all the liquid is in the boundary-phase state. The c_i for h_i of 158-219 mμ is as for the bulk liquid, while in the range 110-158 mμ it is above the bulk value. This may be explained if the boundary phase becomes metastable for a thickness[‡] h greater than about 150 mμ, with the result that on heating in the calorimeter it is converted to bulk liquid, which absorbs heat.

It is unlikely that the rise in c_i in the range in h_i from 110-158 mμ is due to structural features of the film, since the c_i in another series of experiments with h ≈ 153 mμ had the value for the boundary phase. Consequently, if we ascribe the rise to features of the film in the range 153-158 mμ, we may conclude that c_i for this range is about 4.5 cal/g·deg, which is more than 10 times the bulk value.

_____

[†] We did not measure the density of the boundary phase, so the mean thicknesses were calculated from the bulk density and are to be taken as approximate.

[‡] The film thickness h should not be confused with the distance h_i to a point in the film.

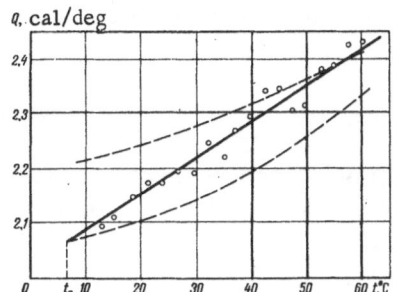

Fig. 3. Thermal capacity Q of the combination of boundary phase and bulk liquid as a function of temperature t; film thickness 158 mμ.

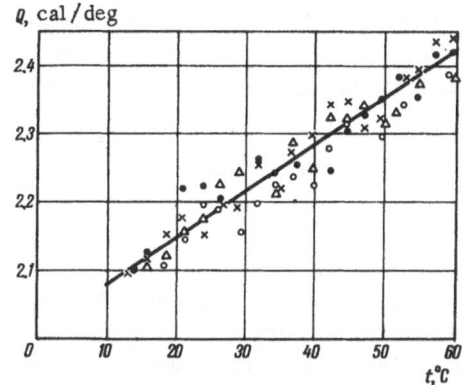

Fig. 4. Thermal capacity of the boundary phase plus bulk nitrobenzene as a function of t for heating rate (deg/min) of: ○, ×) 0.25; △) 0.5; ●) 0.12.

Antoniou [5] obtained similar results for water adsorbed on porous glass, which showed a marked rise in c_i between −30 and −7°C, which was interpreted as due to heat absorption on transition of the boundary phase to bulk water, although other phenomena could complicate the effect below the normal melting point of ice. My measurements were made above the normal melting point of nitrobenzene and are free from this deficiency.

Figure 2 shows the mean specific heat c = $Q/\rho Sm_1 h$ of the boundary phase as a function of temperature for nitrobenzene, where the points correspond to different film thicknesses in the range 50-153 mμ. The broken line is for the bulk liquid.

Figure 3 shows Q for nitrobenzene as a function of t for h = 158 mμ. The upper and lower broken lines have been calculated for the same mass from the c_i for the bulk liquid and the boundary phase. It is seen that Q takes the value for the boundary phase at t $\approx t_0$, i.e., no bulk liquid is present, while Q rises up to 60°C and exceeds the value for the bulk liquid, which agrees with the above assumption as to absorption of heat as the boundary phase is converted to the bulk liquid.

The Q for a 158 mμ film was measured repeatedly over a period of 11 days with various heating rates (Fig. 4). The reproducibility (from repeat experiments after cooling) shows that cooling converts the bulk liquid back to the boundary phase, i.e., the process is reversible. The agreement between the Q for different heating rates shows that there was equilibrium between the boundary and bulk phases, and that this equilibrium is dependent only on t, other factors being unchanged.

In one run, the system contained 6.1 g of nitrobenzene (h $\approx$ 153 mμ), and here all the liquid was present as boundary phase, transition to the bulk phase on heating to 60°C not being observed (Fig. 2).

In another run, the powder had the same total surface and the mass of liquid was 6.31 g, so h was about 158 mμ; here the boundary phase went over to the bulk form on heating (Fig. 3), and the amount converted to the bulk phase (from the qualitative evaluation) was considerably greater than the differences in the masses of liquid in the two runs. The lack of bulk-phase centers in the first run caused superheating of the boundary phase (transition to a metastable state). It would seem that a thickness of about 150 mμ is the limit at which all the nitrobenzene is in the boundary-phase state at t $\approx t_0$.

The transition to the bulk state does not occur at any definite temperature but occurs gradually (by layers) for h > 150 mμ. Fuks [6] also observed thickness reduction in the boundary layer of liquid between solids as t was increased from 25 to 50°C, which he interpreted as melting of the boundary phase.

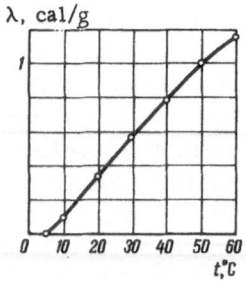

Fig. 5. Latent heat of conversion of boundary phase to bulk liquid as a function of temperature.

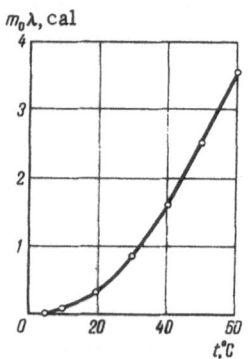

Fig. 6. Heat of formation $m_0\lambda$ for the bulk phase as a function of t.

Kirchoff's formula relates the heat of transition to the specific heats of coexisting phases:

$$\frac{d\lambda}{dT} = c_1 - c_2. \tag{1}$$

Substitution for the specific heats (Fig. 2) and integration gives the results shown in Fig. 5. As the constant of integration is unknown, the curve has been placed arbitrarily.

Consider the heat-balance equation for a temperature T at which the two phases coexist, taking into account that a temperature rise dT produces transition of part of the boundary phase to the bulk phase. Then

$$dW = (M - m_0)c_2 dT + m_0 c_1 dT + \lambda dm_0, \tag{2}$$

in which M is the total mass of nitrobenzene, m_0 is the mass of nitrobenzene in the bulk state at temperature T, c_1 is the specific heat of the bulk phase at temperature T, c_2 is the same for the boundary phase, and λ is the latent heat of the transition at temperature T.

Division of (2) by dT and manipulation gives

$$\frac{dW}{dT} - Mc_2 = m_0(c_1 - c_2) + \lambda \frac{dm_0}{dT}, \tag{3}$$

in which $dW/dT = Q^*$ is the measured thermal capacity (Fig. 3), i.e., the apparent thermal capacity, while $Mc_2 = Q_2$ is the thermal capacity of the same mass of nitrobenzene in the boundary-phase state (lower broken line in Fig. 3).

We transform (3) to

$$\frac{d(m_0\lambda)}{dT} = Q^* - Q_2. \tag{4}$$

Integration of (4) gives

$$m_0\lambda = \int (Q^* - Q_2)\, dT + \text{const.} \tag{5}$$

The constant of integration in (5) may be deduced from the condition $m_0 = 0$ at $t = t_0$, while t_0 is readily determined as the point of intersection (Fig. 3) of the curves for Q^* and Q_2.

Figure 6 shows the results from (5), where λ is positive throughout the temperature range, so the lower limit in λ is estimated by putting the constant of integration in (1) as zero. Then λ for 60°C is ~1.15 cal/g, and $m_0 \approx 3.1$ g at this temperature, so h is about 75 mμ.

Unambiguous determination of the constant of integration in (1) requires measurement of Q^* and Q_2 up to a t such that the transition to the bulk liquid is almost complete, i.e., m_0 is known; but this involves overcoming various difficulties, which has recently been done [7], which has yielded the true value of λ.

Conclusions

1. The specific heat of nitrobenzene mixed with glass powder is independent of film thickness below 150 mμ for temperatures of 10-60°C and is less than the bulk value, i.e., nitrobenzene exists in a boundary-phase state under these conditions.

2. A mean film thickness >150 mμ in this system causes the apparent c of nitrobenzene to increase much more rapidly with temperature, which is due to absorption of heat on transition of the boundary phase to the bulk phase.

3. The equilibrium between the boundary phase and the bulk liquid is dependent only on the temperature if the other parameters are constant.

I am indebted to B. V. Deryagin, Associate Member, Academy of Sciences of the USSR, for a discussion of the results and for valuable advice.

References

1. Yu. M. Popovskii, in: Research in Surface Forces, Vol. 2, Consultants Bureau, New York (1966), p. 149.
2. Yu. M. Popovskii and B. V. Deryagin, Dokl. AN SSSR, 159(4):897 (1964).
3. B. V. Deryagin, Zav. Lab., 17:324 (1951).
4. L. S. Leibenzon, Movement of Natural Gases and Liquids in Porous Media, Gostekhizdat, Moscow-Leningrad (1947).
5. A. A. Antoniou, J. Phys. Chem., 68:2754 (1964).
6. G. I. Fuks, Kolloidn. Zh., 20:748 (1958).
7. B. V. Deryagin and Yu. M. Popovskii, Dokl. AN SSSR, 175:385 (1967).

DERIVATION OF THE EQUILIBRIUM CONTACT ANGLE
FROM THE PARAMETERS OF A SMALL DROP
HANGING ON A WIRE

V. S. Baibakov, P. P. Ryazantsev,
V. P. Safronov, and L. M. Shcherbakov

Tula Polytechnical Institute

A method is described for determining equilibrium contact angles via approximate solution of the differential equation for a small drop resting on a wire. A description is given of an apparatus for measuring contact angles over the range 0 to 100°C.

The sessile drop method is usually considered to be among the best methods of measuring contact angles, and this gives the best result with large drops. However, such a drop has a substantial area of contact, equilibrium is established very slowly, and hysteresis is fairly pronounced. This means that the result is not always the equilibrium angle.

Deryagin [1] proposed measurement with thin wires in order to minimize the effects of roughness and so on; the liquid is taken as a small drop suspended on the wire [2], where equilibrium is established much more rapidly and hysteresis is minimized. Moreover gravitational effects are almost eliminated if the drop is small enough, so the formulas are simplified. Here we must consider what is meant by small.

A drop is considered small if the gravitational forces can be neglected, i.e., if the hydrostatic pressure is small relative to the Laplace pressure, $\rho g h_m \ll 2\sigma/R$, in which ρ is density, σ is surface tension, g is the acceleration due to gravity, R is drop radius, and h_m is the largest distance from the equator; or putting $h_m \approx R$,

$$R^2 \ll a^2 = \frac{2\sigma}{\rho g},\tag{1}$$

in which a^2 is the Laplace capillary constant. If $\rho \approx 1 \text{ g/cm}^3$, $\sigma \approx 30 \text{ erg/cm}^2$, and g = 980 cm/sec², the test of (1) is satisfied for drops less than 1 mm in diameter.

It is hardly possible to measure directly the contact angle from the slope of the tangent to the meridional curve at the wetting perimeter, since this curve has a complicated shape (Fig. 1)

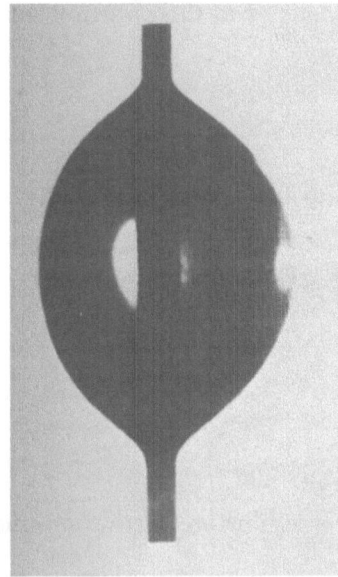

Fig. 1. Drop of distilled water on a copper wire (r_m = 0.285 mm, r_0 = 0.036 mm).

near the wire, and it is difficult to draw the tangent. We consider, therefore, the theoretical shape of a small drop on a wire.

Previous studies [3] have dealt in detail with the general variational method of solving such problems in thermodynamic equilibrium, so we give here only the main results and omit details.

A solid has [4, 5] an adsorption layer that serves as the immediate substrate for the drop. Then the varying part of the thermodynamic potential is as follows for a system composed of a small drop (subscript 1) on a wire in a vapor (subscript 2)[†]

$$\Phi = \mu_1 g + \mu_2 (N - g - \Gamma A) + \sigma S + \omega'(\Gamma) Q + \omega(\Gamma)(A - Q), \quad (2)$$

in which μ_1 and μ_2 are the chemical potentials (per particle) for the bulk liquid and vapor (referred to the vapor pressure P_2), g is the number of particles in a drop, Γ is the number of particles adsorbed per unit area of the wire, σ is the specific surface free energy of the free surface, S is the area of that surface, ω is the specific excess free energy of the free part of the adsorption layer on the wire, $A - Q$ is the area of that part, and ω' and Q are the same for the part covered by the drop.

Variation of Φ with P_2, N, T, and A constant, followed by equating $\delta \Phi$ to zero, gives us

$$\left. \begin{array}{l} \dfrac{\partial \omega}{\partial \Gamma} = \dfrac{\partial \omega'}{\partial \Gamma}, \\[2mm] (\mu_1 - \mu_2) \delta g + \sigma \delta S + (\omega' - \omega) \delta Q = 0. \end{array} \right\} \quad (3)$$

The first of these equations shows that $\omega - \omega'$ = constant throughout the stability range of the adsorption layer.[‡] The second equation represents the variational problem for the equilibrium form of the surface, which may be formulated explicitly as follows.

Let $z = z(r)$ be the equation for the meridional section of the drop (Fig. 2). Then for S, Q, and the drop volume V_1 we have

$$S = 2\pi \int_{r_0}^{r_m} r \sqrt{1 + p^2 dr}, \quad Q = 2\pi r_0 \int_{r_m}^{r_0} p\, dr, \quad V_1 = 2\pi \int_{r_0}^{r_m} zr\, dr$$

(here p = $\partial z / \partial r$). After substitution, the second equation in the above system becomes

$$\delta \int_{r_0}^{r_m} [\Delta\mu \cdot n_1 z \cdot r + \sigma r \sqrt{1 + p^2} - (\omega' - \omega) p]\, dr = 0, \quad (4)$$

in which $\Delta \mu = \mu_1 - \mu_2$, while n_1 is the molecular concentration in the liquid.

[†] All parts of the system are assumed to be bounded by equimolecular separating surfaces corresponding to absence of autoadsorption.

[‡] In the case of a polymolecular adsorption layer, the first equation represents equality of the outward pressures in the various parts of the equilibrium layer.

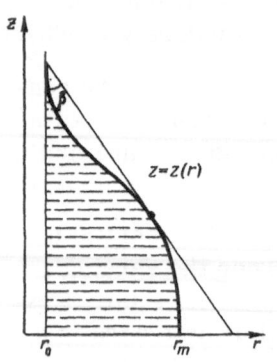

Fig. 2. Derivation of the equation for the section of a drop.

This is a problem with adjustable ends, and the solution may be reduced to that of the Euler equation

$$\frac{\partial F}{\partial z} - \frac{d}{dr} \cdot \frac{\partial F}{\partial p} = 0,$$

in which F is the integrand in (4) subject to the additional conditions at the ends.

The condition at the upper end (Fig. 2), where $\delta r = 0$ and $\delta z \neq 0$, gives us

$$\frac{p_0}{\sqrt{1 + p_0^2}} = \frac{\omega' - \omega}{\sigma}, \tag{5}$$

in which p_0 is the value of $\partial z / \partial r$ at $r = r_0$.

But $p = \cot \beta$, in which β is as shown in Fig. 2. For a point $r = r_0$ on the wetting perimeter, this angle is the contact angle θ, so we have from (5) that

$$\cos \theta = \frac{\omega - \omega'}{\sigma}. \tag{6}$$

This is the equation for the equilibrium contact angle. The condition at the lower end ($\delta z = 0$, $\delta r \neq 0$) gives $p_m = 0$, which shows that the tangent to the curve is parallel to the wire if we take a point on the equator.

Substitution for F in the Euler equation gives

$$\sigma \frac{1}{r} \frac{d}{dr} \left(\frac{rp}{\sqrt{1 + p^2}} \right) = n_1 \Delta \mu$$

or, since $n_1 \Delta \mu = \Delta P$ (in which ΔP is the Laplace pressure difference at the phase interface),[†]

$$\frac{1}{r} \frac{d}{dr} \left(\frac{rp}{\sqrt{1 + p^2}} \right) = \frac{\Delta P}{\sigma} = \text{const} = 2B. \tag{7}$$

This is the differential equation for the meridional curve for a small drop on a thin wetted wire.

Integration of (7) gives

$$r \frac{p}{\sqrt{1 + p^2}} = Br^2 + \text{const} \tag{8}$$

or, introducing β (Fig. 2),

$$r \cos \beta = B r^2 + \text{const}. \tag{9}$$

Putting $r = r_0$, we have for the constant of integration that

$$\text{const} = r_0 \cos \theta - Br_0^2. \tag{10}$$

To eliminate $B = \Delta P / 2\sigma$, which is difficult to measure, we put $r = r_m$ in (9). Then, since $\beta = 0$ at the equator, we have

---

† See [5] for details.

TABLE 1. Values of θ for Double-Distilled Water on a 0.10 mm
Copper Wire for Various Drop Diameters at 17.8°C

Photo No.	Point on photo	Diameter of drop, mm	x	x_m	β	$\theta°$
3—92	1	0.86	3.839	8.750	55 00'	42
3—92	4	0.86	2.678	8.750	58 20'	41
5—94	1	0.75	2.695	7.495	56 30'	44
5—94	2	0.75	3.386	7.405	54 00'	45
5—94	3	0.75	4.000	7.495	50 00'	43
5'—94'	1	0.61	3.598	6.089	45 50'	43
5'—94'	2	0.61	3.848	6.089	43 50'	44
5'—94'	3	0.61	3.152	6.089	48 50'	41

$$B = \frac{r_m - r_0 \cos \theta}{r_m^2 - r_0^2}. \tag{11}$$

Integration of (8) and use of (10) and (11) reduces the formula to one containing elliptic integrals. The final equation is

$$\zeta = z/r_m = E(k, \varphi) + k'F(k; \varphi), \tag{12}$$

in which F and E are elliptic integrals of the first and second kinds, whose modulus k and amplitude φ are defined by

$$k' = \sqrt{1 - k^2} = \frac{\cos \theta - 1/x_m}{x_m - \cos \theta},$$
$$\sin \varphi = \frac{1}{k} \sqrt{1 - \left(\frac{x}{x_m}\right)^2}. \tag{13}$$

Here we have introduced the dimensionless variable $x = r/r_0$ ($x_m = r_m/r_0$).

The above relations provide the basis for several new methods of determining the contact angle from the parameters of a small drop on a wire. One of these [2] employs (9). Using (10) and (11) with $x = r/r_0$, we have

$$\cos \theta = \frac{(x_m^2 - 1) x \cos \beta - x_m(x^2 - 1)}{x_m^2 - x^2}. \tag{14}$$

This defines θ via the auxiliary angle β. It is difficult to draw the tangent directly at the wetting perimeter, but it is much easier to do so at other points, and it is always possible to choose a point such that the operation may be performed with sufficient precision.

To use this method, we must choose a suitable point on the curve and draw the tangent, then measuring β, r, r_m, and r_0. As the right side of (14) contains the dimensionless quantities $x = r/r_0$ and $x_m = r_m/r_0$, as well as β, we can choose the scale for the measurements arbitrarily, i.e., all measurements may be made on a photograph of arbitrary magnification. The result can be tested by using the value for θ to plot the theoretical curve. Of course, the method is applicable only to drops that satisfy (1).

Table 1 illustrates this method via θ for water at 17.8°C on copper in an apparatus previously described [6]. The result from 24 measurements[†] is $\theta = 44 \pm 2°$, which agrees well

---

† Only some of the results are given in Table 1.

TABLE 2. Comparison of Theoretical and Experimental Profiles
for Double-Distilled Water on Copper
(17.8°C, d = 0.07 mm, D = 0.71 mm)

Photo No.	Point on photo	x	x_m	$2r_m$	$2z$ (theor)	$2z$ (exp)	Deviation, %
1—90	1	4.578	10.13	91.20	90.33	90.25	0.1
1—90	3	3.667	10.13	91.20	95.57	95.70	0.1
1—90	A	8.300	10.13	91.20	56.27	55.65	1.1
1—90	B	6.472	10.13	91.20	76.61	75.70	1.2
1—90	C	2.667	10.13	91.20	100.78	100.10	0.7

Note: θ = 44° was assumed in the calculations.

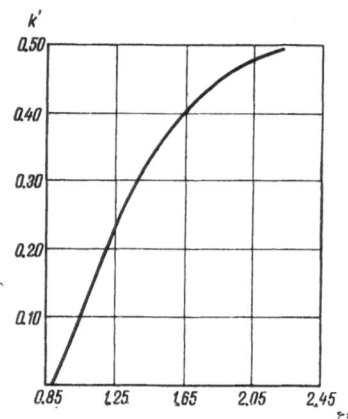

Fig. 3. Relation of ζ^* to the additional modulus k' for m = 1/2.

with published values. Table 2 compares the calculated profile with the measured one; the discrepancies do not exceed 1%.

This method has the slight disadvantage that angle measurements are required. A modified method is very promising in that it requires only linear measurements. Consider a point on the curve such that $r/r_m = m$, in which m is some specified fraction. Then we have from the second equation in (13) that

$$\sin \varphi^* = \frac{1}{k} \sqrt{1 - m^2}, \qquad (15)$$

i.e., the amplitude φ^* for this point is determined by k.

Substitution into the equation for the curve gives the ordinate z* of this point as

$$\zeta^* = z^*/r_m = E(k, \varphi^*) + k'E(k, \varphi^*). \qquad (16)$$

This formula relates the dimensionless ordinate ζ^* to the modulus k or $k' = (1 - k^2)^{1/2}$, e.g., as in Fig. 3 for m = 1/2. We may therefore measure z* from a photograph and deduce $\zeta^* = z^*/r_m$, using the curve to find the corresponding k', which is substituted into the first equation in (13) to give the contact angle as

$$\cos \theta = \frac{k'x_m + 1/x_m}{1 + k'}. \qquad (17)$$

Measurement of θ by this method thus reduces[†] to determination of the distance 2z* between points at which r is some given fraction (e.g., 1/2) or r_m. We also need to know the diameter of the wire and r_m, and to have available tables of ζ^* as a function of k'. The formulas contain only dimensionless quantities, so all measurements may be made directly on photographs. Preliminary measurements have been made in this way on a sealed glass apparatus and have been found to give satisfactory reproducibility.

Both methods are essentially micro methods, so they are especially suitable for measuring equilibrium contact angles, since small drops equilibrate rapidly. They are also suit-

[†] 2z* is the most convenient for practical measurement.

able for examining the effects of local inhomogeneities in the substrate. Also, the volume required is small, so the methods can be used to advantage at high pressures and in thermostats.

References

1. B. V. Deryagin, Dokl. AN SSSR, 51(7):517 (1946).
2. L. M. Shcherbakov and P. P. Ryazantsev, in: Surface Effects in Melts and Derived Solid Phases, Nal'chik (1965), p. 152, 230.
3. L. M. Shcherbakov and P. P. Ryazantsev, Zh. Fiz. Khim., 34:2120 (1960); Research in Surface Forces, Vol. 2, Consultants Bureau, New York (1966), p. 33.
4. B. V. Deryagin and L. M. Shcherbakov, Kolloidn. Zh., 23(1):40 (1961).
5. L. M. Shcherbakov and P. P. Ryazantsev, in: Surface Effects in Melts and in Powder Metallurgy Processes, Izd. AN UkrSSR, Kiev (1963), p. 152.
6. P. P. Ryazantsev, Summaries of Papers at the Machine Design Conference, Tula Polytechnical Institute (1965), p. 108.

MODELS FOR THE TRANSITION LAYER
APPLIED TO THE SURFACE CHARACTERISTICS
OF A LIQUID—VAPOR SYSTEM

T. I. Antonenko

Kishinev University

Detailed models are considered for the structure of the transition layer between a liquid and a saturated vapor. The thickness of the layer is estimated. The ellipticity of light reflected from the layer is related to the surface tension.

A liquid with its saturated vapor is considered on the assumption that between the homogeneous phases there lies a transition layer with distinct properties. The coordinate system is placed with the z axis normal to the interface and directed from the vapor into the liquid, the entire transition zone lying in the region of positive z, while the density at all points with z < 0 is that of the homogeneous vapor.

The methods of the statistical theory of liquids [1, 2] allow the surface characteristics of a two-phase system to be expressed via the unitary and binary correlation functions, whose detailed form is dependent on the model for the structure of the liquid and the form of the molecular interaction potential. The formulas in the superposition approximation are

$$U = \frac{\pi}{v^2} \int_0^\infty F_1(z)\, dz \left[\int_{-z}^\infty F_1(z+t)\, dt \int_{|t|}^\infty \Phi(r)\, g(r)\, dr - 2 \int_0^\infty \Phi(r)\, g(r)\, r^2 dr \right], \tag{1}$$

$$\sigma = \frac{\pi}{v^2} \int_0^\infty F_1(z)\, dz \left[\int_{-z}^\infty F_1(z+t)\, dt \int_{|t|}^\infty \Phi'(r)\, g(r)\, r^2 dr - 3 \int_{-z}^\infty t^2 F_1(z+t)\, dt \int_{|t|}^\infty \Phi'(r)\, g(r)\, dr \right], \tag{2}$$

$$\rho = \frac{4\pi\beta}{\lambda v} \sqrt{n^2+1} \int_0^\infty F_1(z)\, dz \left[1.04 - 0.75 \int_{-z}^\infty F_1(z+t)\, dt \int_{|t|}^\infty g(r) \frac{r^2 - 3t^2}{r^4}\, dr \right], \tag{3}$$

in which U is the specific total surface energy (per unit area of interface), σ is the specific free surface energy (the surface tension), ρ is the ellipticity of light reflected[†] from the

---

[†] It is assumed that the light is polarized in a plane lying at 45° to the plane of incidence and falls on the interface at the Brewster angle.

liquid–vapor interface [3], v is the volume per molecule in the homogeneous liquid, β is the mean polarizability of the molecules, λ is the wavelength of the incident light, n is the refractive index for this wavelength, $F_1(z)$ is the unitary distribution function, g(r) is the radial distribution, and $\Phi(r)$ is the potential for pair interaction of the molecules.

I have estimated [4, 5] ρ and the thickness for two very simple models of the transition zone, which have the following relations between the local density and the position in the surface layer:

$$F_1(z) = \frac{F_1(\infty) - F_1(0)}{d} \cdot z \text{ and } F_1(z) = 1 + Ae^{-\alpha \cdot z}.$$

The constants are deduced from (1) via experimental values for U and via the boundary conditions at z = 0 and z = ∞, while g(r) is represented as a step function: g(r) = 0 for r ≤ a and g(r) = 1 for r > a, in which a acts as the diameter of the molecule. The Lennard–Jones potential is used for the pair interaction:

$$\Phi(r) = 4\varepsilon \left[\left(\frac{a}{r} \right)^{12} - \left(\frac{a}{r} \right)^{6} \right]. \tag{4}$$

The thickness of the transition zone is found to be two to three times a in the linear model; in the case of the second model, the density even in the second molecular layer must be virtually that of the homogeneous liquid in order to give agreement with the observed U and ρ. These results are only estimates, in view of the crudeness of the models.

It appears better to use radial distribution functions obtained by numerical integration [6] for simple liquids with spherical molecules (liquid argon or krypton). These radial functions may be applied with an accuracy sufficient for practical purposes to organic liquids whose molecules differ only slightly from spherical, provided that the asphericity factor f of the molecule is incorporated as in calculating virial coefficients [7].

I have used these functions [8] to calculate the molar potential energy for several liquids via

$$E^l = \frac{24 f v_m \, \varepsilon N}{v} \int_0^\infty \Phi(x) \, g(x) \, x^2 \, dx, \tag{5}$$

in which x = r/a. Comparison with experiment is made via the latent heat of vaporization; initially [8] the agreement was not entirely good, the reason being as follows.

The heat of the transition is defined by the difference between the enthalpies. We neglect the interaction between the vapor molecules far from the critical point and put $PV^{(va)} = RT$ for the vapor to get

$$L = (E^{(va)} + K^{(va)} + PV^{(va)}) - (E^{(l)} + K^{(l)} + PV^{(l)})$$
$$\cong -E^{(l)} + (K^{(va)} - K^{(l)}) + RT,$$

whence

$$-E^{(l)} = L - (K^{(va)} - K^{(l)}) - RT.$$

The original comparison [8] employed, as usual, the assumption that $K^{(va)} - K^{(l)}$, the difference between the molecular kinetic energies of the vapors and liquid, is zero. Although this is true for simple monatomic liquids, it is not obvious for polyatomic ones.

In fact, the specific heat C_v of the vapor is different from that of the liquid. The potential energy is not explicitly dependent on the temperature, so the difference in the C_v must be

TABLE 1. Values of the Molar Potential Energy
of Benzene

°K	$-E^{(l)}/RT$ from (5)	$-E^{(l)}/RT$ from (6)
283	12.0	12.1
353	8.7	8.8
423	6.4	6.3

TABLE 2. Unitary Distribution Function for Benzene

| z/a | $F_1(z)$ | | z/a | $F_1(z)$ | |
	$\lambda^* = 30$ $\beta^* = 1.5$	$\lambda^* = 27.4$ $\beta^* = 1.25$		$\lambda^* = 30$ $\beta^* = 1.5$	$\lambda^* = 27.4$ $\beta^* = 1.25$
1.0	0.74	0.80	2.2	1.13	1.06
1.2	1.02	0.98	2.4	1.08	1.04
1.4	0.98	0.98	2.6	0.97	1.00
1.6	0.88	0.86	2.8	0.93	0.95
1.8	0.89	0.86	3.0	0.98	1.00
2.0	1.02	0.96			

due to difference in the temperature dependence of the K, so the K themselves should differ. Ultrasonic data [9] indicate that $C_v^{(va)}$ and $C_v^{(l)}$ vary almost linearly with T near the critical point in such a way that the second and higher derivatives with respect to T are equal. Then, expanding $C_v^{(va)}$ and $C_v^{(l)}$ as series with respect to $\Delta T = T_c - T$, we get

$$C_v^{(l)} - C_v^{(va)} = \left[\left(\frac{dC_v^{(va)}}{dT} \right)_{cr} - \left(\frac{dC_v^{(l)}}{dT} \right)_{cr} \right] \Delta T,$$

in which $(dC_v^{(va)}/dT)_{cr}$ and $(dC_v^{(l)}/dT)_{cr}$ are derivatives taken at the critical point (on the low-temperature side). This expression is readily integrated to give

$$K^{(va)} - K^{(l)} = \frac{1}{2} (C_v^{(l)} - C_v^{(va)}) \cdot \Delta T.$$

Then $E^{(l)}$ can be expressed in terms of directly measured quantities:

$$-E^{(l)} = L - \frac{1}{2} (C_v^{(l)} - C_v^{(va)}) \Delta T - RT. \tag{6}$$

Comparison of results from (5) and (6) (Table 1) shows good agreement.

On this basis, I have used the same radial functions to find the unitary distribution function and the surface characteristics. The method of [10] was used for these functions at various T on the saturation line:

$$F_1(z) = 1 - \frac{2\pi}{v} \int_{-z}^{\infty} dt \int_{|t|}^{\infty} [g(r) - 1] r dr. \tag{7}$$

Table 2 gives $F_1(z)$ for benzene for two values of the parameters $\lambda^* = 4\pi a^3/v$ and $\beta^* = \varepsilon/kT$, which characterize the volume of the liquid and the temperature. The calculations were performed only for z/a from 1.0 to 3.0, because (7) becomes meaningless[†] for $z/a < 1$.

I used an asymptotic expression [1] for g(r) for $z/a > 3$:

$$g(r) = 1 + A \frac{e^{-\alpha_1 \cdot r} \cdot \sin(\beta_1 r + \delta)}{r},$$

for which the integral in (7) may be calculated exactly; in this approximation

$$F_1(z) = 1 + \frac{2\pi A e^{-\alpha_1 \cdot z}}{v(\alpha_1^2 + \beta_1^2)} [(\alpha_1^2 - \beta_1^2)\sin(\beta_1 z + \delta) - 2\alpha_1\beta_1\cos(\beta_1 z + \delta)].$$

To calculate U, $F_1(z)$ was initially taken in the form[‡] $F_1(z) = 0$ for $z < 0$ and $F_1(z) = 1$ for $z \geq 0$. Then (1) simplifies to

$$U = -\frac{\pi}{2v^2} \int_0^\infty \Phi(r) g(r) r^3 dr. \tag{8}$$

This formula gives U close to the observed value for temperatures near the melting point, but it gives very low values for higher temperatures, so the subsequent calculations were done with (7) for the ratio of the U at 283 and 353°K. Formula (1) gives $U_{283}/U_{353} = 1.09$, whereas the experimental value is 1.07, and (8) gives 1.18.

For benzene, $\rho = 1.05 \times 10^{-3}$, which is somewhat below the experimental values [3], which range from 1.08×10^{-3} to 1.32×10^{-3}.

These results show that the correlation function method, as developed for simple liquids, can be applied to derive physical quantities for normal liquids either as homogeneous phases or as liquid—vapor systems.

References

1. I. Z. Fisher, Statistical Theory of Liquids, Gosfizmatizdat, Moscow (1961).
2. S. Ono and S. Kondo, The Molecular Theory of Surface Tension in Liquids [Russian translation], IL, Moscow (1963).
3. D. V. Sivukhin, Zh. Éksp. Teor. Fiz., 18:976 (1948); 21:367 (1952).
4. T. I. Antonenko, Uch. Zap. Kishinevsk. Gos. Univ., 69:23 (1964).
5. T. I. Antonenko, Uch. Zap. Kishinevsk. Gos. Univ., 75:21 (1964).
6. J. Kirkwood, V. Lovinson, and B. Alder, J. Chem. Phys., 20:929 (1952).
7. J. Hirschfelder, C. Curtiss, and R. Bird, Molecular Theory of Gases and Liquids, Wiley, New York (1964).
8. V. I. Rykov, T. I. Antonenko, T. I. Golomoz, and G. S. Yakovleva, Uch. Zap. Kishinevsk. Gos. Univ., 80:133 (1965).
9. Kh. I. Amirkhanov, A. M. Kerimov, and B. G. Alibekov, in: Use of Ultrasound in Examination of Materials, Vol. 9, Gostekhizdat, Moscow (1959), p. 17.
10. F. M. Kuni, Vestnik LGU, No. 4, p. 11 (1965).

---

† It has been shown [10] that the function should become zero at z = 0 and that the values $F_1(0)$ and $F_1(1)$ may be linked up via any monotonic function (e.g., a linear one). My results show that this link is best made via $\exp[\alpha(z - z_0)/z]$, in which α is determined from two values of $F_1(z)$ taken near $z = a$.

‡ Essentially this means that the thickness of the transition layer is taken as zero.

THERMODYNAMICS OF SMALL BUBBLES

V. A. Matveev and L. M. Shcherbakov

Tula Polytechnical Institute

The thermodynamic theory of perturbations is applied to calculate the excess free energy of a small cavitation bubble or cavity filled with vapor. The concept of capillary effects of the second kind is used to consider the equilibrium conditions for such a bubble, and relationships for the thermodynamic parameters are deduced.

There are fairly many papers on the formation of aerosols by condensation, but little has been published on the nucleation of vapor bubbles in a homogeneous liquid. Some major aspects of boiling and cavitation have been neglected.

There may be two mechanisms for bubble nucleation. In cavitation we have local stretching of the liquid, and nucleation starts as a process of cracking, the subsequent expansion of the cracks and filling with vapor leading to bubble nucleation. In ordinary boiling, the bubbles are nucleated by a fluctuation mechanism.

Independent interest attaches to the thermodynamics of small bubbles; as for other small objects, there are major difficulties in applying thermodynamic principles, which have been overcome in Hill's approach [1] and by the use of the concept of capillary effects of the second kind [2] (see [7, 8] for fuller details).

Excess Free Energy of a Small Bubble

The thermodynamic quantities characterizing a small bubble may be defined as statistical means over an ensemble of nucleus + medium type, e.g., the excess free energy and the surface. The spatial localization of the nucleus is determined by introducing as the boundaries the geometrical equimolecular separating surface, which (as for a small drop) may be defined by

$$v''(P'') \cdot g + v'_\infty (N - g) = V,$$ (1)

in which a single prime refers to the liquid, while a double prime refers to the bubble (containing g particles), while N is the total number of particles, V is the total volume of the system, v'_∞ is the specific volume (per particle) in the bulk liquid, and v'' is the specific volume of the vapor. The latter will be determined by the pressure P'' in the bubble[†] and so in turn

---

[†] We have $v'' = kT/P''$ if the vapor is treated to a first approximation as an ideal gas.

will be dependent on the size of the bubble. This feature makes it difficult to relate the radius r to the number of particles g.

It is somewhat easier to relate the changes in r and g. Consider the virtual evaporation of δg particles from the liquid within the bubble. Then the change in volume of the liquid is

$$- v'_\infty \delta g = - 4\pi r^2 \delta r,$$

and so

$$\frac{\partial g}{\partial r} = \frac{1}{v'_\infty} 4\pi r^2. \tag{2}$$

The liquid here may be considered as incompressible, but it is necessary to allow for the compressibility (and hence for the dependence of v'_∞ on r) if we wish to integrate this relation in order to find the final relation between g and r, since the total number of particles in the liquid is large, and so even slight changes in v'_∞ can lead to appreciable changes in the total volume.[†]

The concept of capillary effects of the second kind indicates that the excess free energy ψ (or other potential) of a system including a small object is associated with the system as a whole, not merely with the interface. As standard state for ψ we take the state corresponding to a system of one of the bulk phases, which allows us to evaluate ψ by the methods of the thermodynamic theory of perturbations. It can then be shown [3] that the excess free energy of medium + small drop may be found via the mean value of the perturbation energy, which is due to isolation of the individual parts of the system from the corresponding bulk phases. The averaging is performed via the unperturbed distribution function corresponding to the bulk system.

There are two components in the perturbation energy for a vapor bubble in a liquid: 1) the change $U_{12}^{lv} - U_{12}^{ll}$ in the interaction energy[‡] due to removal of liquid from the volume of the bubble and filling with vapor, 2) the change in the energy of g particles on transfer from the bulk vapor to the bubble, which is $U_{12}^{lv} - U_{12}^{vv}$. We can neglect terms describing interaction with vapor particles if we are far from the critical point, so

$$\psi = - \langle U_{12}^{ll} \rangle. \tag{3}$$

The excess free energy of this system is then calculated as the mean interaction energy of the regions into which the space is divided by the separating surface. The averaging may be performed via the inhomogeneous binary correlation function

$$F^{II}(r_{ij}, \xi) = \rho(r_{ij}) \cdot \delta(\xi), \tag{4}$$

in which $\rho(r_{ij})$ is the usual radial function for the bulk liquid and $\delta(\xi)$ is a δ function of parameter ξ,[§] with $\delta(\xi) = 0$ for $\xi \le a$ and $\delta(\xi) = 1$ for $\xi > a$.

This representation of F^{II} allows us to put (3) as

$$\psi = - \frac{1}{2} \int_{V_1'} \int_{V_2} n_\infty \cdot u(r_{ij}) \rho(r_{ij}) \, dV_1 \, dV_2, \tag{5}$$

[†] This difficulty does not arise for a small drop, because the number of particles here is small.

[‡] The subscripts denote the interacting regions, while the superscripts denote the phases filling them.

[§] ξ is the distance along the normal characterizing the position of a particle in region 1 with respect to the boundary of the region.

TABLE 1. Dimensionless Thermodynamic Characteristics
of a Nonpolar Liquid as Functions of the Relative Size of
a Cavity

$x = r/a$	$\varphi(x) = \sigma/\sigma_\infty$	Ratio of nucleus radii $\varkappa = r/r_c$	Deviation coefficient $\zeta = W/W_c$
∞	1	1	1
100	1.0094	1.0049	1.029
50	1.0182	1.0097	1.055
30	1.0288	1.0161	1.089
20	1.0409	1.0273	1.134
10	1.0707	1.0450	1.226
8	1.0826	1.0547	1.266
6	1.1023	1.0731	1.337
4	1.1222	1.0959	1.410
3	1.1352	1.1170	1.462
2	1.1417	1.1468	1.488

in which n_∞ is the particle concentration in the bulk liquid, $u(r_{ij})$ is the model interaction potential, V_1' is the volume of the liquid less the surface zone of width a, V_2 is the volume of the bubble, and the 1/2 appears because each particle is counted twice in the integration.

The method itself means that ψ is calculated for the liquid bubble system as a whole, but the result may be referred to unit area of the bubble surface in order to facilitate comparison with the classical theory of surface phases, i.e., we introduce formally the specific free surface energy $\sigma = \psi/4\pi r^2$. We represent the radial function as unit step function: $\rho(r_{ij}) = 0$ for $r_{ij} \leq a$ and $\rho(r_{ij}) = 1$ for $r_{ij} > a$. Then the Lennard−Jones potential is used to derive from (5) the following expression after identifying[†] the energy constants with the corresponding macroscopic quantities:

$$\sigma = \sigma_\infty\left[\left(1 + \frac{a}{r}\right) + 0.007917\,\frac{a^2}{r^2} + 1.0526\left(1 + \frac{a}{r}\right)\left(\frac{a}{2r+a}\right)^2 - 1.0526\,\frac{a^2}{r^2}\ln\frac{2r+a}{a}\right]. \qquad (6)$$

Table 1 gives $\varphi = \sigma/\sigma_\infty$ as a function of the dimensionless bubble radius $x = r/a$.

Equilibrium Conditions for a Small Bubble

in a Homogeneous Liquid

Consider a small spherical bubble (denoted by ″) in a homogeneous liquid (denoted by ′). The thermodynamic potential of this system is

$$\Phi = \mu_\infty'(P'', T)(N - g) + \mu_\infty''(P'', T)\cdot g + \psi(g), \qquad (7)$$

in which μ_∞' and μ_∞'' are the chemical potentials (per particle) of the bulk liquid and vapor as referred to the same pressure P'', g is the number of particles in the bubbles, and N is the total number of particles in the system.

The condition for thermodynamic equilibrium is found to be as follows on variation with P'', T, and N constant:

$$\mu_\infty'(P'', T) - \frac{\partial \psi}{\partial g} = \mu_\infty''(P'', T). \qquad (8a)$$

† This is done by passing to the limit $r \to \infty$.

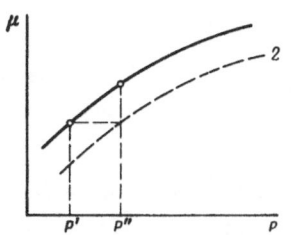

Fig. 1. Deduction of the equilibrium conditions for a small bubble in a homogeneous liquid: chemical potential μ as a function of pressure P for: 1) homogeneous bulk liquid; 2) liquid including a cavity.

The expression $\mu'_\infty(P'', T) - \partial\psi/\partial g = \mu(g)$ on the left may be called the chemical potential of the liquid including the cavity. Then (8a) simply represents equality between this chemical potential and the chemical potential of the vapor.

Figure 1 shows μ as a function of P, which shows that the $\mu(g)$ curve is obtained by displacing the μ'_∞ curve downwards by $\partial\psi/\partial g$. Further

$$\mu(g) = \mu'_\infty(P', T).$$

This means that the reduction in μ for the liquid due to the cavity may be interpreted as the result of a discontinuity (from P'' to P') in the phase pressure on going from the vapor to the liquid. To determine the magnitude of this discontinuity, we substitute for $\mu(g)$ and expand $\mu'_\infty(P'', T)$ as a power series in (P'' − P'), taking only the first two terms in the expansion to get

$$P'' - P' = \frac{1}{v'_\infty}\frac{\partial\psi}{\partial g}.$$

These results allow us to put (8a) in the equivalent form

$$\mu'_\infty(P', T) = \mu''_\infty(P'', T), \quad P'' - P' = \frac{1}{v'_\infty}\frac{\partial\psi}{\partial g}. \tag{9a}$$

Here the chemical potential of the liquid is referred to P', not to P".

The theoretical value for ψ is expressed in terms of the radius of the bubble, not in terms of the number of particles, so it is desirable to convert from g to r for equilibrium conditions by putting $\psi = \sigma(r) \cdot 4\pi r^2$. We then use (2) to put (8a) and (9a) as

$$\mu'_\infty(P'', T) - v'_\infty\left(\frac{2\sigma}{r} + \frac{\partial\sigma}{\partial r}\right) = \mu''(P'', T), \tag{8b}$$

and

$$\mu'_\infty(P', T) = \mu''_\infty(P'', T); \quad P'' - P' = \frac{2\sigma}{r} + \frac{\partial\sigma}{\partial r}, \tag{9b}$$

The equilibrium conditions allow one to examine the effects of bubble size on the properties of the system, in particular the phase pressures. We differentiate with respect to r for T constant in (9b) and use the fact that $\partial\mu_\infty/\partial P = v_\infty$ to get

$$\left.\begin{array}{l} v''_\infty dP'' - v'_\infty dP' = 0 \\ dP'' - dP' = d\left(\frac{2\sigma}{r} + \frac{\partial\sigma}{\partial r}\right) \end{array}\right\}. \tag{10}$$

This system is solved for dP' to give

$$dP' = -\frac{v''_\infty}{v''_\infty - v'_\infty}\,d\left(\frac{2\sigma}{r} + \frac{\partial\sigma}{\partial r}\right) \cong -d\left(\frac{2\sigma}{r} + \frac{\partial\sigma}{\partial r}\right).$$

Integration from r = ∞ to r = r gives us the following expression for the change in the phase pressure of the liquid on account of the production of a bubble of radius r:

$$P' - P_\infty = -\left(\frac{2\sigma}{r} + \frac{\partial\sigma}{\partial r}\right), \tag{11}$$

in which P_∞ is the phase pressure at a planar surface of the liquid.

This equation is one of the forms of the corrected Laplace equation [4].[†] We may similarly solve (10) for dP″ on the assumption that the vapor acts as an ideal gas to get the corrected Kelvin equation

$$kT \ln P''/P_\infty = -v'_\infty\left(\frac{2\sigma}{r} + \frac{\partial\sigma}{\partial r}\right), \tag{12}$$

which defines[‡] the vapor pressure P″ in a bubble of radius r.

Use of the equilibrium condition in the form of (8b) allows us to relate r to the superheating. We differentiate this equation with respect to r with P″ fixed and use $\partial\mu_\infty/\partial P = -\eta_\infty$ (in which η_∞ is specific entropy) to get

$$-\eta_\infty dT - d\left[v'_\infty\left(\frac{2\sigma}{r} + \frac{\partial\sigma}{\partial r}\right)\right] = -\eta''_\infty dT$$

or

$$\lambda_\infty d(\ln T) = d\left[v'_\infty\left(\frac{2\sigma}{r} + \frac{\partial\sigma}{\partial r}\right)\right],$$

in which $\lambda_\infty = (\eta''_\infty - \eta'_\infty)$. T is the specific (per particle) heat of evaporation of the bulk liquid.

We integrate this equation from $r = \infty$ to $r = r$ and neglect the temperature dependence of λ_∞ in a first approximation to get

$$\lambda_\infty \ln T/T_\infty = v'_\infty\left(\frac{2\sigma}{r} + \frac{\partial\sigma}{\partial r}\right), \tag{13}$$

which relates the r of an equilibrium bubble to the superheating, in which T and T_∞ are the temperatures at which the vapor pressure P″ is attained in the bubble and over a plane surface of the liquid. Analogous relationships can be established for other properties, and the method has already been published [6].

It is best to rewrite these equations in terms of the dimensionless quantities $\varphi = \sigma/\sigma_\infty$ and $x = r/a$ for the purposes of comparison with the equations of the classical theory:

$$P' - P_\infty = -\frac{2\sigma_\infty}{r}\cdot\chi, \tag{14}$$

$$kT \ln P''/P_\infty = -v'_\infty\frac{2\sigma_\infty}{r}\cdot\chi, \tag{15}$$

$$\lambda_\infty \ln T/T_\infty = v'_\infty\frac{2\sigma_\infty}{r}\chi, \tag{16}$$

[†] This relation should not be confused with the Gibbs condition, i.e., the second condition in (9b): P″ − P′ = $2\sigma/r + d\sigma/dr$, which relates the pressures of the different phases.

[‡] This deduction implies that (11) and (12) are consequences of the equilibrium conditions of (9b), so Kelvin's equation is sometimes [5] set in contrast to the Gibbs condition, but this is devoid of physical significance (thermodynamic equilibrium of phases always presupposes mechanical equilibrium).

in which the correction coefficient $\chi = \varphi + \frac{1}{2} x \cdot \partial\varphi/\partial x$ can be interpreted as the ratio $\chi = r/r_c$ of the actual radius r to the radius r_c given by the classical equations.

Table 1 gives values of χ for a nonpolar liquid[†] for various $x = r/a$.

Work of Nucleation for a Vapor Bubble within a
Homogeneous Liquid

This work for a bubble of radius r in a liquid having a pressure P' is best calculated via the potential Ω. We have for the initial and final states of the system that

$$\Omega_0 = P' \cdot V,$$

$$\Omega = - P' \left(V - \frac{4}{3}\pi r^3\right) - P'' \cdot \frac{4}{3}\pi r^3 + \sigma(r) \cdot 4\pi r^2;$$

and so the work of nucleation is found as

$$W = \Omega - \Omega_0 = - (P'' - P') \cdot \frac{4}{3}\pi r^3 + \sigma(r) \cdot 4\pi r^2$$

or

$$W = - [(P'' - P_\infty) - (P' - P_\infty)] \frac{4}{3}\pi r^3 + \sigma(r) \cdot 4\pi r^2.$$

Then (11) and (12) show that the first term in the square brackets may be neglected for a nuclear bubble, and so

$$W = (P' - P_\infty) \frac{4}{3}\pi r^3 + \sigma(r) \cdot 4\pi r^2. \tag{17}$$

This implies that nucleation is possible only in a stretched liquid, i.e., for $P' < P_\infty$.

Equation (17) is convenient for calculating W for cavitation, since here we can assume as given the negative pressure $\Delta P' = -(P' - P_\infty)$. In the case of the fluctuation mechanism, it is better to consider the superheating; (11) and (13) allow us to put (17) as

$$W = - \frac{\lambda_\infty}{v'_\infty} \ln\frac{T}{T_\infty} \cdot \frac{4}{3}\pi r^3 + \sigma(r) \cdot 4\pi \cdot r^2. \tag{18}$$

This shows that nucleation can occur only in a superheated liquid, i.e., for $T > T_\infty$.

This W represents an activation energy to be provided before a bubble becomes viable, the magnitude $W = W^*$ corresponding to a critical nucleus ($r = r^*$) in labile equilibrium with the medium.

The turning point in (17) and (18) gives

$$W^* = \frac{4}{3}\pi r^{*2} \left(\sigma - r^* \frac{\partial\sigma}{\partial r^*}\right), \tag{19}$$

in which r^* is deduced from (11) or (13) in accordance with the nucleation mechanism.

To express W^* in terms of directly measured quantities, we transform (19) via (11) and (13), and introduce the dimensionless quantities $x = r/a$ and $\varphi = \sigma/\sigma_\infty$. Then the W^* for the fluctuation mechanism is expressed via the superheating as

---

[†] Equation (6) may be used to approximate $\sigma = \sigma(r)$ for such a liquid.

$$W^* = \frac{16\pi}{3} \frac{\sigma_\infty'^3 \cdot v_\infty'^2}{\lambda_\infty^2 (\ln T / T_\infty)^2} \cdot \zeta, \tag{20}$$

in which $\zeta \doteq \chi^2 \left(\varphi - x \frac{\partial \varphi}{\partial x} \right)$ is the deviation coefficient.

The W* for the cavitation mechanism is defined in terms of the negative pressure $\Delta P = -(P' - P_\infty)$ by

$$W^* = \frac{16\pi}{3} \frac{\sigma_\infty^3}{(\Delta P)^2} \cdot \zeta. \tag{21}$$

Table 1 gives ζ as a function of x = r/a for a nonpolar liquid, and it shows that W* = 0 is not attained for any accessible r,[†] in contrast to generation of a drop of liquid in a supersaturated vapor.

References

1. T. L. Hill, J. Chem. Phys., 36:3182 (1962).
2. L. M. Shcherbakov, Tr. Tul'sk. Mekhan. Inst., Vol. 7, Tula (1955), p. 117; Research in Surface Forces, Vol. 1, Consultants Bureau, New York (1963), p. 19.
3. L. M. Shcherbakov, in: Research in Surface Forces, Vol. 2, Consultants Bureau, New York (1966), p. 26.
4. L. M. Shcherbakov, Kolloidn. Zh., 20:502 (1958).
5. E. I. Nesis, Usp. Fiz. Nauk, 87:615 (1965).
6. L. M. Shcherbakov and V. I. Rykov, Kolloidn. Zh., 23:221 (1961); L. M. Shcherbakov, in: Surface Phenomena in Melts and in Powder Metallurgy, Izd. AN SSSR, Kiev (1963), p. 38; L. M. Shcherbakov and S. A. Pilyus, Ibid., p. 74.
7. L. M. Shcherbakov and V. M. Matveev, this volume, p. 418.
8. V. A. Tereshin and L. M. Shcherbakov, this volume, p. 413.

[†] In our approximation, $r \geq 2a$.

DISTRIBUTION FUNCTIONS AND THERMODYNAMICS
OF THIN FILMS OF AN ELECTROLYTE SOLUTION
ON UNCHARGED SURFACES

V. P. Smilga and V. N. Gorelkin

Kurchatov Institute of Atomic Energy, Moscow

Bogolyubov correlation functions are used to take account of the detailed structure of a double layer, and expressions are derived for the correlation functions as series in the Debye parameter, which allow one to calculate the adsorption integral, the free energy, the differential capacitance, and the change in surface tension of an electrolyte film.

Here we calculate the free energy and pressure of a thin film of electrolyte solution enclosed between two different media. Problems of this type are extremely important as regards the stability of thin films.

The discussion is almost entirely restricted to the case where the interface has no free charges. It is found that there is then a very substantial electrostatic repulsive pressure, which can substantially exceed the van der Waals attraction under certain conditions. In the self-consistent-field approximation, the ion density in a thin film is constant throughout the cross section and is the same as the density in the bulk solution. The effect is therefore due to the discrete structure of the double layer, or rather to the image forces near the boundary, which affect the potential energy of the ions there.

Onsager [1] was the first to consider this effect for the surface of a semiinfinite volume of solution in contact with vacuum; others [2] have since considered it in more detail. Overlap of Onsager layers is responsible for the change in the free energy of a thin film and the corresponding repulsive forces.

Bogolyubov correlation functions [3] will be used here to determine the ion-density distribution in a thin film, as they allow the most logical and rigorous treatment of the problem.

1. Consider a thin film of an electrolyte solution in contact with a massive volume of the solution (infinite thermal reservoir). The solution has a dielectric constant ε_0, while the media adjoining the film have dielectric constants ε_1 and ε_2.

We employ subsequently the average reciprocal Debye length in the film, which is defined as

$$\varkappa^2 = \frac{4\pi e^2 \sum\limits_{i}^{M} z_i^2 \cdot \overline{N}_i}{\varepsilon_0 \theta V},$$

in which $\theta = kT$, $\overline{N}_i$ is the mean number of particles of type i in the film, and V is the volume of the film, the other quantities being standard.

Here we envisage only a symmetrical electrolyte with two types of ion. It is clear that $\overline{N}_i$ and $\varkappa$ are unknown quantities to be deduced from the conditions for equality of the chemical potentials for each type of particle in the film and in the reservoir. These conditions enable us to find $\varkappa$ as a function of $\varkappa_\infty$, the value for the thermal reservoir, and so the final result is defined by the parameters for the bulk solution.

It will be clear from the above that the ions in the film form an open ensemble (large canonical ensemble); but, as Bogolyubov's methods are to be used, we will regard the system as a closed ensemble containing $\overline{N}$ particles. As a canonical ensemble is equivalent to a large canonical ensemble, this replacement is completely in order, provided that we are not interested in the fluctuations in the mean number of particles in a film. All the correlation functions for a large canonical ensemble differ from those for a canonical ensemble by terms of order $1/\overline{N}_i$. Our problem is then to calculate the binary correlation functions $F_{a_1 a_2}(q_1 q_2)$.

2. The binary correlation function for a system with pair interactions contains information sufficient to define the free energy and all the thermodynamic potentials. In general, the mean value of the quantity A dependent as follows on the dynamic variables

$$A = \sum_{1 \leqslant i_1 < i_2 < \cdots < i_s \leqslant N} f_{a_{i_1}, \ldots, a_{i_2}}(q_{i_1} q_{i_2} \cdots q_{i_s}) \tag{2.1}$$

is defined [3] via the correlation functions as follows:

$$\overline{A} = \frac{1}{v^s S!} \sum_{\substack{1 \leqslant a_j \leqslant M \\ 1 \leqslant j \leqslant S}} n_{a_1} \ldots n_{a_s} \int\limits_V \int\limits_V f_{a_1 \ldots a_s}(q_1 \cdots q_s) F_{a_1 \ldots a_s}(q_1, \ldots, q_s) \, dq_1 \ldots dq_s. \tag{2.2}$$

Here M is the number of different kinds of particle, $\nu = V/M$, N is the total number of particles, $n_{aj} = N_{aj}/N$ is the relative number of particles of type a_j, and q_j is the set $q_j^1 q_j^2 q_j^3$ of three coordinates of a particle.

A knowledge of the distribution functions allows us to determine any thermodynamic quantity for the system. As we envisage a system with Coulomb interaction between the particles, we can follow a standard procedure [3] and calculate in place of the correlation functions $F_{a_1 \ldots a_s}(q_1 \ldots q_s)$ the functions $g_{a_1 \ldots a_s}(q_1 \ldots q_s)$:

$$\left. \begin{aligned} F_{a_1}(q_1) &= g_{a_1}(q_1), \\ F_{a_1 a_2}(q_1 q_2) &= g_{a_1}(q_1) \cdot g_{a_2}(q_2) + v \cdot g_{a_1 a_2}(q_1 q_2), \\ F_{a_1 a_2 a_3}(q_1 q_2 q_3) = g_{a_1}(q_1) \cdot g_{a_2}(q_2) \cdot g_{a_3}(q_3) + v \left[g_{a_1}(q_1) \cdot g_{a_2 a_3}(q_2 q_3) + \cdots \right] + v^2 g_{a_1 a_2 a_3}(q_1 q_2 q_3) \text{ etc.} \end{aligned} \right\} \tag{2.3}$$

The functions $g_{a_1 \ldots a_s}(q_1 q_2 \ldots q_s)$ satisfy the chain of Bogolyubov integrodifferential equations [3]

$$\left. \begin{aligned} \frac{\partial g_{a_1}(q)}{\partial q^\alpha} + \frac{\partial U_{a_1}(q)}{\partial q^\alpha \cdot \theta} \cdot g_a(q) + \frac{1}{v\theta} \int\limits_V \sum_{b=1}^{M} n_b \frac{\partial \Phi_{ab}(qq')}{\partial q^\alpha} \left[g_a(q) g_b(q') + v g_{ab}(qq') \right] dq' &= 0, \\ \frac{\partial g_{ab}(qq')}{\partial q^\alpha} + \frac{1}{\theta} \frac{\partial U_{ab}(qq')}{\partial q^\alpha} \cdot g_{ab}(qq') + \frac{1}{v\theta} \frac{\partial \Phi_{ab}(qq')}{\partial q^\alpha} \cdot g_a(q) \cdot g_b(q') & \\ + \frac{1}{v\theta} \int\limits_V \sum_{C=1}^{M} n_C \frac{\partial \Phi_{ac}(qq'')}{\partial q^\alpha} \left[g_a(q) g_{bc}(q'q'') + g_{ab}(qq') g_c(q'') + v g_{abc}(qq'q'') \right] dq'' &= 0. \end{aligned} \right\} \tag{2.4}$$

The conditions for normalization and decline in correlation for $F_{a_1 \dots a_s}(q_1 \dots q_s)$ mean that the $g_{a_1 \dots a_s}(q_1 \dots q_s)$ satisfy

$$\frac{1}{V} \int_V g_a(q)\, dq = 1;$$

$$g_{ab}(qq') \to 0 \quad \text{for} \quad |q - q'| \to \infty. \tag{2.5}$$

The $U_{a_1 \dots a_s}(q_1 \dots q_s)$ of (2.4) denotes the potential energy of the system S of isolated particles.

We can put $U_{a_1 \dots a_s}$ in the following form for a system with pair interaction:

$$U_{a_1 \dots a_s} = \sum_{1 \leqslant i < j \leqslant s} \Phi_{a_i a_j}(q_i q_j) + \sum_{i=1}^{s} \Phi_{a_i}(q_i). \tag{2.6}$$

The second term in (2.6) shows that the potential energy can vary from point to point, an obvious example being a system in an external field. In the present case, the terms dependent on the coordinates of an individual particle have a not entirely trivial form; detailed expressions for these will be derived below.

3. We now need the potential set up by a single point charge in the plane-parallel slot filled by an insulator between two massive insulators differing in nature, the dielectric constants being ε_0, ε_1, and ε_2, as above. The literature appears to carry no general solution to this problem, so one is given below.

The origin is placed at the boundary of the body whose dielectric constant is ε_1, with the x axis perpendicular to the interface and the positive direction into the slot. The y and z axes lie in the plane of the boundary. The width of the slot is l, and the charge has coordinate q. We find the potential $\varphi(qq')$ by solving Poisson's equation:

$$\Delta_q \varphi(qq') = -\frac{4\pi\delta}{\varepsilon_0}(q - q') \tag{3.1}$$

subject to the boundary conditions

$$\left. \begin{array}{l} \varepsilon_1 \dfrac{\partial \varphi_1(qq')}{\partial x} = \varepsilon_0 \dfrac{\partial \varphi_0(qq')}{\partial x} \\[2mm] \varphi_1(qq') = \varphi_0(qq') \end{array} \right\} \quad \text{for} \quad x = 0, \tag{3.2}$$

$$\left. \begin{array}{l} \varepsilon_2 \dfrac{\partial \varphi_2(qq')}{\partial x} = \varepsilon_0 \dfrac{\partial \varphi_0(qq')}{\partial x} \\[2mm] \varphi_2(qq') = \varphi_0(qq') \end{array} \right\} \quad \text{for} \quad x = l, \tag{3.3}$$

where a subscript to the potential indicates the medium in which it is taken.

This is solved via a two-dimensional Fourier transformation with respect to the variables $y - y'$ and $z - z'$. The Fourier amplitude of the potential is defined by

$$\varphi(xx'p) = \frac{1}{2\pi} \iint\limits_{-\infty}^{\infty} \exp\{-ip_y(y - y') - ip_z(z - z')\} \varphi(qq')\, d(y - y')\, d(z - z')$$

$$= \int_0^{\infty} J_0(pr)\, \varphi(rxx')\, r\, dr, \tag{3.4}$$

in which $p = \sqrt{p_y^2 + p_z^2}$, $r = \sqrt{(y - y')^2 + (z - z')^2}$, while $J_0(pr)$ is a Bessel function.

From (3.1) we get for $\varphi(xx'p)$ that

$$\frac{\partial^2 \varphi(xx'p)}{\partial x^2} - p^2 \varphi(xx'p) = -\frac{2}{\varepsilon_0} \delta(x - x'). \tag{3.5}$$

The boundary conditions for (3.5) are exactly the same as (3.2) and (3.3).

It is clear that $\varphi(xx'p)$ must be bounded in both external regions, while in the internal region it must be dependent on x and x' in a symmetrical fashion.

We take solutions to (3.7) that satisfy these conditions to get for the slot that

$$\varphi_0(xx'p) = \frac{1}{p\varepsilon_0} \frac{be^{(l-|x-x'|)p} + ae^{-(l-|x-x'|)p} + ke^{(l-x-x')p} + me^{-(l-x-x')p}}{be^{lp} - ae^{-lp}} \tag{3.6}$$

and for the external regions that

$$\varphi_1(xx'p) = \frac{1}{p\varepsilon_1} \frac{ce^{(l+x-x')p} + de^{-(l-x-x')p}}{be^{lp} - ae^{-lp}},$$

$$\varphi_2(xx'p) = \frac{1}{p\varepsilon_2} \frac{fe^{(l-x+x')p} + he^{(-l+x-x')p}}{be^{lp} - ae^{-lp}}. \tag{3.7}$$

Here the coefficients a, b, m, k, c, d, f, and h have to be determined; the denominators have been introduced for reasons of convenience.

The boundary conditions give us a system of algebraic equations:

$$\frac{1}{\varepsilon_1}[ce^{(l-x')p} + de^{-(l-x')p}] = \frac{1}{\varepsilon_0}[(b+k)e^{(l-x')p} + (a+m)e^{-(l-x')p}],$$

$$ce^{(l-x')p} + de^{-(l-x')p} = (b-k)e^{(l-x')p} + (a-m)e^{-(l-x')p},$$

$$\frac{1}{\varepsilon_2}[fe^{x'p} + he^{-x'p}] = \frac{1}{\varepsilon_0}[(b+m)e^{x'p} + (a+k)e^{-x'p}], \tag{3.8}$$

$$fe^{x'p} + he^{-x'p} = (b-m)e^{x'p} - (a-k)e^{-x'p}.$$

As (3.8) applies for any x' satisfying $0 < x' < l$, it is obviously equivalent to the following system of eight equations:

$$\begin{array}{ll}
\frac{1}{\varepsilon_1}c = \frac{1}{\varepsilon_0}(b+k), & \frac{1}{\varepsilon_2}f = \frac{1}{\varepsilon_0}(b+m), \\
\frac{1}{\varepsilon_1}d = \frac{1}{\varepsilon_0}(a+m), & \frac{1}{\varepsilon_2}h = \frac{1}{\varepsilon_0}(a+k), \\
c = b-k, & f = b-m, \\
d = -(a-m), & h = -(a-k).
\end{array} \tag{3.9}$$

Then

$$\begin{array}{ll}
b = (\varepsilon_0 + \varepsilon_1)(\varepsilon_0 + \varepsilon_2), & f = (\varepsilon_0 + \varepsilon_1)2\varepsilon_2, \\
a = (\varepsilon_0 - \varepsilon_1)(\varepsilon_0 - \varepsilon_2), & h = (\varepsilon_0 - \varepsilon_1)2\varepsilon_2, \\
m = (\varepsilon_0 + \varepsilon_1)(\varepsilon_0 - \varepsilon_2), & c = (\varepsilon_0 + \varepsilon_2)2\varepsilon_1, \\
k = (\varepsilon_0 - \varepsilon_1)(\varepsilon_0 + \varepsilon_2), & d = (\varepsilon_0 - \varepsilon_2)2\varepsilon_1.
\end{array} \tag{3.10}$$

In what follows, we need only $\varphi_0(xx'p)$. If the insulating slot is between two metals, we get the well-known result by putting ε_1 and $\varepsilon_2 \to \infty$.

The potential energy of a system of point charges in the film may be put as

$$U_N(q_1 \ldots q_N) = \frac{1}{2} \int_V \Psi(q) \sum_{i=1}^{N} e z_i \delta(q - q_i)\, dq + \frac{1}{2} \int_{S_i + S_l} \Psi(q)\, \sigma(q)\, dq, \qquad (3.11)$$

in which $\psi(q)$ is the potential set up at point q by all the charges:

$$\Psi(q) = \sum_{j=1}^{N} z_i e \varphi(q q_j). \qquad (3.12)$$

If $\sigma(q) = 0$ at the interface (no free charge there), the integrals over the boundary become zero. Similarly, we can make the integrals over the boundary zero by choice of the potential at the surface for an electrolyte film enclosed between two metal surfaces at the same potential. Expression (3.11) contains the infinite self-energy of the point charges. We eliminate the terms corresponding to that energy to get

$$U_N(q_1 \ldots q_N) = \sum_{i<j}^{N} e^2 z_i z_j \varphi(q_i q_j) + \frac{1}{2} \sum_{i=1}^{N} z_i^2 e^2 \lim_{q \to q_i} \left[\varphi(q q_i) - \frac{1}{\varepsilon_0 |q - q_i|} \right]. \qquad (3.13)$$

We follow Bogolyubov's standard procedure in using the potentials $\Psi_{ab}(qq')$, for the binary interaction, which are defined by

$$\frac{\Phi_{ab}(qq')}{\theta} = \frac{e^2 z_a z_b \varphi(qq')}{\theta} = \nu \Psi_{ab}(qq'). \qquad (3.14)$$

Similarly, the part of the potential energy dependent on the coordinates of a single particle is replaced by the potential

$$\nu \Psi_a(q) = \frac{\Phi_a(q)}{\theta} = \frac{e^2 z_a^2}{2\theta} \lim_{q' \to q} \left[\varphi(qq') - \frac{1}{\varepsilon_0 |q - q'|} \right]. \qquad (3.15)$$

As we envisage only a binary symmetrical electrolyte, we have

$$\Psi_{11}(qq') = \Psi_{22}(qq') = -\Psi_{12}(qq') = -\Psi_{21}(qq'). \qquad (3.16)$$

We deduce the potential from the Fourier amplitude of (3.6) and use (3.14) and (3.15) to get

$$\Psi_{11}(qq') = \frac{\varkappa^2}{4\pi} \frac{1}{\sqrt{(x-x')^2 + r^2}} + \frac{\varkappa^2}{4\pi} \sum_{j=0}^{\infty} \left(\frac{a}{b}\right)^j \left[\frac{a}{b} \frac{1}{\sqrt{[2l(j+1) + (x-x')]^2 + r^2}} \right.$$

$$+ \frac{a}{b} \frac{1}{\sqrt{[2l(j+1) - (x-x')]^2 + r^2}} + \frac{k}{b} \frac{1}{\sqrt{[2jl + (x+x')]^2 + r^2}} + \frac{m}{b} \frac{1}{\sqrt{[2(j+1)l - (x+x')]^2 + r^2}} , \qquad (3.17)$$

$$\Psi_1(x) = \Psi_2(x) = \Psi(x),$$

$$\Psi(x) = \frac{\varkappa^2}{4\pi} \sum_{j=0}^{\infty} \left(\frac{a}{b}\right)^j \left[\frac{a}{b} \frac{1}{2(j+1)l} + \frac{k}{b4(jl + x)} + \frac{m}{b4(jl + l - x)} \right]. \qquad (3.18)$$

It is clear that $\Psi(x)$ corresponds to the interaction of a charge with a series of its own images.

4. Consider the distribution function for the electrolyte ions in a thin film. Following Bogolyubov, we seek the correlation functions as a formal series in ν:

$$g_{a_1 \ldots a_s}(q_1 \ldots q_s) = g^0_{a_1 \ldots a_s}(q_1 \ldots q_s) + \nu g^1_{a_1 \ldots a_s}(q_1 \ldots q_s) + \ldots \qquad (4.1)$$

It has been shown [3] that here we actually have expansion with respect to the dimensionless parameter $\nu \varkappa^3 / 4\pi$ (it is naturally assumed that $\nu \varkappa^3 / 4\pi \ll 1$). We substitute (4.1) into (2.4) and equate terms of equal degree in ν to get equations for the corresponding approximations for the correlation functions. The equations take the following form for the zero approximation for the autocorrelation function:

$$\frac{\partial g_a^0(q)}{\partial q^\alpha} + g_a^0(q) \sum_{b=1}^{M} n_b \int_V \frac{\partial \Psi_{ab}(qq')}{\partial q^\alpha} g_b^0(q') \, dq' = 0. \tag{4.2}$$

The usual normalization conditions apply to the functions $g_a^0(q)$:

$$\frac{1}{V} \int_V g_a^0(q) \, dq = 1. \tag{4.3}$$

In the case of zero charge on the surfaces

$$g_1(x) = g_2(x). \tag{4.4}$$

Then, from (3.16), we readily get for $0 < x < l$ that

$$g_a^{0}(x) = 1. \tag{4.5}$$

We now use (4.5) to write the Bogolyubov equation for the zero approximation for the functions $g_{ab}^0(qq')$:

$$\frac{\partial g_{ab}^0(qq')}{\partial q^\alpha} + \frac{\partial \Psi_{ab}(qq')}{\partial q^\alpha} + \int_V \sum_{c=1}^{2} n_c \frac{\partial \Psi_{ac}(qq'')}{\partial q^\alpha} \cdot g_{bc}^0(q'q'') \, dq'' = 0. \tag{4.6}$$

Decay of the correlation implies that $g_{ab}^0(qq') \to 0$ for $|q - q'| \to \infty$, so we can easily eliminate the derivatives with respect to q^α. Then

$$g_{ab}^0(qq') + \Psi_{ab}(qq') + \int_V \sum_{c=1}^{2} n_c \Psi_{ac}(qq'') g_{bc}^0(q'q'') \, dq'' = 0. \tag{4.6a}$$

From (4.6a) and (3.15) we get

$$g_{11}^0(qq') = g_{22}^0(qq') = - g_{12}^0(qq') = - g_{21}^0(qq'). \tag{4.7}$$

In view of the uniformity in the y and z directions, $g_{ab}^0(qq')$ and $\Psi_{ab}(qq')$ can depend only on x, x', and $r = [(y - y')^2 + (z - z')^2]^{1/2}$.

In (4.6) we perform a Fourier transformation with respect to $y - y'$ and $z - z'$; then (4.7) is used to give

$$g_{11}^0(xx'p) + 2\pi \int_0^l \Psi_{11}(xx''p) g_{11}^0(x'x''p) \, dx'' = - \Psi_{11}(xx'p). \tag{4.8}$$

In view of (3.5), for $\Psi_{11}(xx''p)$, we have that

$$\frac{\partial^2 \Psi_{11}(xx''p)}{\partial x^2} - p^2 \Psi_{11}(xx''p) = - \frac{\varkappa^2}{2\pi} \delta(x - x'') \tag{4.9}$$

subject to boundary conditions analogous to (3.4).

We differentiate (4.8) twice with respect to x and use (4.9) to get the following differential equation:

$$\frac{\partial^2 g_{11}^0(xx'p)}{\partial x^2} - (p^2 + \varkappa^2) g_{11}^0(xx'p) = \frac{\varkappa^2}{2\pi} \delta(x - x'). \tag{4.10}$$

The boundary conditions for (4.10) are found by the following procedure. We use (4.9) to replace $\Psi_{11}(qq')$ in the integral in (4.8), integrate twice by parts, and then use (4.10) to get

$$\int_0^l \Psi_{11}(xx''p)\, g^0{}_{11}(x'x''p)\, dx'' = \frac{2\pi}{\varkappa^3}\left[\frac{\partial g^0_{11}(x'x''p)}{\partial x''}\Psi_{11}(xx''p)\right.$$

$$\left.- g^0_{11}(x'x''p)\frac{\partial \Psi_{11}(xx''p)}{\partial x''}\right]_{x''=l} - \frac{2\pi}{\varkappa^2}\left[\frac{\partial g^0_{11}(x'x''p)}{\partial x''}\Psi_{11}(xx''p)\right.$$

$$\left.- g^0_{11}(x'x''p)\frac{\partial \Psi_{11}(xx''p)}{\partial x''}\right]_{x''=0} - g^0_{11}(xx'p) - \Psi_{11}(xx'p). \tag{4.11}$$

We substitute (4.11) into (4.8) to get the condition that the solution to (4.10) must satisfy:

$$\left[\frac{\partial g^0_{11}(x'x''p)}{\partial x''}\Psi_{11}(xx''p) - g^0_{11}(x'x''p)\frac{\partial \Psi_{11}(xx''p)}{\partial x''}\right]_{x''=l} - \left[\frac{\partial g^0_{11}(x'x''p)}{\partial x''}\Psi_{11}(xx''p) - g^0_{11}(x'x''p)\frac{\partial \Psi_{11}(xx''p)}{\partial x''}\right]_{x''=0} = 0. \tag{4.12}$$

It is clear that $g^0_{11}(q'q) = g^0_{11}(qq')$, so the solution to (4.8) is symmetrical with respect to the coordinates x and x'. Further, (4.10) virtually coincides with (4.9), and so, from (3.6), we put the solution to (4.8) as

$$g^0_{11}(xx'p) = \frac{-\varkappa^3}{4\pi\sqrt{\varkappa^3+p^2}}\left[Be^{(l-|x-x'|)\sqrt{\varkappa^3+p^3}} + Ae^{-(l-|x-x'|)\sqrt{\varkappa^3+p^3}}\right.$$

$$\left. + Ke^{(l-x-x')\sqrt{\varkappa^3+p^3}} + Me^{-(l-x-x')\sqrt{\varkappa^3+p^3}}\right]\left[Be^{l\sqrt{\varkappa^3+p^3}} - Ae^{-l\sqrt{\varkappa^3+p^3}}\right]^{-1}. \tag{4.13}$$

Here again, the coefficients A, B, K, and M have to be determined. We substitute (4.13) into (4.12) and remember that (4.12) must be satisfied for all x and x' to get

$$-p(b-k)(B+K) + \sqrt{\varkappa^2+p^2}(b+k)(B-K) = 0,$$
$$p(a-k)(A+K) - \sqrt{\varkappa^2+p^2}(a+k)(A-K) = 0,$$
$$-p(b-k)(A+M) + \sqrt{\varkappa^2+p^2}(b+k)(-A+M) = 0,$$
$$p(a-k)(B+M) - \sqrt{\varkappa^2+p^2}(a+k)(-B+M) = 0,$$
$$-p(-a+m)(A+M) + \sqrt{\varkappa^2+p^2}(a+m)(-A+M) = 0, \tag{4.14}$$
$$p(-b+m)(B+M) - \sqrt{\varkappa^2+p^2}(b+m)(-B+M) = 0,$$
$$-p(-a+m)(B+K) + \sqrt{\varkappa^2+p^2}(a+m)(B-K) = 0,$$
$$p(-b+m)(A+K) - \sqrt{\varkappa^2+p^2}(b+m)(A-K) = 0.$$

Now

$$\begin{array}{ll}
b - m = 2\varepsilon_2(\varepsilon_0+\varepsilon_1), & -a+k = 2\varepsilon_2(\varepsilon_0-\varepsilon_1) \\
b + m = 2\varepsilon_0(\varepsilon_0+\varepsilon_1), & a+k = 2\varepsilon_0(\varepsilon_0-\varepsilon_1), \\
a - m = 2\varepsilon_1(\varepsilon_2-\varepsilon_0), & -b+k = -2\varepsilon_1(\varepsilon_0+\varepsilon_2), \\
a + m = -2\varepsilon_0(\varepsilon_2-\varepsilon_0), & b+k = 2\varepsilon_0(\varepsilon_0+\varepsilon_2).
\end{array} \tag{4.14a}$$

It is clear from (4.14) and (4.14a) that the latter four equations are equivalent to the first four, which arises because each set of four corresponds to the conditions at one of the boundaries. Then

$$B = (\varepsilon_0\sqrt{\varkappa^2+p^2}+\varepsilon_1 p)(\varepsilon_0\sqrt{\varkappa^2+p^2}+\varepsilon_2 p),$$
$$A = (\varepsilon_0\sqrt{\varkappa^2+p^2}-\varepsilon_1 p)(\varepsilon_0\sqrt{\varkappa^2+p^2}-\varepsilon_2 p),$$
$$M = (\varepsilon_0\sqrt{\varkappa^2+p^2}+\varepsilon_1 p)(\varepsilon_0\sqrt{\varkappa^2+p^2}-\varepsilon_2 p),$$
$$K = (\varepsilon_0\sqrt{\varkappa^2+p^2}-\varepsilon_1 p)(\varepsilon_0\sqrt{\varkappa^2+p^2}+\varepsilon_2 p).$$

As A, B, K, and M have been found by solving a homogeneous system of algebraic equations of fourth order, they are defined apart from an arbitrary constant factor. It is clear from (4.15) that this arbitrary element in no way affects the solution. It is readily verified that $g_{ab}^0(xx'p)$ coincides in the limiting case with Bravina's expression [2] for an electrolyte filling a semiinfinite volume.

5. We now show that $g_{11}^0(qq')$ allows us to determine all the thermodynamic characteristics of the system up to the first order in ν inclusive.

The internal energy is [5]

$$E = \frac{3}{2}\overline{N}\theta + \overline{U}_N. \tag{5.1}$$

From (3.13) and (3.16) we get the mean value of the potential energy of the system as

$$\overline{U}_N = \frac{\theta}{2\nu} \iint_{VV} \sum_{1 \leqslant a,b \leqslant M} n_a n_b \Psi_{ab}(qq')\left[g_a(q)g_b(q') + \nu g_{ab}^0(qq')\right]dq\,dq' + \theta\int_V \sum_{1 \leqslant a \leqslant M} n_a \Psi_a(q)g_a^0(q)dq. \tag{5.2}$$

By virtue of (4.4) and (3.16), the first term in the double integral is zero. We transform the second term having made the transition to the limit in (4.6) for $q \to q'$. It can be shown that the transition to the limit (4.6) is permissible, since the integrand increases not more rapidly than as $1/|q - q'|^2$ for $q \to q'$.

We use (3.15) and (3.16) to rewrite the last term in (5.2); then

$$\overline{U}_N = -\frac{\theta}{2}\int_V \sum_{1 \leqslant a \leqslant 2} n_a \lim_{q \to q'}\left[g_{aa}^0(qq') + \Psi_{aa}(qq')\right]dq + \frac{\theta}{2}\int_V \sum_{a=1}^{2}\lim_{q \to q'}\left[\Psi_{aa}(qq') - \frac{\varkappa^2}{4\pi}\frac{1}{|q-q'|}\right]dq. \tag{5.3}$$

Finally,

$$\overline{U} = -\frac{\theta}{2}\int_V \lim_{q' \to q}\left[g_{11}^0(qq') + \frac{\varkappa^2}{4\pi}\frac{1}{|q-q'|}\right]dq. \tag{5.4}$$

The free energy of the film can be deduced from

$$\frac{\partial\left(\frac{F}{\theta}\right)}{\partial\theta} = -\frac{E}{\theta^2}. \tag{5.5}$$

The function $g_{ab}^0(qq')$ also allows us to find the first-order correction to the autocorrelation function, for which the Bogolyubov equations take the form

$$\frac{\partial g_a^1(q)}{\partial x} + g_a^1(q)\int_V \sum_{b=1}^{M} n_b g_b^0(q'')\frac{\partial\Psi_{ab}(qq'')}{\partial x}dq'' + g_a^0(q)\int_V \sum_{b=1}^{M} n_b \frac{\partial\Psi_{ab}(qq'')}{\partial x}g_b^1(q'')dq''$$

$$= -\frac{\partial\Psi(\alpha)}{\partial x}g_a^0(q) - \int_V \sum_{b=1}^{M} n_b \frac{\partial\Psi_{ab}(qq'')}{\partial x}g_{ab}^0(qq'')dq''.$$

For the present case, from (4.4), (4.5), (3.16), and (4.7),

$$\frac{\partial g_a^1(x)}{\partial x} = -\frac{\partial\Psi(x)}{\partial x} - \int_V \sum_{b=1}^{2} n_b \frac{\partial\Psi_{ab}(qq'')}{\partial x}g_{ab}^0(qq'')dq''. \tag{5.6}$$

Functions $\Psi_{ab}(qq'')$ and $g_{ab}^0(qq'')$ have first-order poles at the points $q'' = q$, as is readily established from the expression for their Fourier components, so the integrand in (5.6) also has a singularity at the point q, and the integral has to be taken in the sense of the principal value. The divergence of (5.6) naturally arises because the short-range forces are not taken into account in Bogolyubov's method. We use (4.6) in order to calculate this integral.

We put x = x' in (4.6) and pass to the limit for $r = \sqrt{(y-y')^2 + (z-z')^2} \to 0$. We will show that for this passage to the limit

$$\lim_{r \to 0}\left[\int_V \sum_{b=1}^2 n_b \frac{\partial \Psi_{ab}(qq'')}{\partial x} [g_{ab}^0(x'=x, z', y', q'') - g_{ab}^0(qq'')] dq'' = 0. \tag{5.7}$$

We divide the region of integration into two:

$$\Omega_R \begin{cases} |x''-x| \leqslant R \leqslant \frac{1}{\varkappa}, \\ \rho^2 = \left(z'' - \frac{z'+z}{2}\right)^2 + \left(y'' - \frac{y'+y}{2}\right)^2 \leqslant R^2 \end{cases} \tag{5.8}$$

and $\{V - \Omega_R\}$ is external with respect to Ω_R.

Taking $r < \delta_R < R/4$, the integral over the external region may be made as small as desired, since the integrand is a continuous function that tends to 0 as $r \to 0$. The integrand in the integral over region Ω_R may be expanded as a series in $|q''-q|$ and $|q''-q'|$, in which we retain only the principal terms, since the integral with respect to the regular terms may be made small by taking R sufficiently small.

Consider the contribution made by the principal terms to the integral over Ω_R:

$$\int_{\rho < R} \int_{x-R}^{x+R} \frac{(x''-x)}{[(x''-x)^2 + (y''-y)^2 + (z''-z)^2]^{3/2}}\left[\frac{1}{\sqrt{(x''-x)^2 + (y''-y')^2 + (z''-z')^2}} - \frac{1}{\sqrt{(x''-x)^2 + (y''-y)^2 + (z''-z)}}\right] dq''. \tag{5.9}$$

As the integrand is odd with respect to $t = x'' - x$, while the region of integration is symmetrical with respect to t, the integral of each of these two terms becomes zero.

We have thus shown that

$$\int_V \sum_{b=1}^2 n_b \frac{\partial \Psi_{ab}(qq'')}{\partial x} g_{ab}^0(qq'') dq'' = -\lim_{r \to 0}\left[\frac{\partial \Psi_{aa}(x, x'=x, r)}{\partial x} + \frac{\partial g_{aa}^0(x, x'=x, r)}{\partial x}\right]. \tag{5.10}$$

Now x and x' appear in $\Psi_{ab}(qq')$ and $g_{ab}^0(qq')$ only in the forms $x - x'$ and $x + x'$, so

$$g_a^1(x) = -\Psi(x) + \frac{1}{2}\lim_{q \to q'}[g_{11}^0(qq') + \Psi_{11}(qq')] + \text{const.} \tag{5.11}$$

We use (3.14) and (3.15) to get finally that

$$g_a^1(x) = \frac{1}{2}\lim_{q \to q'}\left[g_{11}^0(qq') + \frac{\varkappa^2}{4\pi}\frac{1}{|q-q'|}\right] + \text{const.} \tag{5.12}$$

The constant is determined by the normalization conditions

$$\int_0^l g_a^1(x)\,dx = 0. \tag{5.13}$$

Detailed calculations via (5.4) and (5.12) using the general expression of (4.13) are fairly complicated, so we consider only two limiting cases.

6. We calculate the free energy and repulsive pressure for two cases: a) a film of electrolyte between two metal surfaces; b) a free film (both surfaces in contact with air). The latter is called the Onsager case.

The Onsager case is defined only by $\varepsilon_1 \approx \varepsilon_2 \ll \varepsilon_0$. Then we have in these cases

$$\Psi_{11}(qq') = \frac{\varkappa^2}{4\pi}\frac{1}{\sqrt{(x-x')^2+r^2}} + \frac{\varkappa^2}{4\pi}\cdot\sum_{k=0}^{\infty}\left[\frac{1}{\sqrt{[2l(k+1)+(x-x')]^2+r^2}}\right.$$

$$+\frac{1}{\sqrt{[2l(k+1)-(x-x')]^2+r^2}} \mp \frac{1}{\sqrt{(2kl+x+x')^2+r^2}} \mp \frac{1}{\sqrt{[2l(k+1)-(x+x')]^2+r^2}}\left]\left(\frac{a}{b}\right)^k.\right. \qquad (6.1)$$
$$(1 > a/b \approx 1)$$

Here and subsequently, the upper sign corresponds to case a) and the lower one to case b).

The zero approximation for the binary function may here be found somewhat more simply than in the general case (Section 4). We apply the Laplace operator to both parts of (4.6a) and use $\Delta_q \Psi_{11}(qq') = -\varkappa^2\delta(q-q')$ to get

$$\Delta_q g_{11}^0(qq') - \varkappa^2 g_{11}^0(qq') = \varkappa^2\delta(q-q'). \qquad (6.2)$$

We use (6.2) with the obvious relation

$$g_{11}^0(qq') = -\frac{1}{\varkappa^2}\int_V g_{11}^0(qq'')\Delta_{q''}\Psi_{11}(q''q')\,dq'', \qquad (6.3)$$

to put (4.6a) in the form

$$\int[\psi_{11}(qq'')\Delta_{q''}g(q''q') - g_{11}(q''q')\Delta_{q''}\psi_{11}(q''q)]\,dq'' = 0. \qquad (6.4)$$

Green's formula transforms (6.4) to

$$-\int_{S_{x''=0}}\left[\Psi_{11}(qq'')\frac{\partial}{\partial x''}g_{11}^0(q''q') - g_{11}^0(q''q')\frac{\partial}{\partial x''}\Psi_{11}(q''q)\right]dS$$

$$+\int_{S_{x''=e}}\left[\Psi_{11}(qq'')\frac{\partial}{\partial x''}g_{11}^0(q''q') - q_{11}^0(q''q')\frac{\partial}{\partial x''}\Psi_{11}(q''q)\right]dS = 0. \qquad (6.5)$$

The solution to (6.2) that satisfies (6.5) is the sum of the screened potentials of all images:

$$g_{11}^0(qq') = -\frac{\varkappa^2}{4\pi}\frac{e^{-\varkappa\sqrt{(x-x')^2+r^2}}}{\sqrt{(x-x')^2+r^2}} - \frac{\varkappa^2}{4\pi}\sum_{k=0}^{\infty}\left[\frac{e^{-\varkappa\sqrt{[(x-x')+2l(k+1)]^2+r^2}}}{\sqrt{[2l(k+1)+(x+x')]^2+r^2}}\right.$$

$$+\frac{e^{-\varkappa\sqrt{[2l(k+1)-(x-x')]^2+r^2}}}{\sqrt{[2l(k+1)-(x-x')]^2+r^2}} \mp \frac{e^{-\varkappa\sqrt{(2kl+x+x')^2+r^2}}}{\sqrt{(2kl+x+x')^2+r^2}} \mp \frac{e^{-\varkappa\sqrt{[2l(k+1)-(x+x')]^2+r^2}}}{\sqrt{[2l(k+1)-(x+x')]^2+r^2}}\left].\right. \qquad (6.6)$$

At the boundaries of the film we either have $\Psi_{11}^0(qq')$ and $g_{11}^0(qq')$ zero simultaneously (for boundaries with a metal) or the derivatives along the normal zero (Onsager case). Of course, we get previous results [2, 4] on passing to the limit $l \to \infty$.

It has previously been shown that, if the free charge at the boundary is zero, we can use $g_{11}^0(qq')$ to determine the internal energy of the system via (5.1) and (5.4). In the present two cases we get:

$$E = \frac{3}{2}\,\overline{N}\theta - \frac{V\theta\varkappa^3}{8\pi} \pm \frac{\theta S\varkappa^2}{8\pi}\,Ei\,(-2\varkappa\delta) - \frac{\theta S\varkappa^2}{8\pi}\,\ln\,(1 - e^{-2\varkappa l}), \tag{6.7}$$

in which $Ei\,(-\varkappa\delta) = -\int\limits_{\varkappa\delta}^{\infty} \frac{e^{-t}}{t}\,dt$, where δ is the cutoff factor.

The first three terms in (6.7) correspond respectively to the internal energy of an ideal gas with $\overline{N}$ particles and to the Debye and Onsager terms, while the last term is related to interaction between the boundaries and decreases exponentially as l increases. The Debye and Onsager terms are dependent on the number of particles only via $\varkappa$. We integrate (5.5) to get the free energy, subject to $\lim\limits_{\theta\to\infty}\dfrac{F - F_{id}}{\theta} = 0$, as

$$F = F_{id} - \frac{2}{3}\,\frac{\theta V\varkappa^3}{8\pi} + F_{ons}(\theta,\,\varkappa,\,\delta,\,S) + \Delta F_l, \tag{6.8}$$

in which

$$F_{ons} = -\frac{S\theta\varkappa^2}{8\pi}\Big[Ei\,(-2\varkappa\delta) - \frac{1}{2}\Big],$$
$$\Delta F_l = \frac{\theta S}{8\pi}\sum_{k=1}^{\infty}\Big[\frac{1}{2k^3l^2} - \frac{e^{-2\varkappa lk}}{2k^3l^2} - \frac{\varkappa e^{-2\varkappa lk}}{k^2l}\Big]. \tag{6.9}$$

The $1/\varkappa$ appearing in all expressions is governed by $\overline{N}_a$ (mean number of particles in the film), which is an unknown to be found from equality of the chemical potentials for the ions in the film and thermal reservoir. The chemical potential in the thermal reservoir is

$$\mu^\infty = \mu_{id}^\infty + \frac{\partial F_D\,(\varkappa_\infty)}{\partial N_a^\infty} = \mu_{id}^\infty - \frac{\theta v_\infty \varkappa_\infty^3}{8\pi},$$
$$F_D\,(\varkappa_\infty) = -\frac{2}{3}\,\frac{\theta V \varkappa_\infty^3}{8\pi},$$

in which $F_D\,(\varkappa_\infty)$ is the Debye contribution to the free energy determined by the particle density $1/v_\infty$ in the thermal reservoir.

As here $\overline{N}_1 = \overline{N}_2 = \overline{N}/2$, the chemical potentials will be equal if we have that

$$\theta\,\ln\frac{v}{v_\infty} = \frac{-\theta v\varkappa^3}{8\pi} + \Big[\frac{\partial F_{ons}\,(\theta,\varkappa,\delta,S)}{\partial\varkappa} + \frac{\partial\Delta F_l}{\partial\varkappa}\Big]\frac{\partial\varkappa}{\partial\overline{N}} + \frac{\theta v_\infty \varkappa_\infty^3}{8\pi}. \tag{6.10}$$

The quantities on the right in (6.10) are of the first order of smallness in v and v_∞, so we may put as follows, omitting terms of higher order in v:

$$v = v_\infty + \frac{v_\infty}{\theta}\Big[\frac{\partial F_{ons}\,(\theta\varkappa\delta S)}{\partial\varkappa} + \frac{\Delta F_l}{\partial\varkappa}\Big]\frac{\partial\varkappa}{\partial\overline{N}} \tag{6.11}$$

with $\varkappa = \varkappa_\infty$ on the right. Then (6.11) defines $\overline{N}$ or the mean concentration in a film of volume V.

The force acting on unit surface is

$$\mathscr{F} = -\frac{1}{S}\frac{\partial F}{\partial l} = \frac{\theta}{v} - \frac{1}{S}\frac{\partial F_D}{\partial l} - \frac{1}{S}\Big[\frac{\partial F_{ons}\,(\theta,\varkappa,\delta,S)}{\partial\varkappa}\frac{\partial\varkappa}{\partial l} + \Big(\frac{\partial\Delta F_l}{\partial\varkappa}\Big)_l\frac{\partial\varkappa}{\partial l} + \Big(\frac{\partial\Delta F_l}{\partial l}\Big)_\varkappa\Big]. \tag{6.12}$$

We substitute for ν from (6.11) into (6.12) and note that $\partial \varkappa / \partial l = -(N/l)(\partial \varkappa / \partial N)$ to get up to terms of the order of ν inclusive that

$$\mathscr{F} = P_\infty - \frac{1}{S}\left(\frac{\partial \Delta F_l}{\partial l}\right)_{\varkappa = \varkappa_\infty}. \tag{6.13}$$

This force is not dependent on δ, which has to be introduced into the Onsager term in the expression for the free energy because the correlation function tends to infinity as $1/\varkappa$ at short distances from the boundary, which leads to a logarithmic divergence in the free energy. A full treatment would require consideration of the short-range forces, but here we have considered only the Coulomb interaction by introducing the cutoff parameter δ, which corresponds to distances of the order of the ionic radius. The pressure is independent of δ, which shows that the precise form of the correlation function near the boundary is unimportant as regards this pressure, which arises from overlap of Onsager adsorption layers.

It can be shown that a fuller consideration of the short-range forces introduces into the free energy additional terms of order δ/l, so the results are virtually unaffected, because this ratio is very small.

The first term in (6.13) equals the pressure in the thermal reservoir and is independent of the distance between the boundaries. The excess pressure is given by (6.9) and (6.13) as

$$\Delta \mathscr{F}_l = \frac{\theta \varkappa^3}{8\pi} \sum_{k=1}^{\infty} \left[\frac{1}{(\varkappa l k)^3} - \frac{e^{-2\varkappa l k}}{(\varkappa l k)^3} - \frac{2e^{-2\varkappa l k}}{(\varkappa l k)^2} - \frac{2e^{-2\varkappa l k}}{\varkappa l k}\right]. \tag{6.14}$$

This enables us to derive the force in the limiting cases of large and small film thicknesses. For large thicknesses ($\varkappa l \gg 1$) we have

$$\Delta \mathscr{F}_l = \frac{\theta}{8\pi l^3} \sum_{k=1}^{\infty} \frac{1}{k^3}. \tag{6.15}$$

Taking T = 300°K, we have

$$\Delta \mathscr{F}_l = \frac{1.7 \cdot 10^{-15}}{l^3} \text{ dyne/cm}^2.$$

This force is inversely proportional to l^3 and is independent of $\varkappa$.

In the other case ($\varkappa l \approx 1$) we have

$$\Delta \mathscr{F}_l \sim -2\Delta P_D = \frac{4}{3}\frac{\theta \varkappa_\infty^3}{8\pi}. \tag{6.16}$$

From (5.12), (5.13), and (6.6) we now get the density distribution for the particles in a thin film as

$$\frac{1}{\nu(x)} = \frac{1}{\nu}\left\{1 \pm \frac{\nu_\infty \varkappa^3}{8\pi}\left[\sum_{k=0}^{\infty} \frac{e^{-2(kl+x)\varkappa}}{2(kl+x)\varkappa} + \sum_{k=0}^{\infty} \frac{e^{-2(k+1)lx+2x\varkappa}}{2(k+1)lx - 2x\varkappa} + \frac{1}{\varkappa l} Ei(-2\varkappa \delta)\right]\right\}. \tag{6.17}$$

7. These expressions for the electrostatic parts of the free energy and pressure show that on a thin film of the electrolyte solution the compensation of van der Waals force attraction has a place, at least in a long-range region. As it is seen from the formula (6.14) the term corresponding to the zero frequency of the fluctuation field [6] is fully compensated.

References

1. L. Onsager and N. T. Samaras, J. Chem. Phys., 2:528 (1934).
2. V. V. Bravina, Vestnik MGU, No. 2, p. 31 (1958); V. G. Levich and V. A. Kir'yanov, Dokl.
    AN SSSR, 131:1134 (1960).
3. N. N. Bogolyubov, Problems of Dynamical Theory in Statistical Physics, Gostekhizdat,
    Moscow (1944).
4. V. N. Gorelkin and V. P. Smilga, Élektrokhimiya, 2:492 (1966).
5. L. D. Landau and E. M. Lifshits, Statistical Physics, Gostekhizdat, Moscow (1951).
6. I. E. Dzyaloshinskii, E. M. Lifshits, and L. P. Pitaevskii, Zh. Éksp. Teor. Fiz., 37:229
    (1959).

SECTION IV

STABILITY IN DISPERSED SYSTEMS

NONINERTIAL VAN DER WAALS CAPTURE
OF PARTICLES BY A SPHERE IN A FLOW OF LIQUID

B. V. Deryagin and L. P. Smirnov †

Surface Phenomena Division, Institute of Physical Chemistry,
Academy of Sciences of the USSR

Analytic expressions are derived for the critical path and capture coefficient for particles suspended in a flow of a viscous liquid flowing around a spherical obstacle. Allowance is made for van der Waals forces (with and without electromagnetic delay) and for the resistance produced by the layer of viscous liquid between the bodies. The limits of application of the results are examined, and the flux of deposited particles in the van der Waals mechanism is compared with that for deposition by diffusion.

A method previously described [1] is applied here to determine the coefficient of noninertial capture of aerosol particles by a spherical obstacle, with reference to uncharged particles deposited under action of van der Waals forces exerted by the obstacle. Approach to the sphere produces resistance related to displacement of the layer of viscous liquid from the gap.

The hydrodynamic conditions are assumed to be the same as in the electrostatic case [1]. We assume that the particles in the incident flow do not distort the velocity distribution and that the Reynolds number of viscous flow around the sphere is such that Stokes flow can be assumed. A particle is acted on by specific forces, which appear on account of close approach of the particle to the sphere, and also by a Stokes force proportional to the local velocity of the particle relative to the medium.

The calculation gives an analytic expression for the limiting path in the meridional plane (the path separating paths leading to the sphere from those avoiding the sphere) and also a formula for the capture coefficient. A distinction from the case previously considered [1] is that there is difficulty in linking directly this limiting path to a stream line to infinity, but this is overcome by examining the paths in an intermediate region fairly remote from the sphere. The capture-coefficient formula applies for spherical particles whose radii lie in a certain range, and the coefficient given by this formula is much less than that implied by the linking

---

† S. S. Dukhin collaborated in the section on the comparison of van der Waals deposition and diffusion deposition.

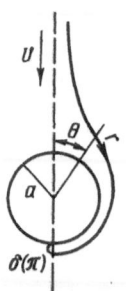

Fig. 1. Flow around
a spherical obstacle
and the critical path.

theory, where no allowance is made for the effects of the viscous
layer. Moreover, the coefficient even becomes zero in the absence
of van der Waals forces, because the viscous resistance prevents the
particles from contacting the sphere.

It seems that the only previous consideration of van der Waals
forces in such a case is Natanson's [2], which deals mainly with electrostatic deposition on a cylinder in a plane-parallel flow. Natanson
[2] corrected errors in an earlier paper [3] (the results then agreed
with even earlier ones [4]) and discussed the deposition on the assumption that the distance from the surface to the center of a particle on
the critical path was much greater than the radius of the particle.
However, it has been shown [5] that in that case a correction must be
made for the electromagnetic delay, which renders incorrect Natan-
son's [2] law of interaction, which is based on extrapolation of Hamaker's formula [4] to the case
of large distances between the bodies.

The equations of motion may be written in spherical coordinates with their origin at the
center of the sphere and with the angle θ reckoned from the vertical directed upwards from the
center (Fig. 1). The incompressible viscous flow approaches the obstacle from above and has
a velocity U at infinity. We neglect the particle's inertia (a discussion at the end of the paper
shows where this is permissible) but incorporate the Stokes force and the specific forces, which
are dependent on the distance δ between the surfaces and on the rate of approach $\dot{r}$. Then,
projecting on the radius and on a direction perpendicular to this, we have

$$0 = -f(\dot{r}, \delta) + k(v_r - \dot{r}),$$
$$0 = v_\theta - r\dot{\theta} \tag{1}$$

(the dot denotes differentiation with respect to time).

The factor k in the Stokes force has the following value for a spherical particle (radius R):

$$k = 6\pi\eta R, \tag{2}$$

in which η is the viscosity of the medium.

The velocity components v_r and v_θ in Stokes flow around a sphere of radius a are [6]

$$v_r = U\cos\theta\left(\frac{3}{2}\frac{a}{r} - \frac{1}{2}\frac{a^3}{r^3} - 1\right),$$
$$v_\theta = U\sin\theta\left(1 - \frac{1}{4}\frac{a^3}{r^3} - \frac{3}{4}\frac{a}{r}\right). \tag{3}$$

Near the sphere (i.e., at r = a + h, in which h ≪ r), these velocity components are approximately as follows:

$$v_r = -\frac{3}{2}U\cos\theta\frac{h^2}{a^2},$$
$$v_\theta = \frac{3}{2}U\sin\theta\frac{h}{a} - \frac{9}{4}U\sin\theta\frac{h^2}{a^2}. \tag{4}$$

Considerations of symmetry make it clear that the resultant of the van der Waals interactions is radial in direction; but the viscous layer produces also a retarding force perpendicular to the radius as well as a torque that tends to rotate the particle, and rotation may affect
the radial force on the particle. Here we neglect these additional effects, which complicate the

situation, and merely note that the retarding force of viscous origin will increase the time spent by the particle near the sphere and so will facilitate deposition under the van der Waals attraction. On the other hand, we may suppose that the action of the retarding force perpendicular to the radius will be less important than that of the radial force, since (4) gives the tangential Stokes force as larger than the radial Stokes force by an order of magnitude.

The literature [4, 5, 7-12] indicates that the force f is

$$f = \frac{\alpha_1}{\delta^2} + \frac{\beta_1}{\delta}\frac{dh}{dt} ; \quad h - R = \delta \ll R \approx h \ll a. \tag{5}$$

The first term represents the resultant of the molecular interactions; coefficient α_1 is dependent [9-11] on the geometry of the surfaces (principal curvatures and mutual disposition of the principal normal sections) and on the absorption spectra of the bodies and of the layer of liquid between them. For a sphere approaching a planar wall, α_1 is proportional to R [4, 12]:

$$\alpha_1 = \frac{AR}{6} , \tag{6}$$

in which A is Hamaker's constant of molecular attraction.

Hamaker deduced the resultant attractive force between bodies by summing the interactions between molecules and derived the following approximate formula [12] for A, which has the dimensions of energy:

$$A = \frac{3}{4}\pi^2 q^2 h\nu_0\widetilde{\alpha}^2, \tag{7}$$

in which q is the number of molecules in unit volume of the interacting bodies, $h\nu_0$ is the characteristic term for the energy (which is evaluated by quantum-mechanical methods or via an experimental formula for the optical dispersion), and $\widetilde{\alpha}$ is the polarizability of the molecules.

Macroscopic treatment of the van der Waals interaction [8] gives the value

$$A = \frac{6\hbar\varkappa}{\pi} , \tag{8}$$

in which $\hbar = h/2\pi$, h is Planck's constant, and $\varkappa$ is a factor in the form of a definite integral, which has the dimensions of a frequency and which is calculated from the frequency dependence of the refractive index and absorption coefficient for the bodies. These results apply if the distance between the surfaces is much less than the wavelength of the electromagnetic radiation emitted and absorbed by the molecules.

The second term in (5) represents the viscous repulsive force, and this has [10] been derived via the perpendicular approach of a nonrotating sphere (radius R) to a plane as:

$$F = \frac{6\pi\eta v R^2}{H} , \tag{9}$$

in which η is viscosity, v is the velocity of approach, and H is the shortest distance between the surfaces.

The analogous result for two spheres of radii R_1 and R_2 is

$$F = \frac{6\pi\eta v}{H}\frac{R_1^2 R_2^2}{(R_1 + R_2)^2}. \tag{10}$$

Then (9) can be used if one sphere is much smaller than the other, R being the radius of the smaller sphere. Then the β_1 in the second term in (5) takes the following value:

$$\beta_1 = 6\pi\eta R^2. \tag{11}$$

For $f \equiv 0$, equations (1) become the equations of motion for the particles of the medium, whose paths coincide with the stream lines around a sphere of radius a, and the exact equation for these in the case of Stokes flow is

$$b^2 = \sin^2 \theta \cdot r^2 \left\{ 1 - \frac{3}{2} \frac{a}{r} + \frac{1}{2} \frac{a^3}{r^3} \right\},$$ (12)

which is written in terms of the variable $h = r - a$ as

$$b^2 = \sin^2 \theta \cdot h^2 \left\{ 1 + \frac{a}{2(a+h)} \right\},$$ (13)

in which b is the radius of the cylindrical current tube representing the asymptotic form of an arbitrary axially symmetric flow surface at infinity.

Equation (13) takes the following simple form for $h/a \ll 1$:

$$b = \sqrt{\frac{3}{2}} \sin \theta \cdot h.$$ (14)

From (5), we write the first equation in (1) as

$$\frac{dr}{dt} = \frac{\delta}{\beta + \delta} \left(v_r - \frac{\alpha}{\delta^2} \right),$$ (15)

where, from (2), (6), and (11)

$$\alpha = \frac{\alpha_1}{k} = \frac{A}{36\pi\eta}; \quad \beta = \frac{\beta_1}{k} = R.$$ (16)

We eliminate the differential dt from (1) and use (15) to get the equation for the paths of the suspended particles:

$$\frac{1}{r} \frac{dr}{d\theta} = \frac{\delta}{\beta + \delta} \cdot \frac{v_r - \frac{\alpha}{\delta^2}}{v_\theta}.$$ (17)

Of course, this becomes the equation of a stream line if $\alpha = \beta = 0$; the same result is obtained by passage to the limit $\delta \to \infty$. This implies that the path of a particle becomes asymptotically a stream line at large distances from the sphere.

From (4), we put (17) as

$$\frac{1}{a} \frac{dh}{d\theta} = -\frac{\delta}{\beta + \delta} \cdot \frac{\frac{3}{2} U \cos \theta \frac{h^2}{a^2} + \frac{\alpha}{\delta^2}}{\frac{3}{2} U \sin \theta \frac{h}{a}}.$$ (18)

Consider the critical path whose elements near the sphere satisfy the inequalities

$$\delta = h - R \ll R \approx h \ll a.$$ (19)

The simplification which results from these assumptions, when the second equation in (16) is used, allows us to put (18) as

$$\sin \theta \frac{d\delta}{d\theta} = -\cos \theta \cdot \delta - \frac{2}{3} \frac{\varkappa a^2}{U R^2} \frac{1}{\delta}.$$ (20)

This is Bernoulli's equation, which with the substitution $\delta^2 = z$ gives a linear inhomogeneous equation having the following general solution:

$$\delta = \left\{ \frac{4}{3} \frac{a^2 \alpha}{U R^2} \frac{\cos \theta + C}{\sin^2 \theta} \right\}^{1/2}.$$ (21)

The limiting path is obtained by putting C = 1, and the equation for this is

$$h = R + \delta = R + \sqrt{\frac{2\alpha}{3U}} \frac{a}{R \sin \frac{\theta}{2}}.$$ (22)

It is readily shown that the limiting path approaches the axis of symmetry orthogonally (Fig. 1) as $\theta \to \pi$, and the distance of the closest point on the limiting surface from the surface of the sphere is

$$h(\pi) = R + \sqrt{\frac{2\alpha}{3U}} \frac{a}{R},$$ (23)

while the equatorial distance of the limiting surface from the sphere is

$$h\left(\frac{\pi}{2}\right) = R + \sqrt{\frac{\alpha}{3U}} \frac{2a}{R}.$$ (24)

It is clear that all particles for which $-1 \le C < 1$ will be deposited on the sphere, since for these δ becomes zero at some $\theta < \pi$.

If there is no van der Waals interaction ($\alpha = 0$), (20) becomes a linear homogeneous equation whose general integral is

$$\delta = \frac{\tilde{c}}{\sin \theta}.$$ (25)

This shows that the paths are then symmetrical with respect to the equatorial plane and have their least distance in that plane; but deposition ($\delta = 0$) is essentially impossible. Particles moving along the line $\theta = 0$ will, for $\alpha = 0$, approach the sphere as follows, as (15) shows:

$$\frac{d\delta}{dt} = -\frac{3}{2} U \frac{R}{a^2} \delta,$$ (26)

whence

$$\delta = \delta_0 e^{-\frac{3}{2} U \frac{R}{a^2} t},$$ (27)

i.e., the approach will take an infinite time.

If pure linkage were to occur (with no van der Waals forces), and if the viscous layer did not oppose approach, the paths of particles trapped noninertially would coincide with stream lines; deposition would occur on the front face of the sphere, and the capture coefficient ζ would be given by (14), in which we should put h = R for $\theta = \pi/2$. Then

$$b = \sqrt{\frac{3}{2}} R$$ (28)

and

$$\zeta = \frac{b^2}{a^2} = \frac{3}{2} \frac{R^2}{a^2}.$$ (29)

It follows from (29) that large particles deposit more readily than small ones.

Consider now how these results are affected by van der Waals and viscous forces. Difficulties arise in direct deduction of the stream line that closely fits the critical path over a finite length, since (22) applies only if the second term on the right is much less than the first; (22) is suitable close to the sphere ($\delta \ll R \ll a$), while the stream lines leading to points on the critical path differ greatly from the latter in that region, and (14) shows that a stream line intersects the critical path at a substantial angle. To link the path with a stream line, the former must therefore be extended to a region further from the sphere. An approximate expression may be obtained for the path in this intermediate region on the basis that $\theta \ll 1$ there.

We introduce the dimensionless quantities

$$\bar{h} = \frac{h}{a}, \ \bar{\delta} = \frac{\delta}{a} \, (\delta = h - R) \tag{30}$$

and use the fact that $\beta = R$ to put (18) as

$$\frac{d\bar{h}}{d\theta} = - \frac{\bar{h}\,\bar{\delta}\cos\theta + \frac{2\alpha}{3Ua^3}\frac{1}{\bar{h}\,\bar{\delta}}}{\bar{h}\sin\theta}. \tag{31}$$

Then (28) for θ small gives the following quadrature:

$$\int \frac{\bar{h}^2\bar{\delta}\,d\bar{h}}{\bar{h}^2\bar{\delta}^2 + \frac{2\alpha}{3Ua^3}} + \ln\theta = \ln C. \tag{32}$$

The integral on the left may be calculated to give

$$\frac{1}{4}\ln\left\{\left[\bar{h}^2 - (\bar{R} + \sqrt{2B})\left(\bar{h} - \sqrt{\frac{B}{2}}\right)\right]\left[\bar{h}^2 - (\bar{R} - \sqrt{2B})\left(\bar{h} + \sqrt{\frac{B}{2}}\right)\right]\right\}$$
$$+ \frac{1}{2}\frac{\bar{R}\sqrt{\frac{B}{2}}}{4B - \bar{R}^2}\ln\frac{\bar{h}^2 - (\bar{R} + \sqrt{2B})\left(\bar{h} - \sqrt{\frac{B}{2}}\right)}{\bar{h}^2 - (\bar{R} - \sqrt{2B})\left(\bar{h} + \sqrt{\frac{B}{2}}\right)}$$
$$+ \frac{\bar{R}}{2}\frac{\sqrt{2B - \bar{R}^2}}{4B - \bar{R}^2}\left(\operatorname{arctg}\frac{\sqrt{2B - \bar{R}^2}}{\bar{R} + \sqrt{2B} - 2\bar{h}} + \operatorname{arctg}\frac{\sqrt{2B - \bar{R}^2}}{\bar{R} - \sqrt{2B} - 2\bar{h}}\right) = \ln\frac{C}{\theta}, \tag{33}$$

in which

$$\bar{R} = \frac{R}{a}; \qquad B = \sqrt{\frac{\bar{R}^4}{16} + \frac{2\alpha}{3Ua^3} + \frac{\bar{R}^2}{4}}. \tag{34}$$

The analytic expression of (33) for the indefinite integral of (32) provides approximations for the integral that correspond to various assumptions about the relative magnitudes of the parameters appearing in the integrand. Such estimates may be made directly with the integrand in (32), but this involves an error because the indefinite integral involves an additive constant. For this reason, the direct estimate may differ by a constant from the value given by (33) for the same range in the parameters.

The problem here is to choose a value for the constant C in (32) or (33) such that the resulting path will be close to the critical path of (22) in the region $\bar{\delta} \ll \bar{h} \approx \bar{R}$. On the other hand, this intermediate path may be linked up with an appropriate stream line in the range of $\bar{\delta}$ sufficiently large, and then ζ can be deduced.

The following approximate relations may be written on the basis of evaluation of the functions on the left in (33):

1) for $\left\{\frac{2\alpha}{3Ua^2}\right\}^{1/4} < \bar{R} \ll \bar{h} \ll 1$

$$\bar{h}\theta = C; \quad h = \frac{Ca}{\theta}; \tag{35}$$

2) for $\bar{h} - \bar{R} \ll \bar{R} \approx \bar{h} \ll 1$

$$\bar{h} = \bar{R} + \left\{\left(\frac{C}{\theta}\right)^2 - \frac{2\alpha}{3R^2U}\right\}^{1/2}; \quad h = R + \left\{\frac{C^2a^2}{\theta^2} - \frac{2\alpha a^2}{3R^2U}\right\}^{1/2}. \tag{36}$$

An important point is that C has the same value in (35) and (36). Of course, the first of these estimates is correct on extrapolating to large $\bar{h}$ and $\bar{\delta}$ the relations for the forces defined by (31) and (32); but the forces at large distances decrease more rapidly than these formulas indicate, so the estimate of (35) applies even earlier (for smaller $\bar{h}$). This feature is not important to us, since it affects only the size of the region where (35) applies. In addition, we note that $\bar{h} \ll 1$ is to be understood in the derivation of (35) and (36), since we have used approximate expressions for the Stokes velocity distribution that apply close to the sphere. The results of (35) and (36) may also be obtained via appropriate approximations for the integrand in (32), i.e., it can be shown that these particular conditions allow us to reverse the sequence of taking the quadrature and making the approximation.

To perform the linkup, we compare (35) and (36) with (14) and (22). Comparison of (36) with (22) gives

$$\frac{C^2a^3}{\theta^2} - \frac{2\alpha a^3}{3R^2U} = \frac{8\alpha a^3}{3R^2U\theta^2}. \tag{37}$$

It is clear that this applies closely for $\theta \ll 1$ if we put

$$C = \frac{2}{R}\sqrt{\frac{2\alpha}{3U}}. \tag{38}$$

On the other hand, (14) and (35) give

$$b = Ca\sqrt{\frac{3}{2}} = 2\frac{a}{R}\sqrt{\frac{\alpha}{U}}. \tag{39}$$

We then get the following expression for ζ:

$$\zeta = \frac{b^2}{a^2} = 4\frac{\alpha}{UR^2} = \frac{A}{9\pi\eta UR^2}. \tag{40}$$

This shows that small particles are deposited more readily than large ones. Two factors in the formulas explain this: 1) the viscous layer produces more resistance to large particles (the resistance is proportional to the square of the radius, while the ratio of the resistance to the Stokes force is proportional to the radius); 2) the centers of small particles come closer to the sphere and enter a region where the flow is slower, and so they are exposed to the van der Waals forces for a longer time.

Also, (40) applies for α/UR^2 sufficiently small, which may have implications for the ζ for very small particles, to which (40) does not apply, since the right part of (20) cannot then become zero for $\theta = \pi$ for the small δ at which the assumed law for the van der Waals forces applies. Therefore, force balance (Stokes and van der Waals forces) for R very small occurs at a critical point lying in a region where the van der Waals forces decrease more rapidly than as the inverse square of the distance from the surface of the sphere. The expression for ζ is correspondingly affected (see below).

More detailed evaluation of the limits of application requires consideration of the conditions that have been imposed on the parameters:

1) $R/a = \bar{R} \ll 1$, in which case we can neglect perturbations due to the particles in the velocity distribution in the medium and can use the relationships for the forces acting on the particles as the sphere is approached.

2) $\text{Re} = Ua\rho/\eta < 1$, which allows us to use the Stokes formulas for the velocity distribution in a medium of density ρ and viscosity η.

3) $\frac{\delta(\pi)}{R} \approx \left\{ \frac{a^2\alpha}{UR^4} \right\}^{1/3} \ll 1$;   a condition that must be obeyed when a particle moves along the critical path near the obstacle, as it insures that we can use the approximations and formulas representing the action of the surface forces on a particle.

4) $\frac{\delta(\pi)}{l} \approx \frac{R}{l} \left\{ \frac{a^2\alpha}{UR^4} \right\}^{1/3} \ll 1$;   in which l is a linear quantity of the order of the wavelength of the electromagnetic waves emitted and absorbed by the molecules of the bodies, and it insures that we can use the δ dependence given for the van der Waals forces.

5) It is shown below that inertial forces can be neglected relative to the other forces near the surface if $\rho_T UR^2/3\eta a \ll 1$, in which ρ_T is the density of a particle.

We thus have the eight parameters R, a, l, U, A, η, ρ, and ρ_T, from which we can produce the five independent dimensionless combinations $\bar{R} = R/a$, $\bar{l} = l/R$, $\text{Re} = Ua\rho/\eta$, $W = A\rho/\eta^2 l$, $\bar{\rho} = \rho_T/\rho$, which should satisfy the above five conditions, which are as follows in terms of the dimensionless quantities:

$$1) \quad \bar{R} \ll 1, \tag{41}$$

$$2) \quad \text{Re} < 1, \tag{42}$$

$$3) \quad \left\{ \frac{W\bar{l}}{\text{Re}\,\bar{R}^3} \right\}^{1/3} \ll 10, \tag{43}$$

$$4) \quad \left\{ \frac{W}{\text{Re}\,\bar{l}\bar{R}^3} \right\}^{1/3} \ll 10, \tag{44}$$

$$5) \quad \bar{\rho}\bar{R}^2 \text{Re} \ll 3. \tag{45}$$

Condition (45) is unimportant, since it is obeyed if (41) and (42) are obeyed.

It is reasonable to take A, η, ρ, l, and a as given and to determine the ranges in $\bar{R}$ and Re, or R and U, for which (40) can be used. We transform (43) and (44) to

$$\frac{A\rho}{\eta^2 a} \ll 10\bar{R}^4 \text{Re}, \tag{43'}$$

$$\frac{A\rho a}{\eta^2 l^2} \ll 10\bar{R}^2 \text{Re}. \tag{44'}$$

Figure 2 shows the hatched region D as that of permissible values of $\bar{R}$ and Re, which is bounded by the straight lines $\text{Re} = 1$ and $\bar{R} = 0.1$ together with parts of the curves

$$\bar{R}^4 \text{Re} = \frac{A\rho}{\eta^2 a}, \tag{46}$$

$$\overline{R}^2 \operatorname{Re} = \frac{A\rho a}{\eta^2 l^2}. \tag{47}$$

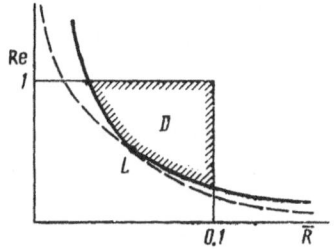

Fig. 2. Region of permissible values of $\overline{R}$ and Re in the absence of electromagnetic delay.

Any point in the hatched region of Fig. 2 represents permissible values of R and U, which can be deduced if we know a and the constants of the liquid. Conversely, the parameters of hyperbolas (46), (47) can be used with a, η, and ρ to test whether (40) is applicable to capture of a particle of radius R from a flow of speed U.

Comparison of (40) with (43') and (44'ʼ) shows that the following inequalities must be obeyed simultaneously:

$$\zeta \ll \frac{R^2}{a^2\pi}, \qquad \zeta \ll \frac{l^2}{\pi a^2}. \tag{48}$$

Then it follows from (41) that ζ is always small at the limits of applicability of (40).

Now (43') and (44') are such that the lower boundary of region D is formed by the hyperbola of (46) or (47) that runs above the other one within the rectangle in Fig. 2. If the point of intersection L of the hyperbolas lies within the rectangle (as in Fig. 2), the boundary to the right of L is given by (47), while that to the left is given by (46). The coordinates of L are defined by

$$\overline{R}_L = \frac{l}{a}, \qquad \operatorname{Re}_L = \frac{A\rho a^3}{\eta^2 l^4}. \tag{49}$$

Point L lies within the rectangle if

$$\frac{l}{a} < 0.1; \quad \frac{A\rho a^3}{\eta^2 l^4} < 1. \tag{50}$$

The inequalities of (50) impose the following restrictions on a:

$$10\,l < a < \left\{\frac{\eta^2 l^4}{A\rho}\right\}^{\frac{1}{3}}. \tag{51}$$

If L lies outside the rectangle in the region with $\overline{R} > 0.1$ and Re < 1, the lower boundary of D is given by (46), which occurs when

$$\frac{l}{a} > 0.1; \quad \frac{A\rho a^3}{\eta^2 l^4} < 1. \tag{52}$$

If the hyperbola of (46) is to intersect the rectangle at all, we must have as follows, with $\overline{R} = 0.1$:

$$\operatorname{Re} = \frac{A\rho}{\overline{R}^4 \eta^2 a} < 1, \text{ i.e., } \frac{10^4 A\rho}{\eta^2 a} < 1. \tag{53}$$

Then a must satisfy

$$\frac{10^4 A\rho}{\eta^2} < a < \begin{cases} 10\,l, \\ \left\{\frac{\eta^2 l^4}{A\rho}\right\}^{\frac{1}{3}}. \end{cases} \tag{54}$$

Finally, if L lies in the region with $\overline{R} < 0.1$ and Re > 1, the lower boundary of D is formed by the hyperbola of (47), and we have

$$\frac{l}{a} < 0.1 ; \; \frac{A\rho a^3}{\eta^2 l^4} > 1. \tag{55}$$

The condition for this hyperbola to intersect the rectangle when $\bar{R} = 0.1$ is

$$\mathrm{Re} = \frac{A\rho a}{\eta^2 l^2 \bar{R}^2} < 1, \; \text{i.e.,} \; \frac{10^2 A\rho a}{\eta^2 l^2} < 1. \tag{56}$$

Then a must satisfy

$$\left. \begin{array}{r} 10l \\ \left\{ \dfrac{\eta^2 l^4}{A\rho} \right\}^{1/3} \end{array} \right\} < a < \frac{\eta^2 l^2}{10^2 A\rho} . \tag{57}$$

If the four quantities $\frac{10^4 A\rho}{\eta^2}$, $10l$, $\left\{ \frac{\eta^2 l^4}{A\rho} \right\}^{1/3}$, $\frac{\eta^2 l^2}{10^2 A\rho}$ (all of which have the dimensions of length) are arranged in increasing order, the intervals of (54), (51), and (57) together form a continuous interval stretching from relatively small a $\left(\frac{10^4 A\rho}{\eta^2} < a < 10l \right)$, via medium a $\left(10l < a < \left\{ \frac{\eta^2 l^4}{A\rho} \right\}^{1/3} \right)$ to large a $\left(\left\{ \frac{\eta^2 l^4}{A\rho} \right\}^{1/3} < a < \frac{\eta^2 l^2}{10^2 A\rho} \right)$. If the order of disposition of the first two numbers is altered, the range for small a is lost; if that of the second and third, the range for medium a is lost; and if that of the last two, the range for a large a is lost.

It is readily seen that for a liquid of low viscosity, e.g., water, for which we suppose $A \approx 10^{-12}$ erg, $\eta = 10^{-2}$ dyn·sec/cm^2, $\rho = 1$ g/cm^3, and $l = 10^{-5}$ cm, we find that all four of the above quantities are about 10^{-4} cm. This means that the hyperbolas meet the rectangle only very near the top right corner and that the only permissible ranges are $a \approx 10^{-4}$ cm and $R \approx 10^{-5}$ cm.

Region D expands if η is increased and ρ is reduced. Moreover, we take the view that the restriction Re < 1 is not absolute and can be replaced by Re$\bar{R}^2 \ll 1$ by considering the inertial and viscous terms in the equation of motion of the liquid near the obstacle, which greatly increases the permissible range in Re. This aspect will be considered in detail in a forthcoming paper.

The above results for a and R for water will apply in the deposition of particles differing little in size when capture is the result of differing speeds of fall. These calculations are not applicable to filtration of a suspension through a filter with comparatively coarse pores, although this case is of considerable technical interest in relation to water purification, wall sealing in irrigation channels, and so on. We consider such cases on the assumption that the critical path almost everywhere lies at distances from the surface such that the molecular attraction is proportional to $1/\delta^3$ (with $\delta \ll R$) on account of electromagnetic delay. It is known [12–15] that for this it is sufficient to have $\delta(\pi) \geqslant 10^{-5}$ cm.

Instead of (5) we now have

$$f = \frac{\mu_1}{\delta^3} + \frac{\beta_1}{\delta} \frac{dh}{dt} ; \; h - R = \delta \ll R \approx h \ll a. \tag{58}$$

where

$$\mu_1 = \frac{2\pi RK}{3} , \tag{59}$$

in which K, which characterizes the attraction, may be given [12–15] the rough value

$$K = 10^{-19} \; \text{erg·cm}. \tag{60}$$

Instead of (20) we get here

$$\sin \theta \cdot \frac{d\delta}{d\theta} = -\delta \cdot \cos \theta - \frac{2\mu a^2}{3R^2 U \delta^2}, \tag{61}$$

in which $\mu = \mu_1 / k$.

The general integral of (61) takes the form

$$\delta = \left\{ -\frac{3}{2} \lambda \frac{\theta - \frac{1}{2} \sin 2\theta + C}{\sin^3 \theta} \right\}^{\frac{1}{3}}, \tag{62}$$

$$\lambda = \frac{2}{3} \frac{\mu a^2}{U R^2}.$$

$C = -\pi$ for the critical path, so (22) is replaced by the following expression for the critical path:

$$h = R + \left\{ -\frac{3}{2} \lambda \frac{\theta - \frac{1}{2} \sin 2\theta - \pi}{\sin^3 \theta} \right\}^{\frac{1}{3}}. \tag{63}$$

Passage to the limit $\theta \to \pi$ gives

$$\delta(\pi) = h(\pi) - R = \lambda^{\frac{1}{3}} = \left\{ \frac{2}{3} \frac{\mu a^2}{U R^2} \right\}^{\frac{1}{3}}. \tag{64}$$

We link the critical path to a stream line by the previous method, which is represented by formulas (30)–(39). We use the symbols used in (30) to get a relation that replaces (32):

$$\int \frac{\bar{h}^2 \bar{\delta}^2 \, d\bar{h}}{\bar{h}^2 \bar{\delta}^3 + \omega} + \ln \theta = \ln C, \tag{65}$$

in which

$$\omega = \frac{2\mu}{3U a^3}.$$

Here it is difficult to obtain a general analytical expression for the quadrature of (65), so we proceed by analogy with the calculations for (35)–(39), for which it may be shown to be permissible to reverse the operations of approximation and taking the quadrature. This gives the following approximate relations, which are suitable for $\theta \ll 1$ and $h \ll a$:

1) for $\left\{ \frac{2\mu}{3U a^3} \right\}^{-\frac{1}{5}} < \bar{R} \ll \bar{h} \ll 1$

$$\bar{h}\theta = C; \; h = \frac{Ca}{\theta}; \tag{66}$$

2) for $\bar{\delta} = \bar{h} - \bar{R} \ll \bar{R} \approx \bar{h} \ll 1$

$$\bar{h} = \bar{R} + \left\{ \left(\frac{C}{\theta} \right)^3 - \frac{2\mu}{3U R^2 a} \right\}^{\frac{1}{3}}; \; h = R + \left\{ \frac{C^3 a^3}{\theta^3} - \frac{2\mu a^2}{3U R^2} \right\}^{\frac{1}{3}}. \tag{67}$$

Comparison of the $\delta = h - R$ given by (63) and (67) for $\theta \ll 1$ gives

$$\frac{C^3 a^3}{\theta^3} - \frac{2\mu a^2}{3UR^2} = \frac{3}{2}\,\pi\,\frac{\lambda}{\theta^3} = \pi\,\frac{\mu a^3}{UR^2}\frac{1}{\theta^3}. \tag{68}$$

This equation is obeyed closely in the region $\theta \ll 1$ if

$$C = \left\{\pi\,\frac{\mu}{UR^2 a}\right\}^{\frac{1}{3}}. \tag{69}$$

On the other hand, comparison of (14) and (66) gives

$$b = Ca\,\sqrt{\frac{3}{2}} = \sqrt{\frac{3}{2}}\left\{\pi\,\frac{\mu a^2}{UR^2}\right\}^{\frac{1}{3}}. \tag{70}$$

The ζ applicable to this case is

$$\zeta = \frac{b^2}{a^2} = \frac{3}{2}\left\{\frac{\pi\mu}{UR^2 a}\right\}^{\frac{2}{3}} = \frac{3}{2}\left\{\frac{\pi K}{9\eta UR^2 a}\right\}^{\frac{2}{3}}. \tag{71}$$

As we have noted, the result of (71) applies if $L < \delta(\pi) \ll R$, in which $L \approx 10^{-5}$ cm. The lower bound to $\delta(\pi)$ arises because only for $\delta(\pi) > L$ can we use the limiting one-term formula for the van der Waals force that follows from complete allowance for the effects of electromagnetic delay, which weakens the attraction. Then, from (64), we must have that

$$L < \left\{\frac{2\mu a^2}{3UR^2}\right\}^{\frac{1}{3}} \ll R, \tag{72}$$

or, in terms of $\overline{R} = \dfrac{R}{a}$, $\mathrm{Re} = \dfrac{Ua\rho}{\eta}$, $\overline{K} = \dfrac{K\rho}{\eta^2 R^2}$, $\overline{L} = \dfrac{L}{a}$, we have

$$\overline{L} < \left\{\frac{2}{27}\frac{\overline{K}}{\mathrm{Re}}\right\}^{\frac{1}{3}} \ll \overline{R}. \tag{73}$$

Conditions (73) may also be put as follows:

$$\mathrm{Re}\,\overline{R}^5 > \frac{2}{27}\cdot 10^3\,\frac{K\rho}{\eta^2 a^2}, \tag{74}$$

$$\mathrm{Re}\,\overline{R}^2 < \frac{2}{27}\,\frac{K\rho a}{\eta^2 L^3}. \tag{75}$$

To these we must add conditions (41) and (42).

Region D (Fig. 3) in the plane of $\overline{R}$ and Re is therefore bounded by the straight lines Re $= 1$ and $\overline{R} = 0.1$ together with the two hyperbolas

$$\mathrm{Re}\,\overline{R}^5 = \frac{2}{27}\cdot 10^3\,\frac{K\rho}{\eta^2 a^2}, \tag{76}$$

$$\mathrm{Re}\,\overline{R}^2 = \frac{2}{27}\,\frac{K\rho a}{\eta^2 L^3}. \tag{77}$$

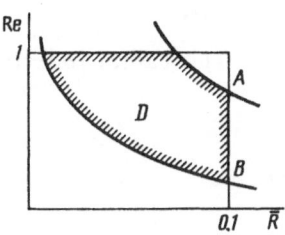

Fig. 3. Region of permissible values of $\bar{R}$ and Re in maximally pronounced electromagnetic delay.

These hyperbolas move apart, and the permissible region extends, as a increases (for given properties of the liquid).

Region D exists if the following two conditions are met:

1) The hyperbola of (76) must intersect the rectangle in Fig. 3, so for $\bar{R} = 0.1$ we have

$$\mathrm{Re} = \frac{2}{27} \cdot 10^3 \frac{K\rho}{\bar{R}^5 \eta^2 a^2} = \frac{2}{27} \cdot 10^8 \frac{K\rho}{\eta^2 a^2} < 1, \tag{78}$$

and so

$$a > 10^4 \left\{ \frac{2}{27} \frac{K\rho}{\eta^2} \right\}^{\frac{1}{2}}. \tag{79}$$

2) The hyperbola of (77) must lie above that of (76), at least in some part of the rectangle in Fig. 3. The ratio of the ordinates of the hyperbolas for a given $\bar{R}$ is

$$\frac{a^3 \bar{R}^3}{10^3 L^3} = \frac{R^3}{10^3 L^3}.$$

The second condition is therefore equivalent to the inequality

$$R > 10L, \tag{80}$$

which coincides with condition (72), $L < \delta \ll R$; it shows that a is subject to the following restriction:

$$a > \frac{10L}{\bar{R}}; \tag{81}$$

as $L \approx 10^{-5}$ cm, $a = 10^{-3}$ cm is the lower limit at which region D still exists, if condition (79) is met.

Condition (79) for water gives $a > 10^{-4}$ cm, so the only remaining restriction is that of (81), and hence region D for water is reasonably large only if

$$a \gg 10^{-3} \text{ cm}, \tag{82}$$

e.g., if $a \geq 10^{-2}$ cm.

In fact, we get from (76) and (77) that the ordinates of points B and A in Fig. 3 for $K \approx 10^{-19}$ erg $\cdot$ cm, $\rho \approx 1$ g/cm^3, and $L \approx 10^{-5}$ cm are

$$\mathrm{Re}_B \approx \frac{10^{-12}}{a^2 \eta^2}, \tag{83}$$

$$\mathrm{Re}_A \approx 10^{-3} \frac{a}{\eta^2}. \tag{84}$$

Then if $\eta = 10^{-2}$ dyn $\cdot$ sec/cm^2 and $a \geq 10^{-2}$ cm we have

$$\mathrm{Re}_B \leqslant 10^{-3}, \quad \mathrm{Re}_A \gg 10^{-2}. \tag{85}$$

Comparison of (71) with (74) and (75) gives us the limits for ζ:

$$\left(\frac{3}{2}\right)^{\frac{5}{3}} \pi^{\frac{2}{3}} \frac{L^2}{a^2} < \zeta < \left(\frac{3}{2}\right)^{\frac{5}{3}} \pi^{\frac{2}{3}} \frac{\bar{R}^2}{10^2}. \tag{86}$$

This means that ζ is always very small within the limits of application of (71).

Finally, we consider the inertial forces, which were ignored above. The radial inertial force j_r and transversal inertial force j_θ are

$$j_r = m\left[\frac{d^2h}{dt^2} - r\left(\frac{d\theta}{dt}\right)^2\right], \qquad j_\theta = m\left(r\frac{d^2\theta}{dt^2} + 2\frac{dh}{dt}\frac{d\theta}{dt}\right),$$

in which the mass of a particle (density ρ_T) is $m = \frac{4}{3}\pi R^3 \rho_T$.

Then (4) and (21) give

$$\frac{dh}{dt} = \frac{d\delta}{dt} = \frac{d\delta}{d\theta}\frac{d\theta}{dt} = \frac{v_\theta}{r}\frac{d\delta}{d\theta} \approx \frac{v_\theta}{a}\delta,$$

$$\frac{d^2h}{dt^2} = \frac{d}{dt}\left(\frac{v_\theta}{a}\right)\cdot\delta + \frac{v_\theta}{a}\frac{d\delta}{d\theta}\frac{d\theta}{dt} \approx \left[\frac{\partial}{\partial\theta}\left(\frac{v_\theta}{a}\right)\frac{d\theta}{dt} + \frac{\partial}{\partial h}\left(\frac{v_\theta}{a}\right)\frac{dh}{dt}\right]\delta + \frac{v_\theta^2}{a^2}\delta$$

$$\approx \left(\frac{v_\theta^2}{a^2} + \frac{v_\theta^2}{a^2}\frac{\delta}{R}\right)\delta + \frac{v_\theta^2}{a^2}\delta \approx \frac{v_\theta^2}{a^2}\delta,$$

$$r\left(\frac{d\theta}{dt}\right)^2 \approx a\frac{v_\theta^2}{a^2} = \frac{v_\theta^2}{a},$$

$$r\frac{d^2\theta}{dt^2} = r\left[\frac{\partial}{\partial\theta}\left(\frac{d\theta}{dt}\right)\frac{d\theta}{dt} + \frac{\partial}{\partial h}\left(\frac{d\theta}{dt}\right)\cdot\frac{dh}{dt}\right] \approx a\left[\frac{v_\theta^2}{a^2} + \frac{v_\theta^2}{a^2}\frac{\delta}{R}\right] \approx \frac{v_\theta^2}{a},$$

$$2\frac{dh}{dt}\frac{d\theta}{dt} \approx 2\frac{v_\theta^2}{a^2}\delta.$$

Then (4) gives

$$mj_r \approx mj_\theta \approx m\frac{v_\theta^2}{a} \approx 3\pi\rho_T U^2 \frac{R^5}{a^3}.$$

Further, as $v_r \approx v_\theta\frac{R}{a}$, $\frac{dh}{dt} \approx v_\theta\frac{\delta}{a}$, we get the correct value for the ratio of the radial inertial and Stokes forces as

$$\frac{mj_r}{kv_r} \approx \frac{\rho_T U R^2}{3\eta a}.$$

It is reasonable to neglect the inertial forces if

$$\frac{\rho_T U R^2}{3\eta a} \ll 1, \tag{87}$$

which is the condition for noninertial deposition during Stokes flow around a spherical obstacle. Inequality (87) becomes as follows in dimensionless quantities:

$$\bar{\rho}\bar{R}^2\,\text{Re} \ll 3.$$

Stokes flow (Re < 1) thus allows noninertial deposition if R is small relative to a.

It is clear from the above that inertial deposition of an aerosol can be neglected for small particles, but there is also the important diffusion (Brownian) mechanism [16], and Brownian motion is very vigorous for small particles. It is thus of interest to compare our results for pure van der Waals deposition with those for diffusion deposition.

The diffusion flux I_D of particles to a sphere of radius a in a Stokes flow of speed U is [17] given by

$$I_D = 8c_0 D^{3/2} U^{1/2} a^{4/3},\tag{88}$$

in which c_0 is the particle concentration far from the sphere and D is the diffusion coefficient; if for the latter we use Einstein's [18] expression,

$$D = \tilde{k}T / 6\pi\eta R,\tag{89}$$

in which $\tilde{k}$ is Boltzmann's constant and T is absolute temperature, then

$$I_D = 8c_0(\tilde{k}T)^{2/3} U^{1/2} a^{4/3}(6\pi\eta R)^{-2/3}.\tag{90}$$

This result may be compared with I_{W_I} and $I_{W_{II}}$, the results for van der Waals deposition with and without electromagnetic delay. Using (40) and (71), we have

$$I_{W_I} = c_0\pi a^2 U\zeta = \frac{c_0 A a^2}{9\eta R^2},\tag{91}$$

$$I_{W_{II}} = c_0\pi a^2 U \cdot \frac{3}{2}\left\{\frac{\pi K}{9\eta U R^2 a}\right\}^{2/3} = c_0\pi^{5/3} \cdot \frac{3}{2} \cdot (9)^{-2/3} a^{4/3} U^{1/3} \eta^{-2/3} R^{-4/3} K^{2/3}.\tag{92}$$

We form the ratios

$$I_{W_I} / I_D = \{(6\pi)^{2/3} A a^{2/3}\}/\{8 \cdot 9\,(\tilde{k}T)^{2/3} U^{1/2} \eta^{1/3} R^{4/3}\},\tag{93}$$

$$I_{W_{II}} / I_D = \left\{\pi^{7/3} \cdot 3 \cdot \left(\frac{2}{3}\right)^{2/3} \cdot K^{2/3}\right\} \Big/ \{2 \cdot 8 \cdot (\tilde{k}T)^{2/3} R^{2/3}\}.\tag{94}$$

These ratios may be evaluated using the permissible a and R found above for water; both of them are around unity if $A \approx 10^{-12}$ erg, $K \approx 10^{-19}$ erg $\cdot$ cm, $\tilde{k}T \approx 4 \times 10^{-14}$ erg, where in the first case we have taken $a \approx 10^{-4}$ cm, $R \approx 10^{-5}$ cm, and Re ≈ 1, while in the second we have taken $R \approx 10^{-4}$ cm. The diffusion and van der Waals effects are thus of the same order, and account must be taken of both.

I_D, $I_{W_{II}}$ and I_{W_I} vary in different ways with R, U, η, and a:

I_D	$I_{W_{II}}$	I_{W_I}
$R^{-2/3}$	$R^{-4/3}$	R^{-2}
$U^{1/2}$	$U^{1/3}$	U^0
$\eta^{-2/3}$	$\eta^{-2/3}$	η^{-1}
$a^{4/3}$	$a^{4/3}$	a^2

The behavior of I_D is almost the same as that of $I_{W_{II}}$, but the dependence on R shows that the van der Waals mechanism becomes relatively more effective as R decreases.

Conclusions

1) Noninertial deposition without surface attraction is impossible, on account of the viscous resistance of the liquid between sphere and particle.

2) Simple analytic expressions can be derived for the critical path near the sphere and for the capture coefficient in the case of van der Waals attraction of uncharged particles in the presence of viscous resistance. These expressions have been derived for two cases, which correspond to different degrees of approach of the critical path to the surface of the obstacle and to incorporation or neglect of electromagnetic delay.

3) In both cases, the capture coefficient for van der Waals attraction increases as the particle size and flow speed decrease for particles in a certain size range, but the coefficient is always small.

4) The permissible range for the parameters is fairly narrow for water and other liquids of low viscosity if the critical path passes so close to the sphere that delay may be neglected. The permissible range increases with the viscosity. If the critical path runs at distances from the sphere such that electromagnetic delay must be considered, the permissible range is fairly wide if the sphere is reasonably large.

5) The van der Waals mechanism has been compared with the inertial and-diffusion mechanisms; it has been found that inertial effects can be neglected for deposition of small particles from a flow having a low Reynolds number. On the other hand, the diffusion and van der Waals fluxes may be comparable when the liquid is water. The performance of the van der Waals mechanism increases more rapidly than that of diffusion as the particle size decreases.

References

1. B. V. Deryagin and L. P. Smirnov, Kolloidn. Zh., 29:400 (1967).
2. G. L. Natanson, Dokl. AN SSSR, 112:696 (1957).
3. N. N. Tunitskii and I. V. Petryanov, Zh. Fiz. Khim., 17:408 (1943).
4. H. C. Hamaker, Physica, 4:1058 (1937).
5. H. B. G. Casimir and D. Polder, Phys. Rev., 73:360 (1948).
6. H. Lamb, Hydrodynamics, 6th edition (1932), Dover, New York.
7. T. Kihara, Adv. Chem. Phys., 1:261 (1958).
8. E. M. Lifshits, Zh. Éksp. Teor. Fiz., 29:99 (1955).
9. B. V. Deryagin, Zh. Fiz. Khim., 6:1036 (1935); Kolloidn. Zh.,69:155 (1934).
10. B. V. Deryagin and N. A. Krotova, Adhesion, Izd. AN SSSR, Moscow (1949).
11. I. E. Dzyaloshinskii, E. M. Lifshits, and L. P. Pitaevskii, Zh. Éksp. Teor. Fiz., 37:229 (1959).
12. B. V. Deryagin, I. I. Abrikosova, and E. M. Lifshits, Usp. Fiz. Nauk, 64:493 (1958).
13. B. V. Derjaguin (Deryagin), J. J. Abrikosova, and E. M. Lifshitz (Lifshits), Quart. Rev., 10:295 (1956); B. V. Derjaguin (Deryagin) and J. J. Abrikosova, J. Phys. Chem. Solids, 5:1 (1958); Zh. Éksp. Teor. Fiz., 30(6):993 (1956); 31(1):3 (1957).
14. J. A. Kitchener and A. P. Prosser, Proc. Roy. Soc., 242:403 (1957).
15. J. H. Schenzel and J. A. Kitchener, Trans. Faraday Soc., 56:161 (1960).
16. N. A. Fuks, Mechanics of Aerosols, Izd. AN SSSR, Moscow (1955).
17. V. G. Levich, Physicochemical Hydrodynamics, Fizmatgiz, Moscow (1959).
18. A. Einstein, Ann. Physik, 17:549 (1905).

COAGULATION OF LYOPHOBIC SOLS
BY ELECTROLYTE MIXTURES

V. M. Barboi and Yu. M. Glazman

Institute of Light-Industry Technology, Kiev

Calculations are presented to show that a purely concentration mechanism does not apply to the coagulation produced by the addition of electrolytes; adsorption effects play a major part.

Symbols

n_1 and n_2	numbers of ions of the first and second electrolytes per unit volume
n_1' and n_2'	the same for the ions of the same charge as the colloid
z_1 and z_2	valencies of ions n_1 and n_2
z_1' and z_2'	valencies of ions n_1' and n_2'
ε	dielectric constant
ψ	potential at any point of the bulk
ψ_a	potential determining the Gouy−Chapman ion distribution
ψ_0	potential at the symmetry plane between two interacting particles
e	electronic charge
$\theta = kT$	product of Boltzmann's constant and absolute temperature

$u = \exp(ez_1\psi/\theta);\; u_a = \exp(ez_1\psi_a/\theta);\; u_0 = \exp(ez_1\psi_0/\theta);$

$f_1(u) = u + (z_1/z_1')\, u^{-z_1'/z_1} - (z_1 + z_1')/z_1';$

$f_2(u) = u^{z_2/z_1} + (z_2/z_2')\, u^{-z_2'/z_1} - (z_2 + z_2')/z_2';$

$f_1(u_0)$ and $f_2(u_0)$ values of $f_1(u)$ and $f_2(u)$ at $u = u_0$;

$B = 4.5\theta^5\, \varepsilon^3 \pi^{-5}\, A^{-2} e^{-6},\;\; m = n_1/n_2$

A   van der Waals constant

$$\rho_1^2(u) = f_1(u) - f_1(u_0); \quad \rho_2^2(u) = f_2(u) - f_2(u_0);$$

$$I = \int_{u_0}^{u_a} \frac{du}{u\sqrt{\rho_1^2(u) - m\rho_2^2(u)}};$$

$$I_1 = \int_{u_0}^{u_a} \frac{u\left[\dfrac{df_1(u)}{du} - \dfrac{df_1(u_0)}{du_0} + m\dfrac{df_2(u)}{du} - m\dfrac{df_2(u_0)}{du_0}\right] + 2\left[\rho_1^2(u) - \rho_2^2(u)\right]}{u^2\left[\rho_1^2(u) + m\rho_2^2(u)\right]^{3/2}}\, du.$$

Subscript zero (above) denotes a quantity relating to the action of an individual electrolyte.

$$I_{10} = \lim_{m=0} I; \qquad u_{00} = \lim_{m=0} u_0;$$

$$I_{20} = \lim_{m=0} I_2; \qquad u_{a0} = \lim_{m=0} u_a;$$

$$I_2 = \int_{u_0}^{u_a} \frac{\rho_2^2(u)\, du}{u\left[\rho_1^2(u) + m\rho_2^2(u)\right]^{3/2}};$$

$F(\varphi_1; k)$ elliptic integral of the first kind

$E(\varphi_1; k)$ elliptic integral of the second kind

$K(k)$ complete elliptic integral of the first kind

$E(k)$ complete elliptic integral of the second kind

Coagulation of lyophobic sols by mixtures of electrolytes has been a fundamental problem in colloid chemistry for many years [1]. The observed regularities are so varied that no interpretation was possible until a quantitative theory of ion–stabilized sytems became available [2, 3]. Deryagin's concepts are used here to consider the concentration mechanism of coagulation for this case.

The curve for coagulation by a mixture of two electrolytes is calculated from the general criterion[†] for the stability of an ion-stabilized system [4]:

$$n_1 = n_2 m^{-1} = Bz_1^{-6}\left[f_1(u_0) + mf_2(u_0)\right]^2 I^6, \tag{1}$$

$$I\frac{d\ln\left[f_1(u_0) + mf_2(u_0)\right]}{du_0} = \frac{3}{2}I_1 + \frac{3}{u_a\sqrt{\rho_1^2(u_a) + m\rho_2^2(u_a)}}. \tag{2}$$

The general nature of the curve, i.e., the course at its ends, may be derived [5] by considering

$$\Phi = \lim_{n_2=0} \frac{dn_1}{dn_2}.$$

As the subscripts 1 and 2 may be assigned to either of the electrolytes, the formula is applicable to both ends of the curve. $\Phi > 0$ corresponds to antagonism, while for superadditivity

$$0 > \Phi > -\frac{n_1^0}{n_2^0},$$

and for synergism

$$\Phi < -\frac{n_1^0}{n_2^0}.$$

_____

[†] Here and subsequently, all quantities refer to the critical state of the system (coagulation value).

Differentation of (1) with respect to m gives

$$\frac{d \ln n_1}{dm} = 6 \, \frac{d \ln I}{dm} + \frac{\dfrac{df_1(u_0)}{dm} + m \dfrac{df_2(u_0)}{dm} + f_2(u_0)}{f_1(u_0) + mf_2(u_0)}. \tag{3}$$

But

$$\frac{dI}{dm} = \frac{\partial I}{\partial m} + \frac{\partial I}{\partial u_0} \frac{du_0}{dm} + \frac{\partial I}{\partial u_a} \frac{du_a}{dm},$$

and so

$$\frac{d \ln n_1}{dm} = 6 \left[\frac{\partial \ln I}{\partial m} + \frac{1}{3} \frac{\partial \ln [f_1(u_0) + mf_2(u_0)]}{\partial u_0} \right] + 6 \frac{\partial \ln I}{\partial m} + 6 \frac{\partial \ln I}{\partial u_a} \frac{du_a}{dm} + \frac{2f_2(u_0)}{f_1(u_0) + mf_2(u_0)}.$$

The expression in square brackets [4] is equal to zero, and

$$\frac{\partial I}{\partial m} = -\frac{1}{2} \int_{u_0}^{u_a} \frac{\rho_2^2(u) \, du}{u \, [\rho_1^2(u) + m\rho_2^2(u)]^{3/2}} = -\frac{1}{2} I_2,$$

$$\frac{\partial I}{\partial u_a} = \frac{1}{u_a \, \sqrt{\rho_1^2(u_a) + m\rho_2^2(u_a)}}; \quad \frac{dn_1}{dn_2} = \frac{1}{m + \dfrac{dm}{d \ln n_1}},$$

so

$$\Phi = \lim_{m=0} \frac{d \ln n_1}{dm} = 2 \, \frac{f_2(u_{00})}{f_1(u_{00})} - \frac{3I_{20}}{I_{10}} + \frac{6 \lim\limits_{m=0} \dfrac{du_a}{dm}}{u_{10} \, \sqrt{f_1(u_{10}) - f_1(u_{00})}}. \tag{3a}$$

The stability of an ion-stabilized system may be lost on addition of electrolytes either solely as a result of compression of the double electrical layer around a colloidal particle (concentration effect) or because this is accompanied by reduction in the charge and potential of the particle (neutralization effect). This means that there can be various reasons for deviation from additivity in the action of electrolyte mixtures: interaction of diffuse ion atmospheres [6, 7] and various adsorption effects [5, 8]. If the potential of a colloid particle is high at the instant of coagulation ($\overline{\psi}_a > 200$ mV) [9], adsorption will not affect the stability, and purely concentration coagulation occurs. Of course, a concentration mechanism also acts at low potentials if no adsorption occurs on addition of the electrolytes. Then (3a) becomes

$$\Phi = \frac{2f_2(u_{00})}{f_1(u_{00})} - \frac{3I_{20}}{I_{10}}. \tag{4}$$

Integrals I_{10} and I_{20} can be reduced to elementary functions and tabulated quadratures for many electrolytes [10].

For instance, we have [6, 11] for the coagulation of a highly charged sol by a mixture of a 2-2 electrolyte with a 1-1 one that

$$\Phi = \frac{3 \, \sqrt{u_{00}} \, (u_{00} + 1)}{(u_{00} - 1)^2 \, K\left(\dfrac{2 \, \sqrt{u_{00}}}{u_{00} + 1}\right)} - \frac{uu_{00}}{(u_{00} - 1)^2} = -0.008.$$

For a mixture of a 2-2 electrolyte with a 1_2-2 one [7, 12]

$$\Phi = \frac{3 \, \sqrt{u_{00}} \, (u_{00}^2 + 1)}{(u_{00}^2 - 1)(u_{00} - 1) K\left(\dfrac{2 \, \sqrt{u_{00}}}{u_{00} + 1}\right)} - \frac{u_{00}^2 + 6u_{00} + 1}{u \, (u_{00} - 1)} = 0.044.$$

TABLE 1. Calculated Values of Φ for the Coagulation of Highly Charged Sols by a Mixture of 1_n-n and $2_{m/2}$-m Electrolytes

m \\ n	1	2	3	4	6	8	∞
1	0.007	0.072	0.122	0.156	0.193	0.212	0.322
2	—0.008	0.044	0.090	0.125	0.162	0.182	0.294
∞	—0.162	—0.147	—0.120	—0.095	—0.050	—0.010	+0.103

Analogous formulas have been derived for mixtures of electrolytes of other types. Table 1 gives results [12] for various mixtures, which indicate that the charges of all the ions are involved for coagulation of highly charged sols.

For mixtures such as $2\text{-}1_2 + 1\text{-}1$, $2\text{-}2 + 1_2\text{-}2$, and $2_{m/2}\text{-}m + 1_m\text{-}m$ for $m \gg 2$, there is always pronounced antagonism for a relatively small concentration of the electrolyte with the smaller coagulating action. The positive deviations from additivity increase as n increases and as m decreases. This result gives a rigorous theoretical basis for the "relieving effect" [13], which is due to the electrostatic effect of the ions having charges of the same sign as those on the colloidal particles. Exact calculations have not proved possible for electrolyte mixtures with $z_1/z_2 > 2$, but an adequate approximation is available in such cases.

If a 1–1 electrolyte is mixed with a 3–3 one, we have

$$I_{20} = \int_{u_{\infty}}^{\infty} \frac{u^{1/3} - u_{00}^{1/3} + u^{-1/3} - u_{00}^{-1/3}}{u + u^{-1} - u_{00} - u_{00}^{-1}} \frac{du}{\sqrt{u(u - u_{00})(u - u_{00}^{-1})}}.$$

It is readily shown that

$$\frac{u^{1/3} - u_{00}^{1/3} + u^{-1/3} - u_{00}^{-1/3}}{u^{1/2} - u_{00}^{1/2} + u^{-1/2} - u_{00}^{-1/2}} < \frac{2}{3} u_{00}^{1/6} \frac{u_{00}^{2/3} - 1}{u_{00} - 1}$$

and so

$$I_{20} < \frac{2}{3} u_{00}^{1/6} \frac{u_{00}^{2/3} - 1}{u_{00} - 1} \int_{u_{\infty}}^{\infty} \frac{u^{1/2} + u^{-1/2} - u_{00}^{1/2} - u_{00}^{-1/2}}{u + u^{-1} - u_{00} - u_{00}^{-1}} \frac{du}{\sqrt{u(u - u_{00})(u - u_{00}^{-1})}}.$$

The integral on the right in this inequality has been calculated [10] and equals $0.131 I_{10}$, so

$$\Phi > 2 \frac{u_{00}^{1/3} + u_{00}^{-1/3}}{u_{00} + u_{00}^{-1} - 2} - 0.262 u_{00}^{1/6} \frac{u_{00}^{2/3} - 1}{u_{00} - 1} = 0.024.$$

Here $n_2^0/n_1^0 = 729$, so it is readily found that such a mixture should show virtually the maximum antagonism, i.e., the coagulation curve virtually coincides with the ordinate axis. Similar results have been obtained for all mixtures of n–n electrolytes with 1_n-n ones (n > 2).

To calculate Φ when the electrolyte is unsymmetrical and has a counterion of high valency, we transform (4) to

$$\frac{z_2}{z_1} \Phi = x(z_1) - y(z_1) - x(z_1') + y(z_1'),$$

TABLE 2. Calculated $x(z_1) - y(z_1)$ and $x(z_1') - y(z_1')$

$\dfrac{z_2'}{z_2}$	$x(z_1) - y(z_1)$		$x(z_1') - y(z_1')$						
	$\dfrac{z_1}{z_2} = 0.5$	$\dfrac{z_1}{z_2} = 1$	$\dfrac{z_1'}{z_2} = 0.5$	$\dfrac{z_1'}{z_2} = 1$	$\dfrac{z_1'}{z_2} = 1.5$	$\dfrac{z_1'}{z_2} = 2.0$	$\dfrac{z_1'}{z_2} = 3.0$	$\dfrac{z_1'}{z_2} = 4.0$	$\dfrac{z_1'}{z_2} \gg 4$
0.5	0.644	−0.371	0.629	0.500	0.400	0.331	0.258	0.220	0.000
0.1	0.588	−0.500	0.604	0.500	0.408	0.339	0.265	0.225	0.000
∞	0.206	−1.00	0.529	0.500	0.468	0.396	0.306	0.227	0.000

in which

$$x(z_1) = \frac{2}{j_1(u_{00})} \frac{u_{00}^k - 1}{k}, \quad y(z_1) = \frac{3}{I_{10}} \int\limits_{u_{00}}^{\infty} \frac{u^2(u^l - u_{00}^l)}{k} \frac{du}{\rho^3 u},$$

$$\rho^2(u) = u(u - u_{00})(u - u_{00}^{-1}); \quad k = z_1/z_2.$$

The integrals $x(z_1)$ and $x(z_1')$ have been calculated [10] for $z_1/z_2 = 0.5$, $z_1/z_2 = 1$, and $z_1/z_2 \gg 1$. Table 2 gives $x(z_1) - y(z_1)$ and $x(z_1') - y(z_1')$, from which Φ is readily determined. For a $n-1_n$ electrolyte mixed with a 1-1 one, there should also be limiting antagonism for $n \geq 3$.

To find the shape of the coagulation curve at the other end, we transform (4) by putting $u = x^{z_1}$ and $u_{00} = x_0^{z_1}$

$$\Phi = 2 \frac{x_0^{z_2} + (z_2/z_2') x_0^{-z_1'/z_1} - (z_2 + z_2')/z_2'}{x_0^{z_1} + (z/z_1') x_0^{-z_1'/z_1} - (z_1 + z_1')/z_1'} - \frac{3z_1}{I_{10}} \int\limits_{x_0}^{\infty} \frac{x^{z_2} - x_0^{z_2} - (z_2/z_2')(x_0^{-z_1'} - x^{-z_2'})}{x[x^{z_1} - x_0^{z_1} - (z_1/z_1')(x_0^{-z_1'} + x^{-z_1'})]^{3/2}} dx,$$

As

$$\lim_{x=\infty} \frac{x^{z_2} - x_0^{z_2} - (z_2/z_2')(x_0^{-z_1'} - x^{-z_2'})}{[(x^{z_1} - x_0^{z_1}) - (z_1/z_1')(x_0^{-z_1'} - x^{-z_1'})]^{3/2}} \begin{cases} > 0 & \text{for } z_2 > 1.5 z_1, \\ = 0 & \text{for } z_2 < 1.5 z_1, \end{cases}$$

the Cauchy integrals I_{20} diverge for $z_2 > 1.5 z_1$. As I_{10} is a convergent integral, we have

$$\Phi = -\infty \quad \text{for } z_2 \geq \frac{3}{2} z_1.$$

Limiting synergism should thus occur for nearly all practical combinations of electrolytes whose counterions differ in valency, and this should be so for a vanishingly small concentration of the component with the greater coagulating action.

We are interested also in the case where the electrolytes differ only in the valency of the auxiliary ions. A mixture of a 1-1 electrolyte with a 1_n-n one gives the following I_{20} for an infinitely small content of the second electrolyte:

$$I_{20} = \int\limits_{u_{00}}^{\infty} \frac{u - u_{00} - \frac{1}{n}(u_{00}^{-n} - u^{-n})}{[u(u - u_{00})(u - u_{00}^{-1})]^{3/2}} u^2 du = \int\limits_{u_{00}}^{\infty} \frac{du}{\rho(u)} - \frac{u_{00}^n - u_{00}^{-n}}{n(u_{00}^2 - 1)} \int\limits_{u_{00}}^{\infty} \frac{du}{(u - u_{00}^{-1})\rho(u)} + \sum_{l=1}^{l=n-1} \frac{u_{00}^{n-l} - u_{00}^{l-n}}{n(u_{00} - u_{00}^{-1})} \int\limits_{u_{00}}^{\infty} \frac{du}{u^2 \rho(u)}.$$

All of these integrals have been determined [10]. For instance, for a 1-1 electrolyte with a 1_2-2 one

$$I_{20} = \int\limits_{u_{00}}^{\infty} \frac{du}{\rho(u)} + \frac{1}{2} \int\limits_{u_{00}}^{\infty} \frac{du}{u\rho(u)} - \frac{(u_{00} - 1)^2}{2u_{00}^2} \int\limits_{u_{00}}^{\infty} \frac{du}{(u - u_{00})\rho(u)} = \frac{u_{00}^2 + u_{00} + 3}{6u_{00}} I_{10},$$

TABLE 3. Values of Φ and $\Phi n_2^0/n_1^0$ for Mixtures of Electrolytes Differing in Valency of the Auxiliary Ions

z'_2/z	0.5	0.5	0.5	1	1	1	1	2	2	2	∞	∞	∞	∞
z'_1/z	0.5	1	∞	0.5	1	2	∞	1	2	∞	0.5	1	2	∞
Φ	−1.000	−0.371	−0.371	−1.104	−1.000	−0.839	−0.568	−1.063	−1.000	−0.588	−1.529	−1.500	−1.396	−1.000
$\dfrac{n_2^0}{n_1^0}\Phi$	−1.00	−0.99	−0.74	−0.97	−1.00	−0.93	−0.84	−0.91	−1.00	−0.85	−0.80	−0.87	−0.97	−1.01

TABLE 4. Calculated Φ and $\Phi n_1^0/n_2^0$ for Coagulation of Lyophobic Sols by 1-1 + 2-2 and 1_2-2 + 2-2 Electrolyte Mixtures

Mixture		highly charged sols	8.1790	5.6511	3.8842	2.8416
				$ez_2\psi'_a/\theta$		
1—1+2—2	Φ	−0.008	−0.031	−0.057	−0.106	−0.149
	$\Phi\dfrac{n_1^0}{n_2^0}$	−0.512	−0.74	−0.81	−0.87	−0.99
1_2—2+2—2	Φ	−0.044	+0.009	−0.001	−0.067	−0.129
	$\Phi\dfrac{n_1^0}{n_2^0}$	+3.29	+0.251	−0.037	−0.71	−0.89

$$\Phi = \frac{u_{00}^2 - 3u_{00} + 1}{2u_{00}}.$$

For n = 3

$$\Phi = -\frac{2}{3}\left(2u_{00}^2 - 8u_{00} + 1 - 8u_{00}^{-1} - 2u_{00}^{-2}\right)$$

and $\lim\limits_{n=\infty}\Phi = -0.5$ in the limit $n \gg 3$.

For a 2-2 electrolyte with a 2-1$_2$ one

$$I_{20} = I_{10} + \frac{4t_0^2}{t_0^4 - 1}\int_{t_0}^{\infty}\frac{dt}{(t+t_0)\rho(t)} - \frac{4t_0^2}{t_0^4 - 1}\int_{t_0}^{\infty}\frac{dt}{(t^2 - t_0^2)\rho(t)} + \frac{2t_0^4 + 2t_0 - 1}{t_0^2(t_0^4 - 1)}\int_{t_0}^{\infty}\frac{dt}{(t^3 - t_0^{-2})\rho(t)};$$

$$I_{20} = \frac{1}{2}I_{10} + \frac{t_0^4 + 4t_0 - 1}{3(t_0^2 - 1)}I_{10} - \frac{8t_0^4}{(t_0^4 - 1)^2};$$

$$\Phi = \frac{12t_0^3}{(t_0^4 - 1)(t_1^2 - 1)K\left(\frac{2t_0}{t_0^2 + 1}\right)} - \frac{t_0^4 + 6t_0^2 + 1}{2(t_0^2 - 1)^2};$$

$$\rho^2(t) = (t^2 - t_0^2)(t^2 - t_0^{-2}); \quad t^2 = u; \quad t_0^2 = u_{00}.$$

Table 3 gives Φ for a $2_{m/2}$-m electrolyte with a $2_{n/2}$-n one.

As $\Phi < 0$ always, we have to calculate $\Phi n_2^0/n_1^0$ (given in Table 3) in order to establish whether there is synergism or superadditivity.

We thus reach the very important conclusion that a great variety of effects (from limiting antagonism to limiting synergism) should be observed in the coagulation of highly charged sols (potential of the colloid particles remaining fairly high at the moment of coagulation), i.e., even when ion adsorption (which always occurs in real cases) cannot affect the stability of the system.

It is necessary to calculate Φ for various constants u_a in order to elucidate the effects of particle potential in the concentration mechanism of coagulation of sols by electrolyte mixtures. Table 4 gives Φ for 2-2 + 1-1 and 2-2 + 1$_2$-2 mixtures for various particle potentials [11, 14]. It is clear that there are only small deviations from additivity in these cases. These calculations were performed from (4), which takes the following forms:

2-2 + 1-1 mixtures:

$$\Phi = \frac{3t_0}{(1-k^2)\sin\varphi_1 F(\varphi_1; k)} - \frac{k^2}{1-k^2}; \quad k = \frac{2t_0}{t_0^2 + 1}; \quad \sin^2\varphi_1 = \frac{\rho^2(t_a)}{t_a^2 - 1}.$$

2-2 + 1$_2$-2 mixtures:

$$\Phi = 0.75\frac{k + k\sqrt{(1-k^2)(1-k^2\sin^2\varphi_1)} - \cos\varphi_1\sqrt{1-k^2}}{(1-k^2)\sin\varphi_1 F(\varphi_1; k)} - \frac{1+k^2}{4(1-k^2)}.$$

These results (and similar ones for other mixtures) show that the deviations from additivity increase with the particle potential if the pure concentration mechanism is involved; they differ from those for highly charged sols even at high but finite potentials. The deviations from additivity are very slight (Table 5) already at $\psi_a = 100$ mV.

However, as n increases in a 1_n-n + $2_{m/2}$-m mixture, the deviations increase for a given potential. For instance, a 1_3-3 + 2-2 mixture shows antagonism for $\psi_a > 70$ mV, while a 1_4-4 + 2-2 one starts to do so at $\psi_a = 50$ mV. No exact calculation has been performed for 1_n-n +

TABLE 5. Calculated Φ_2 for $ez_2\psi_a/\Theta = 3.8842$

Mixtures	n						
	1	2	3	4	6	8	∞
$1_n - n + 2 - 2$	—0.106	—0.067	—0.021	+0.016	+0.083	+0.139	+0.224
$2_{n/2} - n + 2 - 2$	—1.077	—1.000	—0.908	—0.838	—0.704	—0.608	—0.238
$1_n - n + 2 - 1_2$	—0.020	—0.011	—0.004	+0.022	+0.103	+0.164	+0.301
$1_n - n + 2_{m/2} - m$ for $m \gg 2$	—0.102	—0.096	—0.079	—0.064	—0.032	—0.004	+0.023

$3_{m/3}$-m mixtures, but Φ has been estimated for the cases 1_n-n + 3-3 and 1_n-n + 3-1_3 as for highly charged sols. Antagonism here should occur for $\psi_a \geq 25$ mV, so antagonism is possible in isolated cases for concentration coagulation of lyophobic sols even when the particle potential is relatively low.

Of course, the topic is not exhausted by examination of the two ends of the curve. The full curve may be calculated from the general criterion for stability in an ion-stabilized system if I and I_1 can be determined by numerical integration. The calculations involve major difficulties, so it is desirable to consider only cases where I and I_1 can be reduced to tabulated integrals. Calculations have been performed for the following mixtures: 1-1 + 2-2; 1-1 + 2-1_2; 1-1 + $2_{m/2}$-m for m $\gg$ 2; 1_2-2 + 2-2; 1_2-2 + $2_{m/2}$-m (m $\gg$ 2); 1-1 + 1_2-2; 1-1 + 1_m-m and 2-2 + 2-1_2. The method is essentially the same for all of these, so it may be illustrated via a single example.

For a 1_2-2 + 2-2 mixture

$$f_1(u) = u + \frac{1}{2}u^{-2} - \frac{3}{2}, \quad f_2(u) = u^2 + u_0^2 - 2$$

and so

$$I = \int_{u_0}^{u_a} \frac{du}{u\sqrt{m(u - u_0 + 0.5\,u^{-2} - 0.5\,u_0^{-2}) + u^2 - u_0^2 + u^{-2}u_0^{-2}}} = \int_{u_0}^{u_a} \frac{du}{(u - u_0)\sqrt{u^3 + (u_0 + m)u^2 - \dfrac{2+m}{2u_0^2}u - \dfrac{2+m}{2u_0}}} .$$

The polynomial under the root sign has three real roots: one positive (α) and two negative ($-\beta$ and $-\gamma$).

In fact, let

$$u^2 + u^2(u_0 + m) - u\frac{2+m}{2u_0^2} - \frac{2+m}{2u_0} = \varphi(u).$$

It is readily shown that

$$\varphi(u_0) > 0; \quad \varphi(u_0^{-1}) < 0; \quad \varphi(0) < 0; \quad \varphi(-u_0) > 0; \quad \varphi(-u_0 - m) > 0$$

and

$$\varphi\left[\frac{-(u_0 + m) - \sqrt{(u_0 + m)^2 + 2\dfrac{u_0 + m}{u_0}}}{2}\right] < 0,$$

so

$$\varphi(u) = (u - \alpha)(u + \beta)(u + \gamma),$$

in which

$$u_0^{-1} < \alpha < u_0,$$

$$0 < \beta < u_0^{-1},$$

$$\frac{u_0 + m + \sqrt{(u_0 + m)^2 + 2\,\dfrac{u_0 + m}{u_0}}}{2} > \gamma > u_0 + m.$$

The substitution

$$\sin^2 \varphi = \frac{\alpha + \gamma}{u_0 + \gamma} \cdot \frac{u - u_0}{u - \alpha}, \quad k^2 = \frac{\alpha + \beta}{u_0 + \beta}\,\frac{u_0 + \gamma}{\alpha + \gamma}$$

reduces I to an elliptic integral of the first kind:

$$I = \frac{2F(\varphi_1;\, k)}{\sqrt{(u_0 + \beta)(\alpha + \gamma)}},$$

in which

$$\sin^2 \varphi_1 = \frac{\alpha + \gamma}{u_0 + \gamma}\,\frac{u_a - u_0}{u_a - \alpha}.$$

I_1 may be put as

$$I_1 = \int\limits_{u_0}^{u_a} \frac{4u^2 + u\left(2u_0 + 2m + \dfrac{2 + m}{u_0^3}\right)}{(u - u_0)^{1/2}\,[(u - \alpha)(u + \beta)(u + \gamma)]^{3/2}}\, du$$

or after expansion as elementary fractions

$$I_1 = -\frac{4\alpha^2 + \alpha\left(2u_0 + 2m + \dfrac{2 + m}{u_0^3}\right)}{(\alpha + \beta)(\alpha + \gamma)}\,G_1 - \frac{4\beta^2 - \beta\left(2u_0 + 2m + \dfrac{2 + m}{u_0^3}\right)}{(\alpha + \beta)(\gamma - \beta)}\,G_2 + \frac{4\gamma^2 - \gamma\left(2u_0 + 2m + \dfrac{2 + m}{u_0^3}\right)}{(\alpha + \gamma)(\gamma - \beta)}\,G_3,$$

in which

$$G_1 = \int\limits_{u_0}^{u_a} \frac{du}{(u - \alpha)\,\Delta(u)}; \quad G_2 = \int\limits_{u_0}^{u_a} \frac{du}{(u + \beta)\,\Delta(u)}; \quad G_3 \int\limits_{u_0}^{u_a} \frac{du}{(u + \gamma)\,\Delta(u)}.$$

$$\Delta^2(u) = (u - u_0)(u - \alpha)(u + \beta)(u + \gamma).$$

As

$$\frac{1}{u - \alpha} = \frac{u_0 + \gamma}{(u_0 - \alpha)(\alpha + \gamma)}\cos^2\varphi - \frac{1}{\alpha + \gamma},$$

$$\frac{1}{u + \beta} = \frac{u_0 + \beta}{(u_0 - \alpha)(\alpha + \beta)}\Delta^2(u) - \frac{1}{\alpha + \beta},$$

$$\frac{1}{u + \gamma} = \frac{1}{\alpha + \gamma} - \frac{u_0 - \alpha}{(u_0 + \gamma)(\alpha + \gamma)\cos^2\varphi},$$

then

$$\frac{(\alpha + \gamma)\sqrt{(u_0 + \beta)(\alpha + \gamma)}}{2}\,G_1 = \frac{u_0 + \gamma}{u_0 - \alpha}\left[\frac{E(\varphi_1,\, k)}{k^2} - \frac{1 - k^2}{k^2}\,F(\varphi_1;\, k)\right] - F(\varphi_1;\, k),$$

$$(\alpha + \beta)\sqrt{(u_0 + \beta)(\alpha + \gamma)}\,G_2 = 2\,\frac{u_0 + \beta}{u_0 - \alpha}\,E(\varphi_1;\, k) - 2F(\varphi_1;\, k),$$

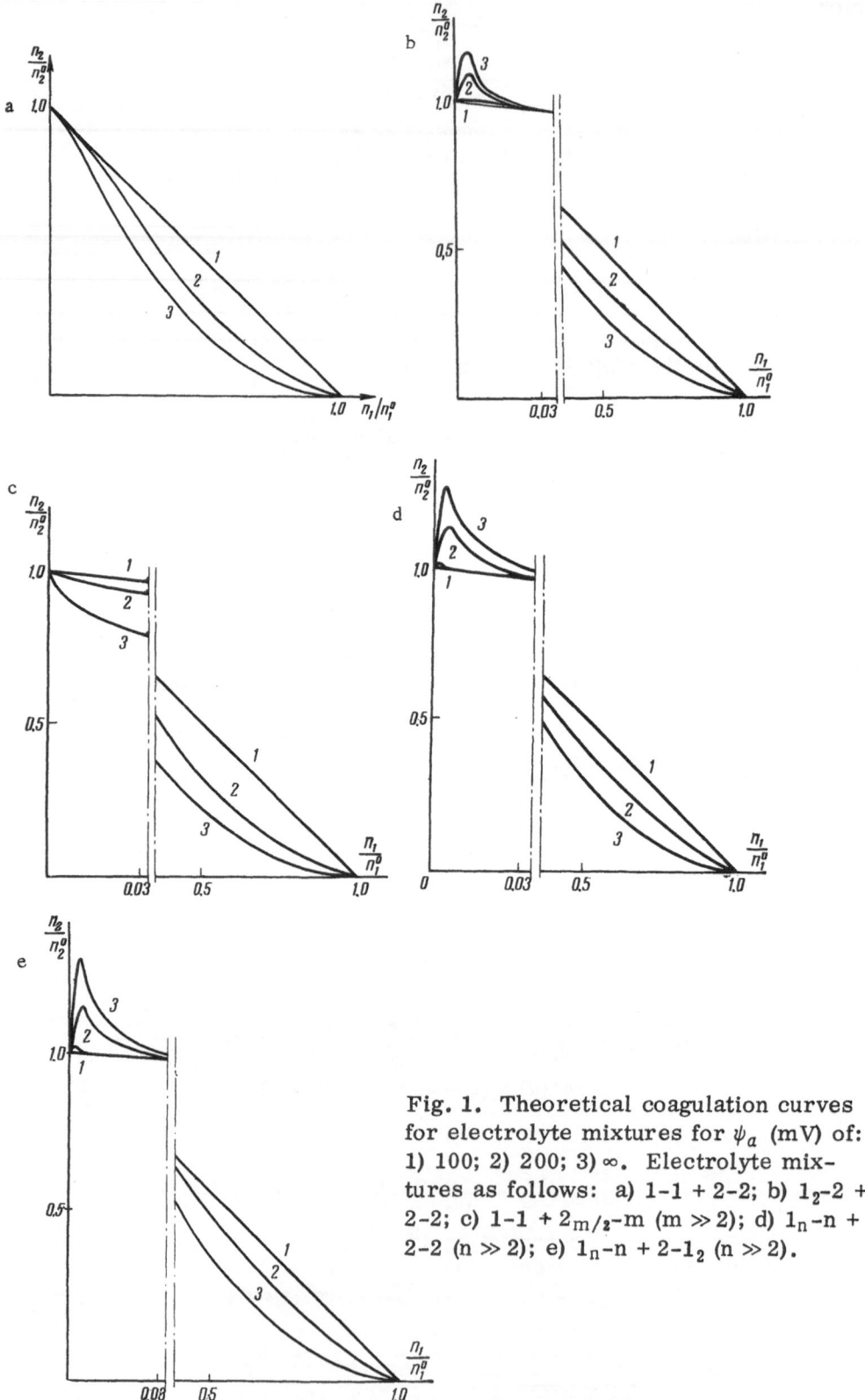

Fig. 1. Theoretical coagulation curves for electrolyte mixtures for ψ_a (mV) of: 1) 100; 2) 200; 3) ∞. Electrolyte mixtures as follows: a) 1-1 + 2-2; b) 1_2-2 + 2-2; c) 1-1 + $2_{m/2}$-m (m $\gg$ 2); d) 1_n-n + 2-2 (n $\gg$ 2); e) 1_n-n + 2-1_2 (n $\gg$ 2).

$$(\alpha + \gamma) \sqrt{(u_0 + \beta)(\alpha + \gamma)} \, G_3 = 2F(\varphi_1; k) \; - 2 \frac{u_0 - \alpha}{u_0 + \gamma} \left[\frac{\Delta \varphi_1 \, \mathrm{tg} \, \varphi_1}{1 - k^2} + F(\varphi_1; k) - \frac{E(\varphi_1; k)}{1 - k^2} \right].$$

These formulas define I and I_1, which are substituted into (2) to derive u_0 and the electrolyte concentrations in the critical state of the disperse system from (1). Figure 1 shows curves calculated in this way for certain mixtures. It is clear that antagonism occurs only at the start, i.e., when there are very small amounts of the electrolyte with the smaller coagulation effect; synergism occurs everywhere else. Even fairly high potentials ($\psi_a = 100$ mV) produce very little deviation from additivity.

These results show that electrical interaction of diffuse ionic atmospheres cannot be the sole cause of marked deviations from additivity in coagulating action at the particle potentials usually observed. We conclude that the actual coagulation of a lyophobic sol by an electrolyte never occurs via the concentration mechanism alone; various adsorption effects play a very important part and largely determine the coagulating action of electrolyte mixtures.

The general stability criterion is inadequate to give the critical electrolyte concentration when the neutralization mechanism of coagulation is applied, and it must be supplemented with an equation defining the relation of particle potential to nature and concentration of the electrolytes.

References

1. Yu. M. Glazman, Research in Surface Forces, Vol. 1, Consultants Bureau (1963), p. 135.
2. B. V. Deryagin, Izv. AN SSSR, Ser. Khim., No. 5, p. 1153 (1937); Kolloidn. Zh., 6:291 (1940); 7:285 (1941); B. V. Deryagin and L. D. Landau, ZhÉTF, 11:802 (1941); 15:663 (1945); B. V. Deryagin, Transactions of the Third All-Union Conference on Colloid Chemistry, Izd. AN SSSR, Moscow (1956), p. 225; Trans. Faraday Soc., 36:203, 730 (1940); Acta Physicochim., URSS, 14:633 (1941).
3. E. J. W. Verwey and J. Th. G. Overbeek, Theory of the Stability of Lyophobic Colloids, New York (1948).
4. V. M. Barboi, Kolloidn. Zh., 26:409 (1964).
5. V. M. Barboi, Yu. M. Glazman, and I. M. Dykman, Kolloidn. Zh., 23:381 (1961).
6. Yu. M. Glazman and I. M. Dykman, Dokl. AN SSSR, 100:299 (1955); Kolloidn. Zh., 18:13 (1956).
7.. Yu. M. Glazman, I. M. Dykman, and Ya. A. Strel'tsova, Dokl. AN SSSR, 117:229 (1957); Kolloidn. Zh., 20:149 (1958).
8. V. M. Barboi, Yu. M. Glazman, and I. M. Dykman, Kolloidn. Zh., 23:376 (1961).
9. V. M. Barboi and Yu. M. Glazman, Kolloidn. Zh., 25:282 (1963).
10. V. M. Barboi, Kolloidn. Zh., 27:151 (1965).
11. V. M. Barboi, Kolloidn. Zh., 26:3 (1964).
12. V. M. Barboi, Kolloidn. Zh., 26:316 (1965).
13. W. Pauli and E. Valko, Elektrochemie der Kolloide, Vienna (1929).
14. V. M. Barboi, Kolloidn. Zh., 25:385 (1963); 27:643 (1965).

EFFECTS OF SURFACE-ACTIVE SUBSTANCES ON THE STABILITY OF LYOPHOBIC SOLS: STABILIZATION OF COLLOIDAL ARSENIC SULFIDE IN RELATION TO NONPOLAR RADICAL AND HYDROXYETHYL CHAIN LENGTH

M. E. Krasnokutskaya and Yu. M. Glazman

Institute of Light-Industry Technology, Kiev

Alkyl polyethyleneglycol ethers have a stabilizing action on lyophobic sols that increases with the length of the nonpolar chain. The results are interpreted in terms of hydrophilization of colloid particles by adsorption on them of SAS molecules.

We have previously [1] examined the effects of surface-active compounds of general formula $C_mH_{2m-1}O(CH_2CH_2O)_nH$ or C_mE_n (n = 18, m of 4, 8, 10, 12, 16, and 18) on the coagulation of arsenic sulfide hydrosol by various electrolytes. It was found that the stabilizing action increases very rapidly with the length of the hydrocarbon radical, the butyl and octyl compounds having hardly any stabilizing action.

The polar part of the molecule of a surfactant also has a marked effect [2] (cf. also [3]) on this stabilization, which increases monotonically with the number of hydroxyethylene groups. This has a natural interpretation in terms of hydrophilization of colloid particles by adsorption on them of SAS molecules [1, 2, 4, 5].

This makes it of interest to examine whether the effect of hydrocarbon chain length is affected by the length of the polar chain, and if so, what are the principal features. For this purpose we used the compounds C_mE_{98} (m of 4, 8, 10, 16, and 18), the methods being largely as previously [1]. Table 1 and Fig. 1 give the results, the abscissa being log C (C is molar concentration of surfactant) and the ordinate the coagulation value for ammonium nitrate. These results show a marked increase in stabilizing action with the length of the nonpolar chain, the general trend resembling Traube's rule, which confirms the view that the stabilization effect of these nonionic surfactants is in some way related to adsorption of individual SAS molecules on the colloidal particles [1].

TABLE 1. Effects of Length of Hydrocarbon Radical in
$C_m E_{98}$ on the Stability of Arsenic Sulfide Sol

Com-pound	Electrolyte	C^*, M	Coagulating concentration, mM		
			initial sol.	sol + surfactant	
				min	max
$C_{18}E_{98}$	NH_4NO_3	$6 \cdot 10^{-5}$	50.0	21.5	5900
	$CaCl_2$		0.40	0.40	4000
	$La(NO_3)_3$		0.026	0.026	>2300
$C_{16}E_{98}$	NH_4NO_3	$7 \cdot 10^{-5}$	50.0	24.0	5800
	$CaCl_2$		0.40	0.40	3500
	$La(NO_3)_3$		0.026	0.026	>2300
$C_{10}E_{98}$	NH_4NO_3	$1.3 \cdot 10^{-3}$	50.0	33.7	4200
	$CaCl_2$		0.40	0.40	3400
	$La(NO_3)_3$		0.026	0.026	>2300
C_8E_{98}	NH_4NO_3	$1.9 \cdot 10^{-3}$	50.0	34.3	3700
	$La(NO_3)_3$		0.026	0.026	>2300
$C_4E_{98}**$	NH_4NO_3	$1.1 \cdot 10^{-2}$	50.0	39.4	3400
	$CaCl_2$	$0.5 \cdot 10^{-2}$	0.40	0.40	3400
	$La(NO_3)_3$	$0.5 \cdot 10^{-2}$	0.026	0.026	>2300

*C is the surfactant concentration corresponding to the sharp rise in
the stability.

**The data for $CaCl_2$ and $La(NO_3)_3$ relate to a surfactant specimen
that had been purified.

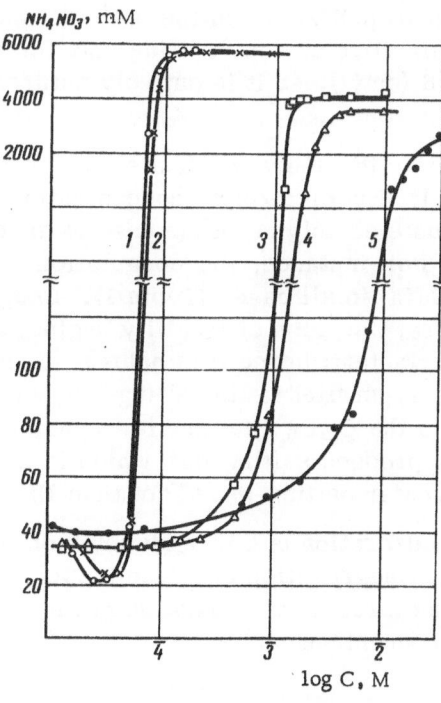

Fig. 1. Effects of surfactants on the coagulation
of arsenic sulfide sol by ammonium nitrate: 1)
$C_{18}H_{37}O(CH_2CH_2O)_{98}H$; 2) $C_{16}H_{33}O(CH_2CH_2O)_{98}H$;
3) $C_{10}H_{21}O(CH_2CH_2O)_{98}H$; 4) $C_8H_{17}O(CH_2CH_2O)_{98}H$;
5) $C_4H_9O(CH_2CH_2O)_{98}H$.

However, the C_mE_{98} do not behave exactly as the C_mE_{18}; the former start to raise sharply the stability at lower C. Also, Fig. 1 shows that all the compounds (instead of merely the higher homologs, as for C_mE_{18}) show a limiting effect.

It is very likely that the surfactants become more readily adsorbed as the length of the E chain increases[†]; but, even if this is not so, the particles should be sufficiently hydrophilic to form a very stable system, an effect that must be dependent on the size of the polar part, being obtained the earlier the greater the degree of hydroxyethylation. This is responsible not only for the above feature but also for some other very important effects. For instance, all the C_mE_{98}, even C_4E_{98}, produce marked stabilization against a variety of electrolytes (Table 1), whereas the C_mE_{18} make the sol hydrophilic only if C_m is octadecyl, hexadecyl, or dodecyl [1]. The decyl compound does produce a considerable increase in stability, but the sol remains lyophobic [4], i.e., is stabilized by a double ionic layer around the particles.

The data indicate that the conversion of hydrophobic sols to hydrophilic ones is not directly related to micelle formation in the solution or in the adsorbed layer on the particles; C_8E_{18} produces no substantial stabilization even at C substantially exceeding the critical C for micelle formation [6], whereas C_4E_{98}, which should not [7] form micelles at accessible concentrations, does produce marked stabilization at C < 0.01 M (Table 1). This confirms the previous views [1, 2, 4, 5] about the stabilizing action of SAS and about the mechanism of conversion of lyophobic sols to lyophilic ones.

Some organic compounds, e.g., saturated monohydric alcohols, do not substantially stabilize hydrophobic sols although they are markedly surface-active. As these are not compounds of semicolloidal type, we might assume that the lack of capacity to form micelles is the reason why they do not stabilize hydrophobic sols; whereas our data show that this is far from being the case. Comparison of the C_mE_{18} with the C_mE_{98} shows that it is not a question of micelle formation but of the degree of hydrophilization of the colloidal particles, which adsorb the surfactant molecules. The C_mE_{18} stabilize arsenic sulfide only for m > 10; in particular, there is no stabilizing action for m = 8 (octyl), so it is entirely natural that octyl alcohol (and more so the lower homologs) has no such action.

There is further evidence for the above mechanism and for the mode of stability loss in response to electrolytes [1, 2, 5]. The electrolyte concentrations required for coagulation are of the same order for the various electrolytes in the case of arsenic sulfide, though there are certain clear-cut differences. For instance, the coagulation value for NH_4NO_3 is [1, 2] somewhat higher than that for $CaCl_2$ in all cases (Table 1). Coagulation is caused by disruption of polymolecular hydration layers, i.e., loss of stability begins, as it were, with reduction in the solubility of the adsorbed surfactant; hence it is natural to assume that the coagulating power of an electrolyte should run parallel to its salting-out action on the surfactant. This is actually so; electrolytes added to the C_mE_n when the latter are at the C necessary for the limiting stabilizing effect do actually produce salting-out, which is most pronounced for the C_mE_{98} series. Here $CaCl_2$ has [8] a greater salting-out effect than NH_4NO_3.

The results all show that adsorption of SAS molecules on the particles makes the latter hydrophilic, with stability against aggregation provided by polymolecular hydration layers. Such a lyophilic sol coagulates as a result of disruption of the stabilizing layers, as from the addition of electrolytes in high concentration.

Conclusions

1. A study has been made of the effects of the C_mE_{98} (m of 4, 8, 10, 16, and 18) on the stability of arsenic sulfide hydrosol in respect of coagulation by NH_4NO_3, $CaCl_2$, and $La(NO_3)_3$.

[†] This aspect is now under study.

2. These compounds differ from the $C_m E_{18}$ in that all of them have a very pronounced stabilizing effect, which increases regularly with the length of the nonpolar chain.

3. These nonionic surfactants greatly stabilize hydrophobic sols by rendering the particles hydrophilic by adsorption of individual SAS molecules; the effect is not directly related to micelle formation.

4. Views presented on the nature of lyophilic sols and coagulation mechanisms are in agreement with the observed parallelism between the precipitating power of electrolytes and their salting-out action on the surfactants.

We are indebted to B. V. Deryagin, Associate Member, Academy of Sciences of the USSR, for constant interest in this work and for reading the draft.

References

1.   M. E. Krasnokutskaya and Yu. M. Glazman, Kolloidn. Zh., 28:847 (1966).
2.   Yu. M. Glazman and M. E. Krasnokutskaya, Kolloidn. Zh., 27:815 (1965).
3.   R. M. Panich, V. V. Kireitsev, D. M. Sandomirskii, and S. S. Voyutskii, Kolloidn. Zh., 24:733 (1962).
4.   Yu. M. Glazman, Kolloidn. Zh., 24:275 (1962).
5.   Yu. M. Glazman and I. P. Sapon, in: Research in Surface Forces, Vol. 2, Consultants Bureau, New York (1966), p. 232; I. P. Sapon and Yu. M. Glazman, Kolloidn. Zh., 27:601 (1965).
6.   J. M. Corkill, J. F. Goodman, and R. H. Ottewill, Trans. Faraday Soc., 57:1627 (1961).
7.   P. H. Elworthy and A. T. Florence, Kolloid.-Z., 204:105 (1965).
8.   M. J. Schick, J. Coll. Sci., 17:801 (1962).

THE MUTUAL POTENTIAL ENERGY
OF TWO COLLOIDAL PARTICLES AND
THE CRITERION FOR RAPID COAGULATION
OF HIGHLY CHARGED LYOPHOBIC SOLS

F. I. Kligman and Yu. M. Glazman

Institute of Light-Industry Technology, Kiev

The physical theory of highly charged lyophobic sols is used to derive an expression for the interaction potential of two spherical particles in a symmetrical electrolyte. The true critical ion concentration is calculated for the onset of rapid coagulation.

It is assumed in the physical theory of coagulation of a highly charged lyophobic sol [1] that the stability limit corresponds to an electrolyte concentration such that, at a certain particle separation H, the electrostatic repulsion F_{el} and the van der Waals attraction F_{vdw} are equal in magnitude, as are the derivatives with respect to H. However, we must distinguish [1] static adhesion of two particles from the kinetic coagulation of a sol, in which aggregation occurs via close approaches produced by Brownian motion. The particles in a sol have thermal energy, so coagulation can occur even if there is a potential barrier with a positive maximum between the particles.

Here we use the force formula [1] to derive an expression for the mutual potential energy of two spherical particles in a highly charged lyophobic sol in the presence of a symmetrical electrolyte. This substantially governs the rate of slow coagulation and the condition for transition to fast coagulation.

We can write the following expression [1] for the resultant force F between two such particles of size $a \gg H$ (in which H is the least distance between the two surfaces):

$$F = F_{el} + F_{vdv} = \pi a \left[\frac{8n\theta}{\varkappa} f_{el}(k) - \frac{A}{12\pi} \frac{1}{H^2} \right], \tag{1}$$

in which

$$f_{el} = \frac{2E(k)}{k} - \left(\frac{1}{k} - k \right) K(k) - 2, \tag{1'}$$

198

$$H = 2kK(k) / \varkappa,$$

(2)

$$\varkappa = \sqrt{\frac{8\pi}{\varepsilon} \frac{e^2 z^2 n}{\theta}} .$$

(3)

Here A is the van der Waals constant, K(k) and E(k) are complete elliptic integrals of the first and second kinds, n is the number of ions of one sign in unit volume, $\theta = kT$ (product of Boltzmann's constant and the absolute temperature), ε is dielectric constant, z is ion valency, and e is the electronic charge.

Substitution for H from (2) into (1) gives

$$F = \frac{8\pi a n \theta}{\varkappa} \left[f_{el}(k) - \frac{1}{8} \frac{A\varkappa^3}{48\pi n \theta} \frac{1}{(kK(k))^2} \right].$$

(4)

We use (4) with (1') and the Deryagin–Landau conditions for the critical state

$$F(H_c) = 0,$$

(5a)

$$\left(\frac{dF}{dH} \right)_{H=H_c} = 0$$

(5b)

(in which subscript c denotes that state) to get

$$4E(k_c) - 3K(k_c)(1 - k_c^2) - 4k_c = 0.$$

(6)

The root $k_c = 0.5425$ of this equation defines $(\varkappa H)_c$ from (2), and hence n_c via (1'), (3), (4), and (5a). We can transform (4) via the dimensionless concentration

$$\delta = \frac{n}{n_c} = \frac{\gamma}{\gamma_c} ,$$

(7)

in which γ is the molar concentration of the electrolyte.

This allows one to examine some features of the interaction without having to specify the constants defining the system. We use (1'), (4), (5a), and (6) to get

$$\frac{1}{8} \frac{A\varkappa_c^3}{48\pi n_c \theta} = \frac{1}{2} \left(\frac{1}{k_c^2} - 1 \right) [k_c K(k_c)]^3.$$

(8)

As $k_c = 0.5425$, we have

$$\frac{1}{2} \left(\frac{1}{k_c^2} - 1 \right) [k_c K(k_c)]^3 = 0.9582.$$

(8')

It follows from (3) and (7) that

$$\varkappa = \varkappa_c \, \delta^{1/2}.$$

(9)

From (8), (8'), and (9), the coefficient to $1/[kK(k)]^2$ in (4) becomes

$$\frac{1}{8} \frac{A\varkappa^3}{48\pi n \theta} = \frac{1}{8} \frac{A\varkappa_c^3}{48\pi n_c \theta} \delta^{1/2} = 0.9582 \, \delta^{1/2}.$$

(9')

Then (4) gives F as

$$F = \frac{8\pi a n \theta}{\varkappa} f(k\delta) = \frac{8\pi a n \theta}{\varkappa} \left[f_{el}(k) - \frac{0.9582 \, \delta^{1/2}}{(kK(k))^2} \right].$$

(10)

From (2), (7), and (9) we have

$$F = \frac{8\pi a n_c \theta}{\varkappa_c} f(\varkappa_c H \delta) = \frac{8\pi a n_c \theta}{\varkappa_c} \left[\delta^{1/s} f_{el}(k) - \frac{4 \cdot 0.9582}{(\varkappa_c H)_i^2} \right]. \tag{10'}$$

From (3), (8), and (8') we get the critical concentration as

$$n_c = \frac{1}{C'} \cdot \frac{\varepsilon^3 \theta^5}{A^2 e^6 z^6}, \tag{11}$$

in which

$$C' = \frac{2\pi}{9} \frac{1}{\left[4 \left(\frac{1}{k_c^2 - 1} \right) (k_c K(k_c))^3 \right]^2} = \frac{2\pi}{9} \frac{1}{(8 \cdot 0.9582)^2} = 1 \cdot 188 \cdot 10^{-2}, \tag{11'}$$

which agrees fully with the earlier result [1].

From (10') we get the mutual potential energy U of two particles as

$$U = \int\limits_H^\infty F \, dH. \tag{12}$$

Integration of (12) with (2), (9), and (10') gives

$$U = U_{el} + U_{vdw} = \frac{16\pi a n_c \theta}{\varkappa_c^2} u(\varkappa_c H \delta) = \frac{16\pi a n_c \theta}{\varkappa_c^2} \left[u_{el}(k) - \frac{2 \cdot 0.9582}{\varkappa_c H} \right], \tag{13}$$

in which

$$u_{el}(k) = \int\limits_k^1 f_{el}(k) \, d(kK). \tag{14}$$

We use (1') to transform the integral in (14) to

$$u_{el}(k) = \left[\int 2E \, dK + \int \frac{2EK}{k} \, dk - \int \frac{K^2}{k} \, dk - \frac{K^2}{2} + \frac{(kK)^2}{2} - 2kK \right]_k^1. \tag{14'}$$

The first term in (14') is integrated by parts, and we use the following relation for elliptic integrals

$$\frac{dE(k)}{dk} = \frac{E(k) - K(k)}{k},$$

which gives

$$u_{el}(k) = \left[2EK - \frac{K^2}{2} + \frac{(kK)^2}{2} - 2kK + \int \frac{K^2}{k} \, dk \right]_k^1. \tag{15}$$

We insert the limits of integration in (14) and note that E = 1 for k = 1 to get

$$u_{el}(k) = \frac{K^2}{2} (1 - k^2) - 2K(E - k) + \int\limits_k^1 \frac{K^2}{k} \, dk. \tag{16}$$

Then (13) and (16) define U for this case.

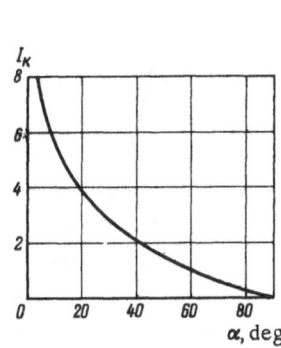

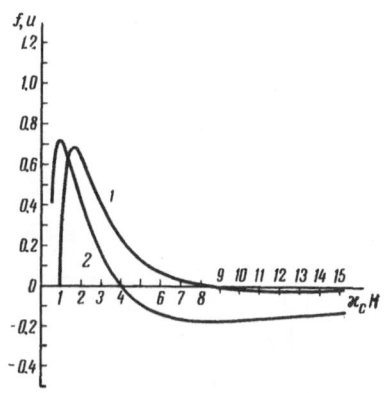

Fig. 1. Dependence of integral
I_k on α.

Fig. 2. Dependence of: 1) f;
2) u on $\varkappa_c H$ for $\delta = 0.25$.

We use (3) to transform the coefficient to u($\varkappa_c H$, δ) in (13) to get U as

$$U = \frac{2ae\theta^2}{e^2 z^2}\, u\,(\varkappa_c H, \delta) = \frac{2ae\theta^2}{e^2 z^2}\left[u_{el}\,(k) - \frac{2 \cdot 0.9582}{\varkappa_c H} \right]. \tag{13'}$$

In order to tabulate $u_{el}(k)$ we need not only the tabulated integrals K(k) and E(k) but also numerical integration of the last term in (16), which we denote by I_k:

$$I_k = \int\limits_k^1 \frac{[K\,(k')]^2}{k'}\, dk' = \int\limits_\alpha^{90°} \frac{[K\,(\alpha')]^2}{\operatorname{tg}\alpha'}\, d\alpha', \tag{17}$$

in which k = sin α and k' = sin α'.

The numerical integration has been performed for various α from 5° to 90° by steps of 1°; Fig. 1 shows I_k as a function of α. The coefficient to f ($\varkappa_c H$, δ) in expression (10') for F is independent of n, as is the coefficient to u($\varkappa_c H$, δ) in (13') for U, so these two dimensionless functions characterize the dependence of F and U on $\varkappa_c H$, i.e., on H in units of $1/\varkappa_c$, which are not dependent on n (for a given δ).

Figure 2 shows f($\varkappa_c H$) and u($\varkappa_c H$) from (1'), (10'), (13'), and (16) for $\delta = 1/4$. As F = $-dU/dH$, maxima and minima in u correspond to zeros in f, and the same applies for other δ. It is clear from (2), (9), (13), and (16) that the shape of u($\varkappa_c H$) is substantially dependent on δ.

The maximum and minimum in u decreases as δ increases, and the turning points come together, the peak shifting to larger $\varkappa_c H$. If $\delta > \delta_0$, in which δ_0 is defined by

$$U = 0, \quad \frac{dU}{dH} = 0, \tag{18}$$

u($\varkappa_c H$) is negative throughout the entire range in $\varkappa_c H$; but if $\delta_0 < \delta < 1$, u($\varkappa_c H$) has a maximum separate from the minimum (both negative), because f ($\varkappa_c H$) has a positive part, i.e., two zero values.

The maximum and minimum in u coincide for $\delta = 1$; if $\delta > 1$, u decreases monotonically in absolute value as $\varkappa_c H$ increases. Figure 3 shows u($\varkappa_c H$) for δ of 1/4, 1/2, δ_0, and 1, and also for u_{el} (k) = 0, when (13) gives

$$u\,(\varkappa_c H) = u_{vdw}(\varkappa_c H) = \frac{2 \cdot 0.9582}{\varkappa_c H}. \tag{19}$$

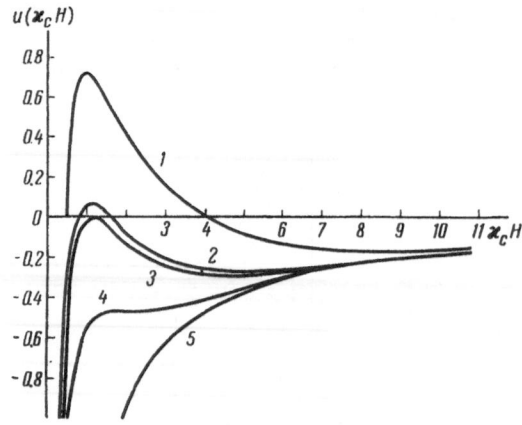

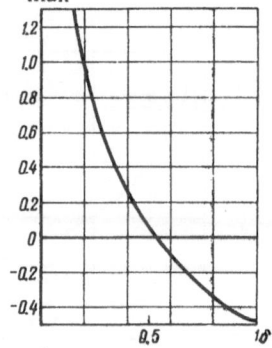

Fig. 3. Relation of: 1-4) u to $\varkappa_c H$ for
δ of: 1) 0.25; 2) 0.5; 3) δ_0; 4) 1; 5) u_{vdw}
to $\varkappa_c H$.

Fig. 4. Relation of u_{max}
to δ.

We have to examine the δ dependence of u_{max}, the maximum value of $u(\varkappa_c H)$, in order to determine the rate of slow coagulation and the condition for transition to fast coagulation.

From (2), (9), and (13)

$$u_{max} = u_{el}(k_m) - \frac{0.9582 \cdot \delta^{1/2}}{k_m K(k_m)}.\tag{20}$$

in which k_m is k for $u = u_{max}$; as F is then zero, the k_m of (20) is given by (1') and (10) as the root of

$$\frac{2E(k_m)}{k_m} - \left(\frac{1}{k_m} - k_m\right) K(k_m) - 2 - \frac{0.9582 \cdot \delta^{1/2}}{[k_m \cdot K(k_m)]^3} = 0\tag{21}$$

for a given value of δ.

Then (21) shows that k_m is a single-valued function of δ, so (20) shows that u_{max} is a function of δ alone. From (16), (20), and (21) we get a formula for calculating u_{max} from the k_m defined by (21):

$$u_{max} = \frac{3}{2}[K(k_m)]^2 (1 - k_m^2) - 4K(k_m)[E(k_m) - k_m] + I_{k_m}.\tag{22}$$

Figure 4 shows values calculated for u_{max}.

Table 1 illustrates the behavior of u_{max} and also $(\varkappa_c H)_m$, the $\varkappa_c H$ for $u = u_{max}$, for various δ.

The curve of Fig. 4 gives us directly the δ corresponding to $u_{max} = 0$, i.e., the condition of (18):

$$\delta = \delta_0 \approx 0.53\dagger.$$

Then the concentration n_0 corresponding to condition (19) is related to the n_c defined by (5a), (5b), (11), and (11') by

$$n_0 = \delta_0 n_c \approx 0.53 n_c.\tag{23}$$

$\dagger$ This numerical value of δ_0 was used in constructing the corresponding curve in Fig. 3.

TABLE 1. Values of k_m and $(x_c H)_m$ for Various δ

δ	k_m	$(x_c H)_m$	δ	k_m	$(x_c H)_m$
1/6	0.1173	0.906	1/2	0.2385	1.079
1/4	0.1500	0.948	2/3	0.3007	1.185
1/3	0.1805	0.990	1	0.5425	1.856

Very general theoretical considerations indicate that (19) should be considered as the criterion for rapid coagulation, i.e., δ_0 is the dimensionless electrolyte concentration above which the coagulation time τ ceases to decrease markedly as the electrolyte concentration is increased further. Now τ is given [2] by

$$\tau = \frac{1}{8\pi R_0 D \nu_0} \cdot w, \quad w = \int_0^\infty e^{\frac{U(H)}{\theta}} \frac{2a \, dH}{(2a + H)^2}, \tag{24}$$

in which $R_0 = 2a$, D is diffusion coefficient, and ν_0 is the number of colloidal particles in unit volume.

From (13') and (24) we get

$$w = \frac{1}{2x_c a} \int_0^\infty e^{b \cdot \frac{a}{z^2} u(x_c H, \delta)} \frac{d(x_c H)}{\left(1 + \frac{H}{2a}\right)^2}, \tag{24'}$$

in which $b = 2\varepsilon\theta/e^2$.

The integrand in (24') is exponentially dependent on $u(x_c H, \delta)$, so w is substantially dependent only on the range of integration corresponding to the highest part of $u(x_c H)$ for a given δ, and for $\delta < \delta_0$ the highest part is that around the maximum. Then (24') with $u_{max}(\delta)$ (Fig. 4) for $\delta > \delta_0$ shows that w decreases rapidly as δ increases. For $\delta > \delta_0$, the highest part is not the region of the turning point but the region of $x_c H$ large, where $u(x_c H) \to 0$, in which U_{vdw} plays the main part in the interaction.[†] Then, although u_{max} continues to fall fairly rapidly (Fig. 4) as δ increases from δ_0 to 1, w and τ cease to decrease substantially as δ increases. This means that electrostatic repulsion ceases to enhance the stability of the sol for $\delta > \delta_0$.

Consider now the limits to the variation in the theoretical values of w and τ for $\delta \geq \delta_0$. The upper limit w_0 to w, which corresponds to $u_{max} = 0$ (i.e., $\delta = \delta_0$), is less than one, since (24) shows that w = 1 when U is zero throughout the region of integration. The lower limit v_{vdw} to w corresponds to $U_{el} = 0$, when $U = U_{vdw}$. The approximation [1] $U_{vdw} = -Aa/12H$ gives us from (24) that

$$w_{vdw} = \int_0^1 e^{-B \frac{v}{1-v}} dy, \tag{25}$$

in which

$$y = \frac{2a}{2a + H}, \tag{25'}$$

---

† Calculations have been performed with a $U_{vdw}(H)$ more accurate than $\sim 1/H$, and these show that $u(x_c H)$ actually tends to zero much more rapidly than Fig. 3 indicates, though the change near the maximum $u(x_c H)$ is unimportant.

$$B = \frac{A}{24\delta}.$$

(26)

The exact expression [3] for U_{vdw} gives

$$U_{vdw} = -\frac{A}{6}\left[\frac{2a^3}{R^2} + \frac{2a^3}{R^2 - 4a^2} + \ln\left(1 - \frac{4a^2}{R^2}\right)\right], \quad R = 2a + H$$

(27)

and from (24)

$$w_{vdw} = \int_0^1 e^{-Bf(y)}\,dy,$$

(28)

in which

$$f(y) = 2y^2 + \frac{2y^3}{1 - y^3} + 4\ln(1 - y^3).$$

Then (26), with $A = 10^{-12}$ erg at 300°K, gives $B \approx 1$; then (25) gives $w_{vdw} \approx 0.4$, while (28) gives $w_{vdw} \approx 0.75$. Also, w_{vdw} is even closer to one if we take a value of A less than 10^{-12} erg.

These estimates of w for $\delta \geq \delta_0$ show that w varies in a ratio not more than 1:0.75 as δ goes from δ_0 to arbitrarily large values, and so τ does the same.† This relatively slight change in τ for $\delta \geq \delta_0$, in conjunction with the rapid variation for $\delta \leq \delta_0$, agrees with our observation that the coagulation rate remains constant above a certain electrolyte concentration, so (18) must be considered as the criterion for the transition to fast coagulation, i.e., it represents the condition for stability of a real sol, and the corresponding ion content defines the true critical electrolyte concentration.

The sharpness of the transition increases with the a/z^2 of (13') which governs τ via (24) and (24'); the larger a/z^2 in (24'), the more rapidly τ decreases as δ increases for $u_{max} \geq 0$ (i.e., for $\delta \leq \delta_0$) and the closer τ to a constant value for $\delta \geq \delta_0$, since the limits of variation in w come together for $\delta \geq \delta_0$: w_0 falls, while w_{vdw} remains unchanged.

Conclusions

1. The physical theory of the stability of highly charged lyophobic sols has been used to derive an expression for the mutual potential energy U of two spherical colloidal particles in a symmetrical electrolyte.

2. The magnitudes and positions of the peaks on the potential curves have been deduced as functions of the "dimensionless" electrolyte concentration.

3. The results indicate that n_0 (true critical electrolyte concentration), corresponding to the beginning of rapid coagulation of the real colloid system, is defined by $U = 0$, $dU/dH = 0$, in which H is the distance between the surfaces of the particles. This n_0 is found to differ from n_c, which corresponds to $F = 0$, $dF/dH = 0$, since $n_0 \approx 0.53n_c$.

We are indebted to B. V. Deryagin for discussion of the results and criticism of the draft.

References

1.   B. V. Deryagin, Izv. AN SSSR, Ser. Khim., No. 5, p. 1153 (1937); Trans. Faraday Soc., 36:203, 730 (1940); B. V. Deryagin and L. D. Landau, Zh. Éksp. Teor. Fiz., 11:802 (1941); 15:663 (1945); Acta Physicochim. URSS, 14:633 (1941).
2.   N. A. Fuks, Z. Physik, 89:736 (1934).

---

† We get $w_{vdw} \approx 0.72$ from (28) even if we take [4] $A = 2 \times 10^{-14}$ erg.

3.   H. C. Hamaker, Physica, 4:1058 (1937).

4.   E. J. W. Verwey and I. Th. G. Overbeek, Theory of the Stability of Lyophobic Colloids, New York (1948).

MAGNETOOPTIC AND ELECTROOPTIC EFFECTS IN RELATION TO STABILITY OF A SUSPENSION

E. E. Bibik, I. S. Lavrov, and O. M. Merkushev

Lensovet Technological Institute, Leningrad

A uniform electric or magnetic field applied to a suspension can cause effects due to aggregation of particles, as well as orientation effects. Magnetite sols are used to show that the stability and the nature of the protective layers affect the optical aggregation effects. Aggregation in response to an electric field explains certain effects observed in the electrophoretic deposition of suspensions.

It is usual to examine electrooptic and magnetooptic effects in colloidal solutions under conditions such as to exclude interaction between particles, i.e., highly dilute solutions. Then a uniform external field will produce only effects due to orientation, which can be used to calculate the particle shape, dipole moment, etc. [1, 2] but which give no direct information on the aggregation stability of the colloidal solution.

More detailed information can be obtained if the conditions are such that interaction between particles is not excluded. If orientation forces act between particles in the absence of external fields, e.g., due to asymmetry in the molecular forces for platy particles [3], then these forces can be deduced from their influence on the orientation effects. Further, while an external field can cause appreciable orientation effects in a dilute colloidal solution, effects of a different kind can occur at high concentrations, e.g., change in transmission from aggregation [4].

The optical effects of aggregation differ from orientation ones in being directly related to the stability, since the density of the aggregates is determined by the position of the minimum on the curve for particle interaction:

$$W(r, H) = P(r) + Q(r) + U(r, H), \qquad (1)$$

while the rate of aggregation is defined by

$$\frac{dW(r, H)}{dr} + F(v) = 0, \qquad (2)$$

in which r is the distance between particles, v is the rate of approach, H is the field strength, F(v) is the frictional force, P(r) is the energy of electrostatic repulsion, and Q(r) and U(r, H)

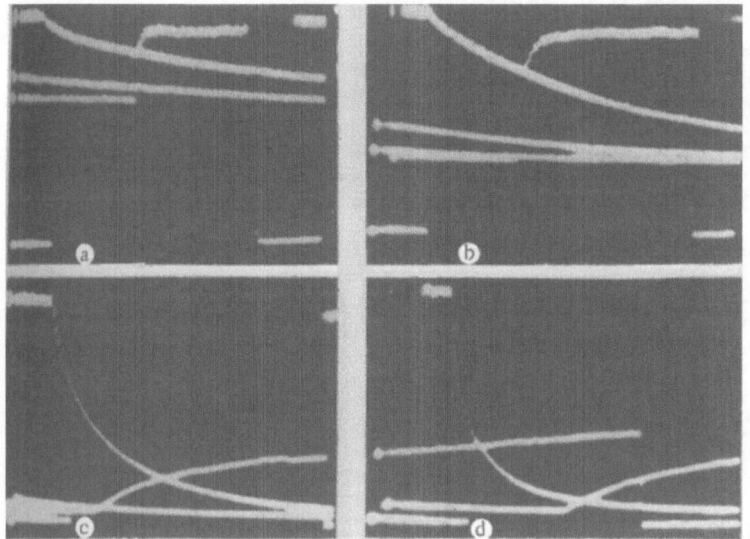

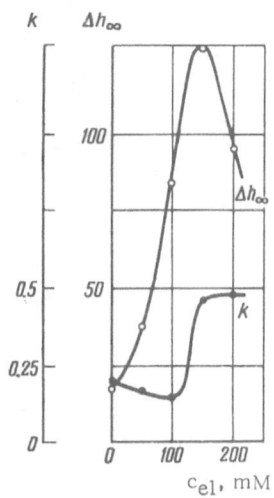

Fig. 1. Change in transmission of a Fe_3O_4 sol in response to a 400 Oe magnetic field for NaCl concentration (mM) of: a) 4; b) 50; c) 100; d) 150.

Fig. 2. Dependence of Δh_∞ and the k of (3) on NaCl concentration in a Fe_3O_4 sol.

are the energies of molecular and dipole attraction respectively. This relation is indicated by the effects of NaCl concentration on the magnetooptic aggregation effects in Fe_3O_4 sols (Fig. 1). The transmission $t = t_0 \exp(-\Delta h)$ of the Fe_3O_4 sol as a function of the time τ of exposure to a magnetic field is expressed by

$$\Delta h = \Delta h_\infty [1 - \exp(-k\tau)], \tag{3}$$

in which Δh is the increment in the extinction of the sol in a magnetic field, Δh_∞ is the limiting value of Δh, e.g., as defined by (1), $k = k(v)$ is a parameter characterizing the rate of change in t, and t_0 is the transmission in the absence of the field.

Parameters k and Δh_∞ characterize the capacity of the repulsive forces to prevent combination into aggregates. Figure 2 shows the dependence of these parameters on the NaCl concentration in an Fe_3O_4 sol. The marked increase in k at about 100 mM indicates that the coagulation threshold is attained. The rise in Δh_∞ indicates displacement of the minimum in $W(r, H)$ to smaller r.

Stability reduction retards the breakup of the aggregates when the field is turned off, and hence the return to t_0 (rising branches in Fig. 1). This indicates directly that this breakup is due to the repulsion. The effects of other compounds on $t(\tau)$ characterize the stabilizing action and sometimes reveal the mechanism. For instance, polyvinyl alcohol (PVA) retards aggregation and breakup, but the density of the aggregates is unaffected (Δh_∞ = constant), i.e., PVA mainly increases the viscosity of the medium and acts via the second term in (2). Detergents such as OP-7 have analogous effects on the ion-stabilized sol $[mFe_3O_4nFe^{3+}3(n-x)Cl^-]^+ \cdot 3xCl^-$. The negatively charged sol $[mFe_3O_4nFe(Ol)_2pOl^-(p-x)Na^+]^-xNa^+$ stabilized by sodium oleate (NaOl) aggregates much more slowly in response to a magnetic field than does the previous sol, but the breakup on switching off the field is also slow. The result is ascribed to anomalous viscosity in the adsorption layer [5], since the bulk viscosity does not differ greatly from that for the ion-stabilized sol.

The rate of breakup for Fe_3O_4 sols on switching off the field indicates that the potential barriers in particle repulsion persist for about 1 min after introduction of sufficient electrolyte

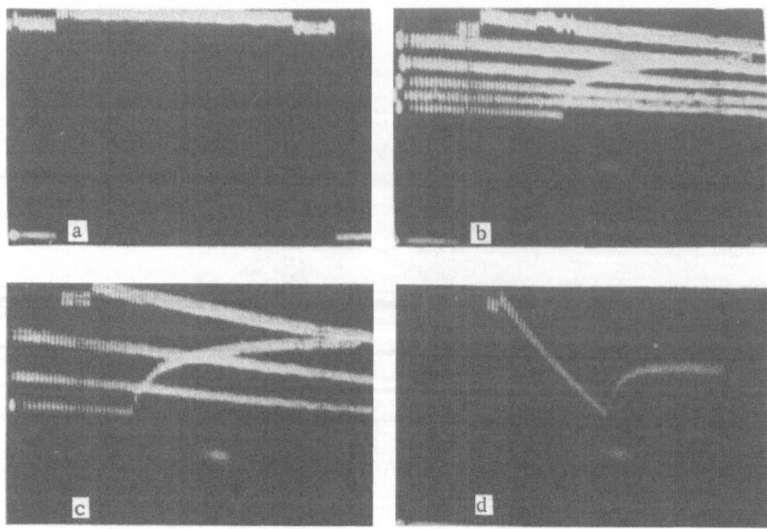

Fig. 3. Transmission change in response to a 400 Oe field
in an Fe_3O_4 sol stabilized by sodium oleate at NaCl concentrations (mM) of: a) 4; b) 150; c) 200; d) 250.

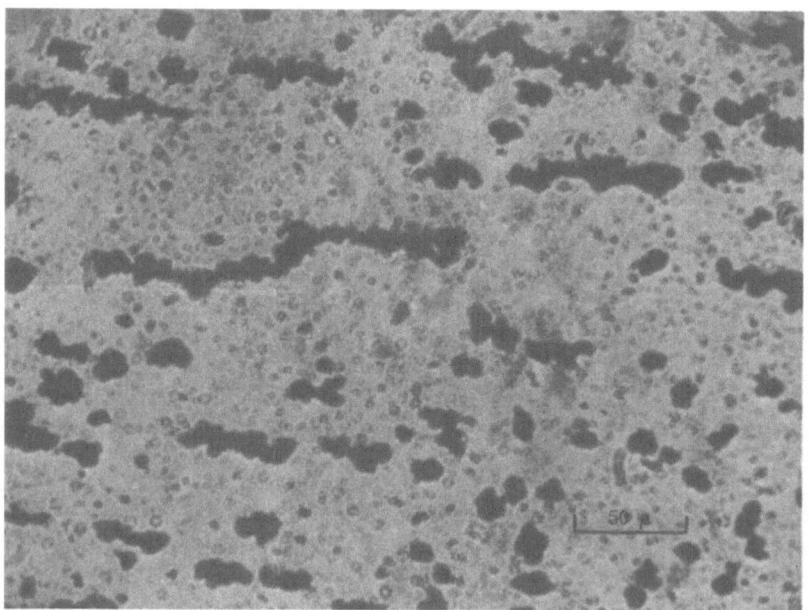

Fig. 4. Aggregation of particles in a Cu_2O suspension in a
50 Hz, 400 V/cm field.

to produce rapid coagulation [6], which points to anomalous properties of the medium in the
double layer, which requires a fairly long time to come to equilibrium.

An electric field produces a substantial charge $\Delta U(r, H)$, and so marked aggregation occurs mainly when the particles are readily polarized [7-9] and when the particles have fixed
dipole moments. The latter is not uncommon [10], so the above method can be extended to suspensions that are not ferromagnetic. In particular, we have observed electrically induced ag-

gregation in suspensions of SiO_2 and Cu_2O, the fact of aggregation being confirmed by the microscope (Fig. 4).

Electrical aggregation explains various effects in electrophoretic deposition. For instance, formation of mixed aggregates is the reason why the deposit from a mixture of suspensions of CuO ($\zeta = 10$ mV) and CoO ($\zeta = 0.5$ mV) reproduces the composition of the suspension, in spite of the large difference in ζ potential. Some suspensions show electrophoretic deposition only at high concentrations (>7% for ZnO, 45% for CuO); at lower concentrations, an electric field causes [11] aggregation and rapid sedimentation, while at higher concentrations the aggregation gives way to structuring of the entire suspension, which prevents sedimentation.

The reversibility of aggregation determines also the strength of fresh electrophoretic coatings; weak coatings are formed when the aggregation is reversible, while strong coatings are produced by irreversible coagulation.

Conclusions

1. Orientation effects can give information on particle interaction when this occurs. These conditions can also produce optical effects due to field-induced aggregation.

2. An empirical equation is given for the reduction in transmission of magnetite sols in response to magnetic fields, and it is shown that the constant in the equation is related to the stability of the sol.

3. Electrical aggregation explains some features of electrophoretic deposition of suspensions.

References

1. V. Tsvetkov and M. Sosinskii, Zh. Éksp. Teor. Fiz., 19:543 (1949).
2. N. A. Tolstoi, Dokl. AN SSSR, 100:893 (1955).
3. I. F. Efremov and S. V. Nerpin, Dokl. AN SSSR, 113:846 (1957).
4. E. E. Bibik, I. F. Efremov, and I. S. Lavrov, in: Research in Surface Forces, Vol. 2, Consultants Bureau, New York (1966), p. 238.
5. P. A. Rebinder and A. A. Trapeznikov, Dokl. AN SSSR, 18:425 (1938); Zh. Fiz. Khim., 12:573 (1938).
6. E. E. Bibik and I. S. Lavrov, Kolloidn. Zh., 26:391 (1964).
7. H. R. Kruyt and J. G. Vogel, Kolloid.-Z., 95:2 (1941).
8. L. G. Gindin and I. N. Putilova, Transactions of the Third All-Union Conference on Colloid Chemistry, Izd. AN SSSR, Moscow (1956), p. 182.
9. J. Stauff, Kolloid.-Z., 143:162 (1955).
10. N. A. Tolstoi, A. A. Spartakov, and A. A. Trusov, in: Research in Surface Forces, Vol. 3, Consultants Bureau, New York.
11. I. S. Lavrov, Kolloidn. Zh., 23:423 (1961).

THE STATISTICAL THEORY OF THE SPECIFIC ADSORPTION
OF IONS FROM ELECTROLYTE SOLUTIONS

G. A. Martynov and A. L. Muler

*Surface Phenomena Division, Institute of Physical Chemistry,
Academy of Sciences of the USSR*

Bogolyubov's equations are used to deduce an isotherm for specific adsorption of ions from an electrolyte
solution. It is found that the electrostatic repulsion causes the ion isotherm to lie below the isotherm for
uncharged particles.

A double electrical layer in a colloidal system or dispersion is usually formed as a result of specific ion adsorption.[†] Such adsorption is ultimately responsible for stabilizing hydrophobic colloids, for producing electrokinetic effects, etc.; it also plays a substantial part in electrochemical processes. Specific adsorption is therefore a basic aspect of the theory of double electrical layers. There are many papers on the topic [2-6], but there is no satisfactory statistical theory of the effect; although an ion adsorption isotherm has been deduced [7] entirely via a canonical Gibbs distribution, the treatment involved various unsound assumptions,[‡] which reduce the value of the results. Moreover, and this is important, this isotherm was derived by expanding the electrostatic terms in the initial quantities as perturbation series, although the parameters used in the expansion are not small for real systems. It is therefore by no means obvious that retention of only the first few terms will provide the required accuracy at substantial degrees of surface coverage. For these reasons, we here deduce the isotherm for specific adsorption of ions by a method not involving series expansion, the method being analogous to ones used in the statistical theory of liquids. For simplicity, we consider only nonlocalized adsorption on a plane.

[†] Image forces can also produce this effect if there are ions differing in valency [1].

[‡] For instance, no allowance was made for electrostatic interaction between the adsorbed ions, or for other interactions of these with the ions in the diffuse part of the double layer, although we show here that both of these are important in specific adsorption.

Derivation of the Basic Equation

1. We start with a chain of Bogolyubov equations for the distribution functions $\mathcal{G}_{a_1,\ldots a_s}$, which is precisely analogous to the canonical Gibbs distribution [8]. In particular, we have as follows for the unary distribution $\mathcal{G}_a(z_1)$, which describes the probability of finding a particle of type a at a distance z_1 from the interface $z = 0$ between the electrolyte and the external medium:

$$\theta \vec{\nabla}_1 \ln \mathcal{G}_a(z_1) + \vec{\nabla}_1 \Phi_a(z_1) + \int_V \sum_{1 \leqslant b \leqslant m} \nu_b \frac{\mathcal{G}_{ab}(\vec{r}_1, \vec{r}_2)}{\mathcal{G}_a(z_1)} \vec{\nabla}_1 \Phi_{ab}(r_{12}) \, d\vec{r}_2 = 0. \tag{1}$$

Here $\theta = kT$ (the temperature in energy units), $\Phi_a = \Phi_a^{(ad)} + \Phi_a^{(el)}$ is the total energy of a particle of type a in the field of the external forces, $\Phi_a^{(ad)}$ is the energy of specific adsorption, $\Phi_a^{(el)} = \frac{4\pi\eta e_a z_1}{\varepsilon}$ is the electrostatic energy of an ion a bearing a charge e_a in the field of an electrode having a charge density η, ε is the dielectric constant of the solvent, $\Phi_{ab} = \Phi_{ab}^{(s)} + \Phi_{ab}^{(el)}$ is the total energy of the pair interaction of particles a and b separated by a distance $r_{12} = |\vec{r}_1 - \vec{r}_2|$; $\Phi_{ab}^{(s)}$ is the short-range component of the previous, $\Phi_{ab}^{(el)} = \frac{e_a e_b}{\varepsilon r_{12}}$ is the electrostatic component, ν_a is the number of particles of type a in unit volume far from the boundary, and $\mathcal{G}_{ab}(\vec{r}_1, \vec{r}_2)$ is the binary distribution function (the probability of finding a and b particles simultaneously at points $\vec{r}_1$ and $\vec{r}_2$).

Equation (1) represents the balance of forces acting on an a particle. The differential term $\nabla_1 \Phi_a$ is the external force exerted on the particle by the adsorbent, while the integral term is the mean force exerted by the other particles in the system. The sum of these two forces in (1) is balanced by a force $\theta \vec{\nabla}_1 \ln \mathcal{G}_a(z_1)$ arising from the thermal motion of the particles. It can be shown that (1) is equivalent to the condition of constancy of the chemical potential μ_a of particle a as written in the form $\vec{\nabla}_1 \mu_a(z_1) = 0$.

2. There are infinitely many solutions to the chain of Bogolyubov equations, but the only ones adequate as regards the Gibbs canonical distribution are those that satisfy the conditions of symmetry, normalization, decrease in correlation, etc. [8]. In other words, these conditions replace the boundary and initial conditions used in solving ordinary differential equations; in both cases, the conditions serve to distinguish the physical solution from the infinite set of nonphysical solutions. We therefore start by giving the Bogolyubov equations a form such that imposition of additional conditions (or at least some of them) will be unnecessary.

The definition of a distribution function [8]

$$\mathcal{G}_{a_1,\ldots,a_s} = \frac{V^s}{\Omega_N} \int_V e^{-u_N/\theta} \, d\vec{r}_{s+1}, \ldots d\vec{r}_N, \quad \Omega_N = \int_V e^{-u_N/\theta} \, d\vec{r}_1, \ldots d\vec{r}_N,$$

$$u_N = \sum_{1 \leqslant i \leqslant N} \Phi_{a_i} + \sum_{1 \leqslant i < j \leqslant N} \Phi_{a_i b_j} \tag{2}$$

implies that $\mathcal{G}_a$ and $\mathcal{G}_{ab}$ can always be put as

$$\mathcal{G}_a = \gamma_a G_a, \quad \mathcal{G}_{ab} = \gamma_{ab} G_{ab}, \quad \gamma_a = e^{-\frac{\Phi_a^{(ad)}}{\theta}}, \quad \gamma_{ab} = e^{-\frac{\Phi_{ab}^{(s)}}{\theta}}, \tag{3}$$

in which G_a and G_{ab} are unknown correlation functions.

The normalization and decay of correlation conditions imply that $\mathscr{G}_a(z) \to 1$ for $z \to \infty$ and $\mathscr{G}_{ab}(z_1, z_2, r_{12}) \to \mathscr{G}_a(z_1) \cdot \mathscr{G}_b(z_2)$ for $r_{12} \to \infty$. As γ_a and γ_{ab} are then unity, we have in place of (3) that

$$\mathscr{G}_a = \gamma_a g_a, \quad \mathscr{G}_{ab} = \gamma_{ab}(\gamma_a g_a)(\gamma_b g_b)(1 + g_{ab}), \tag{4}$$

in which $G_a = g_a$, $G_{ab} = G_a G_b(1 + g_{ab})$, and the new correlation function g_{ab} becomes zero as $r_{12} \to \infty$. It is clear that writing $\mathscr{G}_a$, $\mathscr{G}_{ab}$ in the form of (4) also automatically meets the condition of symmetry provided only that we assume that

$$g_{ab} = g_{ab}(z_1, z_2, r_{12}) \tag{5}$$

is a symmetrical function of the coordinates and charges of particles a and b.

We put

$$g_a = e^{-\frac{\varphi_a}{\theta}}, \quad \varphi_a(z_1) = -\theta \ln g_a \tag{6}$$

(in which φ_a is the potential of the average force) and substitute (4) and (6) into (1) to get

$$\vec{\nabla}_1 \varphi_a(z_1) = \vec{\nabla}_1 \Phi_a^{(el)}(z_1) + \int_V \sum_b \nu_b \gamma_b(z_2) \gamma_{ab}(r_{12}) e^{-\frac{\varphi_b(z_2)}{\theta}} [1 + g_{ab}(z_1, z_2, r_{12})] \vec{\nabla}_1 \Phi_{ab}(r_{12}) d\vec{r}_2. \tag{7}$$

To this equation we must add the condition

$$\varphi_a(\infty) = 0, \tag{8}$$

which follows from the requirement that $g_a(\infty) = 1$.

3. Equation (7) cannot be used directly to determine $\varphi_a(z)$ because it contains the unknown function g_{ab} (as always, we assume that the potentials Φ_a and Φ_{ab}, are given).

There are three possible ways of transforming (7) to a closed equation: 1) series expansion of all the unknown functions (φ_a and g_{ab}) with respect to a small parameter; 2) deletion of g_{ab}; 3) approximation of g_{ab} by some combination of φ_a and φ_b. Each method will be considered separately.

Bogolyubov [8] showed for a system of point charges with $\Phi_{ab}^{(s)} = 0$ and $\gamma_{ab} = 1$, that g_{ab} is a small quantity whose order is higher than that of φ_a, so it can be neglected to a first approximation. However, this result has not yet been extended to particles of finite size. Moreover, no method has yet been devised for solving the Bogolyubov equations for a system consisting of such particles by series expansion of the distribution functions with respect to a small parameter. For this reason, method 1) cannot be used on (7).

It appears physically evident that transition from point charges to particles of finite size in the case of a sufficiently dilute solution should not alter the order of smallness of g_{ab}, since the mean distance between ions is then much larger than the radius of action of the non-Coulomb forces. In other words, we may suppose that the approximation

$$g_{ab} = 0, \quad \mathscr{G}_{ab} = \gamma_{ab}(\gamma_a g_a)(\gamma_b g_b), \tag{9}$$

(which was first used by Kirkwood and Stillinger in relation to the effects of ion volume on the structure of the diffuse part of the double layer)[†] correctly defines the first term in the series $\Phi_{ab}^{(s)} \neq 0$. It remains an open question whether (9) can be used for high concentrations, but we must bear in mind that this approximation correctly describes the behavior of the binary dis-

---

[†] This approximation has since been used in analogous problems [10, 11].

tribution function at large and small distances for any electrolyte concentration; further, g_{ab} appears in (7) only within the integral. Therefore, for the term containing g_{ab} to be negligible it is sufficient for the integral contribution for medium distances to be small relative to the integral contributions from small and large distances, which greatly reduces the demand for smallness[†] in g_{ab}.

The most accurate results might be obtained by putting g_{ab} as some function of g_a and g_b, thereby taking into account that $g_{ab} \neq 0$ at medium distances. However, we know of no statement as to the detailed form of $g_{ab} = g_{ab}[g_a(z_1), g_b(z_2), r_{12}]$ so the arbitrary element in the choice of the form for $g_{ab} = g_{ab}(g_a, g_b)$, substantially reduces the value of any results. On the other hand, the approximation of (9) is unambiguously defined by the conditions that must be satisfied by the binary distribution function. Of course, the simplest possible assumption is that $g_{ab} = 0$, so we will naturally try this first.

We may thus say that the approximation of (9), which gives

$$\frac{d\varphi_a(z_1)}{dz_1} = \frac{d\Phi_a^{(el)}(z_1)}{dz_1} + \int_V \sum_b \nu_b \gamma_b(z_2) e^{-\frac{\varphi_b(z_2)}{\theta}} \gamma_{ab}(r_{12}) \frac{\partial \Phi_{ab}(r_{12})}{\partial z_1} dr_2, \tag{10}$$

is presently the only soundly based method of terminating the infinite chain of Bogolyubov equations for systems of charged particles.

Adsorption of Uncharged Particles

1. Before we consider adsorption from electrolyte solutions, we consider the simpler case of a system of uncharged particles of a single kind. We shall see below that here not only can we solve (10) in a reasonably general way (which is of independent interest) but also we can establish the limits of application of the formulas obtained.

Equation (7) for this case becomes

$$\frac{d\varphi_a(z_1)}{dz_1} = -\nu\theta \int_V \gamma(z_2) e^{-\frac{\varphi(z_1)}{\theta}} [1 + g(z_1, z_2, r_{12})] \frac{\partial f(r_{12})}{\partial z_1} dr_2, \tag{11}$$

in which $f(r_{12}) = \gamma_{12} - 1 = \exp[-\Phi^{(s)}/\theta] - 1$ is the Mayer function.

For simplicity, we consider only the case where the adsorption forces are large but of very short range. Then to high accuracy we have[‡]

$$\gamma(z) = \begin{cases} 0 & \text{for} \quad -\infty < z < 0, \\ 1 + \gamma_0 \delta\left(\frac{z}{h}\right) & \text{for} \quad 0 \leqslant z < \infty, \end{cases} \tag{12}$$

in which γ_0 characterizes the effective depth of the potential well and h is the well width; $\gamma(z) = 0$ for $z < 0$ means that the half-space $z < 0$, which is filled by the adsorbent, is inaccessible to the gas.

The precise values of γ_0 and h in (12) are unimportant, since all subsequent formulas contain only $h\gamma_0$, which is defined by

---

[†] Remember that $g_{ab} \to 0$ for $r_{12} \to \infty$.

[‡] The effects of h and γ_0 on the adsorption are considered at the end of this section.

$$h\gamma_0 = \int\limits_0^\infty [\bar{\gamma}(z) - 1]\, dz, \tag{13}$$

in which $\bar{\gamma}(z)$ is the true value of $\gamma(z)$.

We assume that we always have

$$h\gamma_0 \gg e^{\frac{\varphi(0)}{\theta}} \int\limits_0^\infty [e^{-\frac{\varphi(z)}{\theta}} - 1]\, dz. \tag{14}$$

Now the total number of particles adsorbed per unit surface is

$$N^{(ad)} = \nu \int\limits_0^\infty [\mathscr{G}(z) - 1]\, dz = \nu \int\limits_0^\infty \left\{ \left[1 + \gamma_0 \delta\left(\frac{z}{h}\right) \right] e^{-\frac{\varphi(z)}{\theta}} - 1 \right\} dz$$

$$= \nu \left\{ h\gamma_0 e^{-\frac{\varphi(0)}{\theta}} + \int\limits_0^\infty \left[e^{-\frac{\varphi(z)}{\theta}} - 1 \right] dz \right\} \simeq \nu h\gamma_0 e^{-\frac{\varphi(0)}{\theta}}, \tag{15}$$

so (14) is equivalent to the assertion that the number of particles adsorbed in the layer $(0, h)$ greatly exceeds the number of particles adsorbed in the region $z > h$ (case of monomolecular adsorption). We also assume that the gas density is low far from the surface.

Substitution of (12) into (11) and use of (14) gives

$$\varphi'(z) = -\theta h\gamma_0 \nu e^{-\frac{\varphi(0)}{\theta}} \left\{ \int\limits_{-\infty}^\infty \frac{\partial f_{12}}{\partial z_1}\bigg|_{z_2 = 0} dx_2 dy_2 + \int\limits_{-\infty}^\infty \left[g_{12} \frac{\partial f_{12}}{\partial z_1} \right]_{z_2 = 0} dx_2 dy_2 \right\}.$$

Integration with respect to z from 0 to ∞ is applied with $f_{12}\bigg|_{\substack{z_1 = \infty \\ z_2 = 0}} = f_{12}(r_{12} = \infty) = 0$ to give

$$\varphi(0) = \theta \nu v^{(s)} \gamma_0 e^{-\frac{\varphi(0)}{\theta}} (A + B), \tag{16}$$

in which $v^{(s)} = h s_m$, s_m is the area taken up by one molecule in close packing, and

$$A = -\frac{2\pi}{s_m} \int\limits_0^\infty f_{1,2}(\rho)\rho d\rho;$$

$$B = \frac{1}{s_m} \int\limits_V g(0, z_2, \sqrt{\rho_2^2 + z_2^2}) \frac{\partial f(\sqrt{\rho_2^2 + z_2^2})}{\partial z_2}\, dr_2. \tag{17}$$

Here we have converted to cylindrical coordinates with their center at $x = y = z = 0$, since the symmetry means that subscript 1 to a coordinate in (5) can always be replaced by 2 and vice versa.

Then (15) gives the degree of filling of the monolayer as

$$\tau = \frac{N^{(ad)}}{N^{(ad)}_{\max}} = s_m N^{(ad)} = \nu v^{(s)} \gamma_0 e^{-\frac{\varphi(0)}{\theta}} \tag{18}$$

We express the $\varphi(0)$ of (16) in terms of τ to get the following adsorption isotherm:

$$\tau e^{\tau(A+B)} = v^{(s)}\gamma_0 \nu, \tag{19}$$

which is a rigorous consequence of the canonical Gibbs distribution if (12) and (14) are obeyed, no matter what the form of $\Phi_{12}^{(s)}$.

2. Equation (7) not only permits the calculation of the numbers of particles adsorbed in the z = 0 plane, but also their distribution in the neighborhood of the adsorbent.

We assume for simplicity that $f(r_{12})$ becomes zero[†] for $r_{12} > R$ and consider the region $z > R$. The density is low far from $z = 0$, so g_{12} can always be neglected for $z > R$, so (11) becomes

$$\varphi'(z_1) = v \int\limits_V \varphi(z_2) \frac{\partial f_{12}}{\partial z_1} \vec{dr_2} = v \int\limits_V \varphi(z_1 + r\cos\vartheta) \frac{\partial f(r)}{\partial r} \cdot r\cos\vartheta \cdot 2\pi r^2 dr \sin\vartheta d\vartheta \tag{20}$$

(here we have transferred the origin to $\vec{r_1}$ and use the fact that $\varphi(z) \ll 0$ for z large). We put $\varphi \approx e^{-bz}$ in (20) to get the solution to (20) as

$$\varphi(z) = \sum B_k e^{-b_k z}, \tag{21}$$

in which the b_k are the roots of the transcendental equation

$$b_k = 2\pi v \int\limits_V e^{-b_k r\cos\vartheta} \frac{\partial f(r)}{\partial r} r\cos\vartheta\, d(r\cos\vartheta)\, r dr. \tag{22}$$

In the particular case of hard spheres, for which

$$f = \begin{cases} -1, & 0 \leqslant r < r_0 \\ 0, & r_0 \leqslant r < \infty \end{cases},$$

this equation becomes

$$(b_k r_0)^3 = 4\pi v r_0^3 \{\operatorname{sh} b_k r_0 - b_k r_0 \operatorname{ch} b_k r_0\}. \tag{23}$$

It is readily shown [12] that (23) has only complex solutions, so $\varphi(z) \simeq A_1 e^{-\beta_1 z} \cos\omega_1 z$, if $z > r_0$, but if $\pi r_0^3 v \leq 0.2$,[‡] then $\beta_1 \geqslant 4.5/r_0$,[§] and so we may assume that $\varphi(z) \approx 0$ to high accuracy.

The situation is different in the transition region $0 \leq z \leq R$. Integration of (11) between limits of $z = \infty$ and z, and use of (12) gives that the potential of the mean force is proportional to $h\gamma_0$ in this range, and (14) shows that this is fairly large. It is readily shown that $\varphi(z)$ is positive for a system of hard spheres, so the particle concentration $\nu(z) = \nu \mathcal{G}(z) = \nu \exp[-\varphi(z)/\theta]$ in the transition layer $(0 \leq z \leq r_0)$ is much less than that in the bulk, whereas the converse applies in the adsorption plane $(z = 0)$. The two effects are closely related, since the particles adsorbed at the plane $z = 0$ take up much of the volume of the transition layer, which is thereby made inaccessible to particles incident from the gas (Fig. 1).

3. The adsorption isotherm of (19) contains the unknown function $B = B(\nu, \tau)$, but the magnitude of this is fairly easily estimated for a system of hard spheres, for which

$$f(r) = \begin{cases} -1, & 0 \leqslant r < r_0 \\ 0, & r_0 \leqslant r < \infty \end{cases}; \quad \frac{df}{dr} = \delta\left(1 - \frac{r}{r_0}\right), \quad s_m \simeq r_0^2,$$

$$A = \pi, \qquad B = \frac{2\pi}{r_0^2} \int\limits_0^{r_0} g_{12}(0, t, r_0)\, t dt. \tag{24}$$

[†] The definition of f_{12} implies that R has the meaning of the maximal radius of action of the molecular forces.

[‡] This corresponds to concentrations $\lesssim$ 2-3 M for particles with $r_0 \approx 3$ Å.

[§] The roots of (23) have been numbered [12] so that β_k increases with k.

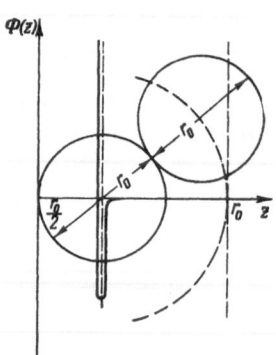

Fig. 1.

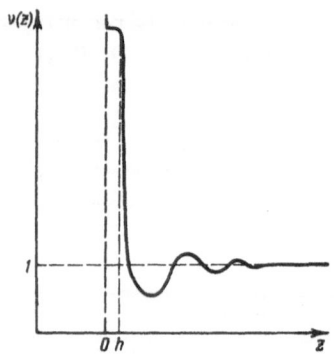

Fig. 2. Density distribution $\nu(z)$ near the surface of the adsorbent.

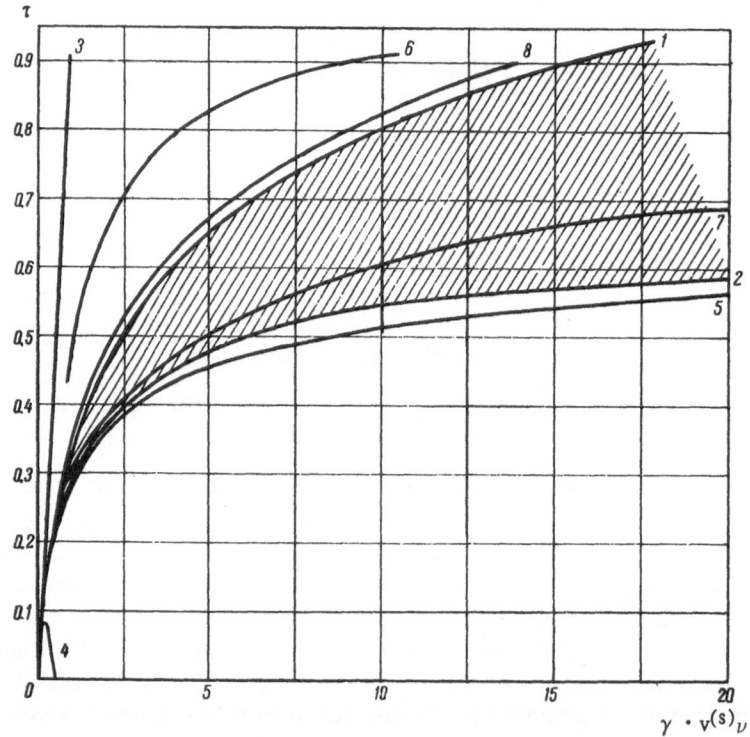

Fig. 3. Adsorption isotherms for neutral particles: 1) calculated from (25); 2) calculated on the assumption that g_{ab} is defined by the superposition approximation; 3) Henry isotherm; 4) isotherm calculated by series expansion with respect to a small parameter; 5) Baff–Stillinger isotherm; 6) Langmuir isotherm for localized adsorption; 7) Frumkin isotherm; 8) isotherm calculated from (31).

As the density is very low in the transition region (Fig. 2), we naturally suppose that g_{12} for an individual particle is substantial only if both molecules are in the adsorption layer (thickness h); $g_{12} = 0$ if only one of them is in this layer. We put $g_{12} = g_{12}^{(0)}$ in (24) to get that

$$B = \pi \frac{h^2}{r_0^2} g_{12}^{(0)}.$$

As h = 0 in our model, B $\equiv$ 0 in this approximation,[†]   and we at once get

$$\tau e^{\pi \tau} = v^{(s)} \gamma_0 v, \tag{25}$$

which is precisely the isotherm derived from (10).

Figure 3 shows the isotherm of (25) (curve 1), and it is clear that the assumption B = 0 leads to the absurd result $\tau > 1$ for $v^{(s)} \gamma_0 \nu \gtrless 23$. This shows that an isotherm of Langmuir type with saturation (and only this is possible for hard spheres) can be obtained only with allowance for the pair correlations.

Another estimate of B may be made on the assumption that the density in the transition layer remains constant at the density in the adsorption layer (Fig. 2). Then

$$g_{12}(z_1, z_2, r_{12}) = g_{12}^{(V)}(r_{12}),$$

in which $g_{12}^{(V)}$ is the value of the correlation function for a homogeneous and isotropic system with density $\nu = r_0^{-3} \tau^{3/2}$.

We take $G_{12}^{(V)} = 1 + g_{12}^{(V)}$ from the numerical solution [13] of the superposition equation to get the following results for $B/A = g_{12}^{(V)}$:

τ	0.21	0.28	0.32	0 36	0.42	0.58	0.77	0.87
$\frac{B}{A}$	0.14	0.20	0.26	0.33	0.45	0.80	1.36	1.66

Curve 2 of Fig. 3 shows the corresponding isotherm. The true isotherm lies somewhere in the hatched region between curves 1 and 2, and we expect for not very high degrees of filling that it will run closer to curve 1 than to curve 2, since the latter was obtained via a gross overestimate of B.

Figure 3 shows that the difference between curves 1 and 2 is very small for $\tau \lesssim 0.4$, and so here the isotherm of (25) can give quantitative agreement with experiment, whereas the agreement is only qualitative for $0.4 \lesssim \tau \lesssim 0.7$.

4. It is of interest to compare the above results with other known isotherms, especially the one obtainable from the Bogolyubov equations by expanding the distribution functions as perturbation series. It can be shown[‡] for a system of hard spheres that, if (12) and (14) are obeyed,

$$\tau = v^{(s)} \gamma_0 v [1 - \pi (v^{(s)} \gamma_0 v) + 1, 2\pi^2 (v^{(s)} \gamma_0 v)^2 \ldots]. \tag{26}$$

If we take only the first term in (26), we get the Henry isotherm:

$$\tau = v^{(s)} e^{-\frac{\eta^{(ad)}}{\theta}} \cdot v, \tag{27}$$

which describes the actual isotherm quite well for $\tau \lesssim 0.2$-0.3 (curve 3 of Fig. 3). However, (27) does not contain the attraction constant A, so the Henry isotherm does not allow us to evaluate the molecular interaction of the adsorbed particles. Incorporation of subsequent terms (curve 4 in Fig. 3) reduces the range of application of (26) to $\tau \lesssim 0.05$; as the contribution from the intermolecular forces is very small at such small τ, it is very difficult to distinguish this contribution from the total effect, whose magnitude is governed mainly by the force field of the adsorbent. It is therefore hardly possible to use (26) in the evaluation of experimental results.

[†] This estimate becomes incorrect for a degree of filling close to one.

[‡] On expanding (25) as powers of $(v^{(s)} \gamma_0 \nu)$ we get $\tau = v^{(s)} \gamma_0 v [1 - \pi (v^{(s)} \gamma_0 v) + \frac{3}{2} \pi^2 (v^{(s)} \gamma_0 v)^2 \ldots]$, which is virtually (26).

The Baff—Stillinger isotherm [7] is the second one derived by calculating the unary distribution for hard spheres, and this in our symbols is

$$v^{(s)}\gamma_0 v = \frac{\tau}{1 - \frac{\pi}{4}\tau} \exp\left[\frac{\frac{\pi}{4}\tau\left(3 - \frac{\pi}{2}\tau\right)}{\left(1 - \frac{\pi}{4}\tau\right)^2}\right]. \tag{28}$$

This may be deduced with exact allowance for the interaction between adsorbed particles in the plane z = 0, but with complete omission of the interaction between the adsorbed particles and those in the transition layer $0 < z < r_0$. Figure 3 shows that curve 5, which is derived from (28), lies very close to curve 2.

Finally, curve 6 in Fig. 3 shows the Langmuir isotherm

$$\tau = \frac{\gamma_0 v^{(s)}\nu}{1 + \gamma_0 v^{(s)}\nu}, \tag{29}$$

which is exact for localized adsorption of particles that do not interact. Also, curve 7 shows the Frumkin isotherm [4]

$$\frac{\tau e^{A\tau}}{1 - \tau} = Cv, \tag{30}$$

which may be considered as a simple approximation to the $e^{B\tau}$ of (19).†

Figure 3 shows that the Frumkin isotherm is the only one that falls in the hatched region, which is so if the attraction constant A is identified with (17), while the adsorption constant C is identified with $v^{(s)}\gamma_0$.

5. Isotherm (25) was deduced from a δ-function form for the potential well corresponding to (12), but the actual adsorption forces have a finite radius of action, so it is of interest to establish the error arising from the approximations of (12) and (13).

We have [14] solved (10) for a system of hard spheres for the case of a potential well of finite width, when

$$\gamma(z) = \begin{cases} \gamma_0, & 0 \leqslant z \leqslant h \leqslant r_0, \\ 1, & h < z < \infty. \end{cases}$$

This solution has the following form for $s_m = r_0^2$ and $A = \pi$:

$$v^{(s)}\gamma_0 v = \frac{\tau\xi}{\int_0^\xi \exp(A\tau x^2)\,dx} \cdot \exp\left[A\tau + \frac{1}{2} - \frac{\xi e^{A\tau\xi^2}}{2\int_0^\xi \exp(A\tau x^2)\,dx}\right]. \tag{31}$$

We have $\xi / \int_0^\xi \exp(A\tau x^2)\,dx \to 1$ if $\xi = h/r_0 \to 0$, and (31) becomes (25). Curve 8 of Fig. 3 shows the isotherm calculated from (31) for the case $h = 0.5r_0$. This almost coincides with the isotherm of (25), so the precise range of the adsorption forces is unimportant in monomolecular adsorption.

---

† Formula (30) was derived on a different basis.

Electrostatic Adsorption of Ions in the Diffuse
Part of a Double Layer

1. The diffuse layer occurs for z > R and plays no important part in the adsorption of uncharged particles at small degrees of filling. The situation is different in the adsorption of electrolyte ions, in which the long-range electrostatic forces cause this layer to make a substantial contribution. We must therefore briefly consider electrostatic adsorption in the diffuse part before we consider specific adsorption in the double layer proper.

We use the approximate Bogolyubov equation (10), because the above study has shown that it gives satisfactory results at not very great degrees of filling. We also restrict consideration to a very simple model of an electrolyte, namely a system of hard spheres bearing charges e_a and moving in a medium of dielectric constant ε. In that case, $f_{ab}(r)$ is defined by (24), while the electrostatic component of the energy of the pair interaction is

$$\Phi_{ab}^{(el)} = \frac{e_a e_b}{\varepsilon |\vec{r_1} - \vec{r_2}|}.$$

We substitute $\Phi_{ab} = \Phi_{ab}^{(s)} + \Phi_{ab}^{(el)}$ into (10) and use Green's formula[†] together with the fact that the derivative of $f_{ab}(r)$ is a δ function, which gives

$$0 \leqslant z \leqslant r_0, \quad \varphi_a'(z) = \frac{2\pi \eta k_a e}{\varepsilon}\left(1 - \frac{z}{r_0}\right) + \frac{k_a \varkappa^2 \theta}{2r_0^3}\int_0^{z+r_0} Q(t)\,dt + 2\pi v\theta \int_0^{z+r_0} q(t)(t-z)\,dt, \tag{32}$$

$$r_0 \leqslant z < \infty, \quad \varphi_a'(z) = \frac{4\pi k_a e}{\varepsilon}\left[\eta + \int_0^\infty \sum_b v_b e_b \gamma_b(z) e^{-\frac{\varphi_b(z)}{\theta}}\,dz\right] + \frac{k_a \varkappa^2 \theta}{2r_0^3}\int_{z-r_0}^{z+r_0} Q(t)\,dt + 2\pi v \int_{z-r_0}^{z+r_0} q(t)(t-z)\,dt, \tag{33}$$

in which $k_a = e_a/e$ is the valency of ion species a (where e is the electronic charge) and

$$q(z) = \sum_b n_b \gamma_b(z) e^{-\frac{\varphi_b(z)}{\theta}}; \quad Q(z) = \int_z^\infty \sum_b n_b k_b \gamma_b(t) e^{-\frac{\varphi_b(t)}{\theta}}\,dt;$$

$$v = \sum_b v_b; \quad n_b = \frac{v_b}{v}; \quad \sum_b n_b = 1; \quad \varkappa^2 = \frac{4\pi v e^2 r_0^2}{\varepsilon \theta}. \tag{34}$$

Putting $z \to \infty$ in (33) and using (8), the boundary condition at infinity, we get that

$$\eta + \int_0^\infty \sum_b v_b e_b \gamma_b(z) e^{-\frac{\varphi_b(z)}{\theta}}\,dz = 0.$$

Here we substitute $\gamma_a(z) = 1 + \gamma_a^{(0)}\delta\left(\frac{z}{h}\right)$ to get

$$\eta + \eta^{(ad)} + \eta^{(dif)} = 0, \tag{35}$$

[†] It is assumed in calculating the integrals via Green's formula that the volume V concerned in the integration is a layer of thickness 2H bounded by two infinite planes. The transition $H \to \infty$ may be made only in the final expression.

in which $\eta^{(ad)} = h \sum_b \nu_b e_b \gamma_b^{(0)} e^{-\frac{\varphi_b(0)}{\theta}}$ is the charge adsorbed in plane $z = 0$ and $\eta^{(dif)}$

$= \int_0^\infty \sum_b \nu_b e_b e^{-\frac{\varphi_b(z)}{\theta}} dz$ is the charge in the diffuse part of the double layer.

Expression (35) has the meaning of a neutrality condition, which arises in the Bogolyubov equations as a consequence of the boundary condition at infinity, and is not an independent requirement.

From (35) we get finally that

$$r_0 \leqslant z < \infty, \quad \varphi_0'(z) = \frac{k_a \varkappa^2 \theta}{2 r_0^3} \int_{z-r_0}^{z+r_0} Q(t)\, dt + 2\pi\nu\theta \int_{z-r_0}^{z+r_0} q(t)(t-z)\, dt. \tag{36}$$

No additional assumptions have been made in passing from (10) to (32) and (36), so (32) and (36) are identical with (10) for this model of an electrolyte.

2. Major mathematical difficulties are encountered in solving (36) for the diffuse part of the double layer [11], so the treatment is simplified by assuming that the solution is so dilute that $\varkappa \ll 1$. Then we convert in (36) to the dimensionless variables $x = z/r_D = \varkappa z/r_0$, $\xi = t/r_D = \varkappa t/r_0$, in which $r_D = \varkappa /r_0$ is the Debye radius, to get

$$\frac{\varphi_a'(x)}{\theta} = \frac{k_a}{2\varkappa} \int_{-\varkappa}^{\varkappa} Q(x+\xi)\, d\xi + \frac{1}{2\varkappa\chi} \int_{-\varkappa}^{\varkappa} q(x+\xi)\, \xi\, d\xi \tag{37}$$

in which $\varkappa^2 = 4\pi\nu r_0^3 \chi$, $\chi = e^2/\varepsilon\theta r_0$.

As $|\xi| \ll \varkappa \ll 1$ always in (37), the functions $Q(x+\xi)$ and $q(x+\xi)$ may be expanded as power series in ξ:

$$\frac{\varphi_a'(x)}{\theta} = k_a \left\{ Q(x) + \frac{\varkappa^2}{3!} Q''(x) + \cdots \right\} + \frac{\varkappa^2}{3\chi} \{ q'(x) + \cdots \}. \tag{38}$$

If we take only terms independent of $\varkappa$ in (38), the equation after differentiation becomes an ordinary Poisson−Boltzmann equation:

$$\psi''(z) = -\frac{4\pi e}{\varepsilon} \cdot \sum_b \nu_b k_b e^{-\frac{k_b e\psi(z)}{\theta}}, \tag{39}$$

whose solution is known, where $\varphi_a = k_a e\psi(z)$. In particular, the total charge in the diffuse layer is

$$\eta^{(dif)} = -\frac{\varepsilon}{4\pi} \sqrt{\frac{8\pi\theta}{\varepsilon} \cdot \sum_b \nu_b (e^{-\frac{k_b e\psi_1}{\theta}} - 1)}, \tag{40}$$

in which $\psi_1 = \psi_1(r_0)$ is the potential in the plane $z = r_0$.

Specific Adsorption of Ions in the Dense Part of a Double Layer

1. Graham's model (Fig. 4a) is generally accepted for this part of the layer on a metal, where it is supposed that the adsorbed ions lie within a molecular capacitor in a layer of liquid

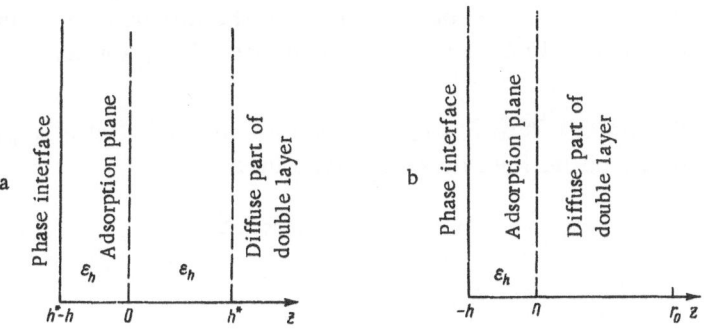

Fig. 4. Models for: a) Graham double layer; b) Stern
double layer.

of thickness h whose dielectric constant $\varepsilon_h \approx 3$ is much less than the dielectric constant ε of the solvent ($\varepsilon \approx 80$). It is also assumed that the region $0 \le z \le h^*$ within this capacitor is not accessible to ions in the solution, though the latter is hardly possible, since exchange between the adsorption plane and the bulk solution must occur through this $(0, h^*)$ layer (see below). It is as yet uncertain whether the region of reduced dielectric constant extends from $z = 0$ to the boundary with the external medium or somewhat further, so we will use the simpler Stern model of Fig. 4b, in which it is assumed that the plane $z = 0$ lies at the outer boundary of the molecular capacitor.

2. We thus use (32) for the potential of the mean force, which is equivalent to the Stern model (Fig. 4b) if we neglect the image forces and if the $\gamma_a(z)$ are given by

$$\gamma_a(z) = \begin{cases} 0, & -\infty < z < 0, \\ 1 + \gamma_a^{(0)}\delta\left(\frac{z}{h}\right), & 0 \le z < \infty. \end{cases} \tag{41}$$

We now use the fact that the definition of (34) and the neutrality condition of (35) imply that

$$Q(z) = Q(0) - \int_0^z \sum_b n_b k_b \gamma_b e^{-\frac{\varphi_b(z)}{\theta}}\, dz = -\frac{\eta}{ve} - \int_0^z \sum_b n_b k_b \gamma_b e^{-\frac{\varphi_b(z)}{\theta}}\, dz. \tag{42}$$

This expression is substituted into (32) to get

$$0 \le z \le r_0, \quad \varphi_a'(z) = -\frac{4\pi\eta e}{\varepsilon} k_a - \frac{k_a \varkappa^2\theta}{2r_0^3} \int_0^{z+r_0} dt \int_0^t \sum_b n_b k_b \gamma_b(\xi) e^{-\frac{\varphi_b(\xi)}{\theta}}\cdot d\xi + 2\pi v\theta \int_0^{z+r_0} q(t)(t-z)\, dt. \tag{43}$$

Substitution from (41) gives

$$\varphi_a'(z) = -\frac{4\pi\eta e}{\varepsilon} k_a - \left\{ \frac{k_a \varkappa^2\theta h}{2r_0^2} \sum_b n_b k_b \gamma_b^{(0)} e^{-\frac{\varphi_b(0)}{\theta}}\left(1 + \frac{z}{r_0}\right) \right.$$

$$\left. + z\cdot 2\pi v\theta h \sum_b n_b \gamma_b^{(0)} e^{-\frac{\varphi_b(0)}{\theta}} \right\} + \left\{ -\frac{k_a \varkappa^2\theta}{2r_0^3} \int_0^{z+r_0} dt \int_0^t \sum_b n_b k_b e^{-\frac{\varphi_b}{\theta}}\, d\xi + 2\pi v\theta \int_0^{z+r_0} \sum_b n_b e^{-\frac{\varphi_b}{\theta}}(t-z)\, dt \right\}. \tag{44}$$

As previously, we consider $h\gamma_b^{(0)}$ as large, so the expression within the braces can be neglected,[†] and this means that we neglect the charge in the part of the diffuse layer in the

---

[†] In principle, allowance is readily made for the discarded integrals (see below).

region $0 \le z \le r_0$; in other words, we assume that near the adsorption plane there is a region of zero charge, which to some extent represents transition from the Stern model to the Graham model.

Integration of (44) from $z = 0$ to $z = r_0$, taking into account that $\varphi_a(r_0) = k_a e \psi_1$, in which ψ_1 is the electrostatic potential defined by (39), gives us that

$$\varphi_a(0) = k_a e \psi_1 + \frac{4\pi e k_a}{\varepsilon}\eta r_0 + \frac{3\pi k_a e r_0 h}{\varepsilon}\sum_b v_b e_b \gamma_b^{(0)} e^{-\frac{\varphi_b(0)}{\theta}} + \pi\theta h r_0^2 \sum_b v_b \gamma_b^{(0)} e^{-\frac{\varphi_b(0)}{\theta}}. \tag{45}$$

We introduce the symbols

$$\tau_a = s_m h \gamma_a^{(0)} v_a e^{-\frac{\varphi_a(a)}{\theta}} = v^{(s)}\gamma_a^{(0)} v_a e^{-\frac{\varphi_a(0)}{\theta}}; \qquad A = \frac{\pi r_0^2}{s_m}; \qquad \tau = \sum_b \tau_b, \tag{46}$$

whose meaning is clear. Substitution of (46) into (45) gives

$$v^{(s)}\gamma_a^{(0)} v_a = \tau_a \exp\left\{A\left[1 + 3\chi k_a \sum_b k_b \frac{\tau_b}{\tau}\right]\tau + \frac{k_a e \psi_a}{\theta} + \frac{4\pi r_0 k_a e}{\varepsilon\theta}\eta\right\}. \tag{47}$$

As we can write such expressions for all M types of ions ($1 \le a \le M$), expression (47) represents a system of M equations for the M unknown degrees of filling τ_a. The other unknown, the electrostatic potential ψ_1 in the plane $z = r_0$, may be deduced from (35), in which $\eta^{(\text{dif})}$ is to be replaced by (40) and $\eta^{(\text{ad})}$ by $(A/\pi r_0^2)e_b\tau_b$.

Analogy with a gas of uncharged particles suggests that the resulting solution will apply only for small and medium τ. It is clear that (47) is similar to the Frumkin isotherm of (30), so we may propose the following generalization, which very probably also applies for τ large:

$$v^{(s)}\gamma_a^{(0)} v_a = \frac{\tau_a}{1-\tau} \exp\left\{A\left[1 + 3\chi k_a \sum_b k_b \frac{\tau_b}{\tau}\right]\tau + \frac{k_a e \psi_1}{\theta} + \frac{4\pi r_0 k_a e}{\varepsilon\theta}\eta\right\}. \tag{48}$$

However, we cannot take this as a soundly based formula.

3. If ions of one kind are adsorbed preferentially (and this is the case of greatest interest), the sum in the exponent in (47) is

$$\sum_b k_b \frac{\tau_b}{\tau} = k_a.$$

Then the total attraction constant is

$$A_a^{(el)} = A\left[1 + 3\chi k_a^2\right]. \tag{49}$$

As χ is 2–3 for aqueous solutions, $A^{(el)}$ for univalent ions is roughly 7–10 times A for uncharged particles, and the factor for divalent ions is 25–40. Charged particles are therefore adsorbed much more weakly than uncharged ones, the increase in τ being the slower the higher the ion charge (Fig. 5). This is as would be expected, because the electrostatic repulsion between the adsorbed particles increases with the charge.

We can also use (35), (40), and (47) to find the potential ψ_1 at the boundary $z = r_0$ of the diffuse layer; Fig. 6 shows the corresponding curves, which indicate that ψ_1 for low concentrations of the potential-determining ions is virtually a linear function of v_a, though this is later followed by saturation and then a fall (all the calculations are for $\eta = 0$). In fact, we know from electrostatics that the potential difference within the capacitor is

$$\Delta\psi = 4\pi\eta L,$$

in which η is the charge on the plates, whose separation is L.

In our case, $\eta^{(ad)} = -\eta^{(dif)}$ acts as η, while r_D (Debye radius) acts as L. If the surface charges are small, $\eta^{(ad)}$ is proportional to ν (this follows from Henry's law), whereas $r_D \sim 1/\sqrt{\nu}$, and so $\Delta\psi = \psi_1$ increases as $\sqrt{\nu}$. The rise in $\eta^{(ad)}$ is less rapid at high τ (Fig. 5), while r_D still continues to fall, which is responsible for the maximum in $\psi_1 = \psi_1(\nu)$.

4. Finally, we consider the potential distribution in the transition layer $(0, r_0)$, for which purpose we put (44) as[†]

$$\varphi_a'(z) = -\frac{4\pi e k_a}{\varepsilon}\eta - \frac{2\pi e k_a}{\varepsilon}\eta^{(ad)}\cdot\left(1 + \frac{z}{r_0}\right) - A\tau\theta\frac{2z}{r_0^2}. \tag{50}$$

Integration of (50) from $z = r_0$ to z gives

$$\varphi_a(z) = k_a e\psi_1 + \frac{4\pi k_a e r_0}{\varepsilon}\eta\left(1 - \frac{z}{r_0}\right) + \frac{\pi k_a e r_0}{\varepsilon}\eta^{(ad)}\left[3 - \frac{2z}{r_0} - \frac{z^2}{r_0^2}\right] + A\tau\theta\left(1 - \frac{z^2}{r_0^2}\right), \tag{51}$$

in which the right part is known, since we have assumed that system (35), (40), and (47) has already been solved.

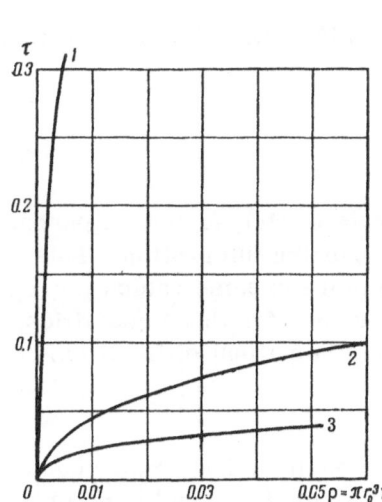

Fig. 5. Adsorption isotherms for $v^{(s)}\gamma_0\nu = 1500$: 1) neutral particles, as calculated from (25); 2) univalent ions ($\chi = 2$); 3) divalent ions ($\chi = 8$).

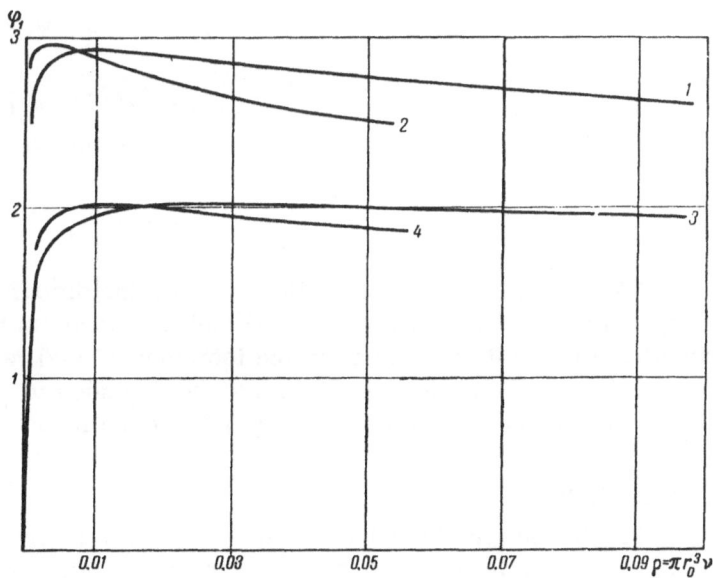

Fig. 6. Relation of electrostatic potential ψ_1 to electrolyte concentration: 1) $h\gamma_0 = 1500$, $\chi = 2$; 2) $h\gamma_0 = 1500$, $\chi = 8$; 3) $h\gamma_0 = 200$, $\chi = 2$; 4) $h\gamma_0 = 200$, $\chi = 8$.

[†] As previously, we neglect the expression in braces.

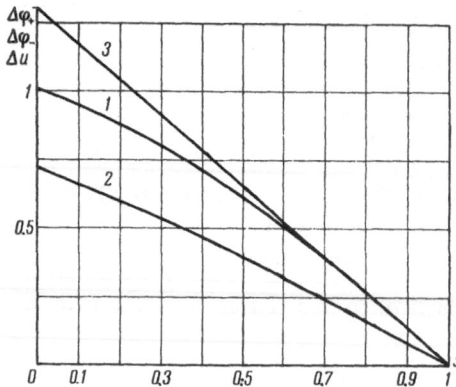

Fig. 7. Distribution of the potential of the mean force $\varphi_\pm$ and of the electrostatic potential u = $e\psi_1/\theta$ in the layer $0 \leq x \leq 1$ (x = z/r_0) for $\pi r_0^3 \nu = 0.01$; $\tau = 0.046$. 1) $\Delta\varphi_+ = \varphi_+(x) - \varphi_+(1)$; 2) $-\Delta\varphi_- = -[\varphi_-(x) - \varphi_-(1)]$; 3) $\Delta u(x) = u(x) - u(1)$.

From $\varphi_a(z)$ we can find the distribution function $\mathcal{G}_a(z) = \exp[-\varphi_a(z)/\theta]$, and use Poisson's equation

$$\psi''(z) = -\frac{4\pi}{\varepsilon}\sum_b \nu_b e_b \mathcal{G}_b(z) \qquad (52)$$

to determine the form of $\psi(z)$. Figure 7 shows the results. These curves show that the charge in the layer $0 \leq z \leq r_0$ has little effect on the distribution of the potential there (provided, of course, that the solution is sufficiently dilute). We can therefore assume without substantial error that the total potential difference between the planes z = 0 and z = ∞ is

$$\psi_{10} = \psi_1 + \frac{\partial\psi}{\partial z}\bigg|_{z=r_0} \cdot r_0 = \psi_1 - \frac{4\pi\eta^{(dif)}}{\varepsilon} \cdot r_0. \qquad (53)$$

In the particular case of low τ, where $A^{(el)}\tau \simeq 0$ and $\eta \simeq -\eta^{(dif)}$, we get from (53) that $\psi_{10} = \psi_1 + (4\pi\eta r_0/\varepsilon)$, so (47) reduces to the Boltzmann isotherm

$$\tau_a = v^{(s)}\gamma_a^{(0)}\nu_a e^{-\frac{l_a e\psi_{10}}{\theta}}. \qquad (54)$$

We can use (51) to make allowance for the discarded integrals in (44), for which purpose we substitute (51) into the solution of (39) and perform the corresponding integration. This gives rise to additional terms in the isotherm of (47), which give more precise values for τ_a and ψ_1. In principle, the operation may be extended to give exact values for these quantities; but it must be borne in mind that this procedure involves major mathematical difficulties.

Conclusions

1. Various methods have been examined for terminating the chain of Bogolyubov equations, and it has been shown that the Kirkwood–Stillinger approximation is the only one presently applicable for a system of charged particles.

2. A solution to the equation is given for a system of uncharged particles with an arbitrary form for the energy of the pair interaction. Evaluations show that the resulting isotherm can give a quantitative description of experimental results for degrees of filling $\tau \lesssim 0.4$ and a qualitatively correct description up to τ of 0.7-0.8.

3. It has been found that the equation goes over to the Poisson–Boltzmann equation in the case of a dilute solution of an electrolyte. The limits of application of the Poisson–Boltzmann equation are evaluated.

4. An isotherm has been derived for specific adsorption of ions from a solution of an electrolyte of arbitrary composition, which applies only for reasonably dilute solutions and not very large degrees of filling. It is shown that electrostatic repulsion of the adsorbed ions causes the adsorption isotherm for charged particles to run much below that for uncharged molecules. The trend in the potential ψ_1 with the concentration of the potential-determining ions has been elucidated.

References

1. G. A. Martynov, in: Research in Surface Forces, Consultants Bureau, New York (1966), p. 75.
2. O. Stern, Z. Elektrochem., 30:508 (1924).
3. D. C. Graham, Chem. Rev., 41:441 (1947).
4. A. N. Frumkin, Z. Phys., 35:792 (1926).
5. B. V. Érshler, Zh. Fiz. Khim., 20:678 (1946).
6. V. A. Krylov and V. G. Levich, Dokl. AN SSSR, 142:123 (1962).
7. N. M. Baff and F. H. Stillinger, Principles of Theoretical Electrochemistry [Russian translation], Izd. Mir, Moscow (1965), p. 141.
8. N. N. Bogolyubov, Problems of Dynamical Theory in Statistical Physics, OGIZ, Moscow—Leningrad (1946).
9. F. H. Stillinger and J. G. Kirkwood, J. Chem. Phys., 33:1282 (1960).
10. V. A. Krylov and V. G. Levich, Zh. Fiz. Khim., 37:106 (1963).
11. G. A. Martynov, in: Research in Surface Forces, Vol. 2, Consultants Bureau, New York (1966), p. 94.
12. G. A. Martynov, Zh. Éksp. Teor. Fiz., 45:656 (1963).
13. I. Z. Fisher, The Statistical Theory of Liquids, University of Chicago (1964).
14. G. A. Martynov and A. L. Muler, Dokl. AN SSSR, 170:1296 (1966).
15. G. A. Martynov and Yu. M. Kessler, Élektrokhimiya, 3(1):76 (1967).
16. G. A. Martynov, V. G. Melamed, and P. T. Ali-Zade, Dokl. AN SSSR, 151:601 (1963).

STABILIZATION MECHANISMS
OF ION-STABILIZED COLLOIDS IN
THE PRESENCE OF SURFACE-ACTIVE SUBSTANCES

G. A. Martynov and D. S. Lychnikov

Surface Phenomena Division, Institute of Physical Chemistry,
Academy of Sciences of the USSR

A study has been made on stabilization of illite clay in suspension by sodium oleate. It is found that the effective force of attraction between particles is inversely related to the thickness of the adsorbed sodium oleate film. The coagulation threshold is dependent on the valency of the electrolyte, which shows that a double electrical layer is present. The sol is stable because this double layer is separated from the surface of the particle by the thickness of the adsorption film, with consequent weakening of the molecular attraction.

There are many papers [1-5] on colloids stabilized by surfactants, but there is no agreed view on the mechanism of the effect. We have examined a system that is stabilized by double electrical layers in the absence of surfactants, which enables one to establish the new features introduced by the surfactant.

Initial Materials

All of the tests were done with an aqueous suspension of the illite clay previously used [6]. The mineral has a fixed lattice and an internal volume porosity not exceeding 8% [9]. The specific surface (by nitrogen adsorption) was 29.6 m^2/g, which agrees well with the value 30 m^2/g calculated from the adsorption isotherm for sodium oleate (see below), but does not agree so well with the 40 m^2/g given by the adsorption of methylene blue. In all cases the clay had a concentration of about 0.450 g/liter. The cation-exchange capacity of illite clay is 5.44 meq per 100 g of dry material [9], which corresponds to a ζ potential of some tens of mV [9]. The surfactant was sodium oleate (NaOl) containing not more than 2% of impurities, mainly oleic acid.

Determination of Deposition Time

1. Methods. The suspension (50–70 ml) was poured into a 100 ml volumetric flask, to which were added the NaOl and electrolyte; the contents were mixed, and the volume was

226

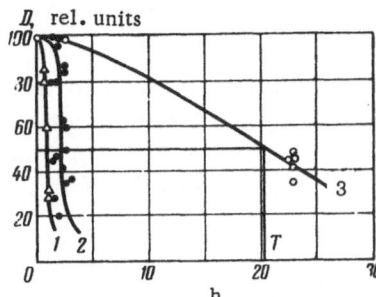

Fig. 1. Optical density D as a function of time for an initial NaOl concentration of 1.33 mM and NaCl concentrations (M) of: 1) 0.02; 2) 0.03; 3) 0.04.

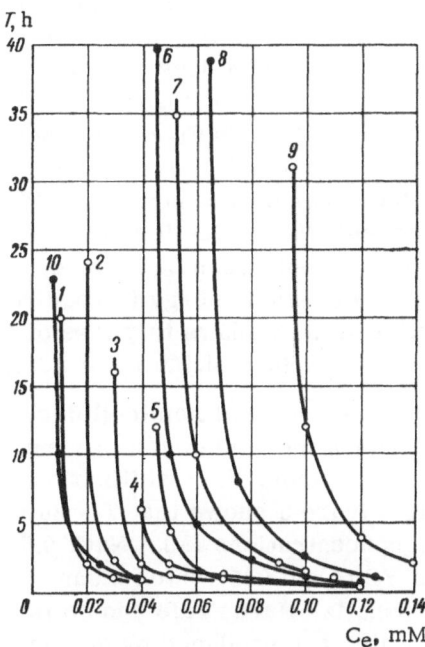

Fig. 2. Time T for half-deposition as a function of electrolyte concentration C_e for initial NaOl concentrations (mM) of: 1) 0.666; 2) 1.333; 3) 2; 4) 2.66; 5) 3; 6) 3.33; 7) 4; 8) 5; 9) 6.66; 10) 0.

made up to 100 ml with the suspension, followed by further mixing for 3-5 min. The mixture was poured into 30 ml cylinders 170 mm high. The deposition rate was deduced from the change in optical density D as recorded by a photoelectric colorimeter used with white light (Fig. 1). The deposition occurs in three stages. In the first, D remains almost constant. In the second, the aggregates grow rapidly and the largest ones are deposited, which is accompanied by an almost linear fall in D. In the third, the small aggregates deposit slowly. The process may be characterized [1, 7] by the time T for D to fall by 50% (deposition of about 50% of the material).

Figure 1 shows that T always falls in the second stage, in which the relative error in the measurement of D is dependent on the deposition rate, being not more than 15% for small rates but rising to 30% for high rates. The corresponding errors in T are 5-10% and 20%.

2. Results. Figure 2 shows T as a function of electrolyte (NaCl) concentration C_e for various initial concentrations of NaOl. There is a rapid fall from an initially almost stable state (T ≫ 20 h) to a state of low stability (T ≈ 5 h) and then a continuing slow fall in T. The NaCl concentration corresponding to loss of stability increases with the initial concentration of NaOl.

We examined the effects of polyvalent electrolytes in order to establish whether the effects are associated with double electrical layers. Figure 3 shows T as a function of concentration for NaCl, Na_2SO_4, $BaCl_2$, and $Y(NO_3)_3$ for an initial NaOl concentration of 2 mM. NaCl and Na_2SO_4 have practically the same effects, whereas the critical concentration of $BaCl_2$ for T = 1 h is about 1/60 of that for NaCl, and the ratio for $Y(NO_3)_3$ is 1/150. Deryagin's theory [8] of highly charged sols explains these results satisfactorily, since it predicts that increases in the counterion (cation) charge by factors of 2 and 3 should reduce the critical concentrations by factors of 64 and 729 (z^6 law), while there should be hardly any effect from increase in the charge of the ion whose sign is the same as that of the colloidal particles. There is thus no doubt that double layers play a major part, but the theory does not explain why the surfactant displaces the critical concentration to higher values (Fig. 2).

Aggregation Mechanism of Micelles in Sodium Oleate Solution

1. Methods. The stability of sol is related to micelle formation and to salting-out of Naol, so we examined the effects of electrolytes on pure soap solution. The NaOl solution was

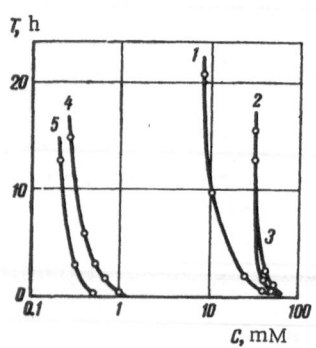

Fig. 3. Time T for half-deposition for an initial NaOl concentration of 2 mM as a function of concentration of: 1, 2) NaCl; 3) Na$_2$SO$_4$; 4) BaCl$_2$; 5) Y(NO$_3$)$_3$. Curve 1 is for no NaOl present.

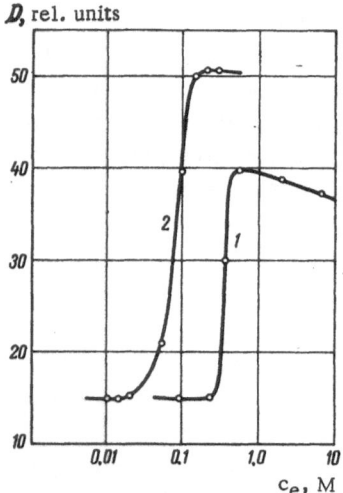

Fig. 4. D for a soap solution as a function of C$_e$ for: 1) NaCl; 2) BaCl$_2$.

appreciably turbid at all concentrations on account of hydrolysis of the NaOl, so D was used as previously.

Test tubes containing the solution (1.3 mM) were treated with the required amount of electrolyte, and D was measured as a function of time to 10% or better. Figure 4 shows D as a function of C$_e$, while Fig. 5 shows T as a function of C$_e$.

2. Discussion. BaOl$_2$ and YOl$_3$ are insoluble in water, so addition of only 0.65 mM BaCl$_2$ to 1.3 mM NaOl would appear capable of precipitating all the soap, but this is not actually the case, since the solution remains stable up to about 13 mM in BaCl$_2$ (Fig. 5), or roughly 20 times the concentration needed to convert NaOl to BaOl$_2$. The same applies on addition of Y(NO$_3$)$_3$.

If the NaOl were in ordinary solution (Na$^+$ and Ol$^-$ ions), the effect would be inexplicable; but most of the molecules in the NaOl solution are present in micelles. On addition of BaCl$_2$, the isolated NaOl molecules react rapidly to give insoluble BaOl$_2$, which should then rapidly precipitate; but measurements extending over more than 4 days at BaCl$_2$ concentrations ≲13 mM revealed no precipitate. Moreover, Fig. 4 shows that D is independent of the BaCl$_2$ content below this concentration. This must mean that the proportion of molecular soap is less than the error of measurement. Molecules of NaOl thus equilibrate with NaOl micelles only very slowly; otherwise insoluble BaOl$_2$ would soon be formed, which means that a substantial activation energy is required in order to disrupt the micelles (Fig. 6 shows that loss of a molecule from a micelle involves rupture of lateral bonds).

A NaOl solution is therefore a typical semicolloidal system[†] in which most of the NaOl molecules are present in aggregates. We thus have to consider the stabilization mechanism for the NaOl sol. Figure 5 shows that T = 20 h corresponds to the following concentrations (M): NaCl 0.5, Na$_2$SO$_4$ 0.2, BaCl$_2$ 0.015, and Y(NO$_3$)$_3$ 0.005, which means that the soap sol obeys the Schultz−Hardy rule and so is stabilized by double layers, which are produced by dissociation of NaOl into Na$^+$ ions (which form the diffuse part of the layer) and Ol$^-$ ions, which are linked into micelles (Fig. 6b).

The effects of ultrasound provide additional evidence for the semicolloidal nature of the system; ultrasound accelerates deposition, as for clay [6], and C$_e$ affects the process also as in the case of clay (Fig. 7), which means that the interaction energy as a function of distance for soap micelles has the second minimum characteristic of ion-stabilized systems.

We can now return to the stabilization of clay by NaOl. Figures 2, 3, and 5 show that the effects of C$_e$ on T are generally much the same for the clay and the soap, but the soap requires a much higher C$_e$, so there is no direct relation between the two effects.

---

† This is also indicated by the resemblance between the T(C$_e$) curves for the clay (Figs. 2 and 3) and the soap (Fig. 5).

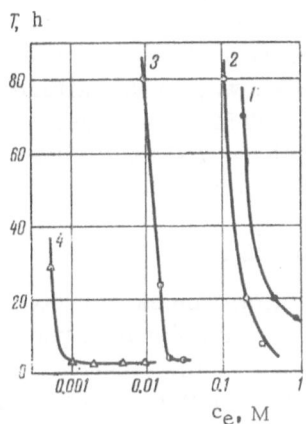

Fig. 5. T for 1.33 mM NaOl as a function of C_e for: 1) NaCl; 2) Na_2SO_4; 3) $BaCl_2$; 4) $Y(NO_3)_3$.

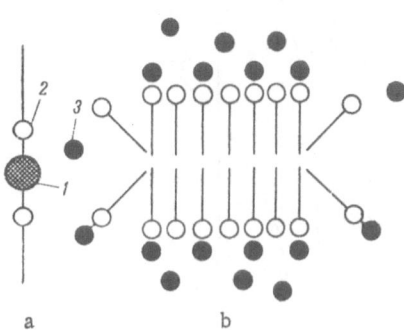

Fig. 6. a) Molecule of $BaOl_2$; b) platy NaOl micelle in a solution containing NaCl: 1) Ba^{2+}; 2) Ol^-; 3) Na^+.

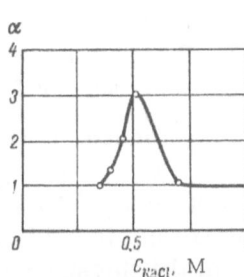

Fig. 7. Effect of NaCl concentration on the relative coagulation rate $\alpha = T_0/T$ for a NaOl emulsion acted on by ultrasound (450 kHz, 1.25 W/cm^2). T_0 is the deposition time for the unexposed solution, and T is the time for the sonicated solution.

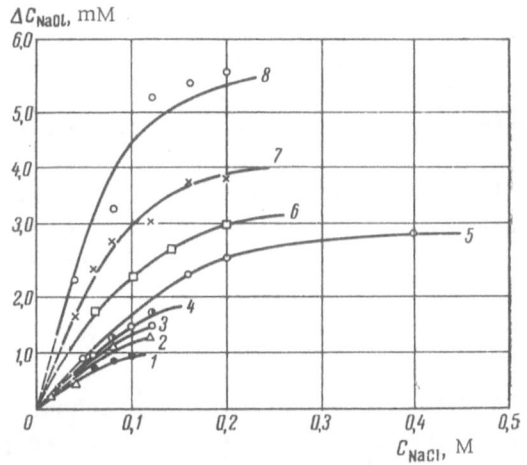

Fig. 8. Adsorbed soap ΔC_{NaOl} on illite clay (0.45 g/liter, 100 ml) as a function of NaCl concentration for initial NaOl concentrations (mM) of: 1) 1; 2) 1.33; 3) 2.0; 4) 2.64; 5) 3.3; 6) 4.0; 7) 5.2; 8) 6.6.

Adsorption of Sodium Oleate on Clay

1. Methods. The adsorption isotherms were recorded optically in the range of soap and electrolyte concentrations in which the clay suspension precipitates fairly rapidly. The suspension was prepared as above, and the measurements were made 24 h after preparation, which allowed suspended particles to settle out and adsorption equilibrium to be reached. If the deposition was not complete, centrifugation for 3 min at 2000-3000 rpm was used in order to

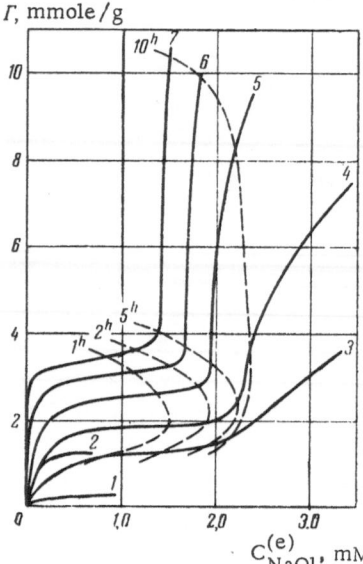

Fig. 9. Adsorption isotherms for NaOl (millimoles/g) for NaCl concentrations (mM) of: 1) 0; 2) 0.06; 3) 0.04; 4) 0.06; 5) 0.08; 6) 0.10; 7) 0.12. Curves 1 and 2 were recorded by the ring method. The broken lines are those of equal time, the times (h) being given on the curves.

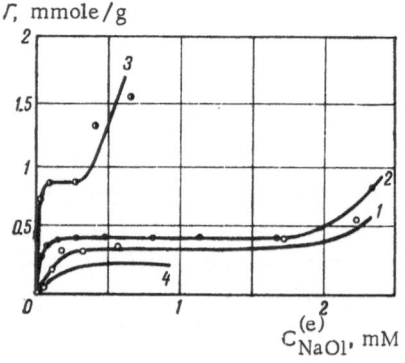

Fig. 10. Adsorption isotherms for NaOl at $BaCl_2$ concentrations (mM) of: 1) 0.8; 2) 1.0; 3) 1.2; 4) 0.0.

complete it. The extinction coefficient at 480 mμ was then measured for the supernatant, which gave the equilibrium concentration $C_{NaOl}^{(e)}$. The initial concentration $C_{NaOl}^{(i)}$ was known, so the difference gives the amount adsorbed. This $\triangle C$ was determined to within 5%. Figure 8 shows $\triangle C$ as a function of NaCl concentration, and the curves were worked up for C_e = constant to give the adsorption isotherms Γ_{NaOl}= $f(C_{NaOl}^{(e)})$ (Figs. 9 and 10).

In the concentration ranges where the suspension was stable, $C_{NaOl}^{(e)}$ was measured via the surface tension σ as determined by the ring-detachment method [11]. Each measurement was performed twice: a few hours after preparation of the specimen and after 24 h. No appreciable change in σ with time was detected. The adsorption isotherms of NaOl for C_e = 0 were recorded in this way, when the suspension is almost indefinitely stable (curve 1 in Figs. 9 and 10). If the horizontal part of the isotherm corresponds to complete filling of a monolayer, and with an area of 28 Å^2 for one NaOl molecule [12], we get S = 30 m^2/g as the specific surface, which agrees very well with the result (S = 29.6 m^2/g) from the adsorption of nitrogen.[†] Curve 2 of Fig. 9 (0.06 M NaCl, ring method) agrees well with curve 4 (photometric method, same NaCl concentration), the relative error not exceeding 20%. This shows that our method gives satisfactory accuracy for Γ.

2. Discussion. Figures 8-10 show that Γ increases rapidly with C_e, e.g., for $C_{NaOl}^{(e)}$ = 1.5 mM, Γ increases from about 0.2 millimole/g to about 10 as C_e increases from 0 to 0.12 M, which corresponds to increase in the thickness of the adsorption layer from monomolecular to about 50 molecules thick (Fig. 9). The effect is probably typical of such systems: 1) in the adsorption of NaOl at an oil—water interface it was found [3] that the film is 29 molecules thick; 2) curves analogous to those of Fig. 8 have been obtained for the effects of electrolytes on the adsorption of hydrolyzed polyacrylonitrile on kaolin [10]; and 3) curves analogous to those of Fig. 9 have been reported [15].[‡]

The curves of Fig. 9 are S shaped. A nearly horizontal part on an adsorption isotherm usually corresponds to formation of a monomolecular layer.

† This measurement was kindly made by G. S. Blyskash in Professor Taubman's laboratory.
‡ See also [13-18] on the polymolecular adsorption of surfactants in the presence of electrolytes.

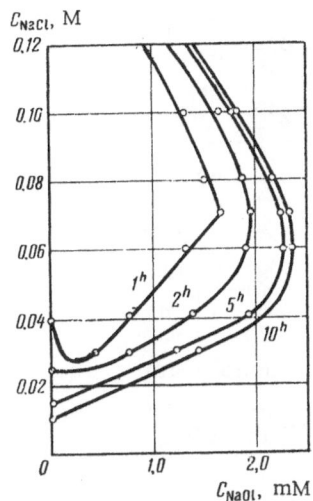

C_{NaCl}, M

0.12

0.10

0.08

0.06

0.04

0.02

0

1.0   2.0

C_{NaOl}, mM

Fig. 11. Relation of NaCl concentration to $C_{NaOl}^{(e)}$ for various T (given in hours on the curves).

In the present case, single NaOl molecules are virtually absent, and their role is played by micelles, which are probably planar (Fig. 6), so the horizontal parts in Fig. 9 correspond to monomicelle coating. The height of the plateau increases with C_e, which may[†] be a result of increase in micelle size. The rise after the plateau represents polymicelle adsorption and cannot be considered as representing filling of internal pores in the clay, since complete filling at a bulk porosity of 8% would require only about 0.08 mM, which is only half the amount of soap needed to produce a layer one molecule thick on the outer surfaces of the particles. The results therefore show that the soap forms a layer many angstroms thick when the soap and electrolyte concentrations are high, although a NaOl molecule is only about 20 Å long.

Discussion

1. The stability data can be replotted as T as a function of $C_{NaOl}^{(e)}$ (instead of the initial concentration, as in Figs. 2 and 3), from which C_e can be related to $C_{NaOl}^{(e)}$ for T = constant (Fig. 11). There are only slight changes in $C_{NaOl}^{(e)}$ and C_e as T increases from 10 h to ∞, so we can assume that the region to the right of the curve for T = 10 h in Fig. 11 corresponds to values of the parameters such that the suspension is absolutely stable, while the region to the left of T = 1 h represents absolute instability.

The critical C_e decreases as $C_{NaOl}^{(e)}$ increases for T of 1 and 2 h at low $C_{NaOl}^{(e)}$, as has previously been found [5]. However, this effect will not be considered in detail, as additional studies are needed to elucidate it.

Figure 11 shows that there are two values of C_e up to $C_{NaOl}^{(e)} \approx 2$ mM, i.e., T at first decreases as C_e increases but then increases again.

The broken lines of Fig. 9 represent the curves of Fig. 11 plotted on the adsorption isotherms. For $C_e \gtrsim 0.06$ M, an increase in T corresponds to an equivalent increase in the film thickness[‡] h (as h is directly proportional to Γ for polymolecular adsorption), whereas there is no clear-cut correlation between h and T for $C_e \lesssim 0.06$ M.

2. Deryagin showed many years ago that an ion-stabilized colloid has molecular attractive forces together with electrostatic repulsion forces. We assume initially that only the molecular attractive forces are present, whose energy can be put in the form U = −Af (H), in which A is the interaction constant and H is the distance between particles. For spherical particles of radius R and for H small we have [8]

$$U = -Af(H) = -\frac{AR}{H}. \tag{1}$$

The particles are in water, not in vacuum, so A must be replaced by $A_{11} + A_{22} - 2A_{12}$, in which A_{11} represents the mutual interaction of the particles, A_{22} is the same for the water, and A_{12} represents the interaction of the particles with the water.

---

† A similar effect has been reported previously [14].

‡ The line for T = 10 h is not known very precisely in Fig. 9, since here a small error in measuring $C_{NaOl}^{(e)}$ produces a large change in Γ.

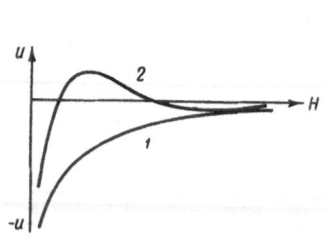

Fig. 12. Total energy U for interacting colloidal particles as
a function of separation H: 1) in
the absence of double layers; 2)
in the presence of double layers.

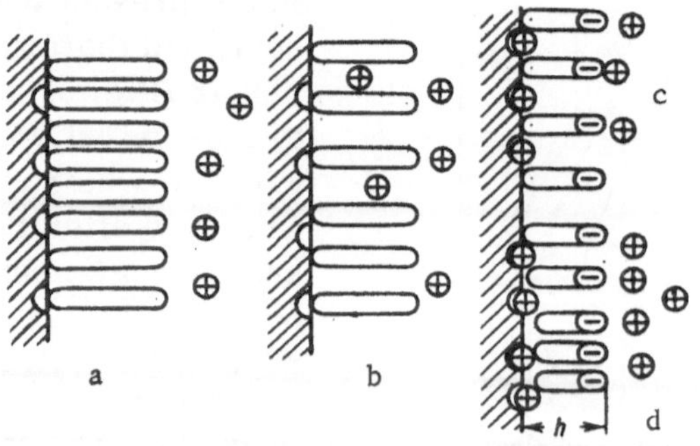

Fig. 13. Formation of a double layer at the surface of a disperse phase
coated with an adsorbed film of
NaOl.

Colloidal particles have a Brownian motion, whose energy per particle is roughly kT, so
two particles whose H is such that U > kT cannot move apart again, which results in an aggregate whose stability is the greater the stronger the inequality $U_{max} \gg kT$, in which U_{max}
is the U corresponding[†] to the smallest H, namely H_{min}. The effective radius R_{eff} of an aggregate of n particles is always greater than R, so the energy $U_{min}^{(n+n')}$ of the interaction of this
aggregate with one consisting of n' particles is certain to be larger than the energy of pair
interactions, $U_{max} = U_{max}^{(1+1)}$. Then, if $U_{max} \gg kT$ for a pair of particles, it certainly must be
more so for aggregates, so the aggregation will continue while any disperse phase is left.
Conversely, the particles cannot form strong aggregations if $U_{max} \ll kT$ and so for

$$U_{max} \ll kT \qquad (2)$$

the system is absolutely stable.[‡]

Absolutely stable systems probably include lyophilic colloids, some macromolecule solutions, and all true solutions. In all other cases, $U_{max} \gg kT$, and so such systems can be stable
only if there is some opposition to approach within the range $H_{min} \le H \le H_0$, in which U(H) > kT.
This H_0 is defined by

$$U(H_0) = kT. \qquad (3)$$

This opposition for an ion-stabilized colloid comes from the electrostatic repulsion,
which produces at a distance $H \gtrsim H_0$ a potential barrier whose height is such that the particles
cannot overcome it during Brownian motion (since this height is substantially greater than kT,
Fig. 12). The barrier is lost when C_e exceeds the critical value and the system becomes unstable.

3. These forces are affected by the addition of NaOl. The double layer on a clay particle is produced mainly by surface dissociation of ions in the exchange complex [9], so a continuous adsorbed film of NaOl displaces the exchange complex (which produces the diffuse part
of the double layer) outwards by a distance h equal to the thickness of the adsorbed film (Fig.
13a). Alternatively, it lies within the film (Fig. 13b) or centers the lattice of the clay particle
(Fig. 13c). The last is the most likely, since it corresponds to minimum electrostatic free

---

[†] There are also Born repulsion forces between particles, so H_{min} cannot be less than several
angstroms.

[‡] Of course, (2) is very rough, since the stability of a colloidal system is also dependent on
other parameters, in particular the particle size and the initial concentration.

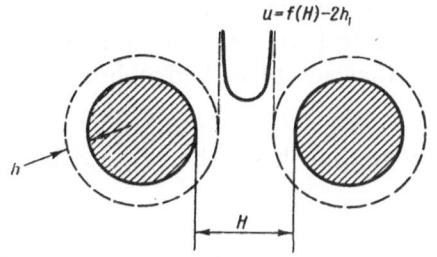

$$u = f(H) - 2h_1$$

Fig. 14. Interaction of two colloidal particles coated with surfactant films of thickness h.

energy; but this does not mean that the NaOl abolishes the double layer, because the film is formed from micelles, each of which bears a charge, so the surface has a second electrical layer (Fig. 13d), which can readily be detected from the electrophoretic mobility. Our experiments show that NaOl added to a clay suspension actually increases the mobility.

The adsorbed film thus moves the double layer outwards by a distance h. Measurement of the ξ potential shows that h in excess of a monolayer produces relatively little alteration in the adsorbed charge, since the ξ potential is always 10-15 mV, as would be expected, since the properties of the surface of a polymolecular NaOl film are not substantially dependent on the thickness.

A continuous surfactant film on the clay particles should affect the law of molecular attraction, because to (1) we must add terms taking account of the interaction between the first particle and the film on the second particle, as well as ones for the forces between the two films. Of course, these additional terms are substantially dependent on A_{NaOl}, the molecular-interaction constant for the adsorbed NaOl.[†] It is reasonable to suppose that this does not differ greatly from $A_{NaOl}^{(m)}$, the constant for micelles, and it can be shown that the latter is substantially less than the constant A_c for interaction of clay particles. The critical electrolyte concentration is proportional to A^{-2} [8], other things being equal, and this is 20 mM for a clay suspension but is ≤ 1 M for NaOl,[‡] so

$$A_c^{-2}/A_{NaOl}^{(m)^2} = \frac{2 \cdot 10^{-2}}{1} = \frac{1}{50},$$

and hence $A_c \simeq 7 A_{NaOl}^{(m)} \approx 7 A_{NaOl}$.

This estimate shows that the additional terms in (1) can be neglected to a first approximation;[§] i.e., without major error we can assume that $U = -Af(H)$ when the particles are coated with NaOl, in which H is the distance between the clay particles, not the distance between the NaOl films, which is $H - 2h$ (Fig. 14).

4. The above experimental results can now be explained, particularly for a clay suspension on the addition of NaOl when the electrolyte concentration is constant.

When the equilibrium NaOl concentration becomes such that a continuous NaOl film (thickness h) is formed on the particles, the stability of the system begins to increase, though the effect should be substantially dependent on the initial electrolyte concentration. If this does not differ too greatly from the threshold value for the clay suspension (i.e., is less than about 0.06 M, Fig. 11), a second double layer of appreciable extent arises at the surface of the soap film, while U(H) remains almost unaltered, and hence H_0 remains as before. However, only the electrostatic forces provide stability in the absence of NaOl, which is possible only if $H_0 \approx 2r_D$, in which r_D is the Debye radius (equal to the thickness of a double layer), whereas the system remains stable at smaller r_D in the presence of the soap, since then (Fig. 14)

$$2(r_D + h) \geqslant H_0. \tag{4}$$

[†] More precisely, on the difference between these constants for NaOl and water.

[‡] The latter is clear also from Fig. 5, which implies that rapid coagulation occurs only for C $\geq$ 1 M.

[§] This is true, of course, only when H is sufficiently large.

Now $r_D \sim 1/\sqrt{C_e}$, so (4) implies that addition of surfactant should increase the critical electrolyte concentration, as is observed (Fig. 11). Also, as the double layer is essential to stability for h not very large, the system should continue to behave as a typical hydrophobic one, as is observed.[†]

If C_e is high (here $\gtrsim 0.06$ M), r_D becomes substantially less than h.[‡] Then r_D can be neglected to a first approximation, and so (4) becomes

$$2h \gg H_0. \tag{5}$$

Figure 9 shows that the clay suspension becomes almost completely stable when h becomes very large, so steric effects play the main part in this concentration range; but the double layers cannot be completely neglected, since then it would be impossible to explain why we do not get coagulation due to attraction between the films alone, as for a pure soap sol. Also, Fig. 9 shows that a given T for $C_e \geq 0.06$ M corresponds to several different Γ, which can be explained only by the effects of double layers.

Now consider what occurs on addition of electrolyte when the equilibrium soap concentration is kept constant. If C_e is small, $r_D \sim 1/\sqrt{C_e}$ is large, and h is small (Fig. 9), and hence the system is stabilized solely by the double layers for C_e small, no matter what $C_{NaOl}^{(e)}$ may be. As C_e increases, r_D decreases, while h increases. The reduction in r_D is more rapid than the increase in h if $C_{NaOl}^{(e)}$ is small (Fig. 9), and so $2(h + r_D)$ at a certain C_e becomes less than H_0, and the system becomes unstable.

However, further increase in C_e for $C_{NaOl}^{(e)} = $ constant causes h to increase further, so that we again have $2h \geq H_0$ and the system is again stable (Fig. 11). If $C_{NaOl}^{(e)}$ is fairly large, the increase in h with C_e is much more rapid (Fig. 9), and so the increase in h more than compensates for the reduction in r_D, and hence $2(h + r_D) \geq H_0$ for all C_e. This explains why the clay suspension is stable for all C_e for $C_{NaOl}^{(e)} \gtrsim 2.33$ mM (Fig. 11).

Conclusions

1. Sodium oleate affects the stability of a suspension of illite clay in the presence of electrolytes. The stability increases with the NaOl concentration at a given C_e, whereas the converse occurs as C_e is increased at a fixed NaOl concentration. A system stabilized by NaOl obeys the z^6 rule.

2. The results for $BaCl_2$ and $Y(NO_3)_3$ in the presence of NaOl show that nearly all the soap molecules are in micelles, provided that the soap concentration is above the critical value for micelle formation, which means that the soap solution is itself a sol stabilized by double layers. This is also indicated by the fact that precipitation of the soap by electrolytes obeys the z^6 rule. There is no quantitative similarity between the precipitation curves for NaOl and for suspended clay stabilized by NaOl.

3. Adsorption of NaOl by the clay in the presence of NaCl and $BaCl_2$ has been examined. At all soap and electrolyte concentrations used, there is polymolecular adsorption, the thickness of the soap film perhaps exceeding 50 layers.

4. A clay suspension containing NaOl is still an ion-stabilized system, but one whose stability in the presence of electrolytes is elevated as a result of reduction in the effective forces of molecular attraction.

[†] Consider, for example, the obedience to the z^6 law for a clay suspension in a soap solution of initial concentration 2 mM.

[‡] Figure 9 shows that Γ is more than 10 times the Γ_{max} for $C_e = 0$ and a monolayer if $C^{(e)} \geq 0.2$ mM and $C_e \geq 0.06$ M. An NaOl molecule has a length of 20 Å, so h for this concentration range exceeds 100 Å. Also $r_D \leq 13$ Å for $C_e > 0.06$ M, so $h \gg r_D$ is obeyed very well.

We are indebted to B. V. Deryagin, Professor A. B. Taubman, Yu. M. Glazman, S. A. Nikitina, and N. N. Serb-Serbina for valuable comments on the paper. We are also indebted to R. M. Panich for assistance with the experiments on surface tension and to N. F. Lobanov for constant assistance in the work.

References

1. P. A. Rebinder, Transactions of the Third All-Union Conference on Colloid Chemistry, Izd. AN SSSR, Moscow (1956), p. 7; Zh. Fiz. Khim., 1:533 (1930); Kolloidn. Zh., 20:527 (1958); P. A. Rebinder and E. K. Venstrem, Zh. Fiz. Khim., 1:163 (1930).

2. D. L. Talmud and S. D. Sukhovol'skaya, Zh. Fiz. Khim., 2:1 (1931); P. A. Rebinder and A. A. Trapeznikov, Zh. Fiz. Khim., 12:573 (1938); Dokl. AN SSSR, 18:185 (1938).

3. S. S. Voyutskii, Usp. Khim., 30:1237 (1961); B. C. Blackey and A. S. C. Lawrence, Disc. Faraday Soc., 18:288 (1954); P. C. Blokker and K. Durham, Ibidem, p. 385; J. H. Schenkel and J. A. Kitchener, Trans. Faraday Soc., 56:161 (1960); A. Lottermoser and R. Stendel, Kolloid, 82:318 (1938); 83:37 (1938); R. H. Ottewill, H. C. Rastogi, and A. Wattanabe, Trans. Faraday Soc., 56:854, 866 (1960); Shinoda, Colloid Surfactants, Academic Press, New York (1963).

4. Yu. M. Glazman, M. E. Krasnokutskaya, and I. P. Sapon, Kolloidn. Zh., 27:290 (1965); I. P. Sapon and Yu. M. Glazman, Kolloidn. Zh., 27:601 (1965); Yu. M. Glazman and M. E. Krasnokutskaya, Kolloidn. Zh., 27:815 (1965); E. M. Aleksandrova and L. A. Shits, Kolloidn. Zh., 24:642 (1962).

5. Yu. M. Glazman and I. P. Sapon, in: Research in Surface Forces, Vol. 2, Consultants Bureau, New York (1966), p. 232.

6. D. S. Lychnikov and G. A. Martynov, Dokl. AN SSSR, 167:855 (1966).

7. V. K. La Mer and R. H. Smellie, Clay and Clay Minerals, 9:253 (1962).

8. H. R. Kruyt, Colloid Science, Elsevier, Barking (England) (1949-1952).

9. R. I. Zlochevskaya, Vestnik MGU, Ser. Geol., No. 3, p. 57 (1965).

10. J. L. Mortensen, Clay and Clay Minerals, 9:530 (1962).

11. P. Lecomte du Noay, Surface Equilibria of Organic Colloids, New York (1926).

12. A. B. Taubman, Dokl. AN SSSR, 71:343 (1950).

13. C. Robinson and H. A. T. Mills, Proc. Roy. Soc. London, A131:576, 596 (1931).

14. A. M. Koganovskii and T. M. Rovinskaya, Kolloidn. Zh., 17:81 (1955).

15. A. M. Koganovskii, Kolloidn. Zh., 28:59, 63 (1966); N. Pilpei, Nature, 204:378 (1964).

16. V. I. Klassen and V. A. Mokrousov, Introduction to the Theory of Flotation, Metallurgizdat, Moscow (1953).

17. L. A. Shits, Kolloidn. Zh., 26:397 (1964); Tr. Mosk. Khim.-Tekh. Inst., No. 41 (1963); P. Lukirskii and A. Echeistova, Zh. Fiz. Khim., 1:353 (1930).

18. N. D. Yakhnin, Dokl. AN SSSR, 165:1107 (1965).

THEORY OF THE STABILITY
OF A HYDROPHOBIC COLLOID

V. M. Muller

Surface Phenomena Division,
Institute of Physical Chemistry,
Academy of Sciences of the USSR

A calculation is presented for: 1) the energy of electrostatic interaction of plane-parallel plates in a solution of a symmetrical electrolyte, 2) the electrostatic energy of spherical particles for medium and high surface potentials. A relationship is derived between the rapid-coagulation potential and the surface potential for planar and spherical colloidal particles.

Introduction

Current theory [1-3] indicates that stability and coagulation in colloidal systems[†] are determined by two types of force: electrostatic repulsion and van der Waals attraction. It has been noted [1] that the mathematical aspect of the stability amounts to calculation of the electrostatic component of the force and energy, since there is no difficulty in calculating the energy of the attractive forces for planar and spherical surfaces.

The repulsion energy is usually calculated subject to the condition of constancy in the surface potential as the configuration changes; but the surface charge and potential are due to the adsorption of potential-determining ions from the electrolyte solution. The radius of action of the forces of specific adsorption is very small (of the order of the ionic radius), while the separation h_{cr} at which attraction begins to exceed repulsion is much larger than that radius, so the specific forces are unaffected as the separation is reduced to h_{cr}, and hence the calculations must be based on constancy of the adsorption energy for the potential-determining ions. It is shown below that obedience to this condition in certain cases corresponds much more to charge constancy than to constancy in the surface potential; this conclusion was reached by Deryagin and Gutop [13] in considering the equilibrium of a free film of an electrolyte solution.

There are two factors that govern the choice of the conditions for transition from a stable state to an unstable one: the shape of the surface and the size of the particles. The simplest

---

[†] It is regrettable that there are cases to be found in the literature where the theory is ascribed to Verwey and Overbeek, although these workers themselves have long accepted that Deryagin [16] has priority and the term DLVO theory has become accepted.

approach, but an unrealistic one, is to treat the surfaces of the particles as infinite parallel planes. We naturally assume in such a case that Brownian motion is absent and that the system goes over to an unstable state when [1]

$$R = 0, \quad \frac{dR}{dh} = 0, \tag{1}$$

in which R is the algebraic sum of the forces of repulsion and attraction per cm² of surface and h is the distance between the surfaces.

It is also permissible to neglect the Brownian motion for large spherical surfaces, since the surfaces will meet when attractive forces predominate at all distances between them, i.e., the critical state is defined by

$$F = 0, \quad \frac{dF}{dh} = 0, \tag{2}$$

in which F is the total force between the surfaces.

There are major difficulties in the exact calculation of the electrostatic interaction between spherical surfaces, but Deryagin [5] has proposed a method of referring this calculation to that of the interaction energy for planar surfaces. Then (2) formally becomes

$$V = 0, \quad \frac{dV}{dh} = 0, \tag{3}$$

in which V is the sum of the repulsion and attraction energies per cm².

However, small spherical particles have Brownian motion, and the system remains stable while the maximum potential energy V_s of the system is greater than the energy kT of the Brownian motion [15], and so in the critical state

$$V_s \approx kT, \quad \frac{dV_s}{dh} = 0. \tag{4}$$

Barboi [8] discussed plane-parallel surfaces and derived a stability criterion for arbitrary potentials subject to constancy in surface potential and to the maximum in the force barrier becoming zero. This criterion for large potentials goes over to Dryagin's expression [1]; Deryagin has also considered certain other limiting cases [2, 3].

A further three possible cases must be considered in a complete discussion of the general case (arbitrary potentials): 1) a potential barrier with a constant surface potential, 2) a potential barrier with constant charge, 3) a force barrier with constant charge. These cases are considered here.

Basic Equations

1. Formulation. As is usual [1, 7, 8], we consider a system consisting of two unbounded plane-parallel plates immersed in a solution of a symmetrical electrolyte and separated by a layer of thickness h. The origin is located on one of the plates, while the z axis is perpendicular to this. We put

$$u = \frac{ez_1\psi}{\theta} \quad \text{and} \quad \varkappa^2 = \frac{8\pi e^2 z_1^2 n}{\varepsilon\theta}, \tag{5}$$

in which e is the elementary charge, z_1 is the valency of the electrolyte, $\theta = kT$, k is Boltzmann's constant, T is absolute temperature, ε is the dielectric constant of the solvent (assumed constant), n is the ion concentration in the bulk solution, $\varkappa$ is the reciprocal of the Debye radius,

238   V. M. MULLER

ψ is the electrostatic potential in the plane z, and u is the corresponding dimensionless potential in that plane.

Then the Poisson−Boltzmann equation for the system can be put as

$$\frac{d^2 u}{dz^2} = \varkappa^2 \operatorname{sh} u .$$
(6)

Let the potentials at the surfaces of the plates be identical at $u_1 = e z_1 \psi_1 / \theta$, and let $u_d = e z_1 \psi_d / \theta$ be the potential in the symmetry plane $z = d = h/2$. Integration of (6) for the region $0 \leq z \leq h/2$ gives

$$\frac{du}{dz} = -\varkappa \sqrt{2 (\operatorname{ch} u - \operatorname{ch} u_d)},$$
(7)

since $u = u_d$ at $z = h/2$ and $\frac{du}{dz} = \frac{e z_1}{\theta} \frac{d\psi}{dz} = 0$ from considerations of symmetry.

Integration of (7) from $z = 0$ to $z = h/2$ gives

$$\frac{\varkappa h}{2} = \int_{u_d}^{u_1} \frac{du}{\sqrt{2 (\operatorname{ch} u - \operatorname{ch} u_d)}} \equiv J(u_1, u_d).$$
(8)

Equation (8) relates u_1 and u_d to h for a given electrolyte concentration. This may be transformed as follows. Let

$$k = \frac{1}{\operatorname{ch} \dfrac{u_d}{2}}$$
(9)

and

$$\cos \varphi = \frac{\operatorname{sh} \dfrac{u_d}{2}}{\operatorname{sh} \dfrac{u}{2}} .$$
(10)

It can then be shown that the following relations apply:

$$\operatorname{cth} \frac{u}{2} d\left(\frac{u}{2}\right) = \operatorname{tg} \varphi \, d\varphi,$$
(11)

$$\frac{\sqrt{2 (\operatorname{ch} u - \operatorname{ch} u_d)}}{\operatorname{sh} \dfrac{u}{2}} = 2 \sin \varphi,$$
(12)

$$\operatorname{ch} \frac{u}{2} = \frac{\sqrt{1 - k^2 \sin^2 \varphi}}{k \cos \varphi} \equiv \frac{\Delta \varphi}{k \cos \varphi} .$$
(13)

Substitution of (11)−(13) into (8) gives [8]

$$\frac{\varkappa h}{2} = k \int_0^{\varphi_1} \frac{d\varphi}{\sqrt{1 - k^2 \sin^2 \varphi}} \equiv k F(\varphi_1, k),$$
(14)

in which $F(\varphi_1, k)$ is an elliptic integral of the first kind and

$$\varphi_1 = \arccos \frac{\operatorname{sh} \dfrac{u_d}{2}}{\operatorname{sh} \dfrac{u_1}{2}} .$$
(15)

2. Forces between the Surfaces: Stability Criterion. Deryagin [3] and Langmuir [4] have shown that the electrostatic repulsion per cm^2 is

$$P = 2n\theta \,(\text{ch}\, u_d - 1), \tag{16}$$

which after substitution from (9) becomes

$$P = 4n\theta \left(\frac{1}{k^2} - 1\right). \tag{17}$$

On the other hand, the attraction per cm^2 between the plates for $h \ll \delta$ (in which δ is plate thickness) is

$$Q = -\frac{A}{6\pi h^3}, \tag{18}$$

in which A is the van der Waals constant. Integration of (18) gives the attraction energy as

$$V_A = -\frac{A}{12\pi h^2}. \tag{19}$$

Also, $V_R = \int\limits_{h}^{\infty} P\,dh$ is the energy of electrostatic repulsion. The total potential energy is $V = V_R + V_A$, so condition (3) is equivalent to

$$V_R = -V_A, \tag{20}$$

$$\frac{dV_R}{dh} = -\frac{dV_A}{dh} \quad \text{or} \quad P = -Q. \tag{21}$$

From (19)–(21) we find that

$$\frac{dV_R}{dh} = -\frac{dV_A}{dh} = -\frac{A}{6\pi h^3} = \frac{2}{h}V_A = -\frac{2}{h}V_R,$$

i.e.,

$$\left(\frac{d \ln V_R}{d \ln h}\right)_{h=h_c} = -2, \tag{22}$$

in which h_c is the h at which (20) and (21) are obeyed simultaneously. This h_c we call the critical distance.

Similarly, it can be shown that (1) for a force barrier becomes [1]

$$\left(\frac{d \ln P}{d \ln h}\right)_{h=h_c} = -3. \tag{23}$$

Adsorption

We noted above that the most logical assumption is that the adsorption energy Φ for the potential-determining ions is constant as h varies. It is of interest to examine how the potential and charge vary under these circumstances. Martynov [9] has considered the adsorption of ions from a mixture of electrolyte on a plane surface. We assume that we have a mixture of two symmetrical electrolytes (valencies z_1 and z_a), only the anions of valency z_d being adsorbed. We also assume that n_a (concentration of the potential-determining ions) is less than n (concentration of the ions of the second electrolyte), which is usually in fact so.

Then the general formula [9] gives the adsorbed charge as

$$\eta^{(ad)} = e z_a n_a r_0 \alpha, \tag{24}$$

in which $\alpha = \alpha(\omega, \varkappa_0)$ is defined by (48) of [9], with

$$\varkappa_0 = \varkappa_a r_0; \quad \omega = e^{-\frac{\Phi}{\theta} - \beta u_1}; \quad \beta = \frac{z_a}{z_1}.$$

Here r_0 is the diameter of the potential-determining ions and u_1 is the dimensionless potential at the boundary of the dense and diffuse regions. We assume that this boundary coincides with the plane $z = 0$, and then the space charge in the region $0 \le z \le h/2$ is

$$\eta^{(dif)} = -\frac{\varepsilon}{4\pi}\left(\frac{d\psi}{dz}\right)_{z=0} = -\frac{\varepsilon\theta}{4\pi e z_1}\left(\frac{du}{dz}\right)_{z=0}.$$

But (7) with $n \gg n_a$ gives

$$\left(\frac{du}{dz}\right)_{z=0} = -\varkappa \sqrt{2(\operatorname{ch} u_1 - \operatorname{ch} u_d)}.$$

Then

$$\eta^{(dif)} = -e z_1 n \frac{2}{\varkappa} \sqrt{2(\operatorname{ch} u_1 - \operatorname{ch} u_d)}. \tag{25}$$

Here z_a, $z_1 < 0$ are the valencies of the anions. The condition for electrical neutrality is $\eta^{(ad)} = -\eta^{(dif)}$, so (24) and (25) give

$$\beta \frac{\varkappa_a^2}{\varkappa} r_0 \alpha = 2\sqrt{2(\operatorname{ch} u_1 - \operatorname{ch} u_d)}, \tag{26}$$

with $\alpha = \alpha(\varkappa_d, \Phi, u_1)$.

Given n_a, n, and Φ cause (8) and (26) together to define u_1 and u_d as functions of h, while (25) gives the charge as a function of h. For Φ large and n_a sufficiently large (which corresponds to a high surface potential), the charge and potential of the surface remain virtually constant in response to change in the configuration of the system; whereas small Φ and n_a (low electrostatic potentials) cause the charge to be much more constant than the potential (the converse is usually assumed in calculating the interaction energy of the surfaces). In that case, (24) becomes Boltzmann's formula:

$$\eta^{(ad)} = e z_a n_a l e^{-\frac{\Phi}{\theta} - \beta u_1}, \tag{27}$$

in which l is the effective thickness of the adsorbed layer of ions. Of course, for $u_1 < 1$, even a change in potential by a factor 2–3 as h is reduced causes only relatively small changes in the adsorbed charge and hence in the space charge.

Case of Small Surface Potential and Charge

It will be seen from the above that it is more reasonable to use the condition $\eta = \text{constant}$ than $\psi_1 = \text{constant}$ in examining the stability of a hydrophobic colloid; but some [6, 7] assume that the difference between these two conditions is small. To test this view, we consider the case of small η and ψ_1, since all operations can then easily be carried to completion. The case $\psi_1 = \text{constant}$ has previously been considered [2, 7], so we only need to consider $\eta = \text{constant}$.

1. Planar Surfaces, Potential Barrier. For $u \ll 1$, we have from (16) that the electrostatic component of the outward pressure is

$$P = n\theta u_d^2. \tag{28}$$

On the other hand, (8) gives

$$\varkappa d = \int\limits_{u_d}^{u_1} \frac{du}{\sqrt{u^2 - u_d^2}} = \text{Arch}\ \frac{u_1}{u_d}\ ,$$

so

$$u_d = \frac{u_1}{\text{ch}\ \varkappa d}\ . \tag{29}$$

It is clear from (25) that the condition η = constant is equivalent to

$$\text{ch}\ u_1 - \text{ch}\ u_d = \text{ch}\ u_\infty - 1, \tag{30}$$

which for small potentials becomes

$$u_1^2 - u_d^2 = u_\infty^2, \tag{31}$$

in which u_∞ is the u corresponding to $h \to \infty$. Then (29) and (31) give

$$u_d = \frac{u_\infty}{\text{sh}\ \varkappa d}\ , \tag{32}$$

and so for η = constant

$$P(d) = \frac{n\theta u_\infty^2}{\text{sh}^2\ \varkappa d} \simeq \frac{16n\theta}{\text{sh}^2\ \varkappa d}\ \gamma^2, \tag{33}$$

in which $\gamma = \tanh(u_\infty/2)$ is independent of h. We are justified in introducing function γ because (see later) the formulas obtained for small potentials are applicable also for large ones (with a certain accuracy). Substitution of (33) into

$$V_R = -\int\limits_\infty^h P(h)\, dh \tag{34}$$

and integration gives

$$V_R^\eta = \frac{32n\theta}{\varkappa}\ \gamma^2\ (\text{cth}\ \varkappa d - 1), \tag{35}$$

in which the superscript η indicates that the energy is calculated for η = constant.

If ψ_1 = constant, we have [7]

$$V_R^\psi = \frac{32n\theta}{\varkappa}\ \gamma^2\ (1 - \text{th}\ \varkappa d). \tag{36}$$

This shows that

$$\frac{V_R^\eta}{V_R^\psi} = \text{cth}\ \varkappa d,$$

i.e., the ratio may become very large as h diminishes, while the ratio is close to one for $\varkappa d > 1$. Substitution of (35) into (22) gives us an equation for the critical state for η = constant:

$$\varkappa d\ (1 + \text{cth}\ \ \varkappa d) = 2. \tag{37}$$

The critical state in terms of the Debye radius is independent of ψ_1 for ψ_1 small and is

$$(\varkappa d)_c^\eta = 0.797, \tag{38}$$

which corresponds to $V_R^\eta = 1.51 V_R^\psi$.

As V and the total force are zero for $h = h_c$, we equate (18) and (33) for the $h = h_c$ given by (38) and use (5) to find the critical concentration as a function of the parameters of the system:

$$n_c^\eta = 155 \frac{\varepsilon^3 \theta^5}{A^2 e^6 z_1^6} \gamma^4. \tag{39}$$

In the same way, it can be shown that for $\psi_1 = $ constant

$$n_c^\psi = 85.5 \frac{\varepsilon^3 \theta^5}{A^2 e^6 z_1^6} \gamma^4. \tag{40}$$

Comparison of (39) and (40) shows that n_c^η, other things being equal, is almost twice as large as n_c^ψ ($n_c^\eta = 1.83 n_c^\psi$).

2. S p h e r i c a l   S u r f a c e s ,   P o t e n t i a l   B a r r i e r .   Deryagin [5] has shown that spherical surfaces of identical radius a whose centers are a distance R apart give, for $\varkappa d \gg 1$ and $H = R - 2a \ll a$, a total interaction energy of

$$V_S(H) = \pi a \int_H^\infty V(h)\, dh, \tag{41}$$

in which V(h) is the interaction energy per cm² for planar surfaces.

Substitution of (35) into (41) and integration gives

$$V_{SR}^\eta = -\frac{8 a \varepsilon \theta^2}{e^2 z_1^2} \gamma^2 \ln(1 - e^{-\varkappa H}). \tag{42}$$

The energy of the attractive force is found similarly from (19) as

$$V_{SA} = -\frac{aA}{12H}. \tag{43}$$

We assume, in accordance with (4), that the colloidal system becomes unstable when

$$V_S = V_{SR} + V_{SA} \approx \theta \text{ and } \frac{dV_S}{dH} = 0, \tag{44}$$

and we repeat the argument leading to (22) to get from (43) and (44) that

$$\left[\frac{d \ln(V_{SR} - \theta)}{d \ln H} \right]_{H=H_c} = -1. \tag{45}$$

But V_{SR} is directly proportional to a, so $V_{SR} \gg \theta$ for large particles at $H = H_c$, and (45) becomes

$$\left(\frac{d \ln V_{SR}}{d \ln H} \right)_{H=H_c} = -1, \tag{46}$$

in which a does not appear. We substitute here from (42) to get a condition for $(\varkappa H)_c$ for $\eta = $ constant:

$$\varkappa H = (e^{\varkappa H} - 1)[\varkappa H - \ln(e^{\varkappa H} - 1)], \tag{47}$$

whose solution is

$$(\varkappa H)_c^\eta = \ln 2 = 0.693. \tag{48}$$

Then $V_{SR}^{\eta} = 1.71 V_{SR}^{\psi}$, since for $\psi_1 = $ constant [2, 7]

$$V_{SR}^{\psi} = \frac{8ae\theta^2}{e^2 z_1^2}\, \gamma^2 \ln\,(1 + e^{-\varkappa H}).\tag{49}$$

As $V_{SR}^{\eta} = -V_{SA}$ for $H = H_c$, we get from (48) the critical concentration for $\eta = $ constant:

$$n_{Sc}^{\eta} = 84.7\, \frac{\varepsilon^3 \theta^5}{A^2 e^6 z_1^6}\, \gamma^4.\tag{50}$$

For $\psi_1 = $ constant

$$n_{Sc}^{\psi} = 36.4\, \frac{\varepsilon^3 \theta^5}{A^2 e^6 z_1^6}\, \gamma^4,\tag{51}$$

which shows that the difference between the cases $\psi_1 = $ constant and $\eta = $ constant is even greater ($n_c^{\eta} = 2.33\, n_c^{\psi}$) for spherical particles than for planes.

    3. Force Barrier. From (23) and (33) we get for plane surfaces with $\eta = $ constant that the critical state is defined by

$$\varkappa d \operatorname{cth} \varkappa d = \frac{3}{2}\,,\tag{52}$$

and so

$$(\varkappa d)_c^{\eta} = 1.290.\tag{53}$$

The corresponding critical concentration is

$$n_c^{\eta} = 213\, \frac{\varepsilon^3 \theta^5}{A^2 e^6 z_1^6}\, \gamma^4.\tag{54}$$

It is readily shown that for $\psi_1 = $ constant

$$n_c^{\psi} = 139\, \frac{\varepsilon^3 \theta^5}{A^2 e^6 z_1^6}\, \gamma^4,\tag{55}$$

i.e., $n_c^{\eta} = 1.53 n_c^{\psi}$, there is a fairly substantial difference between $\eta = $ constant and $\psi_1 = $ constant.

    The total force of interaction for two spherical surface is [5], by analogy with (41),

$$F_S(H) = \pi a \int_H^\infty F(h)\, dh,\tag{56}$$

in which $F(h)$ is the force per cm^2 for planar surfaces.

    This shows that $F_S(H)$ equals $V(H)$ apart from the factor πa, i.e., the stability criterion for spherical surfaces with a force barrier formally coincides with that for planar surfaces with a potential barrier for any value of ψ_1. Then the critical n (that causing rapid coagulation) calculated for $\psi_1 = $ constant is about half of the value calculated for $\eta = $ constant for ψ_1 small, no matter what the shape of the surface and the form of the barrier.

    Deryagin has also considered the interaction of two spherical surfaces differing in radius [10]. The stability criterion is unaffected.

Planar Surfaces with a Potential Barrier for an Arbitrary Constant Potential

    1. V_R^{ψ}. We have already seen that

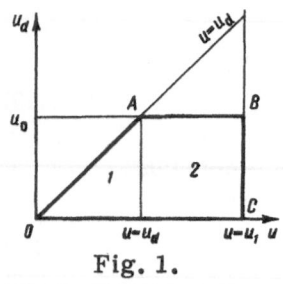

Fig. 1.

$$V_R = -\int_\infty^h P(h)\,dh = -\frac{2}{\varkappa}\int_\infty^h P\,d\left(\frac{\varkappa h}{2}\right). \tag{34}$$

We use (8) and (16) for $\varkappa h/2$ and P, and also the fact that $u_d = 0$ for $h = \infty$, to get that

$$V_R = -\frac{4n\theta}{\varkappa}\int_0^{u_d} (\operatorname{ch} u_d - 1)\,dJ(u_1, u_d). \tag{57}$$

It is readily shown that

$$\lim_{u_d \to 0}[(\operatorname{ch} u_d - 1)\,J(u_1, u_d)] = \lim_{t \to 0}(t \ln t) = 0.$$

Then integration of (57) by parts gives

$$\frac{\varkappa}{5n\theta}V_R = -(\operatorname{ch} u_d - 1)\,J(u_1, u_d) + J_{\dot R}(u_1, u_d), \tag{58}$$

in which

$$J_R(u_1, u_d) = \int_0^{u_d} du_d\left\{\operatorname{sh} u_d \int_{u_d}^{u_1} \frac{du}{\sqrt{2(\operatorname{ch} u - \operatorname{ch} u_d)}}\right\}. \tag{59}$$

As u_1 = constant, the integration in (59) is carried over the area of the trapezium OABC (Fig. 1). We reverse the order of integration to convert the integral of (59) to two integrals over regions 1 and 2 of OABC:

$$J_R^\psi = \int_0^{u_d} du \int_0^u \frac{\operatorname{sh} u_d\,du_d}{\sqrt{2(\operatorname{ch} u - \operatorname{ch} u_d)}} + \int_{u_d}^{u_1} du \int_0^{u_d} \frac{\operatorname{sh} u_d\,du_d}{\sqrt{2(\operatorname{ch} u - \operatorname{ch} u_d)}} = J_1 + J_2. \tag{60}$$

Integration gives

$$J_1 = 4\left(\operatorname{ch}\frac{u_d}{2} - 1\right), \tag{61}$$

$$J_2 = -\int_{u_d}^{u_1} \sqrt{2(\operatorname{ch} u - \operatorname{ch} u_d)}\,du + 4\left(\operatorname{ch}\frac{u_1}{2} - \operatorname{ch}\frac{u_d}{2}\right). \tag{62}$$

We make the substitution x = sinh (u/2) in (62) to show [11] that

$$-\int_{u_d}^{u_1} \sqrt{2(\operatorname{ch} u - \operatorname{ch} u_d)}\,du = 4\left\{\operatorname{ch}\frac{u_d}{2}E(\varphi_1, k) - \operatorname{ch}\frac{u_1}{2}\sin\varphi_1\right\}, \tag{63}$$

in which

$$E(\varphi_1, k) \equiv \int_0^{\varphi_1} \sqrt{1 - k^2\sin^2\varphi}\,d\varphi$$

is an elliptic integral of the second kind, in which k and φ_1 are related to u_1 and u_d by (9) and (15).

Then (9) and (13) give with (60)–(63) that

$$J_R^\psi = \frac{4}{k}\left\{E(\varphi_1, k) - k + \operatorname{tg}\left(\frac{\pi}{4} - \frac{\varphi_1}{2}\right)\sqrt{1 - k^2\sin^2\varphi_1}\right\}. \tag{64}$$

TABLE 1

$\theta_c = \arcsin k_c$	φ_{1c}	u_{1c}	u_{dc}	$(\varkappa h)_c$	$k_c^2(1-k_c^2)^2 F_c^6$	$z_1^{-6} c_{cr}$, mM
90°	53°30′	0	0	2.218	0	0
85°	53°33′	0.295	0.175	2.216	0.0001065	0.00428
80°	54°10′	0.613	0.352	2.210	0.00175	0.0720
75°	55°10′	0.972	0.530	2.198	0.00918	0.378
70°	56°40′	1.243	0.713	2.190	0.03014	1.24
65°	58°43′	1.615	0.902	2.180	0.07921	3.26
60°	61°13′	2.030	1.099	2.160	0.1746	7.18
55°	64°33′	2.528	1.306	2.136	0.3561	14.7
50°	68°16′	3.114	1.525	2.090	0.6433	26.5
45°	73°14′	3.913	1.762	2.044	1.141	47.0
40°	79°14′	5.105	2.022	1.982	1.916	79.0
35°	86°30′	7.693	2 308	1.900	3.066	126
32°52′	90°	∞	2.442	1.858	3.671	151

Note: It has been assumed that $\varepsilon = 80$, $T = 300°K$, and $A = 2 \times 10^{-12}$ erg (column 7).

On the other hand, (9) and (14) give

$$(\operatorname{ch} u_d - 1) J(u_1, u_d) = \frac{2}{k}(1-k^2) F(\varphi_1, k). \tag{65}$$

Substitution of (64) and (65) into (58) gives

$$V_R^{\psi} = \frac{4n\theta}{\varkappa}\left\{-\frac{2}{k}(1-k^2) F(\varphi_1, k) + \frac{4}{k}\left[E(\varphi_1, k) - k + \operatorname{tg}\left(\frac{\pi}{4} - \frac{\varphi_1}{2}\right)\sqrt{1 - k^2 \sin^2\varphi_1}\right]\right\}. \tag{66}$$

The formula, with (14), gives V_R^{ψ} as a function of h.

Verwey and Overbeek [7] calculated the repulsion energy for two planar surfaces for an arbitrary constant potential via the free energy of a plate immersed in an electrolyte solution.† It can be shown that Verwey and Overbeek's formula is precisely the same as (66). The expressions found by Verwey and Overbeek were complicated, so they examined the stability only graphically; but (66) allows the work to be carried to completion.

2. Stability Criterion. From V_R^{ψ} we readily get the equation for the critical state subject to (3); (14), (17), and (22) give

$$-\frac{h}{V_R}\frac{dV_R}{dh} = \frac{2}{\varkappa}\left(\frac{\varkappa h}{2}\right)\frac{P}{V_R} = \frac{2}{\varkappa}\frac{1}{V_R} kF(\varphi_1, k) \cdot 4n\theta\left(\frac{1}{k^2} - 1\right) = 2$$

or

$$\frac{\varkappa}{4n\theta} V_R = \frac{1}{k}(1-k^2) F(\varphi_1, k). \tag{67}$$

Substitution for V_R^{ψ} from (66) gives

$$\frac{3}{4}(1-k_c^2) F_c(\varphi_1, k) = E_c(\varphi_1, k) - k_c + \operatorname{tg}\left(\frac{\pi}{4} - \frac{\varphi_{1c}}{2}\right)\sqrt{1 - k_c^2 \sin^2\varphi_{1c}}, \tag{68}$$

† Deryagin [3] proposed this method of calculating the electrostatic component of the interaction energy.

in which the subscript c denotes that (68) relates k to φ_1 (i.e., u_1 to u_d) for the critical state. Then from (14)

$$\frac{(\varkappa h)_c}{2} = k_c F_c (\varphi_1, k) \qquad (69)$$

and so

$$Q_c = -\frac{A}{6\pi h_c^3} = -\frac{A\varkappa_c^3}{48\pi k_c^3 F_c^3 (\varphi_1, k)} . \qquad (70)$$

From (17), (21), and (70) we get

$$4n_c\theta\left(\frac{1}{k_c^2} - 1\right) = \frac{A\varkappa_c^3}{48\pi k_c^3 F_c^3(\varphi_1, k)} . $$

Squaring and using (5), we have

$$n_c^{\psi} = \frac{72}{\pi} \frac{\varepsilon^2\theta^5}{A^2 e^6 z_1^6} k_c^2 (1 - k_c^2)^2 \, F_c^6 (\varphi_1, k). \qquad (71)$$

This equation has the form of Barboi's equation [8] for n_c for the case of a force barrier, so the dependence of n_c on the surface potential is described by the product $k_c^2(1 - k_c^2)^2 \times F_c^6 (\varphi_1, k)$. in both cases; but in our case the values of k_c and φ_{1c} are defined by (68), which differs substantially from the corresponding equation of Barboi's. Table 1 gives solutions to (68) for arbitrary values of the potential.

For low potentials, (68) becomes

$$\varkappa d\,(1 + \text{th}\,\varkappa d) = 2, \quad (\varkappa d)_c = 1.109, \qquad (72)$$

which may be obtained directly by substituting (36) into (22), the relation between u_1 and u_d being given by (29).

3.   Medium and High Potentials.   We have from (15) that $\varphi_1 \to \pi/2$ as u_1 increases, i.e., $\cos \varphi_1 \to 0$. Also, (9) and (15) give

$$\cos \varphi_1 = \frac{\sqrt{1 - k^2}}{k\,\text{sh}\,\dfrac{u_1}{2}} . $$

We expand (68) in powers of $\cos \varphi_1$ via the relations [12]

$$F (\varphi_1, \ k) = K (k) - F (\varphi_2, \ k), \qquad (73)$$
$$E (\varphi_1, \ k) = E (k) - E (\varphi_2, \ k) + k^2 \sin \varphi_1 \sin \varphi_2, \qquad (74)$$

in which

$$\sin \varphi_2 = \frac{\cos \varphi_1}{\sqrt{1 - k^2 \sin^2 \varphi_1}} , \qquad (75)$$

while K(k) and E(k) are complete elliptic integrals of the first and second kinds. Expansion of (73)-(75) gives

$$F (\varphi_1, k) = K (k) - \frac{\cos \varphi_1}{\sqrt{1 - k^2}} + B_1 (k) \cos^3 \varphi_1 + \dots, \qquad (76)$$

$$E (\varphi_1, k) = E (k) - \sqrt{1 - k^2} \cos \varphi_1 + B_2 (k) \cos^3 \varphi_1 + \dots, \qquad (77)$$

$$\text{tg}\left(\frac{\pi}{4}-\frac{\varphi_1}{2}\right)\sqrt{1-k^2\sin^2\varphi_1}=\frac{1}{2}\sqrt{1-k^2}\cos\varphi_1+B_3(k)\cos^3\varphi_1+\dots. \tag{78}$$

We neglect terms containing $\cos^3\varphi_1$, etc., to get

$$\frac{\varkappa h}{2}=kF(\varphi_1,k)\simeq kK(k)-\frac{k}{\sqrt{1-k^2}}\cos\varphi_1=kK(k)-\frac{1}{\text{sh}\frac{u_1}{2}}, \tag{79}$$

$$V_R^\psi=\frac{8n\theta}{\varkappa}\left\{-\frac{1}{k}(1-k^2)K(k)+\frac{2}{k}E(k)-2\right\}. \tag{80}$$

This shows that V_R^ψ does not contain $\cos\varphi_1$ or $\cos^2\varphi_1$, although it is dependent on u_1 [since V_R^ψ is a function of u_1 and h is defined by (79) and (80)]. Then (68) becomes

$$3(1-k^2)K(k)=4[E(k)-k]+\frac{1-k^2}{k}\frac{1}{\text{sh}\frac{u_1}{2}}. \tag{81}$$

Comparison of the solutions to (81) with the exact equation (68) shows that the error does not exceed 2-3% for $u_1\geq 2$, but that the accuracy falls appreciably for lower values.

We get for $u_1\gg 1$ from (81) that

$$3(1-k^2)K(k)=4[E(k)-k]. \tag{82}$$

The root of this is $k_c=0.5427$, and $K_c(k)=1.7106$, so

$$n_c^\psi=84\frac{\varepsilon^3\theta^5}{A^2e^6z_1^6}\text{ and }(\varkappa h)_c^\psi=1.866. \tag{83}$$

For a force barrier [1, 8]

$$n_c^\psi=122\frac{\varepsilon^3\theta^5}{A^2e^6z_1^6}\text{ and }(\varkappa h)_c^\psi=3.014. \tag{84}$$

We see from (83) that the z_1^{-6} law for the coagulation concentration is obeyed strictly for large potentials, but that the numerical factor is 2/3 of that for the case of a force barrier.

The expansion in powers of $\cos\varphi_1$ is the more accurate the smaller $k=[\cosh(u_d/2)]^{-1}$ for a given φ_1. If $k\to 1$, i.e., $u_d\to 0$, the $B_i(k)$ in (76)-(78) become comparable with $(\cos^3\varphi_1)^{-1}$. But u_d tends to zero either when the surface potential is small or $\varkappa h$ is large; both of these cases have been examined in detail, and the second is of no interest as regards the critical state, for then $\varkappa h$ is always of the order of one. All of the results in this section apply also to spherical surfaces, but only for the case of a force barrier; (82) has previously been derived for this case [1], as has (80) for V_R^ψ.

Plane Surfaces and a Potential Barrier for a

Constant but Arbitrary Charge

1. V_R^η. We have seen above that the expression for the force of electrostatic repulsion has the same form for $\psi_1=$ constant and for $\eta=$ constant:

$$P=2n\theta(\text{ch}\,u_d-1).$$

Then (58) and (59) apply also for $\eta=$ constant, i.e.,

$$\frac{\varkappa}{4n\theta}V_R^\eta=-(\text{ch}\,u_d-1)J(u_1,u_d)+J_R^\eta(u_1,u_d) \tag{85}$$

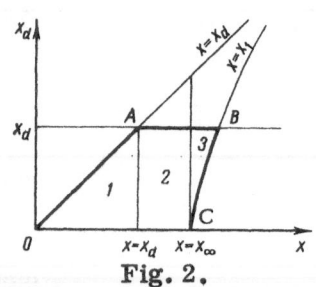

Fig. 2.

and

$$J_R^\eta (u_1, u_d) = \int\limits_0^{u_d} du_d \left\{ \operatorname{sh} u_d \int\limits_{u_d}^{u_1} \frac{du}{\sqrt{2 (\operatorname{ch} u - \operatorname{ch} u_d)}} \right\}. \tag{86}$$

But here u_1 is not independent but is shown by (30) to be related to u_d by

$$\operatorname{ch} u_1 - \operatorname{ch} u_d = 2x_\infty^2 = \text{const.} \tag{87}$$

This same relation applies between u_1 and u_d in (8). We make the substitutions

$$\operatorname{sh} \frac{u}{2} = x, \quad d\left(\frac{u}{2}\right) = \frac{dx}{\sqrt{x^2 + 1}}. \tag{88}$$

Then (87) becomes

$$x_1^2 - x_d^2 = x_\infty^2, \tag{89}$$

and

$$J_R^\eta = 2 \int\limits_0^{u_d} d\left(\operatorname{sh}^2 \frac{u_d}{2}\right) \int\limits_{u_d}^{u_1} \frac{d\left(\frac{u}{2}\right)}{\sqrt{\operatorname{sh}^2 \frac{u}{2} - \operatorname{sh}^2 \frac{u_d}{2}}} = 2 \int\limits_0^{x_d} d(x_d^2) \int\limits_{x_d}^{x_1} \frac{dx}{\sqrt{(x^2 + 1)(x^2 - x_d^2)}}. \tag{90}$$

The integration in (90) is taken over region OABC in Fig. 2. We reverse the order of integration, which splits the integral of (90) into three integrals over regions 1, 2, and 3 of OABC:

$$J_R = \int\limits_0^{x_d} \frac{dx}{\sqrt{x^2 + 1}} \int\limits_0^{x} \frac{d(x_d^2)}{\sqrt{x^2 - x_d^2}} + \int\limits_{x_d}^{x_\infty} \frac{dx}{\sqrt{x^2 + 1}} \int\limits_0^{x_d} \frac{d(x_d^2)}{\sqrt{x^2 - x_d^2}}$$

$$+ \int\limits_{x_\infty}^{x_1} \frac{dx}{\sqrt{x^2 + 1}} \int\limits_{\sqrt{x^2 - x_d^2}}^{x_d} \frac{d(x_d^2)}{\sqrt{x^2 - x_d^2}} = J_1 + J_2 + J_3.$$

We integrate once:

$$J_1 = \int\limits_0^{x_d} \frac{2x\, dx}{\sqrt{x^2 + 1}}; \quad J_2 = \int\limits_{x_d}^{x_\infty} \frac{2x\, dx}{\sqrt{x^2 + 1}} - 2 \int\limits_{x_d}^{x_\infty} \sqrt{\frac{x^2 - x_d^2}{x^2 + 1}}\, dx;$$

$$J_3 = 2x_\infty \int\limits_{x_\infty}^{x_1} \frac{dx}{\sqrt{x^2 + 1}} - 2 \int\limits_{x_\infty}^{x_1} \sqrt{\frac{x^2 - x_d^2}{x^2 + 1}}\, dx.$$

Addition of these integrals gives [11]

$$\frac{1}{4} J_R^\eta = \int\limits_0^{x_\infty} \frac{x\, dx}{\sqrt{x^2 + 1}} - \int\limits_{x_d}^{x_1} \sqrt{\frac{x^2 - x_d^2}{x^2 + 1}}\, dx + x_\infty \int\limits_{x_\infty}^{x_1} \frac{dx}{\sqrt{x^2 + 1}}$$

$$= \sqrt{x_\infty^2 + 1} - 1 + \sqrt{x_d^2 + 1}\, E(\varphi_1, k) - \frac{x_\infty}{x_1} \sqrt{1 + x_1^2} + x_\infty \ln \frac{x_1 + \sqrt{1 + x_1^2}}{x_\infty + \sqrt{1 + x_\infty^2}}, \tag{91}$$

TABLE 2

$\theta_c = \arcsin k_c$	φ_{1c}	$u_{\infty c}$	u_{d_c}	$(\varkappa'_i)_c$	$k_c^2 (1 - k_c^2)^2 \cdot F_c^6$	$z_{1c\,cr}^6$, mM
90°	41°30′	0	0	1.594	0	0
80°	42°40′	0.319	0.352	1.618	0.000272	0.0112
70°	47°00′	0.763	0.713	1.718	0.00705	0.290
60°	55°15′	1.515	1.099	1.892	0.0790	3.25
50°	65°35′	2.743	1.525	1.964	0.479	19.7
45°	72°00′	3.685	1.762	2.002	1.011	41.6
40°	79°00′	5.032	2.022	1.976	1.875	77.2
35°	86°30′	7.688	2.308	1.900	3.066	126
32°52′	90°	∞	2.442	1.858	3.671	151

Note: It has been assumed that $\varepsilon = 80$, $T = 300°K$, and $A = 2 \times 10^{-12}$ erg (column 7).

where, as before

$$\cos \varphi_1 = \frac{x_d}{x_1} = \frac{\operatorname{sh} \dfrac{u_d}{2}}{\operatorname{sh} \dfrac{u_1}{2}} ; \quad k = \frac{1}{\sqrt{1 + x_d^2}} = \frac{1}{\operatorname{ch} \dfrac{u_d}{2}} . \tag{92}$$

Let $k = \sin \theta$. It is then readily shown that

$$x_d = \operatorname{ctg} \theta; \quad x_\infty = \operatorname{ctg} \theta \operatorname{tg} \varphi_1; \quad x_1 = \frac{\operatorname{ctg} \theta}{\cos \varphi_1} ; \tag{93}$$

$$\frac{1}{4} J_R^\eta = \frac{1}{\sin \theta} E(\varphi_1, \theta) - 1 + \operatorname{ctg} \theta \operatorname{tg} \varphi_1 \left\{ \sqrt{1 + \operatorname{ctg}^2 \varphi_1 \operatorname{tg}^2 \theta} - \sqrt{1 + \cos^2 \varphi_1 \operatorname{tg}^2 \theta} - \ln \left[\sin \varphi_1 \frac{1 + \sqrt{1 + \operatorname{ctg}^2 \varphi_1 \operatorname{tg}^2 \theta}}{1 + \sqrt{1 + \cos^2 \varphi_1 \operatorname{tg}^2 \theta}} \right] \right\} . \tag{94}$$

We also make the substitution $k = \sin \theta$ in (65):

$$(\operatorname{ch} u_d - 1) J(u_1, u_d) = 2 \frac{\cos^2 \theta}{\sin \theta} F(\varphi_1, \theta). \tag{95}$$

Then (94) and (95) are substituted into (85):

$$\frac{\varkappa}{16\,n\theta} V_R^\eta = -\frac{1}{2} \frac{\cos^2 \theta}{\sin \theta} F(\varphi_1, \theta) + \frac{1}{\sin \theta} E(\varphi_1, \theta) - 1$$
$$+ \operatorname{ctg} \theta \operatorname{tg} \varphi_1 \left\{ \sqrt{1 + \operatorname{ctg}^2 \varphi_1 \operatorname{tg}^2 \theta} - \sqrt{1 + \cos^2 \varphi_1 \operatorname{tg}^2 \theta} - \ln \left[\sin \varphi_1 \frac{1 + \sqrt{1 + \operatorname{ctg}^2 \varphi_1 \operatorname{tg}^2 \theta}}{1 + \sqrt{1 + \cos^2 \varphi_1 \operatorname{tg}^2 \theta}} \right] \right\} , \tag{96}$$

with

$$\frac{\varkappa h}{2} = \sin \theta \cdot F(\varphi_1; \theta). \tag{97}$$

Then (96) and (97) give V_R^η as a function of h, while u_d, u_1, and u_∞ are given by (93):

$$\operatorname{sh} \frac{u_d}{2} = \operatorname{ctg} \theta; \quad \operatorname{sh} \frac{u_1}{2} = \frac{\operatorname{ctg} \theta}{\cos \varphi_1} ; \quad \operatorname{sh} \frac{u_\infty}{2} = \operatorname{ctg} \theta \operatorname{tg} \varphi_1. \tag{98}$$

2. Stability Criterion. Substitution of (96) into (67) gives

$$\frac{3}{4} \cos^2 \theta F(\varphi_1, \theta) = E(\varphi_1, \theta) - \sin \theta + \cos \theta \operatorname{tg} \varphi_1 \left\{ \sqrt{1 + \operatorname{ctg}^2 \varphi_1 \operatorname{tg}^2 \theta} - \right.$$

$$- \sqrt{1 + \cos^2 \varphi_1 \, \mathrm{tg}^2 \theta} - \ln \left[\sin \varphi_1 \frac{1 + \sqrt{1 + \mathrm{ctg}^2 \varphi_1 \, \mathrm{tg}^2 \theta}}{1 + \sqrt{1 + \cos^2 \varphi_1 \, \mathrm{tg}^2 \theta}} \right] \right\} . \tag{99}$$

From (98) and (99) we get the relation between u_1 and u_d for the critical state, while (97) gives h_c as a function of u_1. Table 2 gives the solution to (99), which becomes (37) for small potentials.

As in the case $\psi_1 = $ constant, n_c is given as a function of surface potential by a formula analogous to (71):

$$n_c^\eta = \frac{72}{\pi} \frac{\varepsilon^3 \theta^5}{A^2 e^6 z_1^6} \, k_c^2 (1 - k_c^2)^2 \, F_c^6 (\varphi_1, k), \tag{100}$$

but $k_c = \sin \theta_c$ and φ_{1c} is defined by (99).

Consider the relation between charge and potential of the surface for the critical state. It follows from (87) and (98) that

$$2 (\mathrm{ch}\, u_1 - \mathrm{ch}\, u_d) = 4 \mathrm{sh}^2 \frac{u_\infty}{2} = 4 \, \mathrm{ctg}^2 \theta \, \mathrm{tg}^2 \varphi_1. \tag{101}$$

Substitution of (100) and (101) in (25) and use of (5) gives

$$\eta_c = \frac{12}{\pi} \frac{\varepsilon^2 \theta^3}{A e^3 z_1^3} \, \mathrm{tg}\, \varphi_{1c} \cos^3 \theta_c F_c^3 (\varphi_1, \theta) = \frac{12}{\pi} \frac{\varepsilon^2 \theta^3}{A e^3 z_1^3} \, \mathrm{tg}\, \varphi_{1c} (1 - k_c^2)^{3/2} F_c^3 (\varphi_1, \theta). \tag{102}$$

3. Medium and High Potentials.

A series expansion in $\cos \varphi_1$ gives

$$\cos \theta \, \mathrm{tg}\, \varphi_1 \left\{ \sqrt{1 + \mathrm{ctg}^2 \varphi_1 \, \mathrm{tg}^2 \theta} - \sqrt{1 + \cos^2 \varphi_1 \, \mathrm{tg}^2 \theta} - \ln \left[\sin \varphi_1 \frac{1 + \sqrt{1 + \mathrm{ctg}^2 \varphi_1 \, \mathrm{tg}^2 \theta}}{1 + \sqrt{1 + \cos^3 \varphi_1 \, \mathrm{tg}^2 \theta}} \right] \right\}$$

$$= \frac{1}{2} \cos \theta \cos \varphi_1 + B_4 (\theta) \cos^3 \varphi_1 + \cdots = \frac{1}{2} \sqrt{1 - k^2} \cos \varphi_1 + B_4 (k) \cos^3 \varphi_1 + \cdots . \tag{103}$$

We substitute (76), (77), and (103) into (96) and neglect terms containing $\cos^3 \varphi_1$, as well as terms of higher orders of smallness, to get

$$V_R^\eta = \frac{8 n \theta}{\varkappa} \left\{ - \frac{1}{k} (1 - k^2) K(k) + \frac{2}{k} E(k) - 2 \right\}, \tag{104}$$

where again

$$\frac{\varkappa h}{2} = k K(k) - \frac{1}{\mathrm{sh} \frac{u_1}{2}}, \tag{105}$$

and (99) becomes

$$3 (1 - k^2) K(k) = 4 [E(k) - k] + \frac{1 - k^2}{k} \frac{1}{\mathrm{sh} \frac{u_1}{2}}, \tag{106}$$

which coincides with (81). Comparison of the solutions to this equation with the solutions to the exact equation (99) shows that (106) provides high accuracy down to u_∞ of 4–5 but that the error increases rapidly below this limit, and that (106) corresponds much better to the exact equation for $\psi_1 = $ constant than for $\eta = $ constant.

This means that there is no difference between the cases $\psi_1 = $ constant and $\eta = $ constant as regards determination of n_c for $u \approx 5$ and above, which agrees with the conclusion drawn in Section 2.

In the limit $u_1 \gg 1$, n_c obeys the z_1^{-6} law strictly, as given by (83).

Plane Surfaces and a Force Barrier at Constant but Arbitrary Charge

1. Stability Criterion.

Here the equation for the critical state is

$$\frac{d \ln P}{d \ln h} = -3.$$

This is transformed via (8) to

$$\frac{d \ln P}{d \ln h} = h \frac{d \ln P}{dh} = J(u_1, u_d) \frac{d \ln P(u_d)}{dJ(u_1, u_d)} = \frac{J(u_1, u_d)}{\frac{dJ(u_1, u_d)}{du_d}} \cdot \frac{d \ln P(u_d)}{du_d} = -3. \tag{107}$$

From (9) and (16) we get that

$$\frac{d \ln P}{du_d} = \operatorname{cth} \frac{u_d}{2} = \frac{1}{k \operatorname{sh} \dfrac{u_d}{2}}. \tag{108}$$

The u_1 and u_d in $J(u_1, u_d)$ are related by (87) because $\eta = \text{constant}$:

$$\operatorname{ch} u_1 - \operatorname{ch} u_d = \text{const},$$

and differentiation gives $\sinh u_1 (du_1/du_d) - \sinh u_d = 0$, i.e.,

$$\frac{du_1}{du_d} = \frac{\operatorname{sh} u_d}{\operatorname{sh} u_1}. \tag{109}$$

On the other hand,

$$\frac{dJ(u_1, u_d)}{du_d} \equiv \frac{d}{du_d} \left\{ \int_{u_d}^{u_1} \frac{du}{\sqrt{2(\operatorname{ch} u - \operatorname{ch} u_d)}} \right\} = \operatorname{sh} u_d \int_{u_d}^{u_1} \frac{du}{[2(\operatorname{ch} u - \operatorname{ch} u_d)]^{3/2}} + \frac{1}{\sqrt{2(\operatorname{ch} u_1 - \operatorname{ch} u_d)}} \frac{du_1}{du_d} - \lim_{u \to u_d} \frac{1}{\sqrt{2(\operatorname{ch} u - \operatorname{ch} u_d)}}. \tag{110}$$

The integral in (110) diverges at the lower limit, and the last term tends to infinity, so we transform it to

$$\lim_{u \to u_d} \frac{1}{\sqrt{2(\operatorname{ch} u - \operatorname{ch} u_d)}} = \int_{u_d}^{u_1} \frac{\operatorname{sh} u \, du}{[2(\operatorname{ch} u - \operatorname{ch} u_d)]^{3/2}} + \frac{1}{\sqrt{2(\operatorname{ch} u_1 - \operatorname{ch} u_d)}}. \tag{111}$$

Substitution of (109) and (111) into (110) gives

$$-\frac{dJ(u_1, u_d)}{du_d} = \int_{u_d}^{u_1} \frac{\operatorname{sh} u - \operatorname{sh} u_d}{[2(\operatorname{ch} u - \operatorname{ch} u_d)]^{3/2}} du + \frac{1 - \dfrac{\operatorname{sh} u_d}{\operatorname{sh} u_1}}{\sqrt{2(\operatorname{ch} u_1 - \operatorname{ch} u_d)}}. \tag{112}$$

We then transform both terms in (112) via (9)-(13) of [11] to

$$\int_{u_d}^{u_1} \frac{\operatorname{sh} n - \operatorname{sh} u_d}{[2(\operatorname{ch} u - \operatorname{ch} u_d)]^{3/2}} du = \frac{1}{2 \operatorname{sh} \dfrac{u_d}{2}} \left\{ \int_0^{\varphi_1} \frac{d\varphi}{\sin^2 \varphi} - \int_0^{\varphi_1} \frac{\operatorname{ctg}^2 \varphi}{\Delta \varphi} d\varphi \right\} = \frac{1}{2 \operatorname{sh} \dfrac{u_d}{2}} \{ \operatorname{ctg} \varphi_1 (\Delta \varphi_1 - 1) + E(\varphi_1, k) \}, \tag{113}$$

TABLE 3

$\theta_c = \arcsin k_c$	φ_{1c}	u_{∞_c}	u_{d_c}	$(\varkappa h)_c$	$k_c^3 (1 - k_c^2)^3 F_c^6$	$z_{1c_{cr}}^6$, mM
90°	59°13′	0	0	2.580	0	0
80°	62°06′	0.655	0.352	2.682	0.00594	0.244
70°	69°57′	1.754	0.713	3.000	0.1982	8.17
60°	79°53′	3.779	1.099	3.134	1.642	67.8
55°	85°08′	5.613	1.306	3.086	3.281	135
50°43′	90°	∞	1.499	3.018	5.270	217

Note: It has been assumed that $\varepsilon = 80$, $T = 300°K$, and $A = 2 \times 10^{-12}$ erg (column 7).

$$\frac{1 - \dfrac{\operatorname{sh} u_d}{\operatorname{sh} u_1}}{\sqrt{2(\operatorname{ch} u_1 - \operatorname{ch} u_d)}} = \frac{1}{2 \operatorname{sh} \dfrac{u_d}{2}} \operatorname{ctg} \varphi_1 \left(1 - \frac{\cos^2 \varphi_1}{\Delta \varphi_1}\right), \tag{114}$$

in which $\Delta \varphi_1 \equiv (1 - k^2 \sin^2 \varphi_1)^{1/2}$. We then substitute (113) and (114) into (102) to get

$$-\frac{dJ(u_1, u_d)}{du_d} = \frac{1}{2 \operatorname{sh} \dfrac{u_d}{2}} \left\{ E(\varphi_1, k) + \frac{(1 - k^2) \sin \varphi_1 \cos \varphi_1}{\sqrt{1 - k^2 \sin^2 \varphi_1}} \right\}. \tag{115}$$

Then (14), (107), (108), and (115) give us the relation between u_d and u_1 for the critical state:

$$\frac{2}{3} F_c(\varphi_1, k) = E_c(\varphi_1, k) + \frac{(1 - k_c^2) \sin \varphi_{1c} \cos \varphi_{1c}}{\sqrt{1 - k_c^2 \sin^2 \varphi_{1c}}}. \tag{116}$$

Again, n_c is defined by (100), in which k_c and φ_{1c} are now solutions to (116), while u_d, u_1, and u_∞ are found from (98), and (102) gives the relation of η_c to u_1. Table 3 gives the numerical values of the roots of (116). Of course, (116) becomes (52) for small potentials.

2. Medium and High Potentials. We again use expansion in powers of $\cos \varphi_1$. Here

$$\frac{(1 - k^2) \sin \varphi_1 \cos \varphi_1}{\sqrt{1 - k^2 \sin^2 \varphi_1}} = \sqrt{1 - k^2} \cos \varphi_1 + B_s(k) \cos^3 \varphi_1 + \dots. \tag{117}$$

Substitution of (76), (77), and (117) into (116) gives

$$K(k) = \frac{3}{2} E(k) + \frac{1}{k \operatorname{sh} \dfrac{u_1}{2}}, \tag{118}$$

(terms in $\cos^3 \varphi_1$ and above have been discarded). Comparison of (105) and (118) shows that in the critical state

$$(\varkappa h)_c = 3k_c E_c(k). \tag{119}$$

Comparison of the solutions to (116) and (118) shows that there is agreement for $u_\infty \geq 4$, while for lower u_∞ the solutions to (118) correspond much better with the exact equation for $\psi_1 = $ constant derived by Barboi [8]. Therefore, here also there is practically no difference between $\psi_1 = $ constant and $\eta = $ constant for $u_\infty \gtrsim 4$. Equation (118) becomes as follows for $u_\infty \gg 1$:

$$2K(k) = 3E(k). \tag{120}$$

This equation was previously derived by Deryagin [1], and it is also a consequence of the equation derived by Barboi for a force barrier at constant potential.

The root of (120) is $k_c = 0.7750$, and then $K_c(k) = 1.9486$ and

$$n_c^\eta = 122 \frac{e^3 \theta^5}{A^2 e^6 z_1^6}.$$

Spherical Surfaces, Potential Barrier, and Medium and High Potentials

1. Electrostatic Repulsion Energy. If $\varkappa a \gg 1$ and $H = R - 2a \ll a$, the energy for two spherical particles is

$$V_{SR}(H) = \pi a \int_H^\infty V_R(h)\, dh = \frac{2\pi a}{\varkappa} \int_H^\infty V_R\, d\left(\frac{\varkappa h}{2}\right), \tag{121}$$

where $V_R(h)$ is the interaction energy per cm^2 for plane surfaces.

We have seen above that we have as follows for $u_1 \geq 4$, no matter whether charge or potential remains constant:

$$V_R = \frac{8n\theta}{\varkappa} \left\{ -\frac{1}{k}(1 - k^2)\, K(k) + \frac{2}{k}\, E(k) - 2 \right\} \equiv \frac{8n\theta}{\varkappa}\, I(k) \tag{122}$$

and

$$\frac{\varkappa h}{2} = kK(k) - \frac{1}{\operatorname{sh}\frac{u_1}{2}} \equiv J(k, u_1). \tag{123}$$

Formula (122) is correct for lower potentials if $\psi_1 = $ constant. Substitution of (122) and (123) into (121) and use of (5) gives

$$V_{SR} = \frac{2a\varepsilon\theta^2}{e^2 z_1^2} \int_k^1 I(k)\, dJ(k, u_1) \equiv \frac{2a\varepsilon\theta^2}{e^2 z_1^2}\, L(k, u_1), \tag{124}$$

since $u_d = 0$ for $h = \infty$ and so $k = 1$.

As (122) corresponds best to the case $\psi_1 = $ constant, we assume that u_1 is constant; the results are correct also for $\eta = $ constant for $u_\infty > 4$. Then it follows from (123) that

$$dJ(k, u_1) = d\left[kK(k) - \frac{1}{\operatorname{sh}\frac{u_1}{2}} \right] = d\,[kK(k)],$$

so

$$L(k, u_1) = L(k) = \int_k^1 I(k)\, d\,(kK). \tag{125}$$

It can be shown that

$$\lim_{k \to 0} [I(k)\, kK(k)] = \lim_{t \to 0} (t \ln t) = 0.$$

Then integration of (125) by parts gives

$$L(k) = - I(k) kK - \int_k^1 kK \frac{dI}{dk} dk,$$ (126)

where from (122)

$$I(k) = -\frac{1}{k}(1-k^2) K + \frac{2}{k} E - 2.$$ (127)

We differentiate I(k) and use the fact [11] that

$$\frac{dK}{dk} = \frac{E}{k(1-k^2)} - \frac{K}{k} \quad \text{and} \quad \frac{dE}{dk} = \frac{E-K}{k},$$

to get

$$\frac{dI}{dk} = -\frac{E}{k^2}.$$ (128)

Then

$$L(k) = - I(k) kK + \int_k^1 \frac{KE}{k} dk,$$ (129)

and the energy of electrostatic repulsion is

$$V_{SR} = \frac{2a\varepsilon\theta^2}{e^2 z_1^2} L(k) = \frac{2a\varepsilon\theta^2}{e^2 z_1^2} \left\{ (1-k^2) K^2 + 2kK - 2KE + \int_k^1 \frac{KE}{k} dk \right\}.$$ (130)

Series expansion of K and E allows one to show that for $k \to 1$

$$\int_K^1 \frac{KE}{k} dk \simeq \frac{1-k^2}{4} \ln \frac{16e}{1-k^2},$$ (131)

while numerical test for $0 < k \le 0.7$ shows that

$$\int_k^1 \frac{KE}{k} dk \simeq -\frac{\pi^2}{4} \ln k + C,$$ (132)

in which C = 0.064.

Formulas (123) and (130) give V_{SR} as a function of h for a given surface potential.

2. Stability Criterion. The equation for transition to the unstable state is found by substituting (130) into (46), after the latter has been transformed via (123), with use of the fact that the definition of L(k) gives [compare (124)]

$$\frac{dL}{dJ} = - I(k),$$

$$\frac{d \ln V_{SR}}{d \ln h} = \frac{h}{V_{SR}} \frac{dV_{SR}}{dh} = \frac{J(k, u_1)}{L(k)} \frac{dL}{dJ} = -\frac{J(k, u_1)}{L(k)} I(k) = 1,$$

i.e.,

$$L_c(k) = I_c(k) J_c(k, u_1).$$ (133)

TABLE 4

k_c	u_{d_c}	u_{1_c}	$(\varkappa H)_c$	$4L_c^2 J_c^2$	$z_{1c_{cr}}^6$, mM	L_c
0.500	2.634	3.57	0.996	0.441	18.2	0.667
0.450	2.874	4.10	0.972	0.620	25.6	0.810
0.400	3.134	4.73	0.934	0.830	34.2	0.975
0.350	3.422	5.62	0.894	1.106	45.6	1.179
0.300	3.748	7.22	0.856	1.512	62.4	1.437
0.280	3.891	8.47	0.840	1.710	70.4	1.559
0.265	4.006	9.77	0.817	1.818	74.9	1.650
0.260	4.045	10.6	0.812	1.867	76.9	1.683
0.251	4.119	∞	0.801	1.958	80.6	1.747

Note: It has been assumed that $\varepsilon = 80$, $T = 300°K$, and $A = 2 \times 10^{-12}$ erg (column 6).

We substitute for L, I, and J from (123), (127), and (129) to get an equation relating u_1 and u_d in the critical state.

First we consider the case $u_1 \gg 1$, where

$$\frac{\varkappa h}{2} = kK,$$

and (133) becomes

$$4K_c E_c - 4k_c K_c - 2(1 - k_c^2)K_c^2 = \int_{k_c}^{1} \frac{KE}{k}\,dk. \tag{134}$$

From (132) we find that

$$k_c = 0.251, \quad K_c = 1.596.$$

We calculate $I_c(k)$ and $L_c(k)$ from (127) and (129) for k_c from 0.251 to 0.5 and then use (123) and (133) to find the corresponding values of u_{1c}. Table 4 gives the solutions to (133).

As we have assumed that $V_{SR} \gg \theta$ in the critical state,

$$V_{SR} \simeq -V_{SA}, \tag{135}$$

where, from (43),

$$V_{SA} = -\frac{aA}{12H_c} = -\frac{\varkappa_c aA}{24J_c(k, u_1)}. \tag{136}$$

From (5), (130), (135), and (136) we get

$$n_c = \frac{72}{\pi}\frac{\varepsilon^3\theta^5}{A^2 e^6 z_1^6}4L_c^2(k)J_c^2(k, u_1). \tag{137}$$

Now $(\varkappa H)_c = 0.801$ for $u_1 \gg 1$ and

$$n_c = 44.9\,\frac{\varepsilon^3\theta^5}{A^2 e^6 z_1^6}.$$

This result agrees with that obtained by Kligman and Glazman [14].

3. Condition for Applicability of the Results. We need to establish the particle sizes that make these relationships correct.

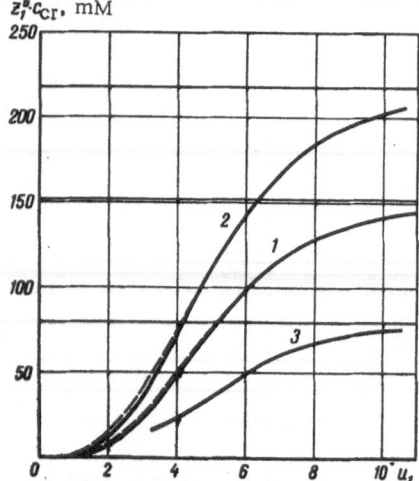

Fig. 3. Relation of $z_1^6 c_{cr}$ to $u_1 =$ $e z_1 \psi_1 / kT$: 1) from Tables 1 and 2; 2) from Table 3 and results of Barboi [8]; 3) from Table 4. The broken lines correspond to $\eta =$ constant; the horizontal lines indicate the asymptotes of curves 1–3 for $u_1 \to \infty$.

Deryagin's method for spherical surfaces is correct if the following apply:

$$\varkappa a \gg 1, \tag{138}$$

$$H = R - 2a \ll a \quad \text{or} \quad \varkappa H \ll \varkappa a. \tag{139}$$

Table 4 shows that $(\varkappa H)_c \approx 1$ for $4 \le u_1 \le \infty$, i.e., (138) and (139) are equivalent. On the other hand, (5), (123), and (137) imply that for conditions close to the critical state

$$\varkappa \approx 24 \frac{e \theta^2}{e^2 z_1^2 A} (\varkappa H)_c L_c (u_d). \tag{140}$$

Now $(\varkappa H)_c L_c \approx 1$ (Table 4), so (138) and (140) give for $\varepsilon = 80$ and $T = 300°K$ that

$$a \gg 3 \cdot 10^{-9} \left(\frac{A}{\theta}\right) z_1^2 \text{ cm}. \tag{141}$$

Then $A = 4 \times 10^{-13}$ erg causes (138) to be obeyed even for small particles ($a > 10^{-7}$ cm). The limiting size of the particles increases with A.

In deducing (46) we assumed that $V_{SR} \gg \theta$ in the critical state; but then, from (136),

$$V_{SR} \simeq -U_{SA} = \frac{aA}{12 H_c} = \frac{A}{12} \frac{\varkappa_c a}{(\varkappa H)_c},$$

and, since $(\varkappa H)_c \approx 1$, we have

$$\varkappa a \approx 10 \left(\frac{\theta}{A}\right) \frac{V_{SR}}{\theta}. \tag{142}$$

This means that the following inequality applies when (138) is obeyed:

$$\frac{V_{SR}}{\theta} \gg \frac{A}{10\theta}. \tag{143}$$

That is, $V_{SR} \gg \theta$ automatically for $A \ge 4 \times 10^{-13}$ erg (since $A/10\theta \ge 1$), and the more so the greater A.

Similarly it may be shown that (141) for small potentials becomes

$$a \gg \frac{2 \cdot 10^{-9}}{\gamma^2} \left(\frac{A}{\theta}\right) z_1^2 \text{ cm}. \tag{144}$$

and the form of the inequality is almost independent of whether $\psi_1 =$ constant or $\eta =$ constant. Moreover, (143) then also applies. It is clear that for high potentials, when $\gamma \approx 1$, (141) and (144) coincide. It follows from (144) that, as ψ_1 decreases, Deryagin's method becomes applicable to larger particles. For instance, for $\psi_1 = 10$ mV we have for all valencies

$$a \gg 2 \cdot 10^{-7} \left(\frac{A}{\theta}\right) \text{cm}.$$

These arguments show that applicability of Deryagin's method automatically provides $V_{SR} \gg \theta$ for $A \ge 4 \times 10^{-13}$ erg, i.e., no real new assumptions have been made in deducing the stability criterion for spherical particles. In the case $A < 4 \times 10^{-13}$ erg, we require a stronger condition on $\varkappa a$, e.g., $\varkappa a \gg 10$, in order to produce $V_{SR} \gg \theta$.

Conclusions

Figure 3 shows $z_1^6 c_{cr}$ as a function of u_1 for all of the cases (it has been assumed that $\varepsilon = 80$, $T = 300°K$, and $A = 2 \times 10^{-12}$ erg). If other values ε_1, A_1, and T_1 are taken, the new curve for $z_1^6 c_{cr}$ is obtained by multiplying the ordinates by $(\varepsilon_1/\varepsilon)^3 (T_1/T)^5 (A/A_1)^2$. The conclusions are as follows.

1. There is only a very small difference between the n_c for $\psi_1 = $ constant and $\eta = $ constant for $u_1 = 2$ and above, so from this viewpoint nothing new is introduced by the condition $\eta = $ constant in places of the usual assumption $\psi_1 = $ constant; but the difference may become much greater if the van der Waals constant is reduced.

2. On the other hand, n_c^η is 1.5-2 times n_c^ψ for small potentials.

3. The curves of Fig. 3 show that the z_1^{-6} limiting law for n_c is more or less correct only for $u_1 \geq 8$; it applies for lower u_1 only on the assumption [8, 17] that $u_1 = $ constant, i.e., $z_1\psi_1 = $ constant, as z_1 varies. This may be explained only by study of the adsorption of the potential-determining ions in the colloidal particles.

4. Any curve in Fig. 3 can be fairly closely approximated by $D_i \gamma^4$ for $2 \leq u \leq \infty$, in which D_i equals the limiting value of $z_1^6 c_{cr}$ for the three exact curves. These limiting values (for a given A) are roughly in the ratio 1:2:3, so any of the three curves may correspond simultaneously to a force barrier or a potential barrier and to a spherical or planar surface, but for slightly different values of the van der Waals constant (c_{cr} is proportional to $1/A^2$).

I am indebted to G. A. Martynov for constant interest in the work and direct participation in the preparation, and to B. V. Deryagin for valuable comments on the draft.

References

1. B. V. Deryagin and L. D. Landau, Acta Physicochim. URSS, 14:633 (1941); Zh. Éksp. Teor. Fiz., 11:802 (1941); 15:662 (1945).
2. B. V. Derjaguin (Deryagin), Trans. Faraday Soc., 36:203, 730 (1940).
3. B. V. Deryagin, Izv. AN SSSR, Ser. Khim., No. 5, p. 1153 (1937); Acta Physicochim. URSS, 10:333 (1939).
4. I. V. Langmuir, J. Chem. Phys., 6:873 (1938).
5. B. V. Derjaguin (Deryagin), Kolloid.-Z., 69:155 (1934).
6. H. R. Kruyt, Colloid Science, Elsevier, Barking (England) (1949-1952).
7. E. J. W. Verwey and J. T. G. Overbeek, Theory of the Stability of Lyophobic Colloids, Amsterdam (1948).
8. V. M. Barboi, Kolloidn. Zh., 25:385 (1963).
9. G. A. Martynov and A. L. Muler, this volume, p. 210.
10. B. V. Deryagin, Zh. Fiz. Khim., 6:1306 (1935).
11. I. S. Gradshtein and I. M. Ryzhik, Tables of Integrals, Sums, Series, and Products, Gos-Fizmatgiz., Moscow (1962).
12. B. I. Segal and K. A. Semendyaev, Five-Figure Mathematical Tables, Fizmatgiz., Moscow (1959).
13. B. V. Deryagin and Yu. V. Gutop, this volume, p. 122.
14. F. I. Kligman and Yu. M. Glazman, this volume, p. 198.
15. N. A. Fuks, Z. Phys., 89:736 (1934).
16. E. J. W. Verwey and J. T. G. Overbeek, Disc. Faraday Soc., 18:180 (1954); Kolloid.-Z., 141:44 (1955); Colloid Sci., 10:224 (1955).
17. V. M. Barboi, Yu. M. Glazman, and I. M. Dykman, Kolloidn. Zh., 24:382 (1962); V. M. Barboi and Yu. M. Glazman, Kolloidn. Zh., 25:282 (1963).

SECTION V

KINETIC EFFECTS IN THIN FILMS OF LIQUID

EFFECTS OF ELECTROKINETIC PHENOMENA AND RHEOLOGICAL PROPERTIES OF BOUNDARY PHASES ON LIQUID FLOW IN PORES AND THIN FILMS

N. V. Churaev and B. V. Deryagin

Surface Phenomena Division, Institute of Physical Chemistry,
Academy of Sciences of the USSR

Deryagin's equation for the potential in overlapping diffuse layers is applied to electrokinetic processes in thin films and pores; the solutions take into account the structural and rheological features of liquid boundary layers. Results are presented for the electroviscous effect and for the filtration anomalies related to the special structure and higher viscosity of boundary layers.

Overlap of diffuse ionic layers near the charged surfaces of walls or pores is an effect that must be taken into account in relation to electrokinetic phenomena in bodies with small pores and in thin wetting films. Consider the flow potential in a wetting film connecting two relatively large volumes having a constant pressure difference (Fig. 1a), which simulates a system in which capillary water is divided by thin liquid films on the solid nonconducting walls of pores. It is assumed that the ion distribution near the surface remains as in a state of rest and that the ion diffusion coefficients are identical. This allows us to neglect the effects of ion didfusion on the flow potential in a first approximation [1]. This flow potential in wetting films has been observed [2].

The calculations have been performed for a planar film (Fig. 1b), which includes the adsorption layer of counterions, a layer of boundary phase (thickness h_0, limiting shear stress θ, and plastic viscosity η_0), and a layer h of liquid with the bulk properties. The potential ψ within the film is described via Deryagin's equation [3], which takes into account the possibility of overlap of diffuse atmospheres. Then we can put as follows the charge density for relatively low potentials ψ_s at the surface separating the boundary phase from the adsorbed layer:

$$\rho = \frac{\varepsilon \varkappa^2 \psi_s}{4\pi} \cdot \frac{\operatorname{ch} \varkappa (h + h_0 - y)}{\operatorname{ch} \varkappa (h + h_0)}, \qquad (1)$$

in which y is the distance from the surface, ε is the dielectric constant (assumed to be the same for the boundary phase and the bulk liquid), and $1/\varkappa$ is the Debye thickness of the double

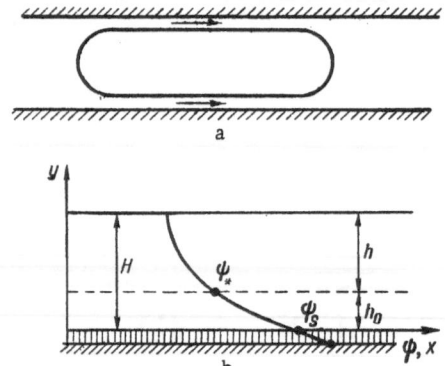

Fig. 1. a) Wetting films in a model for a capillary system; b) calculation scheme for films and slots.

layer. It is assumed [4] that the variation in ψ within the film is as in half the space between two parallel surfaces having the same potential ψ_s. The resulting solutions are then applicable to wetting films and to thin planar pores filled with liquid.

We consider first the case of small pressure gradients, $|P| \leq |P_0|$, when the liquid flows only in the layer h. The Navier–Stokes equations

$$P + \rho E = \eta \nabla v \qquad (2)$$

are to be solved for the steady state where there is a flow-potential gradient E = constant corresponding to the externally imposed constant pressure gradient P = constant; this leads to the following equation for the flow speeds:

$$v = \frac{1}{\eta} \left\{ \frac{P(y - h_0)(y - h_0 - 2h)}{2} + \frac{\varepsilon \psi_s E}{4\pi} \frac{[\operatorname{ch} \varkappa (h + h_0 - y) - \operatorname{ch} \varkappa h]}{\operatorname{ch} \varkappa (h + h_0)} \right\}, \qquad (3)$$

in which η is the viscosity of the bulk liquid.

We can then find a relation between P and E via the condition that the electric current per unit perimeter (A/m) is zero in the steady state:

$$J = \int_{h_0}^{h + h_0} \rho v \, dy - E\lambda (h + h_0)(1 + k) - E u_0 \int_0^{h_0} \rho \, dy, \qquad (4)$$

in which

$$k = u_s \sigma_s / (h + h_0) \lambda. \qquad (5)$$

The first term in (4) takes into account convective transport of ions in the diffuse layer; the second, the electrical conductivity of the neutral electrolyte solution and the surface conductivity of the adsorbed layer of counterions; and the third, the conductivity due to the excess concentration of counterions in the boundary phase. Here λ is the specific conductivity of the solution, σ_s is the surface charge density of the adsorption layer, and u_0 and u_s are the ion mobilities in the boundary phase and adsorption layer respectively.

Substitution for ρ from (1) and for v from (3), with integration within the set limits, gives

$$\left(\frac{E}{P} \right)_{J=0} = -\frac{\varepsilon \psi_s}{4\pi \eta \lambda} \frac{\operatorname{ch} \varkappa h (\varkappa h - \operatorname{th} \varkappa h)}{\operatorname{ch} \varkappa (h + h_0) \left\{ \frac{B (\operatorname{sh} \varkappa h \operatorname{ch} \varkappa h - \varkappa h)}{2 \operatorname{ch}^2 \varkappa (h + h_0)} + \varkappa (h + h_0)(1 + k) + u_0 \varkappa \sqrt{\frac{B\eta}{\lambda}} \frac{\operatorname{sh} \varkappa (h + h_0) - \operatorname{sh} \varkappa h}{\operatorname{ch} \varkappa (h + h_0)} \right\}} \qquad (6)$$

in which

$$B = (1/\eta \lambda)(\varepsilon \psi_s \varkappa / 4\pi)^2. \qquad (7)$$

We replace E in (3) by the value from (6) to get the flow rate (m²/sec) of liquid in the film per unit perimeter:

$$q = \int_{h_0}^{h + h_0} v \, dy =$$

$$= q_0 \left\{ 1 - \frac{3B}{(\varkappa h)^3} \frac{\operatorname{ch}^2 \varkappa h (\varkappa h - \operatorname{th} \varkappa h)^2}{(B/2)(\operatorname{sh} \varkappa h \operatorname{ch} \varkappa h - \varkappa h) + \varkappa (h + h_0)(1 + k) \operatorname{ch}^2 \varkappa (h + h_0) + u_0 \varkappa \sqrt{B\eta/\lambda} [\operatorname{sh} \varkappa (h + h_0) - \operatorname{sh} \varkappa h] \operatorname{ch} \varkappa (h + h_0)} \right\}, \qquad (8)$$

in which $q_0 = -Ph^3/3\eta$ is the flow rate in the absence of electrokinetic effects ($\psi_s = 0$).

By considering the onset of flow in the boundary phase, we get

$$P_0 = -\theta \left| h + h_0 + \frac{\varepsilon \varkappa \psi_s}{4\pi} \left(\frac{E}{P}\right)_{J=0} \text{th } \varkappa (h + h_0) \right|^{-1},\tag{9}$$

as the limit of application of these equations, $|P| \le |P|$. If $h_0 = 0$ (i.e., no boundary phase) or if $h \gg h_0$ (thick film on a lyophilic surface), we have instead of (6) and (8) that

$$\left(\frac{E}{P}\right)_{J=0} = -\frac{\varepsilon \psi_s}{4\pi\eta\lambda} \cdot \frac{\varkappa h - \text{th } \varkappa h}{(B/2) \left[\text{th } \varkappa h - (\varkappa h/\text{ch}^2 \varkappa h)\right] + \varkappa h (1 + k)};\tag{6'}$$

$$\frac{q}{q_0} = 1 - \frac{3B}{(\varkappa h)^3} \cdot \frac{(\varkappa h - \text{th } \varkappa h)^2}{(B/2) \left[\text{th } \varkappa h - (\varkappa h/\text{ch}^2 \varkappa h)\right] + \varkappa h (1 + k)}.\tag{8'}$$

Overlap of the diffuse layers is unimportant for $\varkappa h$ large, and then (6') becomes a classical expression [7] for the flow potential:

$$\left(\frac{E}{P}\right)'_{J=0} = -\frac{\varepsilon \psi_s}{4\pi\eta\lambda}\tag{6''}$$

$$\frac{q}{q_0} = 1 - \frac{3B}{(\varkappa h)^2}.\tag{8''}$$

Figure 2 shows q/q_0 as a function of $\varkappa h$ as calculated from (8') for several values of the dimensionless parameter B with $k \ll 1$. As $\varkappa^2/\lambda$ is only slightly dependent on the electrolyte concentration, B is governed mainly by ψ_s. Use of (1) restricts the application of the solutions to $|\psi_s| \le 150$ mV for a 1-1 electrolyte [5], which corresponds to $B \lesssim 5$ for KCl at 25°C (curve 5 of Fig. 2). Numerical calculations have been performed [6] for $\psi_s = -155$ mV and a 1-1 electrolyte without resort to (1); they agree well with curve 5, which shows that the theory is correct for this range in ψ_s. Curve 3 in Fig. 2 ($B \lesssim 1$) is the limit for a 2-2 electrolyte.

To estimate the possible value of k, we substitute into (5) the expression for σ_s from Stern's equation [7] for a 1-1 electrolyte, putting for the specific adsorption potentials that $\Phi_- = \Phi_+ = 0$. We make the substitution $\lambda = uFc_m \cdot 10^3/v_m$ in (5), in which u is the mobility of the free ions, F is Faraday's number, c_m is the molar fraction of the electrolyte, and v_m is the molar volume of the solution; then

$$k = \frac{u_s}{u} \frac{2zv_m \text{ sh } (F\psi_s/RT)}{(h + h_0) \left[c_m^2 + 2c_m \text{ ch } (F\psi_s/RT) + 1\right]}.\tag{10}$$

Here z is the maximum number of adsorbed ions (mole/cm^2), R is the universal gas constant, and T is absolute temperature.

It has been assumed that $u_s/u = 0.5$ [8] in calculations from (10). If we assume [9] that the surface is 3% covered by the potential-determining ions, each taking up 10 Å^2, then the area per ion is $S \approx 300$ Å^2. This gives $z = 1/NS$, in which N is Avogadro's number. Calculations show that $k \ll 1$ for $|\psi_s| \le 150$ mV for $h + h_0 = H \ge 20$ Å, and so the correction for surface conduction in the adsorbed layers is usually small for a 1-1 electrolyte. In general, this correction (which is especially important for ions of high charge) causes q/q_0 to approach 1.

Figure 2 shows that electrokinetic retardation of the flow can be substantial only for high ψ_s; the maximum reduction in q for sufficiently thin films and small pores does not exceed 55% of q_0. Formally this may be interpreted as an increase in viscosity in thin layers (electroviscosity). This apparent increase in the viscosity should be distinguished from the actual increase due to the change in the structure of the liquid near a solid lyophilic surface [10, 11] and from that produced by the electric field of the surface charge [12, 13].

Deryagin's results [10, 14] confirm directly that there are structural changes in thin layers of liquid, e.g., there is an increase [10] of over fivefold in the viscosity of film water

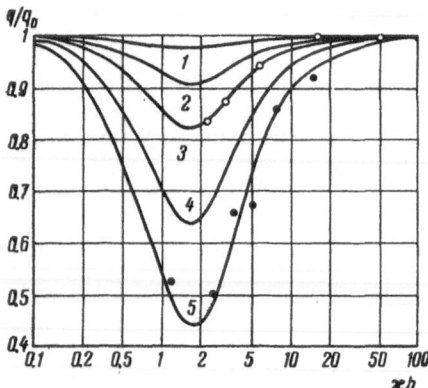

Fig. 2. Variation of q/q_0 with $\varkappa h$ in the electroviscous effect for B of: 1) 0.1; 2) 0.5; 3) 1; 4) 2.5; 5) 5; $\bigcirc$) experiment [12]; $\bullet$) calculation [6].

when $\varkappa h$ is reduced by an order of magnitude, and Fig. 2 shows that this cannot be due to electrokinetic effects even for high ψ_s. Again, the viscosity of pore water was increased by a factor 2.5-3 [14] even for $\varkappa h \geq 100$. On the other hand, electroviscosity accounts for the observed reduction (up to 40%) in the flow rate as the concentration is reduced [12, 15-18] in bodies with relatively broad pores ($h \gg h_0$) at low electrolyte concentrations. The observed points in Fig. 2 [12] agree well with the theory for $\psi_s \approx 50$ mV (curve 3). The curves in Fig. 2 have minima because the convective ion transport is proportional to h^3 for $\varkappa h$ small, whereas the electrical conductivity of a layer is linearly related to the thickness. Development of the flow potential is thus retarded by the electrical conductivity for $\varkappa h$ small.

Similar calculations on q/q_0 from (8) have been made for lyophilic surfaces for $|P| \leq |P_0|$ for various $\alpha = h/(h + h_0)$. These showed that the electroviscous effect is appreciably smaller in this case. The maximal fall in q/q_0 was about 55% for B = 5 and $\alpha = 1$ (curve 5 of Fig. 2), whereas it did not exceed 10% under the same conditions (with $u = u_0 \approx 70$ cm^2/ohm-mole) for $\alpha \leq 0.98$, 1.2% for $\alpha \leq 0.9$, and 0.05% for $\alpha \leq 0.5$. The relation of q/q_0 to $\varkappa h$ for $\alpha = 0.98$ is very similar to curve 2. The effect is reduced on account of the high electrical conductivity of the immobile boundary phase, where the excess concentration of counterions is maximal. This agrees with other results [19, 20]. The electroviscous effect can thus be appreciable for lyophilic surfaces only for $h > 10h_0$. If we take h_0 as 15-20 Å, it is impossible to explain the change in flow speed in films thinner than 150-200 Å or in pores of diameter less than 300-400 Å in terms of electroviscosity.

As the boundary phase does not participate in the flow, the $\zeta = \psi_*$ for this case (Fig. 1) is less than ψ_s:

$$\psi_* = \psi_s \frac{\operatorname{ch} \varkappa h}{\operatorname{ch} \varkappa (h + h_0)} . \qquad (11)$$

For instance, $\psi_* = -32$ mV for $\alpha = 0.5$, $\varkappa h = 1.6$, B=5, and $\psi_s = -150$ mV, as against $\psi_* = -127$ mV for $\alpha = 0.9$ and -145 mV for $\alpha = 0.98$.

Consider now the motion of the liquid in a film at pressure gradients such that the boundary phase flows, the shear extending to the complete layer height h_0 (Fig. 1), which corresponds to $|P| \geq |P_1|$. This P_1 can be found by considering the limiting equilibrium for the surface $y = h_0$:

$$P_1 = -\theta \left[h + \frac{\varepsilon \varkappa \psi_s}{4\pi} \left(\frac{E}{P} \right)_{J=0} \frac{\operatorname{sh} \varkappa h}{\operatorname{ch} \varkappa (h + h_0)} \right]^{-1} . \qquad (12)$$

The Shvedov-Bingham equation gives us for the flow speed in layer h_0 that

$$v_1 = \frac{1}{\eta_0} \left\{ \frac{P}{2} \left[y^2 - 2y \left(h + h_0 + \frac{\theta}{P} \right) \right] + \frac{\varepsilon \psi_s E}{4\pi} \left[\frac{\operatorname{ch} \varkappa (h + h_0 - y)}{\operatorname{ch} \varkappa (h + h_0)} - 1 \right] \right\} . \qquad (13)$$

The speed in layer h can be put as

$$v_2 = v_0 + v, \qquad (14)$$

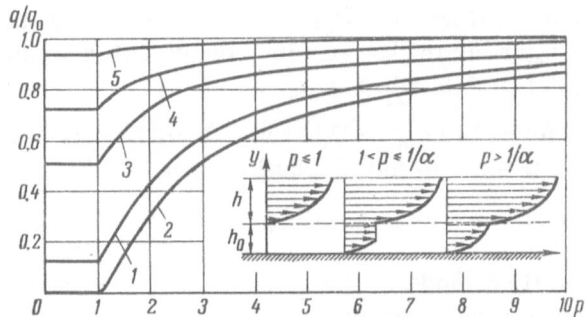

Fig. 3. Reduction in flow speed as a function of $p = PH/\theta$ on account of boundary-layer structure for α of: 1) 0; 2) 0.5; 3) 0.8; 4) 0.9; 5) 0.98. The velocity profiles for three modes of flow are shown in the inset.

in which $v_0 = v_1$ for $y = h_0$, while v is given by (3). The flow current then is

$$J = \int_0^{h_0} \rho v_1 dy + \int_{h_0}^{h+h_0} \rho (v_0 + v) dy - E\lambda (h + h_0)(1 + k), \tag{15}$$

in which the first term characterizes the convective ion transport in the boundary phase, while the second characterizes that in the layer of liquid with ordinary properties. We equate this expression to zero to get, as before, the relation for the flow potential:

$$\left(\frac{E}{P}\right)_{J=0} = -\frac{\varepsilon \psi_s}{4\pi\eta\lambda} \frac{[1 + \gamma (\varkappa h_0)^2/2] \operatorname{sh} \varkappa h - \gamma\varkappa (h + h_0 + \beta) \operatorname{ch} \varkappa (h + h_0) + \varkappa h (\gamma - 1) \operatorname{ch} \varkappa h + \varkappa\beta \operatorname{ch} \varkappa h}{\{(B/2) [\gamma \operatorname{sh} 2\varkappa (h + h_0) - (\gamma - 1) \operatorname{sh} 2\varkappa h - 2\varkappa h(\gamma h_0 + h)/h]/2 \operatorname{ch} \varkappa (h + h_0)\} + \varkappa (h + h_0) (1 + k) \operatorname{ch} \varkappa (h + h_0)} \tag{16}$$

in which $\gamma = \eta/\eta_0$ and $\beta = \theta/P$.

This equation goes over to (6') for $h_0 = 0$, $\gamma = 1$, and $\theta = 0$. Bondarenko and Nerpin [21] were the first to take into account the effects of the rheological parameters of the liquid on electroosmosis in porous media.

Substitution for E from (16) into (13) and (14) gives us the flow rate in the film in the presence of electrokinetic effects:

$$q = \int_0^{h_0} v_1 dy + \int_{h_0}^{h+h_0} v_2 dy. \tag{17}$$

However, it is clear without further calculation that the electroviscous effect in this case cannot exceed the values found from (8') for high mobility of the liquid near the solid, although it is somewhat larger than the values implied by (8) for an immobile boundary phase.

We have already obtained numerical estimates of the electroviscous effect for these limiting cases, so we can now examine the effects of the rheological properties of boundary phases on the flow of liquid in pores and thin films. The scheme is as before (Fig. 3). The solutions for the layers h and h_0 are linked up via the condition $v_1 = v_2$ at $y = h_0$. We consider only the case of steady-state flow for P = constant, and electrokinetic effects are neglected in order to simplify the calculations.

We consider three modes of flow: 1) $|P_0| \leq |P| = \theta/(h + h_0)$, no flow in the boundary phase; 2) $|P_0| < |P| < |P_1|$, shear in the boundary phase extends only to part y_0 of the layer, with $0 < y_0 < h_0$; 3) $|P| \geq |P_1| = \theta/h$, when $y_0 \geq h_0$.

The calculated q are referred to $q_0' = -P(h + h_0)^3/3\eta$ (the rate for a film of liquid whose viscosity throughout the layer $H = h + h_0$ has the bulk value).

Figure 3 shows results[†] for various α with $\gamma = 1$ as a function of $p = PH/\theta$. The value of p for given θ and P characterizes $H = h + h_0$, while it corresponds to the pressure gradient for given H and θ. We get the first mode of flow for $p < 1$, the second for $1 < p \leq 1/\alpha$, and the third for $p > 1/\alpha$. We have $q/q_0' = 1$ when $\alpha = 1$, and the entire film has the properties of the bulk liquid, i.e., $h_0 = 0$. The case $\alpha = 0$, which corresponds to a completely structured liquid when $h = 0$, has been considered in detail elsewhere [22-25], the equations being as follows, which were derived by solution of the equations for steady-state flow of Newtonian and viscoplastic liquids:

$$1) \ q/q_0' = \alpha^3;$$
$$2) \ q/q_0' = \alpha^3 + \gamma \left(1 - \frac{1}{p}\right)^2 \left(1 + \frac{1}{2p}\right);$$
$$3) \ q/q_0' = \alpha^3 + \gamma \left[\left(1 - \frac{1}{p}\right)^3 - \left(\alpha - \frac{1}{p}\right)^3 + \frac{3\gamma}{2p}\left[\left(1 - \frac{1}{p}\right)^2 - \left(\alpha - \frac{1}{p}\right)^2\right]\right].$$

The curves of Fig. 3 show that the layer of structured boundary phase gives a good explanation of the considerable reduction in flow speed observed for pores and thin films. The calculated curves are close to observed ones [26], which also show a region of constant flow speed at low p. If the liquid in the pores or films is considered as structured throughout its volume, the flow stops completely for $p < 1$, which corresponds to an infinite increase in the viscosity. Flow anomalies have now been observed not only for filtration processes [22, 25, 26-29] but also for water transport in unsaturated porous media [22, 30]. Such anomalies may be due solely to viscosity increase in the boundary layers without appreciable structure ($\theta \rightarrow 0$). The flow equations for this case,

$$q/q_0' = \alpha^3 (1 - \gamma) + \gamma$$

imply that the observed reduction in flow speed [10, 14] may be explained by viscosity increase alone. The flow speed must be determined as a function of pressure gradient in order to elucidate the mode of flow in pores and thin films.

The results of Figs. 2 and 3 show that the electroviscosity and the structured sublayer give comparable effects only for thick films and large pores, when $h \geq 50h_0$ ($\alpha \geq 0.98$). The flow anomalies in lyophilic systems are thus mainly due to the special properties of the liquid, whereas those in lyophobic systems, where boundary layers with special properties are thin or absent, may be due also to electroviscosity.

Equations from the thermodynamics of irreversible processes [31] may be used to describe the electrokinetic effects in lyophobic systems:

$$q = L_{11}P + L_{12}E, \tag{18}$$

$$J = L_{21}P + L_{22}E. \tag{19}$$

Solutions (6') and (8') allow us to determine the phenomenological coefficients L_{ik} and to use (18) and (19) to consider a wider range of electrokinetic effects. As $q = q_0$ for $E = 0$, we have $L_{11} = -h^3/3\eta$; and we can deduce L_{21} from (19) by integrating (4) with $E = 0$. Onsager's

———
[†] The ordinates of all the curves for $p > 1$ should be multiplied by γ if $\gamma > 1$.

principle indicates that the mixed coefficients are equal:

$$L_{12} = L_{21} = -\varepsilon\psi_s(\varkappa h - \text{th}\,\varkappa h)/4\pi\eta\varkappa. \tag{20}$$

To find L_{22}, we integrate (4) with $h_0 = 0$ and substitute the v given by (3) with P = 0. The result $J_{P=0} = L_{22}E$ gives

$$L_{22} = -(B/2\varkappa\lambda)\ [\text{th}\varkappa h - (\varkappa h/\text{ch}^2\varkappa h)] - h\,(1+k)/\lambda. \tag{21}$$

Now we can write down some standard relations [31] for electrokinetic processes and use the expressions [32] for the phenomenological coefficients to find appropriate formulas for pores and thin films.

Equation (6') give the relation for the flow potential, $(E/P)_{J=0} = -L_{21}/L_{22}$ which gives the equation for the electroosmotic flow:

$$\left(\frac{q}{J}\right)_{P=0} = \frac{L_{12}}{L_{22}} = \frac{\varepsilon\psi_s}{4\pi\eta\lambda} \cdot \frac{\varkappa h - \text{th}\,\varkappa h}{(B/2)\,[\text{th}\,\varkappa h - (\varkappa h/\text{ch}^2\,\varkappa h)] + \varkappa h\,(1+k)}. \tag{22}$$

We have as follows for the electroosmotic pressure and the mechanoelectric current:

$$-\left(\frac{P}{E}\right)_{q=0} = \left(\frac{J}{q}\right)_{E=0} = \frac{L_{12}}{L_{11}} = \frac{3\varepsilon\psi_s\,(\varkappa h - \text{th}\,\varkappa h)}{4\pi\varkappa h^3}. \tag{23}$$

Equations (22) and (23) for $\varkappa h$ large become standard expressions [7] for the electro-osmotic pressure and electroosmosis in wide pores, so they include the standard solutions for large pores as a special case.

If we insert the L_{12} and L_{22} of (20) and (21) into (22), we get an equation relating the electrokinetic potential $\zeta = \psi_s$ for pores and thin films to the classical value $\zeta' = 4\pi\eta\lambda q/\varepsilon J$

$$\zeta' = \frac{2\eta\lambda\zeta\,(\varkappa h - \text{th}\,\varkappa h)}{(\varepsilon\zeta\varkappa/4\pi)^2\,(\text{th}\,\varkappa h + \varkappa h\,\text{th}^2\,\varkappa h - \varkappa h) + 2\eta\lambda\varkappa h\,(1+k)}. \tag{24}$$

This equation implies that the classical formula (without allowance for interaction on the diffuse layers) causes the electrokinetic potential calculated for $\varkappa h$ small to deviate substantially from the true value, which explains, in particular, the very low ζ' obtained [33] for thin films, as well as the decrease in the zeta potential with $\varkappa h$ for bodies with small pores even after the usual corrections for the surface conductance have been applied [34, 35]. In the case of lyophilic surfaces, allowance must be made for the reduction in the electrokinetic potential at low flow speeds, where the plane for zero velocity moves outwards into the liquid. Then (11) defines $\zeta' = \psi_*$. General solutions can be obtained for electrokinetic processes in films and pores, with allowance for the structured boundary phase, from the equations of the thermodynamics of irreversible processes, by the use of (6) and (8) or (16) and (17) respectively.

The present results show that allowance must be made for overlap of diffuse ion atmospheres and also for the special rheological properties of boundary phases in research on flow in pores and thin films.

References

1.   B. V. Deryagin, Kolloidn. Zh., 22:148 (1960).
2.   O. N. Grigorov and N. K. Barabanshchikov, Dokl. AN SSSR, 93:89 (1953).
3.   B. V. Deryagin, Izv. AN SSSR, OMEN, Ser. Khim., No. 5, p. 1153 (1937).
4.   A. N. Frumkin and A. Gorodetskaya, Zh. Fiz. Khim., 12:511 (1938).
5.   G. A. Elton and F. G. Hirschler, Proc. Roy. Soc., A198:581 (1949).

6. W. D. Kamper, Soil. Sci. Soc. Am. Proc., 24:10 (1960).

7. P. A. Rebinder, ed., Electrokinetic Features of Capillary Systems, Izd. AN SSSR, Moscow (1956).

8. Ling Huang-tsang and D. A. Fridrikhsberg, Vestnik LGU, No. 16, issue 3, 77 (1963).

9. A. A. Baran, D. N. Strazhesko, Yu. M. Glazman, and B. V. Eremenko, Dokl. AN SSSR, 163:125 (1965).

10. B. V. Deryagin and M. M. Samygin, Proceedings of the Conference on the Viscosities of Liquids and Colloidal Solutions, Vol. 1, Izd. AN SSSR (1941), p. 59.

11. B. V. Deryagin and V. V. Karasev, Kolloidn. Zh., 15:365 (1953); B. V. Deryagin, V. V. Karasev, N. N. Zakhavaeva, and V. P. Lazarev, Zh. Éksp. Teor. Fiz., 27:1076 (1957); V. V. Karasev and B. V. Deryagin, Zh. Fiz. Khim., 33:100 (1959).

12. J. C. Henniker, J. Colloid Sci., 7:443 (1952).

13. J. Lyklema and J. T. G. Overbeek, J. Colloid Sci., 16:501 (1961).

14. B. V. Deryagin, N. N. Zakhavaeva, and A. M. Lopatina, in: Research in Surface Forces, Vol. 1, Consultants Bureau, New York (1963), p. 141.

15. N. Street, Producers Monthly, 25:12 (1961).

16. B. F. Ruth, Ind. Eng. Chemistry, 38:564 (1946).

17. J. G. McKelvey and J. H. Milne, Clays and Clay Minerals, Vol. 11, Pergamon Press (1962), p. 248.

18. S. Blaszczynski, Zeszyty Nauk. Politech. slask., 121:141 (1964).

19. D. A. Fridrikhsberg and V. Ya. Barkovskii, Kolloidn. Zh., 26:722 (1964).

20. D. A. Fridrikhsberg and L. V. Pavlova, Kolloidn. Zh., 27:113 (1965).

21. S. V. Nerpin and N. F. Bondarenko, in: Current Views on Bound Water in Rocks, Izd. AN SSSR, Moscow (1963), p. 115.

22. S. V. Nerpin and N. F. Bondarenko, Tr. Leningrad. Inst. Inzh. Vodnogo Transporta, No. 23, p. 36 (1956); A. I. Kotov and S. V. Nerpin, Izv. AN SSSR, OTN, No. 9, p. 106 (1958).

23. S. V. Nerpin, Tr. Leningrad. Inst. Inzh. Vodnogo Transporta, No. 25, p. 37 (1958); S. V. Nerpin and A. I. Kotov, Papers of Soviet Soil Scientists at the Seventh International Congress in the USA, Izd. AN SSSR, Moscow (1960), p. 27.

24. A. M. Gutkin, Kolloidn. Zh., 23:350 (1961).

25. N. V. Churaev, Izv. AN SSSR, Mekhanika i Mashinostroenie, No. 1, p. 136 (1964).

26. G. V. Abelishvili, Tr. Gruz. Nauch.-Issled. Inst. Gidrotekh. i Melior, No. 22, p. 47 (1963).

27. J. V. Nagy and G. Karadi, Oesterreichische Wasserwirtschaft, 13:281 (1961).

28. D. Swartzendruber, Soil Sci., 93:22 (1962).

29. L. S. Ping, Soil. Sci., 95:410 (1963).

30. D. Swartzendruber, Soil. Sci. Soc. Am. Proc., 27:491 (1963).

31. S. R. de Groot, Thermodynamics of Irreversible Processes, Wiley, New York (1951); S. R. de Groot and P. Mazur, Nonequilibrium Thermodynamics (Interscience) Wiley, New York (1962).

32. L. Dresner, J. Phys. Chem., 67:1635 (1963).

33. M. M. Gorokhov, D. M. Lebedev, and N. V. Churaev, Kolloidn. Zh., 29:56 (1967).

34. I. I. Zhukov and D. A. Fridrikhsberg, Kolloidn. Zh., 12:25 (1950).

35. V. V. Ostroumov, Kolloidn. Zh., 13:371 (1951).

APPLICATION OF THERMODYNAMICS
OF IRREVERSIBLE PROCESSES
TO THE ELECTRODIFFUSION THEORY
OF ELECTROKINETIC EFFECTS

B. V. Deryagin and S. S. Dukhin

Institute of Physical Chemistry, Academy of Sciences of the USSR;
Institute of General and Inorganic Chemistry,
Academy of Sciences of the Ukrainian SSR

Methods previously used for internal electrokinetic effects such as electroosmosis and flow potential are extended to external electrokinetic effects (electrophoresis, the Dorn effect) by incorporating the associated diffusion processes. A general theoretical basis is provided for various effects, which include not only the classical electrokinetic effects but also capillary osmosis and diffusophoresis for electrolytes and nonelectrolytes. General formulas are derived for electrophoresis of spherical solid particles and drops (bubbles), with allowance for surface conductivity and diffusion. The mechanism of diffusophoretic slip is elucidated via the conditions for local equilibrium for thin boundary phases.

Introduction

The thermodynamic theory of irreversible processes is based on the Onsager symmetry relations for the kinetic coefficients which appear in the representation of the fluxes in the form of linear combinations of thermodynamic forces. Correct choice of the fluxes and forces requires compilation of an expression for the rate of entropy production, which is represented as a sum of products of the fluxes by the forces. If these fluxes and forces are chosen, the Onsager relations can be written down.

We apply the name internal electrokinetic phenomena[†] to such phenomena occurring in capillary-porous bodies, and a general method has been applied previously [1, 2] to two irreversible processes in a one-component system (thermomolecular pressure difference, the thermomechanical effect). This method is based on compilation of an expression for the rate

---

[†] Electrokinetic phenomena may be classified in this way by analogy with the terminology used in heat and mass transfer. Electrokinetic effects in capillary systems may be termed internal, while electrophoresis and the Dorn effect may be termed external.

of production of entropy in a system consisting of two vessels connected via a porous partition. In the case of electrokinetic effects, we envisage a system of n components bearing charges e_k per unit mass in the presence of a pressure difference Δp and a potential difference $\Delta \varphi$, but no temperature or concentration differences. The equation for entropy production is derived via the equation for entropy change in conjunction with the conservation equations for the energy and the masses of the components.

The laws of conservation of mass and charge may be put as follows:

$$dM_k^I + dM_k^{II} = 0, \tag{1.1}$$

$$\sum_k e_k dM_k^I + \sum_k e_k dM_k^{II} = 0, \tag{1.2}$$

in which M_k^I and M_k^{II} are the masses of component k in vessels I and II, so the electric current flowing from subsystem I to subsystem II is

$$I = \sum_k e_k \frac{dM_k^I}{dt} = -\sum_k e_k \frac{dM_k^{II}}{dt}. \tag{1.3}$$

The law of conservation of energy may be put as

$$dU = dU^I + dU^{II} = dQ - p^I dV^I - p^{II} dV^{II} + (\varphi^I - \varphi^{II})I\, dt, \tag{1.4}$$

in which dQ is the heat given out by the system, U is the energy of the system, and U^I and V^I are the energy and volume of subsystem I.

The equation is as follows for the entropy S^I of subsystem I:

$$TdS^I = dU^I + pdV^I - \Sigma\mu_k^I dM_k^I \tag{1.5}$$

in which μ_k is the chemical potential. The analogous equation for subsystem II may be used with this to find the entropy change for the entire system, which via (1.3)-(1.5) is put as

$$T\,dS = T\,(dS^I + dS^{II}) = dQ - \sum_k (\mu_k^I + e_k\varphi_k^I - \mu_k^{II} - e_k\varphi_k^{II})\, dM_k. \tag{1.6}$$

The entropy charge may be divided into external and internal parts:

$$dS = d_e S + d_i S, \tag{1.7}$$

in which the first is the entropy received from the surroundings:

$$d_e S = dQ/T, \tag{1.8}$$

and so the rate of entropy production is

$$T\,\frac{d_i S}{dt} = -\sum \Delta\mu_k I_k - \sum e_k \Delta\varphi I_k, \tag{1.9}$$

in which $I_k = dM_k/dt$ is the flux of the mass of component k through the diaphragm.

In this theory we envisage isothermal flow with a constant concentration in the whole system, so

$$\Delta\mu_k = v_k \Delta p, \tag{1.10}$$

in which v_k is the partial specific volume of component k.

The bulk velocity J may be expressed in terms of v_k:

$$J = \sum_k v_k \frac{dM_k^I}{dt}, \tag{1.11}$$

and then use of (1.3) allows us to put Td_iS/dt as the sum of products of fluxes and forces:

$$T \frac{d_iS}{dt} = J\Delta p + I\Delta\varphi. \tag{1.12}$$

For this reason, the kinetic coefficients L in the representation of the fluxes as linear combinations of the thermodynamic forces,

$$I = L_{11}\Delta\varphi + L_{12}\Delta p, \tag{1.13}$$

$$J = L_{21}\Delta\varphi + L_{22}\Delta p, \tag{1.14}$$

must satisfy the Onsager relation:

$$L_{12} = L_{21}. \tag{1.15}$$

We now express the characteristics of the various electrokinetic phenomena via L_{11}, L_{12}, L_{21}, and L_{22}. The flow potential is represented via the potential difference $\Delta\varphi$ arising by flow of the liquid under the given pressure difference Δp at zero current, so (1.13) and (1.14) give

$$\left(\frac{\Delta\varphi}{\Delta p}\right)_{I=0} = -\frac{L_{12}}{L_{11}}. \tag{1.16}$$

Electroosmosis may be characterized via the bulk flow rate J for a given current I and $\Delta p = 0$, so (1.13) and (1.14) give

$$\left(\frac{J}{I}\right)_{\Delta p=0} = \frac{L_{21}}{L_{11}}. \tag{1.17}$$

Comparison of the right sides of (1.16) and (1.17) and use of (1.15) will give the following relation between the electroosmosis and the flow potential:

$$\left(\frac{\Delta\varphi}{\Delta p}\right)_{I=0} = -\left(\frac{J}{I}\right)_{\Delta p=0}. \tag{1.18}$$

This relation was derived long before the application of the thermodynamics of irreversible processes and has been repeatedly confirmed. In addition, (1.13) and (1.14) indicate [1] six further electrokinetic effects (transport current, electroosmotic pressure, etc.). Onsager's relation allows one to express all eight electrokinetic effects via the three kinetic coefficients L_{11}, L_{22}, and L_{12}. But L_{11} and L_{22} are related to the electrical and hydraulic resistances, so all eight electrokinetic effects can essentially be characterized via the single specific coefficient L_{12}, which is related to the electrokinetic potential. This means that any of the eight effects should give the same information on the structure of the double layer, and so the choice of some one effect for experimental use is determined not by considerations of principle but by considerations of experimental convenience and accuracy.

De Groot et al. applied these methods also to external electrokinetic effects in order to relate the Dorn effect to electrophoresis [3]; but the treatment for the external effects differs from the above treatment of the internal effects in being very troublesome and unsuitable for extension by allowance for diffusion.

The thermodynamic discussion of the external effects is best approached in a different way [4], which allows the consideration of thermophoresis (external problem) to be referred to thermoosmosis (internal problem). This approach is as follows as regards electrokinetic effects.

Consider electroosmosis in a porous diaphragm consisting of randomly disposed spheres of identical radius whose separations are so large that the local electrical and flow patterns of one sphere are almost unaffected by the presence of other spheres and the liquid has a physical significance.[†] If the spheres are taken as being fixed and at rest, motion of the liquid relative to the diaphragm is to be termed electroosmosis. If the liquid is taken as being at rest, the external electric field moves the spheres, which may be termed electrophoresis. The spheres do not interact, and so the speed of the spheres is not dependent on the concentration. The relation between electrophoresis and electroosmosis becomes more complicated if the diaphragm is more closely packed.

The electrophoresis velocity V_{ef} thus differs from the electroosmosis velocity only in sign for a spheres diaphragm:

$$V_{ef} = -V_{eo}. \qquad\qquad (1.19)$$

The flow potential for a sphere diaphragm is readily related to the Dorn effect. Consider two vessels separated by a spheres diaphragm. The expressions for dS/dt in the Dorn effect and in electrophoresis may be drawn up precisely as above, and so a spheres diaphragm allows one to refer external electrokinetic effects to internal ones.[‡]

A method has also been given [4] for calculating the transfer heat by introducing heat sources and sinks at the boundaries of the diphragm with the vessels, which provide the constant temperature in the vessels needed in deriving the transfer heat. Analogous introduction of mass sources and sinks provides constant concentrations in the extension of the theory by incorporation of diffusion, which provides calculation of the mass transfer in the absence of a concentration difference.

We thus use the method of compiling expressions for the rate of entropy production and for the implied Onsager relations, as first used by de Groot, Mazur, and Overbeek. Our treatment differs from theirs in that we also consider external electrokinetic effects on this basis by using a spheres diaphragm. The method developed for the theory of thermophoresis [4] is applied to the transfer mass via the introduction of mass sources and sinks at the boundaries.

In 1947 it was pointed out [5] in relation to polymolecular boundary phases that there is an analogy with the electrokinetic phenomena for molecular adsorption layers on account of the diffuse structure. The analog of the electric field in this case is the diffusion field, while the analog of the electric field strength is the concentration gradient.

For instance, a concentration gradient arises in a solution of a nonelectrolyte flowing through a porous partition under a pressure difference. This effect is to be considered as the analog of the flow potential that would arise for an electrolyte under the same conditions. The mechanism is simple. The flow involves the diffuse part of the adsorption layer, and so a surface flow arises, which is initiated at the entry of the liquid into the pores as a result of adsorption from the bulk, and this amount of material must be desorbed at the exit. The adsorption at the entry reduces the concentration and vice versa, so the solute concentration should increase in the flow direction. In the steady state, the diffusion flow resulting from this concentration change exactly balances the surface flow. The converse effect was also predicted [5]: a flow of solution must arise in response to a concentration difference across a capillary (diaphragm). The effect is the analog of electroosmosis and has been observed; it is called capillary osmosis. In this observation of the effect it was also found that particles suspended in a solution of a nonelectrolyte should be set in motion by a gradient in the chemical potential. This effect, which is the analog of electrophoresis, was called diffusophoresis. These effects

---

[†] This diaphragm is subsequently termed a spheres diaphragm for brevity.

[‡] However, the method based on a spheres diaphragm does not allow us to include the role of Brownian motion in electrophoresis or the Dorn effect; if this motion is important for some reason, the treatment of de Groot et al. [3] may be of interest.

are of value in research on the diffuse parts of molecular adsorbed layers in the way that electrokinetic effects are useful in research on the diffuse parts of double layers.

In Section 2 we apply this method to the theory of capillary osmosis and diffusophoresis. In Section 3, we apply the thermodynamics of irreversible processes to extend the theory of electrokinetic phenomena by incorporating diffusion. At the same time, the theory of capillary osmosis and diffusophoresis is extended to the case of electrolytes. In Section 4, the usual formulas for electrophoresis are extended to bubbles (drops) and also to solid spherical particles, with allowance for surface electrical conductivity and diffusion. Section 5 deals with the mechanism of capillary osmosis and diffusophoresis for electrolytes and nonelectrolytes.

2. Application of the Thermodynamics of Irreversible Processes to the Theory of Capillary Osmosis and Diffusophoresis

The general method described in Section 1 is applied by considering the rate of entropy production in a system of two vessels joined via a porous diaphragm through which flows a solution of a nonelectrolyte, for which the adsorption layer on the internal surface of the diaphragm has a diffuse structure.

We assume that Δp is accompanied by concentration differences Δc_k for the components, so (1.10) must be generalized:

$$\Delta\mu_k = v_k\Delta p + \Delta'_p\mu_k, \tag{2.1}$$

in which

$$\Delta'_p\mu_k = (\partial\mu_k/\partial c_k)_p\Delta c_k. \tag{2.2}$$

This $\Delta'_p\mu_k$ is substituted into (1.9) with $\Delta\varphi$ put as zero, and (1.11) is used to give

$$T\frac{d_iS}{dt} = J\Delta p + \sum_k I_k\Delta_p\mu_k. \tag{2.3}$$

A standard theorem states that

$$\Sigma c_k\Delta_p\mu_k = 0, \tag{2.4}$$

so it is readily shown that

$$\sum I_k\Delta_p\mu_k = \sum I_k^*\Delta_p\mu_k, \tag{2.5}$$

in which $I_k^* = I_k - c_k V$ characterizes the flux of material less the trivial transfer due to flow of the mixture as a whole, while V is the mean barocentric flow velocity.

Substitution of (2.5) into (2.3) gives us an expression from which we derive the following equations:

$$I_1^* = L_{11}\Delta_p\mu_1 + L_{12}\Delta_p\mu_2 + L_{13}\Delta p, \tag{2.6}$$

$$I_2^* = L_{21}\Delta_p\mu_1 + L_{22}\Delta_p\mu_2 + L_{23}\Delta p, \tag{2.7}$$

$$J = L_{31}\Delta_p\mu_1 + L_{32}\Delta_p\mu_2 + L_{33}\Delta p \tag{2.8}$$

and the Onsager relations

$$L_{12} = L_{21};\ L_{13} = L_{31};\ L_{23} = L_{32}. \tag{2.9}$$

Capillary osmosis may be characterized via the bulk velocity of the liquid due to the gradient in the chemical potential with $\Delta p = 0$. Then, if Δp is put as zero in (2.8), V can be expressed in terms of L_{31} and L_{32} or (if we use the Onsager relations) in terms of L_{13} and L_{23}:

$$V = L_{13}\Delta_p\mu_1 + L_{23}\Delta_p\mu_2. \qquad (2.10)$$

The values of L_{13} and L_{23} in (2.6) and (2.7) can be found by calculating the mass transfer subject to

$$\Delta_p\mu_k = 0, \qquad (2.11)$$

only one of these conditions being independent, in accordance with (2.4).

These quantities I_1^*/c and I_2^*/c can be called transport masses, by analogy with thermal transport [4].

It follows from (2.6) and (2.7) that

$$L_{k3} = I_k^{\cdot}/_c(\Delta p)^{-1}, \qquad (2.12)$$

so

$$V = [I_1^{\cdot}/_c\Delta_p\mu_1 + I_2^{\cdot}/_c\Delta_p\mu_2] (\Delta p)^{-1}. \qquad (2.13)$$

We calculate the transport masses for a spheres diaphragm in order to get the formulas for capillary osmosis and diffusophoresis. It has been shown [6] that an excess concentration δc_k arises around a solid spherical particle in response to flow of the liquid as a result of involvement of the diffuse part of the adsorption layer. The distribution of this concentration is as follows in a spherical coordinate system with its pole at the center of the particle and with the θ axis along the flow direction:

$$\delta c_k(r_1\theta) = M_{ck}\cos\theta/r^2, \qquad (2.14)$$

in which

$$M_{ck} = (3/2)\, uac_k\xi_k/D_k; \quad \xi_k = \frac{1}{c_k}\int\gamma_k(x)\,xdx, \qquad (2.15)$$

in which x is the distance from the given point in the diffuse adsorption layer to the slip plane and $\gamma_k(x)$ is the excess of the local concentration in that layer over the bulk value.

If the particles are far apart, the diffusion field of one is unaffected by the fields of the others, so we may sum these fields via the principle of superposition, by analogy with the formula for averaging the local electric fields in a polarized dielectric, which gives us the Δc_k for flow through a diaphragm as

$$\Delta c_k = 4\pi n M_{ck}l, \qquad (2.16)$$

in which n is the number of spheres in unit volume and l is the thickness of the diaphragm.

These Δc_k either cause (2.11) to be violated or produce inhomogeneity in composition in each vessel. As a concentration difference arises at the ends of the diaphragm, considerations of continuity in the concentrations imply that the concentrations are different, at least in the parts adjoining the diaphragm. Further, a plane perpendicular to the flow direction meets the randomly placed spheres with equal probability if these spheres are sufficiently remote from the ends of the diaphragm. This means that, for any position of the plane, the spheres take up by adsorption precisely the amount released by desorption. If this plane is placed sufficiently close to the inlet end, the parallels produced by intersection with the plane will be characterized by $\theta > \pi/2$ (by $\theta < \pi/2$ similarly for the outlet end). This means that material is preferentially adsorbed at the inlet.

It is clear that, in the steady state, the amount desorbed at the outlet equals that adsorbed at the inlet, and the Δc_k of (2.16) provides the diffusion flux that returns all the material desorbed at the outlet. To meet the condition of homogeneity of the vessels, as well as conditions

(2.11), we introduce mass sources at the inlet and mass sinks of equal capacity at the outlet. Let the capacity of these sources be $(D_k/l)4\pi n M_{ck}l$; then conditions (2.11) will be met, since Fick's law indicates that the Δc_k will compensate the differences of (2.16). For any particle, the flux through a plane due to involvement of the adsorption layers is equal in magnitude and opposite in direction to the total diffusion flux through that plane around it. This result still applies after summation over all particles, so the fluxes will be $D_k\Delta c_k/l$ after introducing the sinks and sources, and thereby when conditions (2.11) are met. In fact, the diffusion flux vanishes, but a flux equal in magnitude and opposite in sign persists, on account of entrainment of the adsorption layers. The transport masses are therefore equal to the capacities of the sources:

$$I_k^{\bullet}/_c = D_k\Delta c_k/l. \tag{2.17}$$

These values are substituted into (2.13) and then (2.4), (2.15), and (2.16) are applied to give

$$V_{c0} = 6\pi\,uanc_1(\xi_1 - \xi_2)\,\Delta_p\mu_1/\Delta p. \tag{2.18}$$

Stokes's law gives a simple relation between Δp and the velocity:

$$\Delta p = 6\pi\eta\,anul, \tag{2.19}$$

and so (2.18) finally gives the following formula for the capillary osmosis velocity[†]:

$$V_{c0} = -c_1(\xi_1 - \xi_2)\,\eta^{-1}\,\mathrm{grad}_p\mu_1. \tag{2.20}$$

The diffusophoresis velocity V_{df} is found simply by changing the sign in (2.20), as shown in Section 1. The following is [7] the Δc_k arising in this way in a diaphragm composed of capillaries of length l and radius a by adsorption and desorption:

$$\Delta c_k = \xi_k c_k \Delta p/\eta D_k, \tag{2.21}$$

and, exactly as for a sphere diaphragm, we can derive the formula for the capillary osmosis in a diaphragm composed of cylindrical capillaries, which is the same as (2.20).

Frumkin [8] and Levich [9] found that a new form of diffusophoresis occurs for large values of the Peclet number Pe; they deduced the corresponding formula. If a concentration gradient in a surfactant is set up in a solution containing drops of another liquid, the adsorption (and hence the surface tension) will vary along the surface of a drop, which produces circulation within the drop and motion with respect to the medium on account of reactive forces. The resulting speed can be calculated either directly (via hydrodynamics) or by the above method (i.e., via the thermodynamics of irreversible processes), and in this way one can verify the correctness of the latter approach.

Consider an idealized diaphragm consisting of randomly placed drops of equal radius, which are assumed to be fixed in some way (they do not move with the liquid). However, the method of fastening must not affect the velocity distribution inside or outside, which is as for a free drop.[‡] The resistance exerted by a drop for Pe $\ll$ 1 is given by the Frumkin–Levich formula [8, 9], if we take the distance between drops to be substantially greater than the size of a drop[§]:

$$F = 6\pi\eta\,anu(2\eta + 3\eta' + 3\gamma)/3(\eta + \eta' + \gamma). \tag{2.22}$$

[†] The sign change in (2.20) is explained by the fact that Δp and Δc were reckoned as between the first and second vessels in deducing (2.3), while the fluxes and corresponding velocities were taken as positive if the corresponding vectors are directed from vessel I to vessel II.

[‡] For instance, each drop might be considered as fixed on a vertical wire through its center and passing through the surface at points where the velocity is zero. It is of no importance that it would be technically difficult to produce such a diaphragm, the only point being that the existence of such a system would not conflict with thermodynamics.

[§] In what follows, η' and η are the viscosities of the liquid and the medium, while γ is the retardation coefficient.

Then

$$\Delta p = n l 6 \pi \eta a u \, (2\eta + 3\eta' + 3\gamma)/3 \, (\eta + \eta' + \gamma). \qquad (2.23)$$

Entrainment of the adsorption layer as a whole by the motion of the drop causes [10] the local concentration distribution of (2.14), where the analog of the dipole moment M_c in this case takes the form

$$M_{ck} = a^2 \, \Gamma_k V_0 / D_k, \qquad (2.24)$$

in which k = 1 corresponds to the solute, k = 2 to the solvent, and Γ_k is the adsorption.

The local concentration field for a drop is entirely similar to that for a solid spherical particle, and the same applies to the transport of adsorbed material along the surface, so the above steps may be repeated, which shows that the transport masses are defined by (2.17), with Δc_k defined by (2.17) and the velocity by (2.18). We substitute for M_{ck} from (2.24) and for Δp from (2.23) in these formulas and neglect Γ_2 to get

$$V = \frac{a\Gamma_1}{2\eta + 3\eta' + 3\gamma} \, \mathrm{grad} \, \mu_1 = \frac{a \, | \, d\sigma/dc_1 \, |}{2\eta + 3\eta' + (\Gamma_1/D_1 | \, d\sigma/dc_1 |)} \, \mathrm{grad} \, c_1. \qquad (2.25)$$

We now derive a formula via hydrodynamics in order to test the above method; we use the analogy between this effect and the motion of a mercury drop in an external electric field E (which is due to the electrocapillary change in surface tension along the surface), which theory [8, 9] gives via the following formula:

$$\Delta\sigma \, (\theta) = [(^3/_2) \, \varepsilon E a - V_0 \varepsilon^2/\varkappa] \cos\theta, \qquad (2.26)$$

in which ε is the charge density at the surface of the drop.

The concentration near a drop must satisfy Laplace's equation and the relatively simple boundary conditions

$$c_1 \, |_{r\to\infty} = c_0 + | \, \mathrm{grad} \, c_1 \, | \, r \cos\theta, \; \frac{\partial c_1}{\partial r} \, (a,\theta) = \frac{2V_0\Gamma_1}{dD_1} \cos\theta, \; \Delta c_1 = 0 \qquad (2.27)$$

for Pe small and for a grad c_1 given at infinity, so c_1 is found without difficulty; a formula analogous to (2.26) may then be derived for the surface tension, assuming a linear dependence of the adsorption on the concentration:

$$\Delta\sigma \, (\theta) = \frac{\partial\sigma}{\partial c_1} \, [c_1 \, (a, \, \theta) - c_0] = \frac{\partial\sigma}{\partial c_1} \, [(3/2) | \, \mathrm{grad} \, c_1/a - V_0\Gamma_1/D_1]. \qquad (2.28)$$

As (2.26) and (2.28) are entirely analogous, the speed may be expressed via the formula for the speed of a mercury drop due to the electrocapillary effect [9]:

$$V = \frac{\varepsilon E a}{2\eta + 3\eta' + \varepsilon^2/\varkappa} \cdot \qquad (2.29)$$

If we make the substitution $a\varepsilon E \to d\sigma/dc_1 \, \mathrm{grad} \, c_1$ in accordance with (2.26) and (2.28), with

$$\varepsilon^2/\varkappa \to (\Gamma_1/D_1) | \, \partial\sigma/\partial c_1 |,$$

we get a formula the same as (2.25), which was derived via the thermodynamics of irreversible processes.

3. Application of the Thermodynamics of Irreversible Processes to the Theory of Electroosmosis, Electrophoresis, Capillary Osmosis, and Diffusophoresis for Electrolytes

Expression (19) gives the entropy production rate for isothermal flow from vessel I to vessel II, between which are maintained finite Δp and $\Delta \varphi$; this still applies in the more general case where the concentrations in the two vessels are different. Here the $\Delta \mu_k$ should be expressed via (2.1), so (1.9) becomes

$$T \frac{d_i S}{dt} = J \Delta p + I \Delta \varphi + \Sigma I_k^* \Delta_p \mu_k, \tag{3.1}$$

where (2.5) has also been used. This expression gives us the following equations:

$$I_1^* = L_{11}\Delta_p\mu_1 + L_{12}\Delta_p\mu_2 + L_{13}\Delta_p\mu_3 + L_{14}\Delta\varphi + L_{15}\Delta p, \tag{3.2}$$

$$I_2^* = L_{21}\Delta_p\mu_1 + L_{22}\Delta_p\mu_2 + L_{23}\Delta_p\mu_3 + L_{24}\Delta\varphi + L_{25}\Delta p, \tag{3.3}$$

$$I_3^* = L_{31}\Delta_p\mu_1 + L_{32}\Delta_p\mu_2 + L_{33}\Delta_p\mu_3 + L_{34}\Delta\varphi + L_{35}\Delta p, \tag{3.4}$$

$$I = L_{41}\Delta_p\mu_1 + L_{42}\Delta_p\mu_2 + L_{43}\Delta_p\mu_3 + L_{44}\Delta\varphi + L_{45}\Delta p, \tag{3.5}$$

$$J = L_{51}\Delta_p\mu_1 + L_{52}\Delta_p\mu_2 + L_{53}\Delta_p\mu_3 + L_{54}\Delta\varphi + L_{55}\Delta p, \tag{3.6}$$

and the Onsager relations, of which we need only the following:

$$L_{51} = L_{15}; \quad L_{52} = L_{25}; \quad L_{53} = L_{35}; \quad L_{54} = L_{45}. \tag{3.7}$$

Electroosmosis may be characterized via the bulk velocity V_{e0} due to $\Delta\varphi$ for $\Delta p = 0$ and $\Delta c_k = 0$; in accordance with this, we put Δp and $\Delta_p\mu_k$ equal to zero in (2.6), so V_{e0} may be expressed via L_{54}, or via L_{45} if we use (3.7). Then

$$V_{e0} = L_{45}\Delta\varphi. \tag{3.8}$$

Capillary osmosis may be characterized via the bulk velocity V_{c0} produced by an ion concentration gradient (and hence by the associated concentration gradient in the solvent) for $\Delta p = 0$ and $\Delta\varphi = 0$, so V_{c0} can be expressed via L_{51}, L_{52}, and L_{53}, or via L_{15}, L_{25}, and L_{35} if we use the Onsager relations:

$$V_{e0} = L_{15}\Delta_p\mu_1 + L_{25}\Delta_p\mu_2 + L_{35}\Delta_p\mu_3. \tag{3.9}$$

To derive the L_{k5} ($k = 1, 2, 3, 4$) we consider (3.2), (3.3), and (3.4) for a given Δp and for

$$\Delta\varphi = 0, \Delta c_k = 0. \tag{3.10), (3.11}$$

The mass fluxes are then the transport masses $I_k^*/c, \dot{\varphi}$. Then (3.2)-(3.4), (3.10), and (3.11) give

$$L_{k5} = (\Delta p)^{-1} I_k^* \big|_{c, \varphi}. \tag{3.12}$$

Even if we do not specify the Δc_k, definite grad c_k arise as the liquid moves within the diaphragm. The diffuse part of the adsorption layer is entrained by the motion of the liquid, which is responsible for the excess fluxes $I_{k\xi}$, which are as follows for a diaphragm composed of cylinders of length l and radius a:

$$I_{k\xi} = (c_k \xi_k \Delta p)/l\eta. \tag{3.13}$$

These excess fluxes arise at the inlet to the capillaries by absorption of material from the bulk, and then the same amount is desorbed at the exit. If these effects are not permanently to violate conditions (3.10) and (3.11), we must introduce sources and sinks whose capacities are equal to the transport masses:

$$I_k^* \big|_{\varphi, \, c} = c_k \xi_k \Delta p / \eta l. \tag{3.14}$$

We substitute these values for $I^*/_{c, \varphi}$ in (3.12) to get

$$L_{k5} = c_k \xi_k / \eta l. \tag{3.15}$$

These values for the L_{k5} are substituted into (3.9), and (2.4) is applied together with the condition for electrical neutrality, which allows us to express $c^{\pm}$ in terms of a single function c, with the result that

$$V_{co} = - \, [z^- (\xi^+ - \xi_3) + z^+ (\xi^- - \xi_3)]^{-1} \, \eta c \, \text{grad} \, \mu \, (c), \tag{3.16}$$

in which $\xi^{\pm}$ and ξ_3 characterize the concentration increase from adsorption for cations, anions, and solvent.

As $I = F(z^+ I^+ + z^- I^-)$, L_{45} is expressible via L_{15} and L_{25}:

$$L_{45} = F \, (z^+ L_{15} - z^- L_{25}) = (F/\eta) \, (z^+ c^+ \xi^+ - z^- c^- \xi^-) = Fcz^+ z^- \, (\xi^+ + \xi^-)/\eta l, \tag{3.17}$$

and also, as $\zeta = 4\pi F c_0 z^+ z^- (\zeta^- - \zeta^+)/\varepsilon$, substitution for L_{45} in (3.8) gives us Smoluchowski's formula for the electroosmosis:

$$V_{eo} = (\varepsilon \zeta / 4\pi \eta) \, \text{grad} \, \varphi. \tag{3.18}$$

We have seen in Section 1 that the theory of electrophoresis may be constructed as for the above theory of electroosmosis if the velocity for the latter is calculated for a spheres diaphragm. The velocities in diffusophoresis and electrophoresis may be calculated from formulas that differ from (3.8) and (3.9) only in sign, while the calculation of the L_{k5} is, from (3.12), reduced to that of the transport masses, which are easily found via the electrodiffusion theory of the Dorn effect.

For Pe small, entrainment of the double layer ions gives rise to a local electrical field $\varphi(r, \, \theta)$ around a sphere together with an excess-concentration distribution $\delta c_k (r, \, \theta)$, the structure of these being as for the electric-dipole field of (2.14). These local fields sum to give the differences Δc_k and $\Delta \varphi$ in concentration, which can be expressed in terms of dipole moments via a formula in the theory of dielectrics:

$$\Delta c^{\pm} = 4\pi n z^{\mp} M_c; \quad \Delta \varphi = 4\pi n \, M_{\varphi}. \tag{3.19}, (3.20)$$

Δc_k and $\Delta \varphi$ either cause violation of (3.10) and (3.11) or upset the composition uniformity of each vessel. The latter is a necessary condition in deducing the equation for the rate of entropy production within the diaphragm via the entropy-balance equations. Mass sources at the inlet and equal mass sinks at the outlet must be introduced in order to maintain composition uniformity and to obey (3.10) and (3.11). If the capacities of these sources are

$$(\mp \chi^{\pm} \, \Delta \varphi - D^{\pm} \Delta c^{\pm})/l,$$

in which $\chi^{\pm} = \lambda^{-1} D^{\pm} z^+ z^- c$, $\lambda = RT/F$, R is the universal gas constant, T is absolute temperature, and F is Avogadro's number, then we get concentration and potential differences that balance out those of (3.19) and (3.20). Exactly as in Section 2, we can show that the transport masses are equal to the capacities of the sources, so from (3.12), (3.19), and (3.20) we get

$$L_{15} = (4\pi n/l \, \Delta p) \, (-\chi^+ M_{\varphi} - D^+ z^- M_c), \tag{3.21}$$

$$L_{25} = (4\pi n/l \, \Delta p) \, (\chi^- M_{\varphi} - D^- z^+ M_c). \tag{3.22}$$

There is only a diffusion flux for the solvent, so L_{35} can be calculated exactly as in Section 2:

$$L_{35} = 4\pi\eta D_3 M_3 / l\,\Delta p. \tag{3.23}$$

These expressions are substituted into (3.9) to give

$$V_{df} = (4\pi/\Delta p)\left\{(D^+ - D^-)\frac{z^+z^-c_0}{\lambda} M_\varphi - (D^+z^- + D^-z^+)\,M_c - (z^+ - z^-)\xi_3\right\}\mathrm{grad}\,\mu. \tag{3.24}$$

We use (3.17), (3.21), and (3.22) to derive an expression for V_{e0}; then a formula differing from (3.8) only in sign gives the general formula

$$V_{ef} = (4\pi nF/\Delta p)\left[(z^+D^+ + z^-D^-)\frac{z^+z^-c}{\lambda} M_\varphi + (D^+ - D^-)z^+z^-M_c\right]\mathrm{grad}\,\varphi. \tag{3.25}$$

General formulas (3.24) and (3.25) may be applied to the particular case of spherical solid particles. We neglect initially the surface conductivity and use previous simplifications [6] to get the following expressions for the effective dipole moments:

$$M_c = -\frac{3ua\,(D^-z^-\xi^+ + D^+z^+\xi^-)}{2D^+D^-\,(z^+ + z^-)}, \tag{3.26}$$

$$M_\varphi = \frac{3ua\lambda\,(D^+\xi^- + D^-\xi^+)}{2D^+D^-\,(z^+ + z^-)}. \tag{3.27}$$

We substitute these expressions and the Δp of (2.19) into (3.24) to get for V_{df} a formula differing from (3.16) for V_{c0} only in sign. The definition of ξ_i allows us to neglect ξ_3 relative to $\xi^\pm$ for the case of a weak solution of an electrolyte, so we have

$$V_{df} = [z^-\xi^+ + z^+\xi^-]c\,\mathrm{grad}\,\mu\,(c)/\eta. \tag{3.28}$$

Analogous substitution in (3.25) gives us Smoluchowski's formula for V_{ef}, which differs from (3.18) only in sign.

4. Extension of the Theory of Electrophoresis

We use the theory of the Dorn effect and the results of the previous section to extend the electrophoresis formulas to the case of a bubble (drop) and to a solid spherical particle, with allowance for surface electrical conduction and diffusion.

First we consider a solid spherical particle whose surface activity is so high that the effects of bulk diffusion can be neglected relative to those of surface electrical conduction. The effective dipole moments for this case can be deduced from formulas [11] for the concentration and potential:

$$M_\varphi = \frac{3}{2}\frac{ua^2\lambda}{D_s^+D_s^-\,(z^+ + z^-)}\,(D_s^+\xi^- - D_s^-\xi^+)\frac{c_0}{\Gamma_0}, \tag{4.1}$$

$$M_c = -\frac{3}{2}\frac{ua^2c_0^2}{D_s^+D_s^-\,(z^+ + z^-)}\,\Gamma_0\,(D_s^-z^-\xi^+ + D_s^+z^+\xi^-), \tag{4.2}$$

in which $D_s^\pm$ are the coefficients of surface diffusion of the ions.

These values and the Δp of (2.19) are substituted into (3.25) to give

$$V_{ef} = -\frac{Fc_0z^+z^-}{\eta}\left(\frac{D^-}{D_s^-}\xi^- - \frac{D^+}{D_s^+}\xi^+\right)\frac{ac_0}{\Gamma_0}\,\mathrm{grad}\,\varphi. \tag{4.3}$$

Henry [12] and Booth [13] have shown that allowance for surface conductivity gives rise to the formula

$$V_{ef} = -\frac{\varepsilon\zeta}{4\pi\eta}\frac{K}{K + K_s/a}\,\text{grad}\,\varphi,$$ (4.4)

in which K and K_s are the specific bulk and surface conductivities.

Formulas (4.3) and (4.4) agree closely, since K should be neglected relative to K_s/a in (4.4) in this comparison, because that simplification was made in deducing (4.3); also, we should put $D^+/D_s^+ = D^-/D_s^-$ in (4.3), since Henry and Booth did not take into account differences in ion mobility. But then $Dc_0/D_s\Gamma_0 = K/K_s$ and, since the ζ potential is related to $\xi^\pm$, we readily see that the simplified formulas coincide

The electrodiffusion theory of the Dorn effect for a drop with an adsorbed layer at $Pe \ll 1$ was first developed [10, 11, 14] for the limiting cases of low surface activity [10]

$$\Gamma_0/c_0 \ll a$$ (4.5)

and high surface activity [11]

$$\Gamma_0/c_0 \gg a$$ (4.6)

with

$$\Gamma_0 = \Gamma^\pm/z^\pm,$$ (4.7)

$$\Gamma^\mp = \Gamma_{St}^\pm + \Gamma_H^\pm,$$ (4.8)

in which $\Gamma_{St}^\pm$ and $\Gamma_H^\pm$ are the uptakes in the Stern and Hui layers.

These results for limiting cases apply also for the general case, in that there is only a small deviation from local electrical neutrality outside the double layer, while the potential and concentration outside that layer satisfy the Laplace equations and are given by

$$c(r_1\theta) = c_0 + \frac{M_c}{r^2}\cos\theta;$$ (4.9)

$$\varphi(r_1\theta) = \frac{M_\varphi}{r^2}\cos\theta.$$ (4.10)

There is only a small relative change in concentration along the surface of a drop for $Pe \ll 1$. We therefore have the following boundary relations between the tangential ion fluxes in the double layer and the normal components of the ion fluxes at the outer boundary of the double layer:

$$\frac{2V_0}{a}\left[z^\mp\Gamma_0 + \frac{2 + 3(\eta' + \gamma)/\eta}{a}z^\mp c_0\xi^\pm\right]$$
$$+ \frac{1}{a^2\sin\theta}\frac{\partial}{\partial\theta}\left\{\sin\theta\left[-D_s^\pm\frac{\partial\Gamma_{St}^\pm}{\partial\theta} - D^\pm\frac{\partial\Gamma_H^\pm}{\partial\theta} \mp \lambda^{-1}z^\pm(D_S^\pm\Gamma_{St}^\pm + D^\pm\Gamma_H^\pm)\frac{\partial\varphi}{\partial\theta}\right]\right\}_{r=a}$$
$$= \left(D^\pm z^\mp\frac{\partial c}{\partial r} \pm \lambda^{-1}D^\pm z^+z^-c_0\frac{\partial\varphi}{\partial r}\right)_{r=a}.$$ (4.11)

Previously [10], only the first terms on the left were retained in (4.11) by virtue of (4.5), while only the second terms were retained in the other case [11], in view of (4.6). As the concentration changes along the surface are relatively small, we can put

$$\frac{\partial\Gamma_{St}^\pm}{\partial\theta} = \alpha_{St}^\pm\frac{\partial c^\pm}{\partial\theta},$$ (4.12)

$$\frac{\partial \Gamma_{\overline{H}}^{\pm}}{\partial \theta} = \alpha_{\overline{H}}^{\pm} \frac{\partial c^{\pm}}{\partial \theta},$$ (4.13)

in which $\alpha_{St}^{\pm}$, $\alpha_{H}^{\pm}$ are constants; then the boundary conditions of (4.11) contain only the two unknown functions c and φ. When (4.9) and (4.10) are used in substituting for these, the trigonometric functions cancel, which gives us a system of two inhomogeneous linear algebraic equations, which is solved to give

$$M_{\varphi} = M_{\varphi\zeta} + M_{\varphi\Gamma} + M_{\varphi\Gamma S},$$ (4.14)

$$M_{c} = M_{c\zeta} + M_{c\Gamma} + M_{c\Gamma S},$$ (4.15)

where each term reflects the contribution of a certain factor to the Dorn effect and to electrophoresis. Components $M_{\varphi\zeta}$ and $M_{c\zeta}$ are related to convective transport of double layer ions on account of difference in the mean tangential velocities

$$M_{\varphi\zeta} = \frac{3ua\lambda}{2R} \{ D^{+}\xi^{-} - D^{-}\xi^{+} + a^{-1} [D_{S}^{+} \alpha_{St}^{+}\xi^{-} - D_{S}^{-} \alpha_{St}^{-}\xi^{+} + D^{+}\alpha_{H}^{+}\xi^{-} + D^{-}\alpha_{H}^{-}\xi^{+}] \},$$ (4.16)

$$M_{c\zeta} = -\frac{3uac}{2} \{ z^{+}D^{+}\xi^{-} - z^{-}D^{-}\xi^{+} + a^{-1} [z^{+}D_{S}^{+} \alpha_{St}^{+}\xi^{-} + z^{-}D_{S}^{-} \alpha_{St}^{-}\xi^{+} + z^{+}D^{+}\alpha_{H}^{+}\xi^{-} + z^{-}D^{-}\alpha_{H}^{-}\xi^{+}] \},$$ (4.17)

in which

$$R = (z^{+} + z^{-}) \left\{ D^{+}D^{-} \left[1 + \frac{\alpha_{H}^{+}+\alpha_{H}^{-}}{a} + \frac{\alpha_{H}^{+}\alpha_{H}^{-}}{a^{2}} \right] + D_{S}^{+}D^{-} \frac{\alpha_{St}^{+}}{a} \left(1 + \frac{\alpha_{H}^{-}}{a} \right) + D_{S}^{-}D^{+} \frac{\alpha_{St}^{-}}{a} \left(1 + \frac{\alpha_{H}^{+}}{a} \right) + \frac{D_{S}^{+}D_{S}^{-}\alpha_{St}^{+}\alpha_{St}^{-}}{a^{2}} \right\}.$$

Components $M_{\varphi\Gamma}$ and $M_{c\Gamma}$ arise from ion transport due to motion of the double layer as a whole with velocity V_0:

$$M_{\varphi\Gamma} = (D^{+} - D^{-}) \Gamma_{0}v_{0}^{2}a^{2}\lambda/c_{0}R,$$ (4.18)

$$M_{c\Gamma} = -v_{0}\Gamma_{0}a^{2} (z^{+}D^{+} + z^{-}D^{-}).$$ (4.19)

Components $M_{\varphi\Gamma S}$ and $M_{c\Gamma S}$ are due to ion transport by tangential diffusion and conduction:

$$M_{\varphi\Gamma S} = [D_{S}^{+} \alpha_{St}^{+} - D_{S}^{-} \alpha_{St}^{-} + D^{+}\alpha_{H}^{+} - D^{-}\alpha_{H}^{-}] v_{0}a\Gamma_{0}\lambda/c_{0}R,$$ (4.20)

$$M_{c\Gamma S} = - [z^{+}D_{S}^{+} \alpha_{St}^{+} + z^{-}D_{S}^{-} \alpha_{St}^{-} + z^{+}D^{+}\alpha_{H}^{+} + z^{-}D^{-}\alpha_{H}^{-}] v_{0}a\Gamma_{0}.$$ (4.21)

We substitute in (3.25) from (4.14)-(4.21) for M_{φ} and M_{c}, and from (2.19) for Δp, to get

$$V_{ef} = -\frac{1}{R^{*}} \left[\frac{\varepsilon\zeta}{4\pi\eta} + \frac{Fcz^{+}z^{-}}{\eta a} \left(\alpha_{H}^{+}\xi^{-} - \alpha_{H}^{-}\xi^{+} + \frac{D_{S}^{+}}{D^{+}} \alpha_{St}^{+}\xi^{-} - \frac{D_{S}^{-}}{D^{-}} \alpha_{St}^{-}\xi^{+} \right) \right.$$
$$\left. + \frac{Fcz^{+}z^{-}}{\eta} \frac{\Gamma_{0}}{c_{0}} \left(\alpha_{H}^{-} - \alpha_{H}^{+} + \frac{D_{S}^{+}}{D^{+}} \alpha_{St}^{+} - \frac{D_{S}^{-}}{D^{-}} \alpha_{St}^{-} \right) \frac{1}{2+3 (\eta'+\gamma)/\eta} \right] \text{grad } \varphi,$$ (4.22)

in which

$$R^{*} = 1 + \frac{\alpha_{H}^{+} + \alpha_{H}^{-}}{a} + \frac{D_{S}^{+}\alpha_{st}^{+}}{D^{+}a} \left(1 + \frac{\alpha_{H}^{+}}{a} \right) + \frac{D_{S}^{-}\alpha_{St}^{-}}{D^{-}a} \left(1 + \frac{\alpha_{H}^{-}}{a} \right) + \frac{D_{S}^{+}D_{S}^{-}\alpha_{St}^{+}\alpha_{St}^{-}}{D^{+}D^{-}a^{2}}.$$

The first two terms are the result of substitution for $M_{\varphi\zeta}$ and $M_{c\zeta}$. The components $M_{\varphi\Gamma}$ and $M_{c\Gamma}$ after substitution make zero contribution to (4.22). The third term in (4.22) results from substitution for $M_{\varphi\Gamma S}$ and $M_{c\Gamma S}$. This term vanishes if $\gamma \rightarrow \infty$, and the other two remaining terms give the electrophoretic mobility of a solid particle, i.e., the third term reflects the effects of surface mobility in a drop on V_{ef}.

To simplify evaluation of the ratio of the second term to the third, we assume that the $D_s^{\ddagger}$ are small relative to the $D^{\ddagger}$ and that only negative ions are adsorbed, the ζ potential being so high that $\alpha_H^- \gg \alpha_H^+$; $\xi^- \gg \xi^+$. Then this ratio is of the order of

$$S_{2,3} = \frac{\xi^- c_0}{a\Gamma_0}\Big[2 + 3\frac{\eta' + \gamma}{\eta}\Big], \tag{4.23}$$

and for $3\gamma \ll \eta$.

$$S_{2,3} \sim \frac{\tilde{\zeta}}{\chi a\,[\exp{(\tilde{\zeta}/2)} - 1]} \ll 1, \tag{4.24}$$

in which $\tilde{\zeta} = F\zeta/RT$ and χ is the reciprocal of the Debye radius, i.e., surface mobility is always more important than surface conductivity under these circumstances.

If $3\gamma \gg \eta$ occurs for the moderate surface activity of (4.5) as a result of a sufficiently high concentration, we can use a standard formula [9, 10] for γ:

$$\gamma = (RT\Gamma_0^2)/(3c_0 D_e), \tag{4.25}$$

which gives us that

$$S_{2,3} \sim \frac{\xi^- c_0}{a\Gamma_0}\frac{3\gamma}{\eta} \sim \frac{\varepsilon}{4\pi\eta D}\Big(\frac{RT}{F}\Big)^2 \frac{\Gamma_0}{c_0 a} \ll 1, \tag{4.26}$$

i.e., surface mobility predominates in this limiting case also. Finally, if we simplify the third term in (4.22) as above and use (4.25), we get the following result for $3\gamma \gg \eta$:

$$S_{1,3} \sim \frac{\varepsilon}{4\pi\eta D}\Big(\frac{RT}{F}\Big)^2\Big(\frac{e\zeta}{kT}\Big). \tag{4.27}$$

The surface mobility still influences V_{ef} even though the surface motion is reduced as 3γ increases relative to η. The deviation from Smoluchowski's formula due to surface mobility is very substantial, at least when ζ is not too large.

5. Mechanism of Capillary Osmosis and Diffusophoresis in Electrolytes and Nonelectrolytes

It has been shown [15] that there is local equilibrium in a thin polarized double layer, which implies that local equilibrium should occur in a molecular diffuse adsorption layer.

The equilibrium conditions for a cross-section of the double layer provide some information on the diffusophoresis mechanism. We can restrict consideration to the case of a dilute solution, which allows us to use the theory of surface molecular forces in a binary solution [16] and also to write down explicitly the conditions for constancy of the chemical potentials of solute and solvent in the double layer. In relation to the latter we must bear in mind that the pressure may change within the double layer and that the molecules have an energy of interaction with the surface. Although the solvent molecules (high concentration c_2) interact with the surface as well as the solute ones (low concentration c_1), the potential energy $U_1(x)$ of the latter is used alone in the theory of surface molecular forces [16]. In the method used in that theory, $U_1(x)$ is an effective quantity that takes into account the interaction of the surface not only with the solute but also with the solvent.[†]

[†] $U_1(x)$ characterizes the energy change in the system on bringing solute molecules from infinity to a distance x from the surface. This process actually does alter the energy, since

Then the conditions for constancy of the chemical potentials within the adsorption layer are as follows:

$$\frac{\partial \mu_1}{\partial x} = \frac{kT}{C}\frac{\partial C}{\partial x} + v_1\frac{\partial p}{\partial x} + \frac{\partial V_1}{\partial x} = 0, \tag{5.1}$$

$$\frac{\partial \mu_2}{\partial x} = -kT\frac{\partial C}{\partial x} + v_2\frac{\partial p}{\partial x} = 0, \tag{5.2}$$

in which

$$C = \frac{C_1}{C_2 + C_1} \cong \frac{C_1}{C_2} \ll 1, \tag{5.3}$$

and v_1 and v_3 are the specific partial molecular volumes, so

$$C_1(x, y)v_1 + C_2(x, y)v_2 = 1. \tag{5.4}$$

Here we have introduced a local rectangular coordinate system (x, y) such that $x = 0$ corresponds to the surface of the particle while y represents displacement along the surface in a plane including the macroscopic concentration gradient. Although v_1 and v_2 in the general case are dependent on the composition of the solution, they can be taken as constant for a weak solution.

Equations (5.1), (5.2), and (5.4) allow us to determine the three functions c_1, c_2, and p. We eliminate the first terms from (5.1) and (5.2) and then apply (5.4), which gives

$$\frac{\partial p}{\partial x}(x, y) = C_1(x, y)\frac{\partial V_1}{\partial x}. \tag{5.5}$$

As

$$C_1 v_1 \ll 1, \tag{5.6}$$

substitution of the $\partial p / \partial x$ of (5.5) into (5.1) shows that the second term in (5.1) may be neglected relative to the third, so (5.1) and (5.3) together give

$$kT\frac{\partial \ln c_1}{\partial x} + \frac{\partial v_1}{\partial x} = 0. \tag{5.7}$$

As $c_1(y)$ is given for the outer boundary of the adsorption layer, we have

$$C_1(x, y) = c_1(y)\exp[v_1(x)/kT]. \tag{5.8}$$

We substitute (5.8) into (5.5) and integrate the latter subject to constancy of p outside the adsorption layer, which gives us the excess pressure in that layer as

$$p(x, y) = kT\,\gamma_1(x)c_1(y)/c_{1_0}, \tag{5.9}$$

in which

$$\gamma_1(x) = c_{10}[\exp(V_1(x)/kT) - 1]. \tag{5.10}$$

We substitute the expression implied by (5.9) for the tangential component of the pressure gradient into the equations of hydrodynamics and integrate to get for the capillary osmosis a formula differing from (2.20) in the absence of the second term:

$$V_{c0} = -\frac{\xi_1 c_1}{\eta}\frac{\partial \mu_1}{\partial y}. \tag{5.11}$$

a solute molecule displaces a solvent molecule to the region outside the adsorption layer, whereas analogous motion of a solvent molecule leaves the energy of the system unchanged.

The difference between these formulas arises because (5.11) is approximate, being derived subject to (5.3), and the error being of the order of c_1/c_2. It follows from (5.4) that

$$[C_1(x, y) - c_1(y)] v_1 + [C_2(x, y) - c_2(y)] v_2 = 0.$$

We multiply this equation by x and integrate to give finally

$$c_{10}v_1\xi_1 + c_{20}v_2\xi_2 = 0,$$

so

$$\xi_2/\xi_1 = \frac{c_{10}v_1}{c_{20}v_2} \sim \frac{c_{10}}{c_{20}} \ll 1. \tag{5.12}$$

Formula (5.5) can also be derived via elementary molecular-kinetic concepts. Consider a volume $S\Delta x$ between two planar areas parallel to the surface, the first being at a distance x from the surface and the second at a distance $x + \Delta x$. The total force on the molecules in this volume is

$$F(x, y) = C_1(x, y) S\Delta x \frac{\partial V_1}{\partial x}.$$

This force must be balanced by the difference between the pressures acting on the two areas:

$$S[p(x + \Delta x) - p(x)] = F(x) \cdot S\Delta x. \tag{5.13}$$

Passing to the limit, we obtain at once (5.5), so capillary osmosis and diffusophoresis are due to the change along the surface in the excess pressure in the diffuse part of the adsorption layer on account of concentration change along the external boundary of the adsorption layer.

Picard [17] states that it is necessary to consider an additional specific force due to diffusion of ions along the surface of a particle. It seems that formula (3.28), which was derived via the thermodynamics of irreversible processes, agrees qualitatively[†] with Picard's concept. We consider only the excess pressure of purely electrical origin in our initial discussion of diffusophoresis in an electrolyte. The nonelectrical component of the excess pressure can be neglected relative to the electrical one for a dilute electrolyte. In deducing (3.28) we used a theory of the Dorn effect [14] in which polarization of the double layer is neglected, so in deriving (3.28) via consideration of the pressure distribution we must assume that the potential $\Phi(x)$ in the double layer remains unaltered. But then the excess concentrations of the ions should be characterized via formulas analogous to (5.10):

$$\gamma^\pm(x, y) = c^\pm(y) [(\mp z^\pm e\Phi/kT) - 1]. \tag{5.14}$$

Repeating the above sequence of arguments, we arrive at (3.28), so diffusophoresis in an electrolyte is also due to the tangential component of the excess pressure in the double layer caused by the action of the spherically symmetrical field of the particle on the charge in the diffuse part of the double layer and by the variation in the latter along the surface of the particle in response to the macroscopic concentration gradient. But this means that the theory of diffusophoresis and electrophoresis should be based on the same system of equations and boundary conditions, only with the difference that the potential gradient at infinity is given for electrophoresis, whereas the concentration gradient at infinity is given for diffusophoresis.

V_{ef} is affected by the concentration change along the outer boundary of the double layer and by the polarization of that layer not only because there is an additional tangential component

---

[†] The agreement is not quantitative, because (3.28) is not the same as Picard's formula, which thus conflicts with the thermodynamics of irreversible processes.

of the electric field but also because the spherically symmetrical field of the particle acts on the deformed ion atmosphere. The last effect has a mechanism analogous to that of diffusophoresis in a nonelectrolyte, and allowance for it eliminates the need to introduce into the electrophoresis formula additional terms proportional to the concentration gradient along the surface of the particle, which Picard does (incorrectly), which implies that diffusophoresis is incorporated twice.

The theory of diffusophoresis may be extended to highly charged particles via the theory of thin double layer polarization, as has been done for electrophoresis [18].

Conclusions

1. The methods of de Groot et al. for internal electrokinetic effects (electroosmosis, flow potential) have been extended by allowance for the accompanying diffusion processes. It is shown that a diaphragm of randomly placed fixed spheres of equal radius may be used with this method to derive a formula relating electroosmosis to the flow potential, which can be transformed to a formula relating electrophoresis to the Dorn effect.

This provides a simple unified method of discussing internal and external electrokinetic phenomena with allowance for diffusion via the thermodynamics of irreversible processes. The value of this result will be clear from the fact that de Groot et al. had to develop a special method in order to consider external electrokinetic effects, which is incomparably more complicated than the one given here and which is unsuitable for extension by allowance for diffusion.

2. Formulas have been derived for capillary osmosis and diffusophoresis in nonelectrolyte solutions. The method and results are shown to be correct by the form of the formula for diffusophoresis of a drop bearing an adsorbed layer in a nonelectrolyte, which is precisely the formula for this effect derived by Frumkin and Levich directly via the equations of physicochemical hydrodynamics.

3. It is found on applying the thermodynamics of irreversible processes to the electrodiffusion theory of electrokinetic phenomena that a new effect (diffusophoresis, motion of particles in response to a macroscopic concentration gradient in the absence of an external electric field) is the analog of the influence of diffusion on the sedimentation potential in the case of electrophoresis. The velocity of diffusophoresis resembles that of electrophoresis in being expressed via a linear combination of the electric and concentration dipole moments, and correspondingly via a linear combination of the cation and anion components of the ζ potential.

4. Smoluchowski's formulas are obtained for electroosmosis and electrophoresis, although allowance is made for bulk diffusion effects in the initial formulas for the flow potential and the Dorn effect.

The method has been applied with allowance for surface conductivity and diffusion in the formulas for the Dorn effect to derive a formula for the electrophoresis of an unpolarized particle, in which the surface diffusion is considered more rigorously than in Booth's and Henry's formulas, which may be derived as special cases from the new formula.

5. A new formula has been derived for the electrophoresis of unpolarized drops and bubbles via the electrical-diffusion theory of the Dorn effect for these, with allowance for surface conductivity and diffusion. The electrophoretic velocity of a drop consists of: 1) that velocity for a solid particle with the same characteristics, 2) an additional component reflecting the role of surface mobility. The latter effect persists in the presence of ionic surfactants at virtually any concentration, at least for moderate values of the ζ potential. This shows that it is incorrect to use Smoluchowski's formula to describe the electrophoresis of drops even when the surface is highly strengthened in this way, and that it is therefore necessary to reconsider some erroneous interpretations of experiments.

6. Capillary osmosis and diffusophoresis in nonelectrolytes and electrolytes involve an excess pressure either in the diffuse molecular layer or the ionic double layer. Concentration change along the outer boundaries of these layers is responsible for the tangential component of the pressure gradient, which leads to diffusophoretic slip. Formulas have been derived via the thermodynamics of irreversible processes for capillary osmosis and diffusophoresis, on the basis of the thermodynamic conditions for local equilibrium in the diffuse layers in the presence of a macroscopic concentration gradient.

References

1. P. Mazur and J. Th. G. Overbeek, Rec. Trav. Pays-Bas (1950).
2. S. R. de Groot, Thermodynamics of Irreversible Processes, Wiley, New York (1951).
3. S. R. de Groot, P. Mazur, and J. Th. G. Overbeek, J. Chem. Phys., 20:1825 (1952).
4. B. V. Deryagin and S. P. Bakanov, Dokl. AN SSSR, 147:139 (1962); J. Coll. Sci., 20:555 (1965).
5. B. V. Deryagin, G. P. Sidorenko, E. A. Zubashchenko, and E. V. Kiseleva, Kolloidn. Zh., 9:335 (1947).
6. B. V. Deryagin and S. S. Dukhin, Dokl. AN SSSR, 129:1328 (1961).
7. B. V. Deryagin, Kolloidn. Zh., 22:148 (1960).
8. A. N. Frumkin, S. Sat'yanarayana, and N. V. Nikolaeva-Fedorovich, Izv. AN SSSR, Otdel. Khim. Nauk, No. 11, p. 1917 (1962); V. G. Levich and A. M. Kuznetsov, Dokl. AN SSSR, 146:145 (1962).
9. V. G. Levich, Physicochemical Hydrodynamics, Fizmatizdat, Moscow (1962), §74.
10. S. S. Dukhin, Kolloidn. Zh., 25:418 (1963).
11. S. S. Dukhin, Kolloidn. Zh., 25:520 (1963).
12. D. S. Henry, Trans. Faraday Soc., 44:1021 (1948).
13. F. Booth, Trans. Faraday Soc., 44:955 (1948).
14. S. S. Dukhin, in: Research in Surface Forces, Vol. 2, Consultants Bureau, New York (1966), p. 54.
15. S. S. Dukhin, this volume, p. 287.
16. B. V. Derjaguin (Deryagin), I. E. Dzyaloshinsky, M. M. Koptelova, and L. P. Pitayevsky, Disc. Faraday Soc., No. 40, p. 246 (1965).
17. W. F. Picard, Kolloid.-Z., 179:117 (1961).
18. S. S. Dukhin, this volume, p. 312.

THEORY OF THE POLARIZATION
OF THE DOUBLE ELECTRICAL LAYER
OF A COLLOIDAL PARTICLE

S. S. Dukhin

Institute of General and Inorganic Chemistry,
Academy of Sciences of the Ukrainian SSR

An analytic theory is presented for a thin double layer on an axially symmetric particle for any value of the ζ potential. Allowance is made for the dense part of the layer and also for the deformation of the diffuse part. The results provide the basis for the theory of various effects: coagulation in an external electric field, the ionic component of the dipole moment, and the effects of double layer deformation on the electrophoretic mobility.

Double-layer polarization is of interest in electrochemistry in relation to the passage of current in an electrode−electrolyte system [1, 2]. The effect was first considered in colloid chemistry in relation to the theory of electrophoresis [3]. The external field and the tangential motion of the liquid alter the distribution of the ions in the diffuse part along the surface; the double layer is deformed (polarized). This polarization gives rise to an additional electric field, which can become comparable with the applied field, which means that the electrophoresis velocity is substantially affected and is dependent on the size and shape of the particle. There is also a very great increase in the difficulty in measuring the ζ potential.

Interest in double-layer polarization has recently increased, on account of various effects related to the permanent dipole moment [4] and to the ionic component of the dipole moment, which arises by polarization of the double layer.[†] Stoilov [5] has described a new method of examining the size, shape, and electrical parameters of anisodiametric particles, which is based on measurement of the light scattering by particles oriented in an external alternating electric field. This method gives access to the ionic component, which makes it necessary to relate the dipole moment to the ζ potential in order to derive the latter by this method.

In electrophoretic coating, it is often found that the particles coagulate into chains on applying the electric field [6]. The structure of the coatings from deposited aggregates is sub-

---

[†] Effects from the ionic component have to be eliminated in order to measure the ionic component, for which purpose we require a general theory that takes into account both components.

stantially different from that from deposition of single particles. For this reason, double layer polarization is of some practical interest.

Overbeek [3] and Booth [7] have developed the theory of double-layer polarization for the important particular case of spherical particles in relation to determination of the electrophoretic mobility. The generality of the argument is restricted by the condition that the $\tilde{\zeta}$ (dimensionless) potential is small ($\tilde{\zeta} = e\zeta/kT$, in which e is electronic charge, k is Boltzmann's constant, and T is absolute temperature), this potential being used as a small parameter. Even so, the derivations are so complicated that it is impossible to extend them to particles of less regular form. Here I show that the theory can be developed in a very general form for axially symmetric particles having any ζ potential, provided that the thickness of the double layer is small:

$$\chi l \gg 1, \tag{1}$$

in which χ is the reciprocal of that thickness and l is the characteristic linear dimension of the particles.

This theory allows one to extend the theory of electrophoresis to particles of any axially symmetric shape and any value of ζ potential. It can also be used as the basis for a theory of the ionic component of the dipole moment and for various related effects.

Equations and Boundary Conditions. The system of equations for a polarized double layer is exactly that used in theoretical electrochemistry for passage of dc in the presence of convective diffusion and space charge outside the double layer†: the equations for convective diffusion of anions and cations, Poisson's equation, and the Navier−Stokes equation:

$$\operatorname{div} \vec{j}^{\pm} = 0; \tag{2}$$

$$\Delta \Psi = -\frac{4\pi F}{\varepsilon}(z^+ C^+ - z^- C^-); \tag{3}$$

$$\eta \Delta V = \operatorname{grad} p + \rho_e \operatorname{grad} \Psi; \tag{4}$$

$$\operatorname{div} \vec{V} = 0. \tag{5}$$

All three possible transport mechanisms for ions in the electrolyte have to be incorporated in the expressions for the ion flux densities $\vec{j}^{\pm}$:

$$\vec{j}^{\pm} = -D^{\pm} \operatorname{grad} C^{\pm} \pm \frac{F}{RT} D^{\pm} z^{\pm} C^{\pm} \vec{E} + C^{\pm} \vec{V}, \tag{6), (7)}$$

in which $C^{\pm}$, $D^{\pm}$, and $z^{\pm}$ are respectively the partial concentrations, diffusion coefficients, and valencies of the ions, $\vec{V}$ is velocity, $\vec{E}$ is electric field strength, F is Faraday's constant, R is

---

† However, the macroscopic motion is controlled independently in an electrolytic cell, whereas that motion in electrophoresis is due to thermodynamic forces, the macroscopic transport being of the same order as the molecular transport. For this reason, we need a rigorous derivation of the macroscopic velocity in this problem. The Navier−Stokes equation describes the macroscopic motion of the mixture if by the velocity of the mixture we understand the momentum of unit mass (the barocentric velocity). The mass flux is then represented by the product $\rho \vec{V}$ (ρ is density), so the liquid can be considered as incompressible (which corresponds to transition from div $\rho \vec{V}$ = 0 to div $\vec{V}$ = 0) only for dilute electrolytes. Moreover, we can use the ordinary definition of ion mobility (from equality of the resistance to the force exerted by the field in the steady state) only to calculate the flow relative to a liquid barocentrically at rest. The velocity in the expression for the resistance is the barocentric one, since it characterizes the relative motion of the body (here an ion) and the mixture, and it appears in the boundary condition for the Navier−Stokes equation that has to be solved in deriving the resistance.

the universal gas constant, T is absolute temperature, ε is dielectric constant, η is viscosity, and p is pressure. In the Navier−Stokes equation we have incorporated the electrical forces acting on the space charge, whose density is ρ_e.

In the general case, a system of equations analogous to (2)-(6) must be written for the region within a particle; but we consider only nonconducting solid particles, so the need for (2), (3), and (5) drops out, while (4) becomes the Laplace equation

$$\Delta \Psi' = 0, \tag{8}$$

in which the prime refers to the internal region.

Two equations analogous to (2) and (3) provide the boundary conditions at the interface:

$$\text{div}_s I_s^\pm = -j_n^\pm, \tag{9}, (10)$$

in which $j_n^\pm$ are the normal components of the bulk ion fluxes at the surface and $I_s^\pm$ are the densities of the surface fluxes:

$$I_s^\pm = -D_s^\pm \, \text{grad}_t \, \Gamma_{St}^\pm \mp \frac{F}{RT} D_s^\pm z^\pm \Gamma_{St}^\pm \, \text{grad}_t \, \Psi, \tag{11}, (12)$$

in which $D_s^\pm$ are the coefficients of surface diffusion of the ions and $\Gamma_{St}^\pm$ is the adsorption in the Stern layer.

The surface analog of (4) is represented by the boundary condition for the electrical induction:

$$\varepsilon E_n - \varepsilon' E_n' = 4\pi\sigma_{St} = 4\pi F \left(z^+ \Gamma_{St}^+ - z^- \Gamma_{St}^- \right), \tag{13}$$

in which ε' is the dielectric constant of a particle.

One further boundary condition takes into account the change in potential on passage through the interface, which equals the potential step in the dense part of the double layer:

$$\Psi - \Psi'|_S = \delta V. \tag{14}$$

This δV may be taken as given if the surface is in a state reasonably close to equilibrium.

Quasi-equilibrium Double Layer. An initial strictly equilibrium double layer can, if (1) is obeyed, be considered as locally planar, so the potential in it must satisfy the Poisson−Boltzmann equation:

$$\frac{d^2(\Phi/\lambda)}{dx^2} = \frac{\dot\chi^2}{z^+ + z^-} \left[\exp\left(-z^+\Phi/\lambda \right) - \exp\left(z^-\Phi/\lambda \right) \right], \tag{15}$$

in which $\chi = [4\pi F^2 z^+ z^- c_0 (z^+ + z^-)/(\varepsilon RT)]^{1/2}$; $\lambda = RT/F$; $c_0 = c_0^+/z^- = c_0^-/z^+$ is the ion concentration in the body of the solution and x is the distance from the surface of the particle.

Integration of (15) gives

$$\frac{df}{dx} = -\frac{\chi l \sqrt{2}}{(z^+ + z^-)^{1/2}} \, \Omega(z^+, z^-, f), \tag{16}$$

in which x = x/l and $f = \Phi/\lambda$, with

$$\Omega(z^+, z^-, f) = \left[\frac{\exp(-z^+ f) - 1}{z^-} + \frac{\exp(z^- f) - 1}{z^-} \right]^{1/2}, \tag{17}$$

and with the boundary condition

$$f|_{x\to\infty} = 0. \tag{18}$$

Deformation of the equilibrium double layer by the field or by motion is ignored in the classical theory of electrokinetic effects [9]. We assume that the potential can be represented as a superposition of Φ and φ_0, the latter being the distribution around a particle in an external field in the absence of a double layer, which must satisfy the Laplace equation and the condition of equality to zero in the normal component of the current, and hence in the field at the surface of the particle:

$$\Delta\varphi_0 = 0; \tag{19}$$

$$\frac{\partial\varphi_0}{\partial n} = 0. \tag{20}$$

In fact, the double layer is deformed, but the deformation may be slight. We term quasi-equilibrium a double layer whose potential may be represented as $\Phi + \varphi_0$ together with a small correction ψ^* (governed by the polarization). The quasi-equilibrium state is realized for a sufficiently small value of the external field for a given Φ. If the thickness of the double layer is relatively small, ψ^* may conveniently be represented as the sum of two functions, the first φ_1 representing the potential outside the double layer and the second Φ_1 representing the change in polarization potential within the double layer. Then the polarization produces a change in potential $\Phi_1 + \varphi_1$ within the double layer, with φ_1 taken as unchanged within the cross-section of the double layer. Similarly, $C^{\pm}$ is best represented as a superposition of $c^{\pm}$ (concentration outside the double layer), $\gamma_0^{\pm}$ (excess ion concentrations in the equilibrium double layer), and $C_1^{\pm}$ (change in ion concentration over the double layer).

There are marked differences between the potential distribution within the double layer and outside it, since the potential within the double layer varies very rapidly within distances of the order of the thickness, whereas the scale of the variation outside is set by the size of the particle. This indicates that any simplifications in the equations and the method of solution must differ basically for the two regions.

Potential and Ion Concentration Outside the Layer. The condition of local electrical neutrality [10] is used to describe the trends in the current outside the layer. If the ion concentrations are not too small, condition (21) should be obeyed outside the double layer, and this can be replaced by condition (22):

$$z^+c^+ - z^-c^- \ll c^{\pm}; \tag{21}$$

$$z^+c^+ - z^-c^- = 0. \tag{22}$$

Condition (22) allows us to express the ion concentrations via a function c, which reduces the number of unknowns and removes the need to use Poisson's equation to determine the potential and concentration outside the double layer. It is sufficient to consider (2) and (3), which for small values of Pe (Peclet number)

$$\text{Pe} = \frac{lV}{D} \ll 1 \tag{23}$$

(in which V is particle velocity) substantially simplifies to

$$\Delta c = 0, \tag{24}$$

$$\Delta\varphi = 0, \quad \varphi = \varphi_0 + \varphi_1. \tag{25}$$

Integral transformation of (2) and (3) gives the boundary conditions for the concentration c and potential φ at the outer boundary of the double layer:

$$\pm \lambda D^{\pm} z^+ z^- c_0 \frac{\partial\varphi}{\partial n} + D^{\pm} z^{\pm} \frac{dc}{dn} = \text{div}_s(I_E^{\pm} + I_D^{\pm} + I_V^{\pm} + I_s^{\pm}), \tag{26), (27)}$$

in which $I_E^\pm$, $I_D^\pm$, $I_V^\pm$ are the densities of the various components of the tangential ion fluxes in the diffuse part of the double layer. The first term takes into account the electrical migration transport; the second, the diffusion; and the third, convective transport. The tangential fluxes are expressed not only via φ but also via Φ_1.

Analog of the Poisson − Boltzmann Equation for a Thin Polarized Double Layer. Here (3), (4), and (21) allow us to draw up the equation for the potential in a quasi-equilibrium double layer (the analog of the Poisson−Boltzmann equation). The sequence of steps is as for an equilibrium double layer. First, we use the equations of continuity to express $C^\pm$ in terms of ψ, and then the formulas for $C^\pm$ are used in Poisson's equation to obtain the desired equation.

We introduce an (x, y) coordinate system such that x = 0 corresponds to the surface of the particle, while y represents displacement along the surface of the particle in a plane including the symmetry axis. The double layer is thin, so the normal components of the electrical migration and diffusion fluxes substantially exceed the tangential components. For this reason, in writing out (2) and (3) in detail it is desirable to distinguish the normal components of the fluxes, with the tangential components and the convective term transferred to the right side of the equation. To relate the distributions inside and outside the double layer, it is convenient to subtract from the left side an analogous expression for the ion flux outside the double layer.

From (2), (3), (24), and (25)

$$\frac{\partial}{\partial \widetilde{x}}\left[\widetilde{C}^\pm\left(\frac{\partial \ln \widetilde{C}^\pm}{\partial \widetilde{x}} \pm z^\pm \frac{\partial \widetilde{\Psi}}{\partial \widetilde{x}}\right)\right] - \frac{\partial}{\partial \widetilde{x}}\left[\widetilde{c}^\pm\left(\frac{\partial \ln \widetilde{c}^\pm}{\partial \widetilde{x}} \pm z^\pm \frac{\partial \widetilde{\varphi}}{\partial \widetilde{x}}\right)\right]$$

$$= \operatorname{div}\left(\widetilde{\gamma}_0^\pm \frac{Ul}{D}\right) - \operatorname{div}_y\left[\operatorname{grad}_y \widetilde{C}_1^\pm \pm z^\pm\left(\operatorname{grad}_y \frac{\Phi_1}{\lambda} + \gamma_0^\pm \operatorname{grad}_y \widetilde{\varphi}\right)\right], \qquad (28),\ (29)$$

in which $\widetilde{\gamma}_0^\pm$ are the excess ion concentrations in the double layer, U is the speed of the particle; $\widetilde{\Psi} = \Psi/\lambda$; $\widetilde{\varphi} = \varphi/\lambda$; $\widetilde{C}^\pm = C^\pm/c_0$, etc. We integrate (28) and (29) on both sides with respect to $\widetilde{x}$ from 0 to $\widetilde{x}$ to get

$$\widetilde{C}^\pm \frac{\partial}{\partial \widetilde{x}}(\ln \widetilde{C}^\pm \pm z^\pm \widetilde{\Psi}) - \widetilde{c}^\pm \frac{\partial}{\partial \widetilde{x}}(\ln \widetilde{c}^\pm \pm z^\pm \widetilde{\varphi}) = \widetilde{I}^\pm(0,\ \widetilde{x}), \qquad (30),\ (31)$$

in which

$$I^\pm(0,\ \widetilde{x}) = \int_0^x \left\{\operatorname{div}\left(\widetilde{\gamma}_0^\pm \frac{Va}{D^\pm}\widetilde{\vec{V}}\right)\right.$$

$$\left. - \operatorname{div}_y\left[\operatorname{grad}_y \widetilde{C}_1^\pm \pm z^\pm(\operatorname{grad}_y \Phi_1/\lambda + \widetilde{\gamma}_0^\pm \operatorname{grad}_y \widetilde{\varphi})\right]\right\} d\widetilde{x}.$$

As $\widetilde{c}^\pm$ and φ vary only slightly over the thickness of the double layer, the second term on the left in (30) and (31) may be neglected relative to the first:

$$\widetilde{C}^\pm \frac{\partial}{\partial \widetilde{x}}(\ln \widetilde{C}^\pm \pm z^\pm \widetilde{\Psi}) = \widetilde{I}^\pm(0,\ \widetilde{x}). \qquad (32),\ (33)$$

We neglect the surface ion fluxes in (9) and (10) to get the left sides of the equations as zero for $\widetilde{x} = 0$. From (32) and (33) with $\widetilde{x} \to \infty$ we get boundary conditions (26) and (27), which may be written in more compact form as follows:

$$\frac{\partial \ln \widetilde{c}^\pm}{\partial \widetilde{x}} \pm z^\pm \frac{\partial \widetilde{\varphi}}{\partial \widetilde{x}} = \widetilde{I}^\pm(0,\infty)/c_0^\pm. \qquad (34),\ (35)$$

Combination of (32)–(35) gives

$$\frac{\partial}{\partial \tilde{x}}\left[\ln C^{\pm} / \tilde{c}^{\pm} \pm z^{\pm}(\Psi - \tilde{\varphi})\right] = \frac{\tilde{I}^{\pm}(0, x)}{\tilde{C}^{\pm}(x)} - \frac{\tilde{I}^{\pm}(0, \infty)}{\tilde{c}_0^{\pm}},$$ (36), (37)

whence it follows that

$$C^{\pm}(x, y) = \tilde{c}^{\pm}(x, y) e^{\pm z^{\pm}(\tilde{\Psi} - \tilde{\varphi}) + \tilde{Q}^{\pm}},$$ (38), (39)

in which

$$\tilde{Q}^{\pm} = \int_{\tilde{x}}^{\infty}\left(\frac{\tilde{I}^{\pm}(0, x)}{\tilde{C}^{\pm}(x)} - \frac{\tilde{I}^{\pm}(0, \infty)}{\tilde{c}_0^{\pm}}\right) d\tilde{x}.$$ (40), (41)

As $c^{\pm}$ differs from $C^{\pm}$ in varying only slightly over the thickness of the double layer, we can put

$$c^{\pm}(x, y) \cong c_0^{\pm} + \delta c^{\pm}(y),$$ (42)

with obedience to the condition of small concentration charge near a particle, $\delta c^{\pm}(y) \ll c_0^{\pm}$. Moreover, if the external field is sufficiently weak we have the conditions

$$f_1^{\pm} = \Phi_1^{\pm}/\lambda \ll 1, \quad Q^{\pm} \ll 1,$$ (43)

which allows us to retain only the first two terms on series expansion of the exponentials. The final result of all these simplifications is

$$z^+C^+ - z^-C^- = z^+c_0^+e^{-z+f} - z^-c_0^-e^{z-f} + z^+\delta c^+(y)e^{-z+f} - z^-\delta c^-(y)e^{z-f}$$
$$- (z^+c_0^+e^{-z+f} + z^-c_0^-e^{z-f})f_1^{\pm} + c_0^+e^{-z+f}Q^+ + c_0^+e^{z-f}Q^-.$$ (44)

We get the following estimates from (34), (35), (40), (41), and (42):

$$\frac{\delta c^+}{c_0}e^{-z+f} + \frac{\delta c^-}{c_0}e^{z-f} \sim e^{|\zeta|}(\tilde{I}^+(0, \infty) + \tilde{I}^-(0, \infty),$$ (45)

$$e^{-z+f}Q^+ + e^{z-f}Q^- \sim \frac{e^{|\zeta|/2}}{\chi l}(\tilde{I}^+(0, \infty) + \tilde{I}^-(0, \infty)).$$ (46)

Condition (1) thus allows us to neglect the complicated integral term in (44). We substitute the simplified right part of (44) into Poisson's equation and use (15) to get the following equation for f_1:

$$\Delta f_1(x, y) \cong \frac{d^2 f_1}{dx^2} = \frac{(\chi l)^2}{z^+ + z^-}\left[(e^{-z+f} - e^{z-f})\frac{\delta c(y)}{c_0} - (z^+e^{-z+f} + z^-e^{z-f})\right]f_1.$$ (47)

Now Ψ coincides with φ for $x \rightarrow \infty$, so

$$f_1|_{x\rightarrow\infty} = 0.$$ (48)

The boundary condition at $x = 0$ follows from (8) and (13):

$$\varepsilon\frac{\partial\tilde{\varphi}}{\partial\tilde{x}} - \varepsilon'\tilde{\psi}(0) = 4\pi\tilde{\sigma}_{st}.$$ (49)

If we follow Overbeek and Booth in neglecting the polarization of the internal sheath in the double layer, and also impose the condition

$$\varepsilon' \ll \varepsilon,$$ (50)

the boundary condition simplifies substantially to

$$\frac{\partial f_1}{\partial \widetilde{x}}\Big|_{\widetilde{x}=0} = -\frac{\partial \widetilde{\varphi}}{\partial \widetilde{x}}\Big|_{\widetilde{x}=0} = -\frac{\partial}{\partial x}(\widetilde{\varphi}_0 + \widetilde{\varphi}_1)\Big|_{\widetilde{x}=0} = -\frac{\partial \widetilde{\varphi}_1}{\partial \widetilde{x}}\Big|_{\widetilde{x}=0}. \tag{51}$$

The $f(x)$ for an unsymmetrical electrolyte and highly charged particles cannot be expressed in terms of elementary functions. The variable coefficients in the equations simplify substantially if we take f as independent variable via (16). This replacement is also desirable because $f(x)$ contains not only z^+ and z^- but also the additional parameter $\widetilde{\zeta}$, which on this substitution vanishes from the equation, which takes the form

$$2\left[\frac{e^{z^- f} - 1}{z^-} + \frac{e^{-z^+ f} - 1}{z^+}\right]\frac{d^2 f_1}{df^2} + (e^{z^- f} - e^{-z^+ f})\frac{df_1}{df} + (z^- e^{z^+ f} + z^+ e^{-z^+ f})f_1 = \frac{\delta c(y)}{c_0}(e^{z^- f} - e^{-z^+ f}). \tag{52}$$

Boundary conditions (48) and (51) now becomes

$$f_1\big|_{f=0} = 0; \tag{53}$$

$$-\frac{\partial f_1}{\partial f}\Big|_{f=\widetilde{\zeta}} = \frac{\partial \widetilde{\varphi}}{\partial \widetilde{x}}\Big|_{\widetilde{x}=0} \frac{\sqrt{z^+ + z^-}}{\sqrt{2}\chi l}\left[\frac{e^{z^- \widetilde{\zeta}} - 1}{z^-} + \frac{e^{-z^+ \widetilde{\zeta}} - 1}{z^+}\right]^{-1/2}. \tag{54}$$

The simplification of (38) and (39), which has a simple physical significance, arises from the possibility of quantitative description of the potential in a thin polarized double layer for arbitrary ζ and z together with an axially symmetric form for the particle. Diffusion and electrical migration in the normal direction are the main processes governing the ion distribution during polarization, as for an equilibrium double layer, and so the concentration has an exponential relation to the potential. Tangential diffusion and electrical migration are of secondary importance, as is convective transfer, which allows us to neglect $Q^\pm$; but these processes affect the concentration and potential outside the double layer.

We can derive (38) and (39) directly, with $Q^\pm$ omitted, by assuming that, although as a whole the double layer deviates from equilibrium locally (i.e., for small parts), there is equilibrium between a part and the adjacent volume of electrolyte. In fact, (38) and (39) (with $Q^\pm$ omitted) may be considered as Boltzmann distributions that take into account the deviations from the mean value in the ion concentration at the outer boundary of the double layer. However, these arguments are not sufficiently rigorous, and so it appears best to present a derivation based directly on the equations of continuity, the more so since condition (1) is obtained as criterion for the applicability of the results.

Grafov and Chernenko [1, 2] considered the one-dimensional polarization of a double layer at an electrode in response to passage of current, and they concluded that a Boltzmann distribution could be used; Chernenko [2] employed the boundary-layer method. Appendix 1 deals with this question, in view of its importance.

Solution of the Equations for a Polarized Double Layer. To eliminate $\delta c(y)/c_0$ from (52) we seek a particular solution in the form $\delta c(y)Y_p/c_0$. Then we get for Y_p an equation containing no physical parameters apart from the z:

$$\left(\frac{e^{z^- f} - 1}{z^-} + \frac{e^{-z^+ f} - 1}{z^+}\right)\frac{d^2 Y_p}{df^2} + \frac{e^{z^- f} - e^{-z^+ f}}{2}\frac{dY_p}{df} - \frac{z^+ e^{-z^+ f} + z^- e^{z^- f}}{2}Y_p = \frac{e^{z^- f} - e^{-z^+ f}}{2}. \tag{55}$$

The general theory of linear differential equations shows that the function should be determined as the superposition

$$f_1 = \frac{\delta c\,(y)}{c_0}\,Y_p + Y_q, \tag{56}$$

in which Y_q is the general solution of the homogeneous equation corresponding to (55):

$$\left(\frac{e^{z-f}-1}{z^-}+\frac{e^{-z+f}-1}{z^+}\right)\frac{d^2Y_q}{df^2}+\frac{e^{z-f}-e^{-z+f}}{2}\frac{dY_q}{df}-\frac{z^+e^{-z+f}+z^-e^{z-f}}{2}\,Y_q = 0. \tag{57}$$

To transform (53) and (54) to boundary conditions for Y_p and Y_q we note that (55) for $f \to 0$ becomes an inhomogeneous Euler equation:

$$\frac{z^+ + z^-}{2}\,f^2\,\frac{d^2Y_p}{df^2}+\frac{z^+ + z^-}{2}\,f\,\frac{dY_p}{df}-\frac{z^+ + z^-}{2}\,Y_p = \frac{z^+ + z^-}{2}\,f, \tag{58}$$

whose solution tends either to 0 or ∞ as $f \to 0$. We rule out the second possibility as having no physical significance and find that for any z^+ and z^-

$$Y_p|_{f\to0} = 0. \tag{59}$$

Comparison of (53), (56), and (59) shows that

$$Y_q|_{f\to0} = 0. \tag{60}$$

Equation (57) for $f \to 0$ becomes the homogeneous Euler equation corresponding to (58). The solutions to this equation tend to 0 or to ∞, so we can at once select one of the solutions, which shows that (60) defines Y_q apart from an undetermined constant factor, which can be deduced via the boundary condition derived by substituting the f_1 of (56) into (54):

$$f_1 = \frac{\delta c\,(y)}{c_0}\,Y_p\,(f) + K Y_q\,(f), \tag{61}$$

in which

$$K = -\frac{1}{Y_q'(\widetilde{\zeta})}\left[-\frac{\partial\widetilde{\varphi}}{\partial x}\,(0y)\,\frac{\sqrt{z^+ + z^-}}{\sqrt{2}\chi l}\times\left(\frac{e^{-z+\widetilde{\zeta}}+1}{z^+}+\frac{e^{z-\widetilde{\zeta}}-1}{z^-}\right)+\frac{\delta c\,(y)}{c_0}\,Y_p'(\widetilde{\zeta})\right].$$

As $\delta c(y)/c_0$ and $(\partial\widetilde{\varphi}/\partial\widetilde{x})(0, y)$ are to be determined by solution of (26) and (27), it is best to put (61) in the form

$$f_1 = \frac{\delta c\,(y)}{c_0}\,Y_c\,(f) + \frac{\partial\widetilde{\varphi}}{\partial x}\,Y_\varphi\,(f), \tag{62}$$

in which

$$Y_c\,(f) = Y_p\,(f) - [Y_p'(\zeta)/Y_q'(\zeta)]\,Y_q\,(f); \tag{63}$$

$$Y_\varphi\,(f) = -\frac{\sqrt{z^+ + z^-}}{\sqrt{2}\chi l}\left[\frac{e^{-z+f}-1}{z^+}+\frac{e^{z-f}-1}{z^-}\right]^{-1/2}Y_\varphi Y_q(f). \tag{64}$$

We put (57) in the canonical form

$$Y_q'' + d_1\,(f)\,Y_q' + d_2\,(f)\,Y_q = 0. \tag{65}$$

The coefficients in this equation have a singular point for $f = 0$, but this is a regular singular point, since d_1 has a first-order pole and d_2 has a second-order pole. Equation (57) does not satisfy the analogous conditions for an infinitely remote point and so is not entirely regular [11]. We therefore draw up the defining equation

$$r(r-1) + a_0 r + b_0 = 0,$$

in which a_0 and b_0 are the first coefficients in the power-series expansions of $A(f)$ and $B(f)$, which have the following simple relations to $d_1(f)$ and $d_2(f)$:

$$A(f) = f \, d_1(f) = \sum_{n=0}^{\infty} a_n f^n; \quad B(f) = f^2 \, d_2(f) = \sum_{n=0}^{\infty} b_n f^n. \tag{66}$$

In particular, $a_0 = 1$ and $b_0 = 1$ for $z^+ = z^- = 1$, so

$$r_1 = r = 1, \quad r_2 = -r = -1. \tag{67}$$

Near the regular singular point, the differential equation has a fundamental system of solutions, which have an especially simple form if $s = r_1 - r_2$ is not an integer. This is not the case if $z^+ = z^- = 1$, and the fundamental system then has the form

$$Y_1 = f^{r_1} \sum_{l=0}^{\infty} D_l f^l; \tag{68}$$

$$Y_2 = Y_1 [B \ln f + \Psi(f)], \tag{69}$$

in which $\Psi(f)$ is an analytic function expressible as a series and having a pole of order s at $f = 0$.

The radii of convergence for the series in Y_1 and Y_2 are [11] not less than the distance from the special point to the nearest of the other singularities. As (57) has no finite singular points other than $f = 0$, we may conclude that the above series have an unrestricted region of convergence. As Y_2 increases without limit as $f \to 0$, Y_q should be identified with Y_1.

To demonstrate the monotonic character of Y_q we transform (57) for $z^+ = z^- = 1$ by the variable replacement

$$\xi = 2 \ln \mathrm{th} \, f/4 \tag{70}$$

which gives

$$\frac{d^2 Y_q}{d\xi^2} = \Phi_1(\xi) Y_q, \tag{71}$$

in which

$$\Phi_1(\xi) = \frac{1 + e^{2\xi} + 6e^{\xi}}{4(1 + e^{\xi})^2} > 0. \tag{72}$$

It follows from (71) that Y_q is monotonic. We substitute the Y_q of (68) into (57) for $z^+ = z^- = 1$ to get the following recurrence formula for the coefficients in the series:

$$Y_q = \sum_{l=0}^{\infty} B_{2l+1} f^{2l+1}, \tag{73}$$

$$B_{2l+1} = \frac{1}{(2l+1)^2 - 1} \left\{ \sum_{m=1}^{l-1} \frac{4(1+l)(3m-l)}{(2l-2m+2)!} B_{2m+1} - \frac{2l}{2l+1} B_1 \right\}. \tag{74}$$

The following are the first six coefficients in this series:

$$B_1 = \frac{1}{2}; \quad B_3 = \frac{1}{48}; \quad B_5 = \frac{1}{48 \cdot 80}; \quad B_7 = \frac{1}{48^2 \cdot 280};$$
$$B_9 = \frac{1}{9! \cdot 16^2 \cdot 2}; \quad B_{11} = \frac{1}{12! \cdot 2 \cdot 16^2}. \tag{75}$$

The particular solution of (55) can be expressed via Y_1, Y_2, and the coefficients of (55) via the quadrature

$$Y_p = \frac{1}{2} \int^f \frac{e^{f'/2} + e^{-f'/2}}{e^{f'/2} - e^{-f'/2}} \frac{Y_1(f) Y_2(f') - Y_2(f) Y_1(f')}{Y_1(f') Y_2'(f') - Y_2(f') Y_1'(f')} df'. \tag{76}$$

We express the Wronskian in the denominator of the integrand in (76) in terms of the coefficients of (55) to convert this integral to the power series

$$Y_p = \ln f \sum_{l=0}^{\infty} B_{2l+1} f^{2l+1} + \sum_{l=0}^{\infty} C_{2l+1} f^{2l+1}. \tag{77}$$

The dependence of C_{2l+1} on the coefficients of the series for Y_1 and Y_2 is complicated, so we derive a recurrence formula by substituting (77) into (55):

$$C_{2l+1} = \frac{1}{(2l+1)^2 - 1} \left\{ \frac{l}{(2l+2)!} - \sum_{m+1}^{l-1} \frac{4(1+l)(3m-l)}{(2l-2m+2)!} C_{2m+1} - \sum_{m=1}^{l} \frac{6m+2l+4}{(2l-2m+2)!} B_{2m+1} \right\}. \tag{78}$$

This gives the first six coefficients as

$$C_1 = 0; \quad C_3 = -\frac{1}{96}; \quad C_5 = -\frac{1}{96^2}; \quad C_7 = -\frac{23}{48^3 \cdot 30}; \quad C_9 = \frac{1211}{9! \, 16^3 \cdot 4}; \quad C_{11} = -\frac{81347}{10! \cdot 12^2 \cdot 16^3 \cdot 5}. \tag{79}$$

It is of interest to establish the range in f for which we need take only the first six terms in (73) and (77). For this purpose, Milne's numerical method was applied to solve (71) and the corresponding inhomogeneous equation. It is found that Y_q can be expressed as the sum of the first six terms to 0.1% in the range $0 < f < 6$, while Y_p can be so represented for $0 < f < 3.5$. For this reason, Y_p was calculated by Milne's method for $3.5 < f < 6$.

Appendix 2 shows that we have the following asymptotic representations for f large:

$$Y_q = k_q e^{f/2}, \tag{80}$$

$$Y_p = k_p e^{f/2}, \tag{81}$$

in which the values $k_p = 0.67$ and $k_q = 0.49$ have been derived by comparison with the numerical values of Y_p and Y_q given by Milne's method. Figure 1 shows Y_p and Y_q calculated in various ways, including via (80) and (81). The error of the asymptotic representation increases as f becomes small very rapidly for Y_p and much less rapidly for Y_q.

<u>Structure of a Thin Polarized Double Layer.</u> The first term in (62) is the main one, since $\delta c(y)/c_0$ and $\partial \widetilde{\varphi}/\partial \widetilde{x}$ are of the same order, while the large quantity χl appears in the denominator in (64). The derivatives $Y_p'(\widetilde{\xi})$ and $Y_q'(\widetilde{\xi})$ are required in order to calculate Y_c. In calculating $Y_q'(\widetilde{\xi})$ we can use the part series

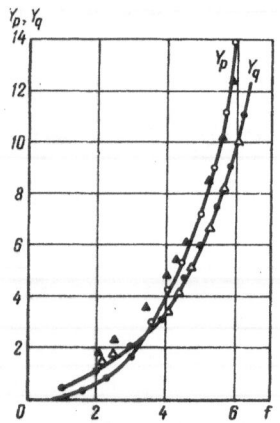

Fig. 1. Curves for $Y_q(f)$ and $Y_p(f)$: ●) taking the first six terms of the generalized power series; ○) from Milne's numerical method; ▲) from asymptotic formula (80); △) from asymptotic formula (81).

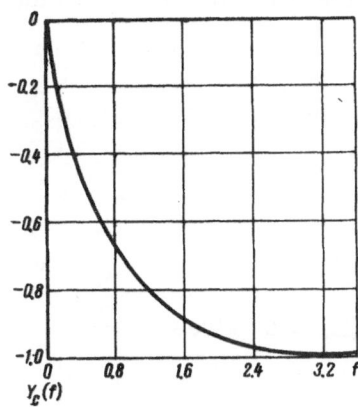

Fig. 2. Graph of $Y_c(f)$.

$$Y'_q(\widetilde{\zeta}) = \sum_{n=0}^{6} (2n+1) B_{2n+1} \widetilde{\zeta}^{2n} \qquad (82)$$

for $\widetilde{\zeta} < 6$, as is demonstrated by the results from numerical differentiation. The ratio of the derivatives varies only slightly in the range $4 < \widetilde{\zeta} < 6$ and is close to $k_p/k_q \approx 1.35$. This result is not accidental, since Y_p and Y_q are represented satisfactorily by the asymptotic formulas in this range. For this reason, the first and second terms in (63) balance out to a considerable extent, and Y_c is nearly constant (close to one) for $f = 2.5$ and $\widetilde{\zeta} > 4$, being independent of $\widetilde{\zeta}$. Also, Y_p and Y_q are satisfactorily represented by the first two terms in (73) and (77) for $f < 2.5$, which allows us to use the following simple representation for Y_c:

$$Y_c = \begin{cases} -1, & \widetilde{\zeta} > 4, \ f > 2.5 \\ Y_c(\alpha), & f < 2.5, \end{cases} \qquad (83)$$

in which

$$Y_c(\alpha) = \left(\alpha + \frac{\alpha^3}{6}\right)(\ln \alpha + q(\alpha_0)) - \frac{\alpha^3}{12}; \quad \alpha = f/2;$$

$$q(\alpha_0) = \ln 2 - p(\alpha_0); \quad p(\alpha_0) = Y''_p(\widetilde{\zeta})/Y'_q(\widetilde{\zeta}); \quad \alpha_0 = \widetilde{\zeta}/2.$$

This asymptotic behavior of Y_c as a function of $\widetilde{\zeta}$ is exceptionally important, so the question is considered analytically in Appendix 3. It is found that Y_c for any z^+ and z^- decreases monotonically from zero to a value slightly exceeding $1/z^-$ for $\widetilde{\zeta} > 0$ (Fig. 2), while Y_c for $\widetilde{\zeta} < 0$ increases monotonically to a value somewhat less than $1/z^+$. For $\widetilde{\zeta} > 0$, the polarization is due mainly to anion redistribution:

$$C_1^-(x, y) = c_0^- e^{z^- f}(z^- Y_c + 1)\frac{\delta c(y)}{c_0}. \qquad (84)$$

These features of $Y(f)$ correspond to a marked reduction in the relative change in the counterion concentration on polarization

$$C_1^-(x, y)/C_0^-(x) = \frac{\delta c(y)}{c_0}(z^- Y_c + 1) \qquad (84')$$

as the surface is approached.

Electroosmosis on an Axially Symmetric Particle. If we neglect the polarization of the thin double layer, the distribution of the electroosmosis velocity over the cross section of the layer is described by Smoluchowski's formula:

$$u(x) = \frac{\varepsilon(\zeta - \Phi(x))}{4\pi\eta}\frac{\partial\Psi}{\partial y}\bigg|_{x=0}. \qquad (85)$$

Polarization alters the pressure distribution within the double layer, and there arises a tangential component in the pressure gradient, which is not taken into account in (85). To derive the equation for the pressure distribution in the polarized layer, we take the divergence in (4):

$$\Delta p = -\operatorname{div}(\rho\operatorname{grad}\psi). \qquad (86)$$

This equation can be simplified substantially by neglecting the tangential derivatives relative to the normal ones and by omitting terms nonlinear in E, after which the double integration is readily performed. As boundary conditions we take the requirement that the pressure outside the double layer does not vary, and so

$$\frac{\partial p}{\partial y} = \frac{\varepsilon}{4\pi} \frac{\partial}{\partial y}\left(\frac{\partial \Phi}{\partial x} \frac{\partial \psi}{\partial x}\right). \tag{87}$$

This is substituted into the equations of hydrodynamics to give

$$\eta \frac{\partial^2 u}{\partial y^2} + \frac{\partial p}{\partial y} + \rho \frac{\partial \psi}{\partial y} = \eta \frac{\partial^2 u}{\partial y^2} + \frac{\varepsilon}{4\pi} \frac{\partial}{\partial y}\left(\frac{\partial \Phi}{\partial x} \frac{\partial \psi}{\partial x} - \psi \frac{\partial^2 \Phi}{\partial x^2}\right) = 0. \tag{88}$$

Double integration with the boundary conditions $u|_{x=0} = 0$ and $\frac{\partial u}{\partial x}\big|_{x\to\infty} = 0$ gives

$$u(\widetilde{x}, \widetilde{y}) = \frac{\varepsilon\lambda^2}{4\pi\eta l} \frac{\partial}{\partial \widetilde{y}} \int_0^{\widetilde{x}} \left[2\int_{\widetilde{x}}^{\infty} \widetilde{\Psi} \frac{\partial^2 f}{\partial \widetilde{x}^2} d\widetilde{x} - \widetilde{\psi}\frac{\partial f}{\partial \widetilde{x}}\right] d\widetilde{x}. \tag{89}$$

Boundary Conditions for the Potential and Concentration Outside the Layer. The above results for the double layer allow one to express the tangential ion fluxes appearing in (26) and (27) in terms of the tangential components of the gradients in the concentration and potential at the outer boundary of the double layer. Here we need the formula for the ion adsorption in the diffuse part of the double layer,

$$\Gamma_0^{\pm} = \int_0^{\infty} \gamma_0^{\pm}(\widetilde{x}) d\widetilde{x} = \frac{c_0^{\pm}}{\chi} I_1^{\pm}(\widetilde{\zeta}), \tag{90}$$

in which

$$I_1^{\pm}(\widetilde{\zeta}) = \sqrt{\frac{z^+ + z^-}{2}} \int_{\widetilde{\zeta}}^{0} \frac{e^{\mp z^{\pm} f} - 1}{\Omega(z^+, z^-, f)} df. \tag{91}$$

The results represented by (61) and (90) then give

$$I_E^{\pm} = \mp D^{\pm} z^{\pm} \operatorname{grad}_y\left[\Gamma_0\widetilde{\varphi} + \int_0^{\infty} \gamma_0(\widetilde{x}) f_1 dx\right] = \mp \frac{z^+ z^- D c_0}{\chi l} \operatorname{grad}_y[I_1^{\pm}(\widetilde{\zeta})\widetilde{\varphi} - I_{1c}^{\pm}\widetilde{c}], \tag{92}, (93)$$

in which

$$I_{1c}^{\pm}(\widetilde{\zeta}) = \sqrt{\frac{z^+ + z^-}{2}} \int_{\widetilde{\zeta}}^{0} \frac{e^{\mp z^{\pm}} - 1}{\Omega(z^+, z^-, f)} Y_c(f) df, \tag{94}$$

$$I_D^{\pm} = -D^{\pm}(\Gamma_0^{\pm} \operatorname{grad}_y \widetilde{C} + \operatorname{grad}_y C_1) = \frac{D^{\pm} c_0 z^{\mp}}{\chi l}[I_1^{\pm}(\widetilde{\zeta}) \mp z^{\pm} I_{1l}^{\pm}(\widetilde{\zeta})] \operatorname{grad}_y \widetilde{C}. \tag{95}, (96)$$

Use of (85) and (89) gives

$$I_V^{\pm} = \frac{\varepsilon\lambda^2}{4\pi\eta} \frac{c_0 z^{\mp}}{\chi l} \operatorname{grad}_y[I_2^{\pm}(\widetilde{\zeta})\widetilde{\varphi} + I_{2c}^{\pm}(\widetilde{\zeta})\widetilde{c}], \tag{97}, (98)$$

in which

$$I_2^{\pm} = \sqrt{\frac{z^+ + z^-}{2}} \int_{\widetilde{\zeta}}^{0} \frac{(e^{\mp z f} - 1)(\widetilde{\zeta} - f)}{\Omega(z^{\pm}, f)} df, \tag{99}$$

$$I_{2c}^{\pm} = \sqrt{\frac{z^+ + z^-}{2}} \int_{\tilde{\zeta}}^{0} \left\{ \frac{e^{\mp z^{\pm}f} - 1}{\Omega(z^+, z^- f)} \int_{f}^{\tilde{\zeta}} \left[\frac{1}{\Omega \cdot (z^+, z^-, f)} \int_{0}^{f} \frac{e^{z^- f} - e^{-z^+ f}}{\Omega(z^+, z^-, f)} Y_c df - Y_c \right] df \right\} df. \tag{100}$$

The terms containing $I_{1c}^{\pm}(\tilde{\zeta})$ condense when these expressions for the tangential fluxes are substituted into (26) and (27):

$$\frac{1}{\chi l} \operatorname{div}_v \left\{ \left[\frac{3m^{\pm}}{2} I_2^{\pm}(\tilde{\zeta}) \mp z^{\pm} I_1^{\pm}(\tilde{\zeta}) \right] \operatorname{grad}_v \tilde{\varphi} + \left[\frac{3}{2} m^{\pm} I_2^{\pm}(\tilde{\zeta}) - I_1^{\pm}(\tilde{\zeta}) \right] \operatorname{grad}_v \tilde{c} \right\}_{x=0} = \left(\frac{\partial \tilde{c}}{\partial x} \pm z^{\pm} \frac{\partial \tilde{\varphi}}{\partial x} \right)_{\tilde{x}=0},$$
$$\tag{101}, \ (102)$$

in which

$$m^{\pm} = \varepsilon \lambda^2 / 6\pi \eta D^{\pm}.$$

Then it is sufficient to solve (24) and (25) in conjunction with (101) and (102) in order to determine the total potential distribution characterizing the polarization of a thin double layer on an axially symmetric particle.

Polarization of the Dense Part of the Double Layer. This topic is considered in most papers on the theory of double-layer polarization, but only Booth's largely neglected paper [12] has a comment on the method of approach. Booth considered that the polarization of this part must be caused by the change in electrolyte concentration along the polarized diffuse sheath and that a quantitative characterization could be obtained by assuming equilibrium in the double layer. This approach appears more justified for a thin double layer, in view of the results obtained above, for it has been shown that local equilibrium persists between the diffuse part of a thin polarized double layer and the adjacent volume of solution. It remains to demonstrate equilibrium between the dense and diffuse parts.

Consider the processes that may cause polarization in the dense part. If the ion surface-diffusion coefficients are not zero, the surface fluxes $I_s^{\pm}$ produce an altered distribution; but if these fluxes are zero, the concentration in the dense part may alter as a result of change in the conditions of ion exchange between the dense and diffuse parts. If the state is steady and $I_s^{\pm} \neq 0$, boundary conditions (9) and (10) show that the flux from the dense part to the diffuse part is not zero. It is clear that the deviation from true equilibrium between these parts increases with $I_s^{\pm}$ and hence $j_n^{\pm}$; but we may take these parts as being in equilibrium for $I_s^{\pm} = 0$ and hence $j_n^{\pm} = 0$. We may therefore assume for a thin layer with $I_s^{\pm} = 0$ that the two parts are in equilibrium one with another and with the solution. A Boltzmann distribution applies to the diffuse part, while for the dense part we can use one of the formulas characterizing the internal sheath of an equilibrium double layer. The effect of the polarizing factor is seen in that the ion concentration and the potential vary along the outer boundary of the double layer.

We need to characterize the equilibrium double layer in some definite way in order to describe the deformation of the dense part of the double layer. We consider only particles whose double layers arise by physical adsorption of ions, which we assume to be negative. The monolayer of adsorbed potential-determining ions on the internal Helmholtz plane is accompanied by a monolayer of counterions localized on the external plane if the ζ potential is sufficiently high. We use Stern's formulas to relate the ion adsorption in the dense part to the concentration and potential (which vary along the surface). We assume that the specific adsorption potential need be borne in mind only in the formula for the adsorption of the potential-determining ions, while the adsorption on the external plane is due only to the electrostatic factor, i.e.,

$$\Gamma_i^- = \frac{\Gamma_{i\infty}^-}{1 + \frac{1}{c} e^{\theta^- - z^- \psi_i^* / \lambda}} \cong \Gamma_{i\infty}^- (c_0^- + \delta c^-) e^{-\theta^- + z^- (f_i^+ / i_1)}, \tag{103}$$

$$\Gamma_e^+ = \frac{\Gamma_{e\infty}^+}{1 + \frac{1}{c} + e^{z + \psi_i^{\bullet}/\lambda}} \cong \Gamma_{e\infty}^+ (c_0^+ + \delta c^+) e^{-z^+ (f_e + f_{e1})}. \tag{104}$$

Here subscripts i and e relate respectively to the internal and external planes, $\Gamma_{i\infty}^+$ and $\Gamma_{e\infty}^-$ are the limiting uptakes in adsorption, $c^{\pm}$ is expressed in different units (this does not affect the results, since these are dependent on the relative change in concentration), $\psi^* = \Phi + \Phi_1$, because ψ^* is reckoned from the outer boundary of the double layer, f_i, f_{i1}, f_e, and f_{e1} are the values of the corresponding potentials on the internal and external planes, and θ^- is the adsorption potential.

The relative changes in concentration and potential are small, so (103) and (104) give

$$\Gamma_i^- = \Gamma_{i0}^- + \delta\Gamma_i^-; \quad \Gamma_e^+ = \Gamma_{e0}^+ + \delta\Gamma_e^+, \tag{105}$$

in which Γ_{i0}^- and Γ_{e0}^+ are the uptakes in the initial equilibrium double layer, while for $\delta\Gamma_e^+$ and $\delta\Gamma_i^-$, which characterize the polarization of the Stern layer, we have that

$$\delta\Gamma_i^- = \Gamma_{i0}^- \left(\frac{\delta c}{c} + z^- f_{i1}\right); \quad \delta\Gamma_e^+ = \Gamma_{e0}^+ \left(\frac{\delta c}{c_0} - z^+ f_{1e}\right). \tag{106}$$

The equations of electrostatics relate the space charges to the field, and in this way (106) gives a boundary condition relating f_{1e} and $(\partial f_{1e}/\partial x)_e$. The treatment is simplified by restricting consideration to an electrolyte concentration low enough for the adsorption of the counterions in the Stern layer to be negligible relative to that elsewhere. Then, as the double layer as a whole is electrically neutral,

$$\delta\Gamma_e^+ \ll \delta\Gamma_i^-, \quad f_{i1} \cong f_{e1}. \tag{107}$$

Allowance for the polarization of the Stern layer, i.e., for the change on the right in (13), converts this boundary condition, subject to the simplifying assumptions for (106), to the form[†]

$$\frac{\partial f_1}{\partial x}(\zeta) = \frac{4\pi F^2}{\varepsilon RT} z^- \Gamma_{i0}^- (\widetilde{\delta c} + z^- f_1(\widetilde{\zeta})). \tag{108}$$

Here we have also used the possibility of neglecting $\partial\widetilde{\varphi}_1/\partial x$ relative to $\partial f/\partial x$ for a small thickness of the double layer.

If the absolute ζ are high and the concentrations are low, the adsorption of the potential-determining ions can be expressed in terms of Γ_H^+, and hence directly via ζ:

$$\Gamma_{St}^- \cong (z^+/z^-) \Gamma_H^+, \tag{109}$$

whence, from (90), with the integral of (91) calculated with $z^+ = z^- = 1$,

$$\Gamma_{St}^- = \Gamma_H^+ = (c_0/\chi) [\exp(-\widetilde{\zeta}/2) - 1]. \tag{110}$$

Subject to the above simplifying assumptions, (108) becomes

$$\frac{\partial f_1}{\partial x} = \chi [\exp(-\widetilde{\zeta}/2) - 1](\widetilde{\delta c} - f_1(\widetilde{\zeta})). \tag{111}$$

[†] Here, as previously, we identify the ζ potential with the Stern potential, which is justified only for a sufficiently low electrolyte concentration, when the thickness of the mobile part of the double layer substantially exceeds the thickness of the immobile part. The formulas are readily extended to the case when this is not so, it being sufficient at certain points to replace ζ by the Stern potential. For instance, the limits of integration in the calculation of $I_E^{\pm}$ and $I_D^{\pm}$ should be the Stern potential, while ζ should remain the limit for $I_V^{\pm}$.

The coefficient on the right arises from the polarization of the Stern layer, and it is large, so it is impossible to ignore the change in the surface charge in the theory of deformation of the double layer.

Disturbed Equilibrium between the Internal and Diffuse Parts of a Polarized Double Layer.

If ion exchange between these two parts is very slow, the equilibrium between them may be substantially disturbed on polarization. The flux of ions from the bulk to the surface is proportional to the bulk concentration directly at the surface $C^{\pm}|_{x=0}$, while the flux of desorbed ions is proportional to the adsorption, so

$$\alpha^{\pm} (\psi|_{x=0})\, \Gamma^{\pm} - \beta^{\pm}\, C^{\pm}|_{\widetilde{x}=0} = \operatorname{div}_y \operatorname{grad}_y (-D_s^{\pm} \Gamma^{\pm} \pm D_s^{\pm}\, z^{\pm}\, \Gamma^{\pm} \psi)_{x=0}. \tag{112}$$

But desorption involves a potential barrier dependent on the surface potential, and so the coefficients are dependent on the surface potential $\psi|_{x=0}$. There are only small relative changes in adsorption and potential along the surface, so $\alpha^{\pm} \Gamma^{\pm}$ can be linearized. Solution of (112) allows us to express $\Gamma^{\pm}$ via $C^{\pm}|_{x=0}$. The results for $\Gamma^{\pm}$ are substituted into (106) to get the boundary condition relating the normal derivative of f_1 to the value at the surface. This condition allows us to determine the angular dependence $Z(y)$ of the second term in the general formula for f_1:

$$f_1 = \delta \widetilde{c}(y) Y_c(f) + Y_q(f) Z(\alpha^{\pm}, D_s^{\pm}, y). \tag{113}$$

Deviation of the internal part from the equilibrium state should be taken into account also in drawing up the expression for the $I_s^{\pm}$, which appear in (26) and (27), if the $I_s^{\pm}$ are comparable in magnitude with the other surface fluxes.

Deviation from equilibrium between the parts thus does not introduce further complications into the mathematical part of the theory, but it involves the physical parameters $\alpha^{\pm}$ and $D_s^{\pm}$ about which little is known. In this connection it is of interest to note that the coefficients of surface diffusion are very small for adsorbed potential-determining ions. If the simultaneous filling of the Stern layer with counterions is negligible, the right sides in the boundary conditions of (112) are small, so the dependence of the double-layer deformation on $\alpha^{\pm}$ and $D_s^{\pm}$ may be negligible even for very small values of $\alpha^{\pm}$.

We can still retain the conclusion that a Boltzmann distribution applies to a polarized double layer even when the tangential ion transport occurs mainly in the internal part of the layer, i.e., $I_s^{\pm} \gg I^{\pm}$, for in that case the relative change in j_x over the cross-section is small, since the ions from the bulk are transported mainly into the internal part of the layer. From this it follows directly that a Boltzmann distribution applies.

Ionic Component of the Dipole Moment.

There is no need to consider separately the diffuse and dense parts of the double layer as regards polarization if the layer is thin; it is sufficient to know the total excess charge on a given part due to the polarization of the dense and diffuse sections, which is denoted by $\sigma(y)dy$. We apply Gauss's theorem to this part of the double layer and ignore the tangential induction fluxes to get

$$4\pi\sigma(y) = \varepsilon\frac{\partial\varphi}{\partial x}(0, y) - \varepsilon'\frac{\partial\Psi}{\partial x}(0, y), \tag{114}$$

whence (50) gives

$$\sigma(y) = \frac{\varepsilon}{4\pi}\frac{\partial\varphi}{\partial x}(0, y). \tag{115}$$

Let $R(y)$ be the distance from a given point on the surface of a particle to some specified point on the symmetry axis, with $S(y)$ as the projection of $R(y)$ on the axis and $r(y)$ the distance from the given point on the surface to the axis. Then we get for the dipole moment

$$d = \int\limits_0^{y_0} \sigma(y)\, S(y)\, 2\pi r(y)\, dy = \frac{\varepsilon}{2} \int\limits_0^{y_0} S(y)\, \frac{\partial \varphi}{\partial x}(0, y)\, r(y)\, dy. \qquad (116)$$

The charge system as a whole is neutral, so the dipole moment is independent of the choice of the point on the axis.

The theory thus allows one to calculate the ionic component of the dipole moment on the basis of the potential distribution outside the double layer.

Conclusions

1. Deformation of the double layer on a colloidal particle has a bearing on many effects in colloidal systems that are of interest in relation to the double layer and to the interaction and coagulation of the particles. An analytic theory has been developed for the deformation of the diffuse part of a thin double layer on an axially symmetric particle, and this can serve as the basis for the theory of various effects such as the effects of double-layer polarization on the electrophoretic mobility, coagulation in an external electric field, the ionic component of the dipole moment, etc.

2. It has been found that a Boltzmann distribution still applies to a slightly deformed thin double layer, which occurs because there is local equilibrium between an area of the layer and the adjacent volume of solution. This provides an analog of the Poisson–Boltzmann equation for a slightly deformed thin double layer.

3. Allowance has been made for the deformation of the dense part in the theory of double-layer polarization. It is shown to be necessary to take account of the polarization of the dense part.

Mathematical Appendices

Appendix 1.   Derivation of the Boltzmann Distribution for the Ions in a Polarized Double Layer by the Boundary-Layer Method

To reveal the presence of a small parameter we convert to the dimensionless variables

$$n^{\pm} = C^{\pm}/c_0; \quad \varphi = \psi/\lambda, \quad \widetilde{x} = x/l, \quad \widetilde{y} = y/\lambda l.$$

System (2)-(4) takes the following form in these variables[†]:

$$\Delta n^{\pm} \mp \operatorname{div}(n^{\pm} \nabla \psi) = 0; \quad \varepsilon_0^2 \Delta \psi = n^+ - n^-, \qquad (A.1.1)-(A.1.3)$$

in which $\varepsilon_0 = (RT\varepsilon/4\pi F^2 c_0)^{1/2}/l \ll 1$.

The boundary-layer method [12] indicates that the solution should be sought as the sum of two functions, the first varying smoothly in the main region while the second varies rapidly and is localized in the thin boundary layer:

$$\psi = \psi_0 + \psi^*; \quad n^{\pm} = n_0^{\pm} + n^{*\pm}. \qquad (A.1.4)-(A.1.6)$$

To determine $n_0^{\pm}$ and ψ_0 we put $\varepsilon_0 = 0$ in the equations; then $n_0^+ = n_0^-$, and from (A.1.1) and (A.1.2) we get $\operatorname{div}(n_0^{\pm} \nabla \psi_0) = 0$; but the deviation of the concentration from the mean value is proportional to the applied electric field, so we may put $n_0^{\pm} = 1$ in this expression if the field is weak. Then

$$\Delta n_0^{\pm} = 0; \quad \Delta \psi_0 = 0. \qquad (A.1.7)-(A.1.9)$$

To derive the equations for $n^{*\pm}$ and ψ^* we substitute (A.1.4)-(A.1.6) into (A.1.1)-(A.1.3). The derivatives along the normal for the rapidly varying functions $n^{*\pm}$ and ψ^* substantially exceed

---

[†] The expressions have been simplified by omitting the convective term and putting $z^+ = z^- = 1$. Allowance for these factors does not affect the results.

the corresponding derivatives of $n_0^{\pm}$ and ψ_0 as well as the tangential derivatives, so the first terms in the equations may be isolated. For this purpose we formally extend the independent variable $x = \varepsilon_0 \xi$, with the result that $n^{*\pm}$ and ψ^* in the variables y and ξ become as smooth as $n_0^{\pm}$ and ψ_0 in y and x. Then ε_0 tends to zero, and thus the secondary terms in the equations are eliminated. This procedure gives

$$\frac{\partial^2 n^{*\pm}}{\partial \xi^2} \mp \frac{\partial}{\partial \xi} \left(n_0^{\pm} \frac{\partial \psi^*}{\partial \xi} \right) \mp \frac{\partial}{\partial \xi} \left(n^{*\pm} \frac{\partial \psi^*}{\partial \xi} \right) = 0. \qquad \text{(A.1.10), (A.1.11)}$$

As

$$n^{*\pm} |_{\zeta \to \infty} = 0; \quad \psi^* |_{\zeta \to \infty} = 0, \qquad \text{(A.1.12)-(A.1.14)}$$

the constant of integration may be equated to zero:

$$\frac{\partial n^{*\pm}}{\partial \xi} \mp n^{*\pm} \frac{\partial \psi^*}{\partial \xi} \mp n_0^{\pm} \frac{\partial \psi^*}{\partial \xi} = 0. \qquad \text{(A.1.15), (A.1.16)}$$

The integrals of these equations may, in the light of (A.1.12) and (A.1.13), be put as

$$n^{*\pm} = e^{\psi^*} \int_{\xi}^{\infty} n_0^{\pm} \frac{\partial \psi^*}{\partial \xi} e^{-\psi^*} d\xi. \qquad \text{(A.1.17), (A.1.18)}$$

As n_0 varies only slightly within the thickness of the double layer, we can use the following simplifying representation for $n_0^{\pm}$, which resembles (42):

$$n_0^{\pm} = 1 + \delta n^{\pm}(y) + \delta n^{\pm}(y) \, kx / l, \qquad \text{(A.1.19), (A.1.20)}$$

in which k is a factor of the order of unity.

These expressions are substituted into (A.1.17) and (A.1.18); integration then gives

$$n^{*\pm} = n_0^{\pm} (1 + \delta n^{\pm}(y))(e^{f+f_1} - 1) + Q^{*\pm}, \qquad \text{(A.1.21), (A.1.22)}$$

in which

$$Q^{*\pm} \cong \frac{\delta n^{\pm}(y)}{\chi l} e^{\mp f} \int_{f}^{\infty} \frac{\ln \text{th} \, f / 4}{\text{sh} \, f / 4} e^{\pm f} df.$$

It is readily seen that the integrals $Q^{*\pm}$ are analogs of the integrals $Q^{\pm}$ in (3) and (39), so they can, for analogous reasons, be deleted if (1) is obeyed. If, in addition, we delete the unity in the second bracket in (A.1.21) and (A.1.22), we get a Boltzmann distribution. This unity is retained in drawing up the expression for $n^+ - n^-$, i.e., (A.1.21) and (A.1.22) may be replaced by Boltzmann distributions in drawing up the Poisson-Boltzmann equation.

Appendix 2. It has been shown [11] that the asymptotic solution to (57) may be expressed via the coefficients of the asymptotic series for the variable coefficients in (65):

$$d_1(f) = a_0 + \frac{a_1}{f} + \cdots; \quad d_2(f) = b_0 + \frac{b_1}{f} + \cdots \qquad \text{(A.2.1)}$$

and this solution takes the form

$$Y_1 = e^{\alpha_1 f} f^{\rho_1} z_1(f), \qquad Y_2 = e^{\alpha_2 f} f^{\rho_2} z_2(f), \qquad \text{(A.2.2)}$$

in which α_1 and α_2 are the roots of the characteristic equation

$$\alpha^2 + a_0 \alpha + b_0 = 0, \qquad \text{(A.2.3)}$$

$$\rho_1 = -\frac{a_1\alpha_1 + b_1}{2\alpha_1 + a_0}, \qquad \rho_2 = -\frac{a_1\alpha_2 + b_1}{2\alpha_2 + a_0}, \qquad\qquad (A.2.4)$$

$$z_1(f) = 1 + \frac{C_1}{f}, \qquad z_2(f) = 1 + \frac{C_2}{f}. \qquad\qquad (A.2.5)$$

The detailed forms of $d_1(f)$ and $d_2(f)$ give us that $a_0 = 1/2$ and $b_0 = -1/2$, while the other coefficients in the series are zero, so

$$\rho_1 = 0; \ \rho_2 = 0, \ \alpha_1 = 1/2; \ \alpha_2 = -1, \qquad\qquad (A.2.6)$$

and hence

$$Y_{1as} = \exp(f/2), \quad Y_{2as} = \exp(-f). \qquad\qquad (A.2.7)$$

These two functions have been chosen for the asymptotic expression for Y_q because Y_q increases monotonically. To obtain the asymptotic form for Y_p we represent the true values Y_1 and Y_2 by the sums

$$Y_1 = Y_{1as} + \delta Y_1, \quad Y_2 = Y_{2as} + \delta Y_2. \qquad\qquad (A.2.8)$$

We express the Wronskian appearing in the integrand in (76) via Liouville's theorem:

$$W(f) = W(f_0) \exp\left[-\int_f^{f_0} d_1(f_1) \, df'\right], \qquad\qquad (A.2.9)$$

and transform this formula to the simpler form

$$Y_p = Y_1(f) \int \mathrm{ch}\,\frac{f'}{2}\, Y_2(f')\, df' - Y_2(f) \int \mathrm{ch}\, f'/2\, Y_1(f')\, df'. \qquad\qquad (A.2.10)$$

We replace the functions outside the integrals by their asymptotic values, while we use (A.2.8) to represent the functions within the integrands. Then

$$Y_{pas} = Y_{1as} \int^f \mathrm{ch}\,\frac{f'}{2}\, \delta Y_2(f')\, df' - Y_{2as} \int^f \mathrm{ch}\,\frac{f'}{2}\, \delta Y_1(f')\, df'. \qquad\qquad (A.2.11)$$

The integrals in (A.2.11) tend to constant values as $f \to \infty$, so (81) follows from (A.2.11).

Appendix 3. We combine (55) and (57), and use the definition of (13) for Y_c, to get an equation for the latter that may be put in the form:

$$A(f)Y_c'' + \frac{1 - e^{-(z^+ + z^-)f}}{2} Y_c' - \frac{z^- + z^+ e^{-(z^+ + z^-)f}}{2} Y_c = \frac{1 - e^{-(z^+ + z^-)f}}{2}, \qquad\qquad (A.3.1)$$

$$Y_c|_{f=0} = 0; \ Y_c'|_{f=\tilde{\zeta}} = 0, \qquad\qquad (A.3.2), (A.3.3)$$

$$A(f) = \frac{1 - e^{-z^-f}}{2} + \frac{e^{-(z^+ + z^-)f} - e^{-z^-f}}{2} > 1, \ \text{where } z^-f > 1. \qquad\qquad (A.3.4)$$

Y_c, as expressed via the first terms on the series for Y_q and Y_p, decreases monotonically for small and medium values of f. In this way it can be shown for $z^+ = z^- = 1$ that Y_c for $f > 2.5$ decreases from 0 to a value slightly exceeding -1. We shall show that Y_c decreases monotonically towards $-1/z^-$ for $z^-f > 1$, and hence throughout the range of variation of f.

We can simplify (A.3.1) via (A.3.3) at the point $f = \tilde{\zeta}$:

$$2A(f)Y_c'' = z^- Y_c + 1 + e^{-(z^+ + z^-)f}(z^+ Y_c - 1). \qquad\qquad (A.3.5)$$

We assume that $Y_c(\widetilde{\zeta}) < -1/z$. Then the right side in (A.3.5) is negative, and from (A.3.4) this means that $Y_c'' \ll 0$. A negative value for the second derivative to the left of a turning point means that the function increases monotonically, so Y (f) has a minimum for $f < \widetilde{\zeta}$, where (A.3.5) is again obeyed. The right part of (A.3.5) is then negative at the minimum; but the second derivative should be positive at a minimum, so we have a contradiction. Consequently,

$$Y_c(\widetilde{\zeta}) > -1/z^- \tag{A.3.6}$$

and Y_c decreases monotonically near $f = \widetilde{\zeta}$. Monotonic decrease in Y_c at the ends of an interval means that the number of minima within the latter must be equal to the number of maxima. Then (A.3.5) implies that minima are possible only for $Y_c > -1/z^-$ and maxima for $Y_c < -1/z^-$, which is absurd. Therefore, Y_c everywhere decrease monotonically.

References

1. B. M. Grafov and A. A. Chernenko, Dokl. AN SSSR, 146:135 (1962); Zh. Fiz. Khim., 37:666 (1963).
2. A. A. Chernenko, Dokl. AN SSSR, 153:1129 (1963).
3. J. Th. G. Overbeek, Koll. Beihefte, 54:287 (1943).
4. N. A. Tolstoi and P. P. Feofilov, Dokl. AN SSSR, 66:617 (1949); N. A. Tolstoi, A. A. Spartakov, and G. I. Khil'ko, Kolloidn. Zh., 22:705 (1960).
5. St. Stoilov, in: Proceedings of the 20th International Congress of Pure and Applied Chemistry (1965).
6. O. G. Us'yarov, I. S. Lavrov, and I. F. Efremov, Kolloidn. Zh., 28:596 (1966).
7. F. Booth, Proc. Roy. Soc., A203:514 (1950).
8. L. D. Landau and E. M. Lifshits, Mechanics of Continuous Media, Gosfizmatizdat, Moscow (1957), p. 57.
9. M. W. Smoluchowski, Z. Phys. Chem., 92:129 (1918).
10. V. G. Levich, Physicochemical Hydrodynamics, Gosfizmatizdat, Moscow (1959).
11. F. G. Trikomi, A Textbook of Differential Equations, GITTL, Moscow (1962).
12. F. Booth, J. Coll. Sci., 6:549 (1951).
13. S. N. Carrier, Advances in Applied Mathematics, 3:137 (1953).
14. D. A. Fridrikhsberg, Nauchn. Byul. LGU, No. 26, p. 6 (1950).

THE ELECTRODIFFUSION THEORY OF
THE DORN EFFECT FOR SPHERICAL SOLID PARTICLES

B. V. Deryagin and S. S. Dukhin

Institute of Physical Chemistry, Academy of Sciences of the USSR;
Institute of General and Inorganic Chemistry,
Academy of Sciences of the Ukrainian SSR

The formula for the sedimentation potential for small values of the Peclet number Pe is refined with allowance for surface conductivity and deformation of the double layer, and it is applied to the theory of the Dorn effect for large values of Pe. It is shown that the range of application of Booth's theory for the Dorn effect is restricted by the conditions of smallness of Pe and the ζ potential. A method is proposed for testing the theory of electrokinetic phenomena and the double layer via measurements of the Dorn potential at small and large values of Pe.

There are difficulties in testing the theory of electrokinetic effects, which hinders development of the theory and also application of electrokinetic measurements in research in double layers. For example, the measured electrophoretic mobility allows one to calculate the ζ potential, but test of the formula requires some independent method of measuring the ζ potential, which at present is lacking.

Here we may note a situation encountered in the electrodiffusion theory of the Dorn effect [1, 2]. Different formulas apply for the sedimentation potential for small and large values of Pe. Measurements at large and small Pe on the same object (fixed ζ potential) allow one, in principle, to derive the ζ potential from one set of conditions and then to test the theory for the other set of conditions. Moreover, the formula for Pe small contains not ζ itself but its cationic and anionic components, whose dependence on ζ can be calculated from double-layer theory, so the above measurements at different Pe can provide the more detailed information on ζ^+ and ζ^-, which should allow one to evaluate the limits of application of the theory.

Here we show that the theory[†] of electrokinetic effects can conveniently be tested via the sedimentation potential for spherical solid particles.

[†] The Dorn effect for droplets and bubbles is substantially complicated by the second double layer and the convective-diffusion potential. For this reason, various parameters (values largely unknown) in addition to the ζ potential appear in the formulas for the sedimentation potential in such systems, e.g., ion adsorption and speed of the surface of the drop. This makes it difficult to test the theory if the particles have mobile surfaces.

The electrodiffusion theory of the Dorn effect for such particles was given in 1961 for small Pe. The main result is the following formula for the distribution of the potential:

$$\varphi(r, \theta) = M_\varphi \cos \theta / r^2, \tag{1}$$

in which

$$M_\varphi = \frac{3vaRT}{2F} \cdot \frac{D^+\xi^- - D^-\xi^+}{D^+D^-(z^+ + z^-)}; \tag{2}$$

v is particle velocity, a is particle radius, r and θ are the coordinates of a spherical system with its origin at the center of the particle, $D^\pm$ are ion diffusion coefficients, $z^\pm$ are ion charges, F is Faraday's number, R is the universal gas constant, T is absolute temperature, and

$$\xi^\pm = \frac{RT\varepsilon}{8\pi F^2} \int_0^{\widetilde{\zeta}} \frac{[\exp(\pm z^\pm f) - 1] \int_f^{\widetilde{\zeta}} \{c_0^+ [\exp(-z^+f) - 1] + c_0^- [\exp(z^-f) - 1]\}^{-1/2} df}{\{c_0^+ [\exp(-z^+f) - 1] + c_0^- [\exp(z^-f) - 1]\}^{1/2}} df; \tag{3}$$

$$f = e\Phi/kT.$$

Formula (2) shows that neglect of diffusion in Smoluchowski's theory is justified only when the diffusion coefficients are equal; Smoluchowski's formula may be deduced [3] from (2) only for the particular case $D^+ = D^-$.

The theory of the Dorn effect for this case for Pe $\gg 1$ and Re $\ll 1$ (in which Re is the Reynolds number) can be developed via the general method previously described [1], as in the theory of a dynamic adsorption layer and a diffusion boundary layer [4].

The concentration change in a diffusion boundary layer on such a particle directly at the surface is

$$\delta c(0, \theta) = c_0 - c(0, \theta) = AF(\theta), \tag{4}$$

in which

$$F(\theta) = \int_0^\theta \frac{\sin \theta' \cos \theta' \, d\theta'}{\left[\theta - \theta' - \frac{1}{2}(\sin 2\theta - \sin 2\theta')\right]^{2/3}}, \tag{5}$$

$$A = 2^5 3^{-5/3} \Gamma(1/3) \, \text{Pe}^{2/3} a^{-1} \xi c_0 \tag{6}$$

where Γ is the gamma function. For a binary electrolyte (the case envisaged here) we must make the substitution

$$\xi \to (z^- D^+ \xi^+ + z^+ D^- \xi^-)(z^+ D^+ + z^- D^-)^{-1}.$$

As $F(\theta) \approx 1$, the above formulas give us the estimate

$$\delta c(0, \theta)/c_0 \sim \text{Pe}^{2/3} (\chi a)^{-2}, \tag{7}$$

in which χ is the reciprocal of the Debye thickness of the double layer.

The condition Re $\ll 1$ restricts Pe to values of about 10^3, so $\text{Pe}^{2/3} \approx 10^2$ and $a > 10^{-2}$ cm. Approximate electrical neutrality outside the double layer is always assumed in the theory of the Dorn effect, and this is so for electrolyte concentrations exceeding 10^{-8} to 10^{-7} M, so $(\chi a)^{-2} < 10^{-4}$. Then the relative changes in concentration and adsorption are virtually always small in the sedimentation of solid particles at Re < 1, so a single value for the ζ potential may be used for the entire surface of the particle.

As for a drop or bubble, the sedimentation potential for Pe $\gg 1$ is proportional to the sum of the surface ion fluxes: I_ζ due to the equilibrium double layer, I_{V_1} and I_{V_2} due to the secondary double layer, and I_{D_s} due to the tangential components of the diffusion and electrical migration fluxes. I_{V_1} is proportional to the speed of the surface of a drop and is zero for a solid particle. Formula (92) of [1] gives the following estimate for the relative contribution from the two double layers to the total sedimentation potential:

$$I_{V_2}/I_\zeta = \frac{v_D}{\zeta} \frac{\delta c\,(0,\,\theta)}{c_0},\tag{8}$$

in which

$$v_D = \frac{RT}{F} \frac{D^+ - D^-}{z^+ D^+ + z^- D^-}.$$

If we exclude from consideration the region near the isoelectric point, the contribution from the secondary double layer is negligible here.

The tangential component of the field of the secondary double layer may be neglected in evaluating the tangential fluxes due to diffusion and migration in the electric field. This allows us to ignore the contribution from the electrical migration flux to the Dorn effect. This simplification is permissible if the surface divergence from this flux is much less than the electrical migration flux from the volume to the surface. The tangential and radial components of the field are of the same order of magnitude in this case, so we have the condition

$$\Gamma_0/c_0 \ll a.\tag{9}$$

This condition has been derived [1] similarly for the case Pe $\ll 1$, so [1] may be consulted for the more detailed derivation. This condition is not always obeyed for Pe $\gg 1$; e.g., the tangential fluxes for a drop or bubble are determined by the convective-diffusion potential, and the importance of the electrical migration fluxes may increase. If we express the adsorption Γ_0 in the diffuse part of the double layer via the ζ potential, we may convert (9) to

$$2\,[\exp{(\mp z\widetilde{\zeta}/2)} - 1] \ll \chi a.\tag{10}$$

If $a > 10^{-2}$ cm and if the electrolyte concentration exceeds 10^{-7} M, this condition is violated only for $\zeta > 150$ mV in conjunction with very low concentrations.

Smoluchowski's formula thus applies with small corrections for the sedimentation potential of a solid particle for Pe $\gg 1$; these corrections reflect the effects of the secondary double layer and the surface electrical conductivity. The latter effect becomes appreciable if $\zeta > 150$ mV in conjunction with $c_0 \approx 10^{-7}$ M, while the effect of the secondary double layer becomes appreciable if $\zeta < 1$ mV together with $c_0 \approx 10^{-7}$ M.

The theory of the sedimentation potential for a solid particle for Pe $\ll 1$ may be refined as follows. In deducing (2) we took into account only the convective surface ion fluxes due to the Stokes velocity distribution in the liquid. If we are to incorporate surface conductivity, we have to allow for the deformation of the double layer and the change in the tangential component of the field within the cross section of the double layer. Here we can use the theory of a thin polarized double layer [5]. The distribution of the potential $\psi*$ for this case differs from that for electrophoresis [formula (17) of [6]] in the absence of the potential φ_0:

$$\widetilde{\psi}_1^\bullet = \widetilde{\varphi}_1 + f_1,\tag{11}$$

in which

$$\widetilde{\varphi}_1 = -A \cos\theta/\widetilde{x}^2; \quad f_1 = \delta c Y_c\,(f) \cos\theta;\tag{12}$$

where A is a constant to be determined and δc is the concentration change along the external boundary of the double layer, so that $c = c_0(1 + \delta c \cos \theta/\widetilde{x}^2)$. Equations (101) and (102) of [5] for the balance of the ion fluxes should differ from those for the case of electrophoresis [6] [Equation (22)] first in that $A + \frac{3}{2}$ is replaced by A (which corresponds to absence of φ_0) and second in the additional term $\frac{3}{2}v\xi^\pm/D^\pm$, which reflects the tangential convective transfer by the Stokes flow of liquid:

$$\frac{3m^\pm}{2\chi a}[I_2^\pm A - I_{2c}^\pm \widetilde{\delta c}] \mp \frac{z^\pm}{\chi a} I_1^\pm A + \frac{\delta \widetilde{c}}{\chi a} I_1^\pm + \frac{3}{2}\frac{v}{a}\frac{\xi^\pm}{D^\pm} = -\delta\widetilde{c} \pm z^\pm A, \tag{13}$$

in which the integrals $I_1^\pm$, $I_2^\pm$, $I_{2c}^\pm$ and the numbers $m^\pm$ have been determined [6].

The results of solving this system is

$$M_\varphi = a^2\lambda A = \frac{3}{2}va\frac{RT}{F}\frac{(\xi^+/D^+ - \xi^-/D^-) + (\chi a)^{-1}(\xi^+I_1^-/D^+ - \xi^-I_1^+/D) - 3(2\chi a)^{-1}(\xi^+m^-I_{2c}^-/D^+ - \xi^-m^+I_{2c}^+/D^-)}{(z^+ + z^-) - (\chi a)^{-1}[(z^+ + z^-)(I_1^+ + I_1^-) + 3m^+(I_2^+ + z^-I_{2c}^+) - 3m^-(I_2^- - z^+I_{2c}^-)]}. \tag{14}$$

The integrals $I_1^\pm$, $I_2^\pm$, $I_{2c}^\pm$ may be expressed in terms of elementary functions in the case of a symmetrical electrolyte. If (10) is obeyed, we can neglect all terms except the first in the numerator and denominator in (14), which then becomes (2), and so it is necessary to consider the effects of surface conductivity and deformation of the double layer only if (10) is violated.

The sedimentation potential may be measured for Pe small and large by use of motion under gravity and in a centrifugal field. Pe may also be adjusted via the particle size, but this complicates the experiment somewhat, since the ζ potential must not alter. A centrifugal field is thus more convenient, since the measurements at large and small Pe may be made with the same particles by varying the centrifugal field. Condition (10) must be obeyed if Smoluchowski's formula is to apply for large Pe, and formula (2) for Pe small. If this condition is obeyed for Pe $\ll 1$, it is automatically obeyed for Pe $\gg 1$.

We can insure that this condition is obeyed by making c_0 not too small, and we will assume in what follows that the condition is obeyed. We take the ratio of the sedimentation potentials at large and small Pe; on subtracting unity, we get the correction to Smoluchowski's formula due to the difference between the diffusion coefficients of the ions:

$$D = \frac{4\pi Fcz^+z^-}{\varepsilon\zeta D_e}(D^+\xi^- - D^-\xi^+) - 1. \tag{15}$$

Formulas (3) for $\xi^\pm$ may be simplified if $\zeta < 25$ mV:

$$\xi^\pm = \mp\frac{\varepsilon\zeta}{4\pi Fcz^+z^-(z^+ + z^-)} \tag{16}$$

while for high ζ

$$\xi^- = \frac{\varepsilon\zeta}{4\pi Fcz^+z^\mp} \quad , \quad |\xi^+| \ll \xi^-(\zeta \gg 25 \text{ mV}), \tag{17}$$

$$\xi^+ = -\frac{\varepsilon\zeta}{4\pi Fcz^+z^\mp} \quad , \quad |\xi^-| \ll \xi^+(-\zeta \gg 25 \text{ mV}). \tag{18}$$

Exact integration is possible if $z^+ = z^- = z$:

$$\xi^\pm = \frac{\varepsilon RT}{2\pi F^2 c_0 z^3}\{\ln[1 + \exp(\mp ze\zeta/2kT] - \ln 2\}. \tag{19}$$

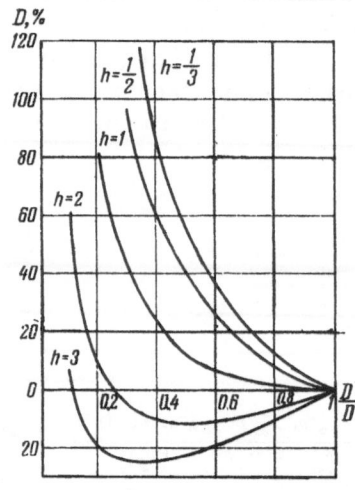

Fig. 1. Relative deviation D from Smoluchowski's formula given by (20) as a function of D^+/D^- for various $h = z^+/z^-$.

We express $\xi^\pm(\zeta)$ in (15) via the above formulas to get

with $\zeta \ll 25\,\mathrm{mV}$ $\quad D = \dfrac{(D^+z^+ + D^-z^-)^2}{D^+D^-(z^+ + z^-)^2} - 1,$ (20)

with $\zeta \gg 25\,\mathrm{mV}$ $\quad D = \dfrac{z^+}{z^+ + z^-}\left(1 - \dfrac{D^+}{D^-}\right),$ (21)

and for $z^+ = z^- = z$

$$D = \frac{1}{2\alpha}\left\{\left(1 + \frac{D^-}{D^+}\right)[\ln(1 + e^{\alpha/2}) - \ln 2]\right.$$

$$\left. - \left(1 + \frac{D^+}{D^-}\right)[\ln(1 + e^{-\alpha/2}) - \ln 2]\right\} - 1,$$ (22)

in which $\alpha = ze\zeta/kT$.

Figures 1–3 show the relationships represented by (20)–(22).

Booth's [7] electrodiffusion theory of the Dorn effect employs the above system of equations and boundary conditions. His theory is somewhat more general, since no restriction is placed on the ratio of the double-layer thickness to the radius of a particle; but Booth's theory is found to have only a restricted range of application on comparing his formula in the simplified form he gives for the limiting case of a thin double layer, and this restriction occurs not only as regards the magnitude of the ζ potential but also that of Pe. Booth imposed a restriction on Re in his formulation, but none on Pe, so Booth's formula should apply for Pe $\ll 1$ and Pe $\gg 1$; but that formula does not contain Pe, so there should be no qualitative differences between the Dorn effects in these limiting cases. We must thus conclude that Booth's method fails to reveal the most important aspect of the Dorn effect, namely the qualitative differences for Pe $\ll 1$ and Pe $\gg 1$, which represent exceptionally good scope for testing the theory of electrokinetic phenomena. All quantities in Booth's theory are represented as series expan-

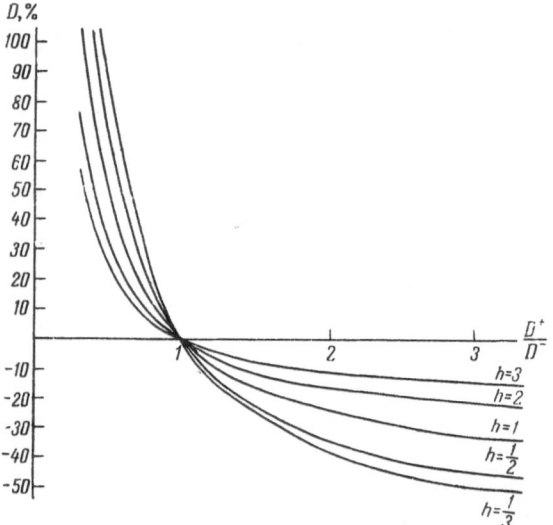

Fig. 2. $D(D^+/D^-)$ for $-\zeta \gg 25$ mV as given by (21).

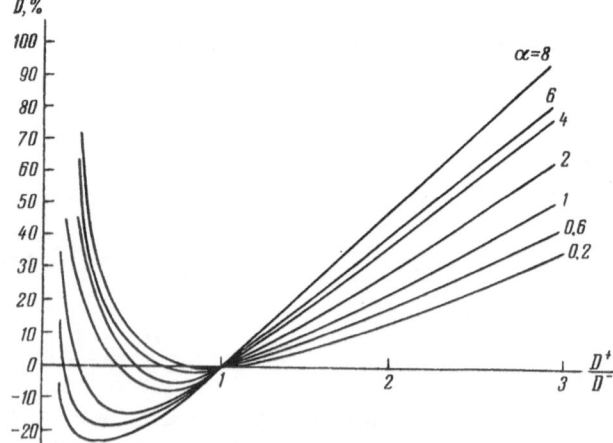

Fig. 3. $D(D^+/D^-)$ from (22) for $z^+ = z^- = z$ (symmetrical electrolyte) for several $\alpha = ze\zeta/kT$.

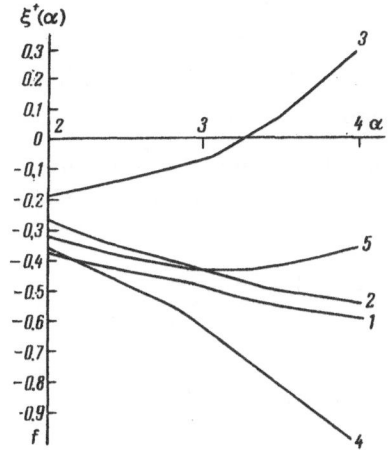

Fig. 4. ξ^+ and related functions characterizing the limits of application of Booth's theory: 1) $\xi^+(\alpha)$; 2) $f_1^{(\alpha)}$; 3) $f_2^{(\alpha)}$; 4) $f_3^{(\alpha)}$; 5) $f_4^{(\alpha)}$.

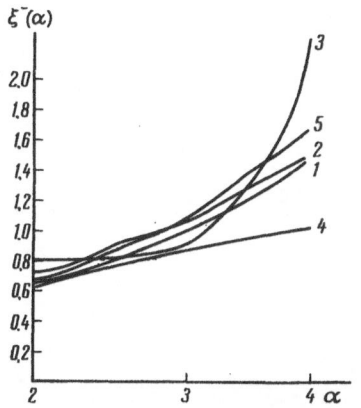

Fig. 5. ξ^- and related functions characterizing the limits of application of Booth's theory: 1) $\xi^-(\alpha)$; 2) $g_1(\alpha)$; 3) $g_3(\alpha)$; 4) $g_4(\alpha)$; 5) $g_5(\alpha)$.

sions in $e\Phi/kT$ (including the effective dipole moment of a particle), so we used series expansion on the integrands in (2) and (3), retained only the first few terms, integrated, and derived formulas for $\xi^\pm$ as power series in $e\zeta/kT$. Subsitution for $\xi^\pm$ in the formula for the potential of a moving particle for Pe $\ll 1$ shows that Booth's formula follows from our formula (2) as the particular case in which $\xi^\pm(\zeta)$ is expanded as a series and then the terms in ζ^3 and above are discarded.

Figure 4 shows ξ^+ for a symmetrical electrolyte and representations in terms of series in which only the first terms have been retained, in order to determine the limits in ζ for the applicability of these series. Curve 1 of Fig. 4 shows $\xi^+ = 2\pi Fcz/\varepsilon RT = \ln(1 + e^{-\alpha/2}) - \ln 2$. Curve 2 shows $f_1 = e^{-\alpha/2} - \ln 2$, which is obtained by series expansion of the logarithm and retention of the first two terms. This expansion is incovenient to continue further for $\alpha > 2$, so we have taken only f_1; Fig. 4 shows that this differs from ξ^+ by not more than 10% for $2 < \alpha < 4$. Curves 3-5 represent partial sums of the power series for f_1: $f_2 = 1 - (\alpha/2) + (\alpha^2/8) - \ln 2$, $f_3 = 1 - (\alpha/2) + (\alpha^2/8) - (\alpha^3/48) - \ln 2$ and $f_4 = f_3 + \alpha^4/24 \cdot 16$. Curve 1 in Fig. 5 shows

$$\hat{\xi}^- = \frac{2\pi F^2 cz^3 \xi^-}{\varepsilon RT} = [\ln(1 + e^{\alpha/2}) - \ln 2]$$
$$= \frac{\alpha}{2} + [\ln(1 + e^{-\alpha/2}) - \ln 2].$$

Curve 2 in Fig. 5 shows function g_1, which is obtained by series expansion of the logarithm and retention of the first two terms: $g_1 = f_2 + (\alpha/2)$. The expansion is inconvenient beyond this point, so the expansion has been restricted to g_1. Figure 5 shows that g_1 differs from ξ^- by not more than 10%. Curves 3-5 show partial sums of the power series for g_1: $g_2 = (\alpha/2) + f_2$; $g_3 = (\alpha/2) + f_3$; $g_4 = (\alpha/2) + f_4$.

It is clear that f_2 differs substantially from ξ^+ throughout the range $2 < \alpha < 4$, while g_2 differs substantially from ξ^- for $\alpha > 3$. This means that Booth's formula cannot be used for $\zeta > 50$ mV. If we could overcome the very great mathematical difficulties and incorporate the terms in α^3 and α^4 in Booth's method, we would still not get the desired accuracy, as the figures show. Booth himself considered the ratio of the second term to the first and concluded that his theory could not be accurate for NaOH, for example, for $\zeta > 30$-40 mV. Booth's theory is thus restricted to small ζ and small Pe.

Conclusions

1. A theory for Pe large has been given for the Dorn effect for solid spherical particles. It is found that Smoluchowski's formula retains its value even when the ions differ in diffusion coefficient. There are corrections to this formula for the secondary double layer and for surface

conductivity, but these are important only if the electrolyte concentration is very low and simultaneously ζ is very large or very small.

2. The formula previously derived for the sedimentation potential for Pe small has been corrected for surface conductivity and deformation of the double layer.

3. A method is proposed for testing the theory of electrokinetic phenomena and double layers via measurement of the Dorn potential for Pe large and small. The choice of conditions for such measurements is considered.

4. It is shown that Booth's theory is restricted to small values of Pe and ζ.

References

1.   S. S. Dukhin, in: Research in Surface Forces, Vol. 2, Consultants Bureau, New York (1966), p. 54.
2.   S. S. Dukhin, Kolloidn. Zh., 25:418, 520, 524 (1963); 26:36 (1964).
3.   B. V. Deryagin and S. S. Dukhin, Dokl. AN SSSR, 129:1328 (1961).
4.   S. S. Dukhin and B. V. Buikov, Zh. Fiz. Khim., 39:913 (1965).
5.   S. S. Dukhin, this volume, p. 287.
6.   S. S. Dukhin, this volume, p. 312.
7.   F. Booth, J. Chem. Phys., 22:1956 (1954).

THE ELECTRODIFFUSION THEORY
OF ELECTROPHORESIS

S. S. Dukhin

Institute of General and Inorganic Chemistry,
Academy of Sciences of the Ukrainian SSR

The theory of polarization in a thin double layer is used to derive a three-term formula for the electrophoretic mobility; the first term coincides with Smoluchowski's formula, while the second and third terms reflect the effects of the polarization.

Introduction

Application of a steady electric field causes particles to acquire a constant velocity corresponding to stationary hydrodynamic, electrical, and diffusion fields near the particles. Figure 1 shows schematically the forces that determine the speed v of a particle in response to a field $\vec{E}$. Force $\vec{K}_1$ is due to the action of the field on the charge Q of the particle:

$$\vec{K}_1 = Q\vec{E}. \tag{1}$$

The second force is given by Stokes's formula and is due to viscous resistance:

$$\vec{K}_2 = -6\pi\eta a v. \tag{2}$$

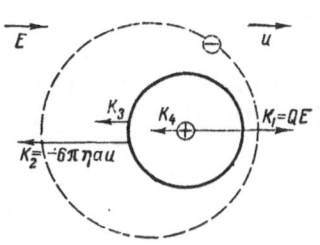

Fig. 1. Schematic representation of the forces acting on a particle in electrophoresis [12].

The other two forces are due to the ionic sheath of the particle. The force from the field on the ions in the diffusion part of the double layer is transmitted to the liquid; this motion of the liquid is responsible for the hydrodynamic force $\vec{K}_3$. The sign of the net charge of the counterions is the reverse of that for the charge on the particle, so $\vec{K}_3$ reduces the electrophoretic mobility (electrophoretic retardation). The fourth force is due to asymmetry in the ion atmosphere, because the ions are displaced by the field and by the motion of the liquid. The spherically symmetrical field of the particle tends to produce a symmetrical structure in the diffusion atmosphere, and this actually occurs when the external field is removed, though a certain relaxation time is involved.

The unsymmetrical component of the ion distribution in the steady state produces a field which acts on the charge of the particle to produce a force $\vec{K}_4$ (relaxation effect). Forces $\vec{K}_3$ and $\vec{K}_4$ are taken into account in the Debye−Hückel theory of strong electrolytes [1], where they are calculated separately and considered as additive. Overbeek has shown that the interaction between these effects cannot be neglected for colloidal particles; $\vec{K}_3$ must be calculated with allowance for the symmetry in the ion distribution, while $\vec{K}_4$ must be calculated with allowance for the velocity distribution near the particle. Severe mathematical difficulties arise from the need for simultaneous consideration of the hydrodynamic and diffusion aspects. The sum of these four forces must be zero in the steady state, so

$$\vec{v} = \frac{1}{6\pi\eta a}(Q\vec{E} + \vec{K}_3 + \vec{K}_4). \tag{3}$$

$\vec{K}_3$ and $\vec{K}_4$ are complicated functions of ζ and of the other parameters of the colloidal solution (particle size, valency, concentration, ionic mobility). As these parameters are usually known, ζ can be calculated from experimental data on the electrophoretic mobility via the formulas for $\vec{K}_3$ and $\vec{K}_4$.

Early work by Smoluchowski [2], Hückel [3], and Henry [4] on the theory of electrophoresis did not incorporate relaxation, although electrical retardaion was considered; see [5] for a detailed discussion of these papers. No allowance was made for polarization of the double layer, so the speed in electrophoresis was proportional to the ζ potential:

$$\vec{v} = \frac{e\zeta}{4\pi\eta}\vec{E} \text{ (Smoluchowski);} \tag{4}$$

$$\vec{v} = \frac{e\zeta}{6\pi\eta}\vec{E} \text{ Hückel} \tag{5}$$

Henry [4] explained the apparent conflict between these formulas. In Smoluchowski's theory it is assumed that the external field does not penetrate into the particle, which thus deforms the potential distribution, whereas Hückel neglected this factor and assumed that the field lines remain straight within the double layer. This approach is justified as an approximation only when the particle size is much less than the thickness of the double layer:

$$\chi a \ll 1. \tag{6}$$

Smoluchowski assumed that the strength of the applied field remained unchanged throughout the thickness of the double layer, which is approximately so if that thickness is much less than the particle radius:

$$\chi a \gg 1. \tag{7}$$

The differences between the various formulas arise from the different ways of considering K_3. Henry's treatment is the most rigorous as regards deformation of the applied field by the surface of the particle, as no simplifying assumptions were made about the field structure. Henry's formula

$$\vec{v} = \frac{\varepsilon_0\zeta}{6\pi\eta a} F_1(\chi a) \vec{E} \tag{8}$$

is applicable for any χa if it is in order to neglect the polarization of the double layer. In the limit $\chi a \to \infty$, $F_1(\chi a) \to 3/2$, and (8) becomes Smoluchowski's formula; if $\chi a \to 0$, $F_1(\chi a) = 1$, and Hückel's formula follows from (8).

The Debye−Hückel−Onsager theory [6] implies that the relaxation effect for a strong electrolyte is of the same order as K_3, which led to several discussions [7] of the relaxation

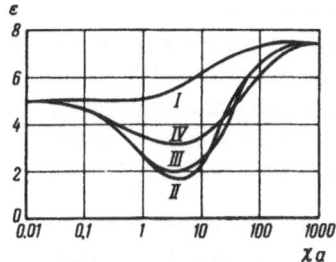

Fig. 2. Relation of ε to χa for $\zeta = 125$ mV, $z^+ = z^- = 1$ as given by: I) Henry's formula; II) Overbeek's formula; III) Booth's formula; IV) Wiersema's numerical calculations.

effect in colloidal solutions between 1928 and 1938. Overbeek [5] showed that these workers could not overcome the mathematical difficulties and that a new approach to the problem was necessary.

Relatively recently, Overbeek [5] and Booth [8] independently discussed in detail the relaxation effect for a spherical particle with a Chapman double layer, with no restriction on χa. The potential distribution in the equilibrium double layer was in both cases described via a solution of the Poisson−Boltzmann equations in terms of functional series [9]. This gives $\vec{v}$ as a power series in ζ. The mathematical difficulties restrict the discussion to the first few terms of this series. Overbeek's and Booth's results agree well and may be represented as

$$\vec{v} = \frac{\varepsilon_0 \zeta}{6\pi\eta} F(\chi a, \zeta)\, \vec{E}. \tag{9}$$

$F(\chi a, \zeta)$ approximates to Henry's $F_1(\chi a)$ for the limiting cases of (6) and (7), and also when $\zeta < 25$ mV. This means that the effects of the polarization can be neglected if one of these conditions is met. Overbeek has demonstrated this by an independent method [10] for the case of a thin double layer. The analytic methods developed by Overbeek and Booth do not allow one to derive information on the effects of the polarization for highly charged particles, so Wiersema has recently used numerical methods in the theory of polarization effects. The numerical methods were first applied to give the distributions of the potential and charge in the equilibrium double layer; a monograph [11] deals with these characteristics for the double layer of a spherical solid particle for various χa, z^+, z^-, and ζ. Then Wiersema applied these methods to Overbeek's relaxation problem [12], with computer assistance. Wiersema's approach is precisely the same as Overbeek's, but the use of numerical methods allowed Wiersema to obtain results for high values of ζ.

Figure 2, derived from [12], shows the dimensionless electrophoresis velocity

$$\varepsilon = \frac{6\eta\pi e}{\varepsilon_0 kT}\, \frac{v}{E} \tag{10}$$

as a function of χa for $\zeta = 125$ mV with $z^+ = z^- = 1$ and equal values of the diffusion coefficients, with

$$m^\pm = \varepsilon_0 \lambda^2 / 6\pi\eta D^\pm = m = 0.124 \tag{11}$$

taken as the numerical value corresponding to the normal mobility of inorganic ions in water at room temperature. Curve I is from Henry's formula and curve II is from Overbeek's formula; curve III is from Booth's paper, and curve IV is from Wiersema's numerical calculations.

It is clear that the polarization effects are unimportant only for $\chi a > 10^3$, being maximal for $\chi a \sim 5$. Wiersema's curve lies closer to Henry's curve than it does to Overbeek's or Booth's.

Wiersema in his dissertation gives a full survey of the experimental studies that can be used to test the theory of polarization effects, and he indicates two ways in which the theory could be tested. The first is to measure the electrophoretic mobility as a function of ζ for fixed χa by varying the concentration of the ions that determine the potential. As these ions determine not ζ but the total potential discontinuity in the double layer, and the relation between these characteristics of the double layer is unknown, it is difficult to test the theory. However,

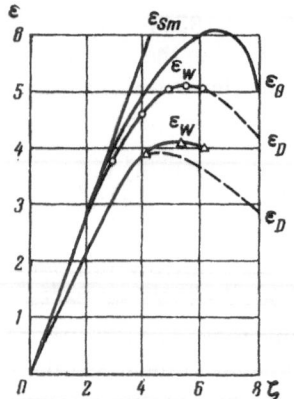

Fig. 3. Relation of ε to $\tilde{\zeta}$, $z^+ = z^- = 1$. For $\chi a = 50$: ε_{sm}) from Smoluchowski's formula; ε_0) from Overbeek's formula; $\bigcirc$) from Wiersema's numerical calculations; $---$) from (25). For $\chi a = 20$: $\triangle$) from Wiersema's numerical calculations; $---$) from (25).

the theory does indicate that either $\varepsilon(\zeta)$ has a maximum or that $\varepsilon(\zeta)$ tends to a limit as ζ increases. Wiersema correctly observed that this lack of information on the relation of ζ to the potential discontinuity should not interfere with the test for a maximum in $\varepsilon(\zeta)$. The second method is to determine the dependence of ε on particle radius for a fixed ζ.

There have been many measurements of electrophoretic mobility, but only a few have been designed to test the theory, and these have been concerned mainly with Henry's theory, so it is difficult to use them to test the theory of the relaxation effect. From a large number of papers Wiersema selected a few concerning electrophoresis of highly charged sols with nearly spherical particles. In every case, the form of $\varepsilon(\zeta)$ agreed qualitatively with theory, but the limiting (or maximal) values of the electrophoretic mobility differed somewhat from the theoretically predicted values. As the observed maximum only slightly exceeded the calculated value in certain instances, Wiersema concluded that experiment confirms the theory to a first approximation, though it could not be said that the verification was of the necessary accuracy. In all cases, the experimental data agreed better with Wiersema's numerical calculations than with the Overbeek−Booth analytic formulas.

Picard [13] has recently considered the theory of electrophoresis, but the results will not be considered in detail here, because he did not take into account the polarization of the double layer. However, we may note Picard's observation that it is necessary to take diffusion forces into account, which arise from change in the ion concentrations in the double layer along the surface. The liquid exerts a viscous resistance to the motion of the ions and so produces a force proportional to the flux density. The ion flux has diffusion and electromigration components, so the force has analogous components, the value given [13] for the diffusion component being

$$f_D \, dV = kT \, (\text{grad} \, c^+ + \text{grad} \, c^-) \, dV. \tag{12}$$

However, the formula for diffusophoresis cannot be derived from this formula for the bulk forces [14].

We thus conclude that (12) conflicts with the thermodynamics of irreversible processes. Picard's conclusion is unfounded because electrical forces are not analogous to diffusion forces; electrical forces are external to the volume of liquid, so there is no objection to applying Picard's conclusion concerning bulk forces to this component. But the diffusion component of the ion flux is not caused by external forces, and we cannot mechanically transfer the conclusion correct for electrical forces to the diffusion ones.

Experiment thus shows that there is a marked effect from polarization on the electrophoresis of highly charged particles. The analytic solution for this effect is of value in that it allows one to define the conditions under which the effect gives only a small correction to Henry's formula. The required accuracy is not provided by Overbeek's and Booth's formulas under conditions where relaxation leads to a substantial deviation from Henry's formulas, as is readily seen by comparing the results from these formulas with Wiersema's numerical results (Fig. 3).

Numerical methods have thus so far provided the only effective method of overcoming the mathematical difficulties in the theory of the relaxation effect. In spite of the value of the numerical results, there are at least three reasons why the theory of the polarization effect should be developed further: 1) For $\zeta > 150$ mV, Wiersema could not produce convergence in the process used in the numerical calculations, and so Wiersema does not give results for such ζ. Also, the calculations in the range 100-150 mV were troublesome, and so Wiersema was unable to establish clearly whether there is a maximum on the $\varepsilon(\zeta)$ curve or whether ε takes a constant value as ζ increase (a question very important to the treatment of the experimental data). 2) A second disadvantage is common to all numerical methods. The theory contains many parameters, whereas the calculations have to be performed selectively for certain combinations of these. 3) It is now necessary to formulate the problem of electrophoresis more precisely, and also to extend the theory to a wider class of highly charged particles. For instance, a theory has been proposed [15] for the electroviscous effect, in which the reduced rate of increase in v with ζ, or even a reduction in v, is ascribed to increase in the viscosity of the boundary layers of liquid in response to the increasing field of the double layer. It proved impracticable to take into account simultaneously the layer polarization and the electroviscous effect.

The mathematical difficulties have meant that only the polarization of the diffusion part of the double layer has been taken into account in the theory of electrophoresis. We are naturally interested in the role of the polarization of the inner part. The formula for the electrophoretic mobility may alter if the structure of the double layer is described via a theory more rigorous than Chapman's.

Systems of direct emulsion type have provided [16] the best data for testing the theory of layer polarization in relation to electrophoresis rate. This is no accident, because methods have been developed for examining the double layers at oil−water interfaces, and here it is possible to test the theory of electrophoresis. Moreover, the particles are here spherical, as is assumed in Overbeek's, Booth's, and Wiersema's treatments. Polarization theory must be extended to a drop with a mobile surface. Also, highly charged dispersed particles often have very regular but not spherical shapes.

It is extremely difficult to examine these topics within the framework of Wiersema's numerical method, but the theory of a thin polarized double layer [17] allows one to examine them. As the theory of double-layer polarization is approximate, and its accuracy decreases if condition (7) is not obeyed closely, we have first to establish whether the results from Wiersema's numerical calculations agree with formulas for the electrophoretic mobility derived from the theory of a thin polarized double layer. We therefore retain the formulation used in Overbeek's, Booth's, and Wiersema's theories.

Derivation of Formulas for the Electrophoretic

Mobility

Here, as noted above, we retain the usual formulation (in particular, we ignore the polarization of the inner part of the double layer), so we make extensive use of the general results in order to simplify the exposition. The system of equations and boundary conditions in the Overbeek−Booth−Wiersema theory is the same as the system adopted in the theory of a polarized double layer [17] (without allowance for polarization of the Stern layer).

Booth [8] and Wiersema [12] showed by different methods that one can represent the functions for the spatial distributions of the quantities in electrophoresis theory for spherical particles as functions dependent only on the radius and multiplied by $\sin\theta$ or $\cos\theta$. Overbeek used this form for the distributions of the charge and potential in the double layer, and in this way he obtained a general solution to the equations of hydrodynamics and expressed the velocity distribution in general form via the potential distribution. He then gave a detailed form to the equation for the balance of forces to obtain a formula for the electrophoresis rate, which simplifies substantially for the present case (thin double layer) and becomes

$$v = \frac{\varepsilon_0 a}{6\pi\eta} \int\limits_x^1 \xi \, dx, \tag{13}$$

where the expression for ξ simplifies substantially subject to condition (7):

$$\xi = \frac{\lambda^2}{a^2} \Big[2\int\limits_\infty^x \widetilde{\psi}^* \frac{d^2 f}{d\widetilde{x}^2} \, d\widetilde{x} - \widetilde{\psi}^* \frac{df}{dx} + 2\int\limits_\infty^{\widetilde{x}} \widetilde{\psi}^* \frac{df}{d\widetilde{x}} \, d\widetilde{x} \Big]. \tag{14}$$

Here $f(x)$ is the potential distribution in the equilibrium double layer in dimensionless form and $\widetilde{\psi}^*$ is the unsymmetrical component in the potential distribution. The third term in (14) may be neglected, by virtue of condition (7).

We alter the variable of integration in accordance with formula (16) of [17] to get

$$v = \frac{\varepsilon_0 \lambda^2}{6\pi\eta a} \Big[2\int\limits_f^{\widetilde{\zeta}} \frac{df}{\Omega(z^+, z^-, f)} \int\limits_0^f \frac{e^{z-f} - e^{-z+f}}{2\Omega(z^+, z^-, f)} \widetilde{\psi}^* \, df - \int\limits_f^{\widetilde{\zeta}} \widetilde{\psi}^* \, df \Big]. \tag{15}$$

The earlier results [17] for double-layer polarization were detailed only for $z^+ = z^- = 1$, so we apply (15) only for this case. The dimensionless electrophoretic mobility is

$$\mathbf{8} = \frac{2kT}{aeE} \int\limits_0^{\alpha_0} \Big[\frac{2}{\sh\alpha} \int\limits_0^\alpha \ch\alpha\widetilde{\psi}^* \, d\alpha - \widetilde{\psi}^* \Big] d\alpha, \tag{16}$$

in which

$$\alpha = f/2.$$

The results [17] show that we have

$$\widetilde{\psi}^* = \widetilde{\varphi}_0 + \widetilde{\varphi}_1 + f_1. \tag{17}$$

For the case of a spherical particle

$$\widetilde{\varphi}_0 = -\Big(\widetilde{x} + \frac{1}{2x^2}\Big) \Delta\widetilde{\varphi}_0 \cos\theta; \quad \widetilde{\varphi}_1 = -A\frac{\Delta\widetilde{\varphi}_0}{x^2} \cos\theta; \quad f_1 = \Delta\widetilde{c} Y_c(f), \tag{18}$$

in which $\Delta\widetilde{\varphi}_0 = Ea/\lambda$, while A and $\Delta\widetilde{c}$ are constant coefficients, which will be found below.

We use (17) and (18) to reduce (16) to

$$\mathbf{8} = \frac{3}{2}\frac{e\zeta}{kT} + \frac{e\zeta}{kT} A - \frac{\Delta\widetilde{c}}{\Delta\widetilde{\varphi}_0} \frac{2}{z} \int\limits_0^{\alpha_0} \Big[\frac{2}{\sh\alpha} \int\limits_0^\alpha \ch\alpha Y_c(\alpha) \, d\alpha - Y_c(\alpha) \Big] d\alpha. \tag{19}$$

In the more general case of a binary electrolyte of arbitrary valency composition

$$\mathbf{8} = \frac{3}{2}\frac{e\zeta}{kT} + A\frac{e\zeta}{kT} - \frac{\Delta\widetilde{c}}{\Delta\widetilde{\varphi}_0} \int\limits_0^{\zeta} \Big[\frac{1}{\Omega(Z^\pm, f)} \int\limits_0^{\widetilde{f}} \frac{e^{z-f} - e^{-z'f}}{\Omega(Z^\pm, f)} Y_c(z^\pm, f) \, df - Y_c(z^\pm, f) \Big] df. \tag{20}$$

The concentration distribution outside the thin double layer is

$$\widetilde{c} = 1 + \frac{\Delta\widetilde{c}}{x^2} \cos\theta, \tag{21}$$

with $\Delta\widetilde{c} = \delta c / c_0$.

Expressions (18) and (21) allow us to put equations (26) and (27) of [17] for the balance of the ion fluxes as

$$\frac{3m^\pm}{2\chi a}\left[I_2^\pm\left(\frac{3}{2}+A\right) - I_{2c}^\pm\frac{\widetilde{\Delta c}}{\Delta\widetilde{\varphi}_0}\right] \mp \frac{z^\pm}{\chi a}I_1^\pm\left(\frac{3}{2}+A\right) + \frac{1}{\chi a}I_1^\pm\frac{\widetilde{\Delta c}}{\Delta\widetilde{\varphi}_0} = -\frac{\widetilde{\Delta c}}{\Delta\widetilde{\varphi}_0} \pm z^\pm A. \tag{22}$$

This system is solved to give

$$A = -\frac{2}{3}\frac{\beta}{1+\beta}; \quad \beta = \frac{(^3/_2)(m^- I_2^- - m^+ I_2^+) + (z^+ I_1^+ + z^- I_1^-)}{(z^+ + z^-)\chi a + z^+ I_1^+ + z^- I_1^- - (^3/_2)(z^- m^+ I_{2c}^+ + z^+ m^- I_{2e}^-)}, \tag{23}$$

$$\frac{\widetilde{\Delta c}}{\Delta\widetilde{\varphi}_0} = -\frac{3}{2}\frac{z^+ z^-(I_1^- - I_1^+) + (^3/_2)(Z^+ m^- I_2^- + Z^- m^+ I_2^+)}{[(z^+ + z^-)\chi a + z^+ I_1^+ + z^- I_1^- - (^3/_2)(z^+ m^+ I_{2c}^+ + z^- m^- I_{rc}^-)](1+\beta)}, \tag{24}$$

in which

$$I_1^\pm = \left(\frac{z^+ + z^-}{2}\right)^{1/2}\int\limits_{\widetilde{\zeta}}^0 \frac{e^{\mp z^\pm f}-1}{\Omega(z^\pm, f)}df,$$

$$I_2^\pm = \left(\frac{z^+ + z^-}{2}\right)^{1/2}\int\limits_{\widetilde{\zeta}}^0 \frac{(e^{\mp zf}-1)(\widehat{\zeta}-f)}{\Omega(z^\pm, f)}df,$$

$$I_{2c}^\pm = \left(\frac{z^\pm + z^-}{2}\right)^{1/2}\int\limits_{\widetilde{\zeta}}^0\left\{\frac{e^{\mp z^\pm f}-1}{\Omega(z^\pm, f)}\int\limits_f^{\widetilde{\zeta}}\left[\frac{1}{\Omega(z^\pm, f)}\int\limits_0^f \frac{e^{z^- f}-e^{-z+f}}{\Omega(z^\pm, f)}Y_c\,df - Y_c\right]df\right\}df.$$

We substitute for $\widetilde{\Delta c}/\Delta\widetilde{\varphi}$ and A from (23) and (24) into (20) to get ε as a function of $\widetilde{\zeta}$. For the case of a symmetrical electrolyte we have

$$\varepsilon = \frac{3}{2}\widetilde{\zeta} - \frac{3\beta}{2(1+\beta)}\widetilde{\zeta} + \frac{3}{2}\frac{[(2+6m/z)\,\text{sh }z\,\widetilde{\zeta}/2 - 2m\widetilde{\zeta}]\,L(0,\alpha_0)}{[\chi a + 4\,\text{sh}^2\,z\widetilde{\zeta}/4 - (^3/_4)m(I_{2a}^+ + I_{2c}^-)](1+\beta)}, \tag{25}$$

in which

$$\beta = \frac{(4+12m/z^2)\,\text{sh}^2\,z\widetilde{\zeta}/4}{\chi a + 4\,\text{sh}^2\,z\widetilde{\zeta}/4 - (^3/_4)m(I_{2c}^+ + I_{2c}^-)}, \tag{26}$$

$$L(\alpha,\alpha_0) = \frac{2}{z}\int\limits_\alpha^{\alpha_0}\left[\frac{2}{\text{sh }\alpha}\int\limits_0^\alpha \text{ch }\alpha Y_c\,d\alpha - Y_c\right]d\alpha, \tag{27}$$

$$I_{2c}^+ + I_{2c}^- = 4\int\limits_0^\alpha L(\alpha,\alpha_0)\,\text{sh }\alpha\,d\alpha, \tag{28}$$

where we have put $m^+ = m^- = m$.

It is necessary to express $L(0,\alpha_0)$, I_{2c}^+, and I_{2c}^- in terms of elementary functions in order to put these formulas in a form convenient for numerical calculations.

Integral $L(0, \alpha_0)$

I have derived [17] the following approximation for the function $Y_c(\alpha)$ for a symmetrical univalent electrolyte for $\widetilde{\zeta} \gg 1$:

$$Y_c = \begin{cases} y_c(\alpha), & \alpha < a = 1.25, \\ -y_{cm} = 1, & \alpha > a, \ \widetilde{\zeta} > 4, \end{cases} \tag{29}$$

in which

$$y_c(\alpha) = \left(\alpha + \frac{\alpha^3}{6}\right)(\ln\alpha + q(\alpha_0)) - \frac{\alpha^3}{12},$$

$$q(\alpha_0) = \ln 2 - p(\alpha_0); \quad p(\alpha_0) = Y'_p(\tilde{\zeta})/Y'_q(\tilde{\zeta}).$$

Functions Y_p and Y_q have been defined [17]. We divide the range of integration into two parts to get

$$L(0, \alpha_0) = L(0, a) + L(a, \alpha_0), \tag{30}$$

in which

$$L(0, a) = 2\int\limits_0^a \left[\frac{2}{\mathrm{sh}\,\alpha}\int\limits_0^\alpha \mathrm{ch}\,\alpha\,y_c(\alpha)\,d\alpha - y_{cm}\right]d\alpha, \tag{31}$$

$$L(a, \alpha_0) = 2\int\limits_a^{\alpha_0}\left\{\frac{2}{\mathrm{sh}\,\alpha}\left[\int\limits_0^a \mathrm{ch}\,\alpha y_c(\alpha)\,d\alpha + \int\limits_a^\alpha \mathrm{ch}\,\alpha y_{cm}\,d\alpha\right] - y_{cm}\right\}d\alpha$$

$$= -2y_{cm}(\alpha_0 - a) + 4y_{cm}\int\limits_a^{\alpha_0}\left(\frac{1}{\mathrm{sh}\,\alpha}\int\limits_a^\alpha \mathrm{ch}\,\alpha\,d\alpha\right)d\alpha + 4\int\limits_0^a \mathrm{ch}\,\alpha y_c(\alpha)\,d\alpha\int\limits_a^\alpha \frac{d\alpha}{\mathrm{sh}\,\alpha}$$

$$= 2y_{cm}(\alpha_0 - a) + 4\left(\ln\mathrm{th}\,\frac{\alpha_0}{2} - \ln\mathrm{th}\,\frac{a}{2}\right)\left(\int\limits_0^a \mathrm{ch}\,\alpha y_c(\alpha)\,d\alpha - y_{cm}\,\mathrm{sh}\,a\right). \tag{32}$$

The following approximate expressions may be used for the hyperbolic functions in the range $(0, a)$:

$$\mathrm{sh}^{-1}\alpha = \frac{1}{\alpha} - \frac{\alpha}{6}, \quad \mathrm{ch}\,\alpha = 1 + \frac{\alpha^2}{2}.$$

Then

$$\mathrm{ch}\,\alpha y_c(\alpha) = \alpha q(\alpha_0) + \alpha^3 q_1 + \alpha^5 q_2 + \ln\alpha\left(\alpha + \frac{2}{3}\alpha^3 + \frac{\alpha^5}{12}\right),$$

in which

$$q_1 = \frac{2}{3}q(\alpha_0) - \frac{1}{12}; \quad q_2 = \frac{1}{12}q(\alpha_0) - \frac{1}{24}.$$

We can now readily perform the integration:

$$\int\limits_0^\alpha \mathrm{ch}\,\alpha y_c(\alpha)\,d\alpha = \alpha^2 q_3 + \alpha^4 q_4 + \alpha^6 q_5 + \ln\alpha\left(\frac{\alpha^2}{2} + \frac{\alpha^4}{6} + \frac{\alpha^6}{72}\right), \tag{33}$$

in which

$$q_3 = \frac{q(\alpha_0)}{2} - \frac{1}{4}; \quad q_4 = \frac{1}{6}q(\alpha_0) - \frac{1}{16}; \quad q_5 = \frac{q(\alpha_0)}{72} - \frac{1}{3\cdot 36}.$$

The substitution $\alpha = a$ gives the formula for the integral appearing on the right in (32).

To find $L(0, a)$ we transform the integrand in (31) via (33) to

$$\frac{1}{\mathrm{sh}\,\alpha}\int\limits_0^\alpha \mathrm{ch}\,\alpha\,y_c(\alpha)\,d\alpha - y_c(\alpha)/2 = -\frac{\alpha}{4} + \frac{\alpha^3}{48} + \alpha^5 q_7 + \alpha^7 - \frac{q_5}{6} - \ln\alpha\left[\frac{\alpha^5}{72} + \frac{\alpha^7}{12\cdot 36}\right],$$

in which

$$q_7 = -\frac{q(\alpha_0)}{72} + \frac{1}{12 \cdot 72}.$$

Integration gives

$$L(0, a) = -\frac{a^2}{2} + \frac{a^4}{48} + a^6 q_8 + a^8 q_9 - \ln a \left[\frac{a^6}{6 \cdot 18} + \frac{a^8}{12 \cdot 72}\right], \tag{34}$$

in which

$$q_8 = -\frac{1}{9 \cdot 12} q(\alpha_0) + \frac{1}{6 \cdot 72}; \quad q_9 = \frac{q(\alpha_0)}{36 \cdot 24} + \frac{19}{9 \cdot 36 \cdot 64}.$$

We calculate I_{2c}^+ and I_{2c}^-:

$$\frac{1}{4}(I_{2c}^+ + I_{2c}^-) = \int_0^{\alpha_0} \operatorname{sh} \alpha L(\alpha, \alpha_0) \, d\alpha = \int_0^a \operatorname{sh} \alpha L(\alpha, \alpha_0) \, d\alpha + \int_a^{\alpha_0} \operatorname{sh} \alpha L(\alpha, \alpha_0) \, d\alpha$$

$$= L(0, a)(\operatorname{ch} a - 1) - \int_0^a \operatorname{sh} \alpha L(0, \alpha) \, d\alpha - L(a, \alpha_0)(\operatorname{ch} a - 1) + \int_a^{\alpha_0} \operatorname{sh} \alpha L(\alpha, \alpha_0) \, d\alpha. \tag{35}$$

The derivation of $L(\alpha, \alpha_0)$ is analogous to that of $L(a, \alpha_0)$ and gives a formula resembling (32):

$$L(\alpha, \alpha_0) = 2 \left\{ y_{cm}(\alpha_0 - \alpha) + 2 \left[\int_0^a y_c \operatorname{ch} \alpha \, d\alpha - y_{cm} \operatorname{sh} a\right] \left[\ln \operatorname{th} \frac{\alpha_0}{2} - \ln \operatorname{th} \frac{\alpha}{2}\right] \right\}, \tag{36}$$

whence we have

$$\int_a^{\alpha_0} L(\alpha, \alpha_0) \operatorname{sh} \alpha \, d\alpha = 2 y_{cm} [\operatorname{sh} \alpha_0 - \operatorname{ch} a (\alpha_0 - a) - \operatorname{sh} a]$$

$$+ 4 \left[\int_0^a y_c(\alpha) \operatorname{ch} \alpha \, d\alpha - y_{cm} \operatorname{sh} a\right] \left(\ln \frac{\operatorname{sh} \alpha_0}{\operatorname{sh} a} - \operatorname{ch} a \ln \frac{\operatorname{th} \frac{\alpha_0}{2}}{\operatorname{th} \frac{a}{2}}\right). \tag{37}$$

There are many similar terms in summing the penultimate and last parts in (35), as (32) and (37) show. We finally get

$$I_{2c}^+ + I_{2c}^- = 8 y_{cm} [\operatorname{sh} \alpha_0 - (\alpha_0 - a) - \operatorname{sh} a] + 32 \left[\int_0^a y_c(\alpha) \operatorname{ch} \alpha \, d\alpha \right.$$

$$\left. - y_{cm} \operatorname{sh} a\right] \ln \frac{\operatorname{ch} \alpha_0/2}{\operatorname{ch} a/2} + 4 L(0, a)(\operatorname{ch} a - 1) - 4 \int_0^a \operatorname{sh} \alpha L(0, \alpha) \, d\alpha. \tag{38}$$

The last integral in (38) is calculated as follows. We replace a by α on the right in (31) to get $L(0, \alpha)$. We use the approximate value of $\sinh \alpha$; multiplication and collection of similar terms allows us to represent the integrand as the sum of a polynomial and the product of $\ln \alpha$ and a polynomial. Integration gives

$$\int_0^a L(0, \alpha) \operatorname{sh} \alpha \, d\alpha = -\frac{a^4}{8} - \frac{a^6}{96} + q_{12} a^8 + q_{13} a^{10} + q_{14} a^{12} - \ln a \left[\frac{a^8}{18 \cdot 48} + \frac{7 a^{10}}{5 \cdot 72^2} + \frac{a^{12}}{12 \cdot 72^2}\right], \tag{39}$$

in which

$$q_{12} = -\frac{q(\alpha_0)}{24 \cdot 36} + \frac{1}{32 \cdot 48}; \quad q_{13} = -\frac{7 q(\alpha_0)}{5 \cdot 72^2} + \frac{163}{200 \cdot 72^2}; \quad q_{14} = -\frac{q(\alpha_0)}{12 \cdot 72^2} + \frac{7}{96 \cdot 72^2}.$$

TABLE 1. Electrophoretic Mobilities: ε_w from Wiersema's
Calculations and ε from (25)

$\tilde{\zeta}$	$\chi a = 10$		$\chi a = 20$		$\chi a = 50$	
	ε_w	ε	ε_w	ε	ε_w	ε
4		3.05	3.963	3.88	4.79	4.86
5	3.47	2.84	4.098	3.84	5.25	5.23
6	3.347	2.66	3.937	3.53	5.266	5.15
8				2.41		3.85
10						3.45

Comparison with Theory

This formula is compared with Overbeek's formula by expanding the functions in (25) as series in $\tilde{\zeta}$, with isolation of terms† proportional to $\tilde{\zeta}$ and $\tilde{\zeta}^3$:

$$\varepsilon = \frac{3}{2}\frac{e\zeta}{kT} - \frac{9}{16}\frac{z^2}{\chi^2}\left(\frac{e\zeta}{kT}\right)^3 - \frac{3\varepsilon_0\lambda^2}{16\pi\eta D}\frac{1}{\chi a}\left(\frac{e\zeta}{kT}\right)^3. \tag{40}$$

Here the first term agrees with Smoluchowski's formula. The third term agrees exactly with the corresponding term in Overbeek's formula ($\chi a \gg 1$), while the second term differs only in the magnitude of the numerical factor (9/16 instead of 1/2) from the corresponding term in Overbeek's formula. The agreement with Overbeek's formula is therefore almost exact.

Overbeek and Booth were not convinced that their formulas were applicable for $e\zeta/kT \approx 1$, since this quantity was used as a small parameter in the theory. It is therefore somewhat unexpected that Wiersema concluded that Overbeek's formula takes fairly good account of the double-layer polarization up to ζ of 75-100 mV. Our formula (25) agrees with this conclusion; the hyperbolic functions appearing there do not depend directly on ζ, but on $\zeta/2$ and $\zeta/4$, which means that the later terms in the power series in ζ are relatively unimportant for values exceeding one. $L(0, \alpha_0)$, I_{2c}^+, and I_{2c}^- have a similar feature. The following is the reason why the Overbeek−Booth formula applies even when $e\zeta/kT$ is not small.

I have shown [17] that the relative concentration change of the ions in the double layer on polarization is the less the nearer the part of the layer to the surface. In other words, the outer part of the double layer, which corresponds to $f \approx 1$, is much more deformed than the diffusion part, which adjoins the surface and where $f \gg 1$. Therefore, the range of application of the Overbeek−Booth method is somewhat greater than was originally believed. However, the Overbeek−Booth theory does not agree satisfactorily with Wiersema's numerical calculations for fairly high $\tilde{\zeta}$ in conjunction with not very large χa, as Fig. 2 shows.

On the other hand, the present formula agrees well with Wiersema's calculations for χa reasonably large. Table 1 gives ε from (25) for several $\tilde{\zeta}$ and χa for which Wiersema performed numerical calculations. The discrepancy between the ε derived in the different ways does not exceed 1% for $\chi a = 50$, as against 10% for $\chi a = 20$; but the discrepancies rise to 20% in certain cases for $\chi a = 10$. We may therefore say that (25) can be used for $\chi a > 50$ or $\chi a > 20$, in accordance with the accuracy needed.

A very important point is that there is no substantial disagreement between Wiersema's numerical calculations and the theory based on a thin polarized double layer. We may conclude

† As $L(0, \alpha_0)$, I_{2c}^+, and I_{2c}^- appear in the terms of the $\tilde{\zeta}$ series with powers >3, formula (40) may be considered for $z \neq 1$.

that the approximations made in that theory do not substantially affect the accuracy if χa is sufficiently large.

Formula (25) contains information not available from Wiersema's calculations, which did not extend to $\chi a > 50$, and which involved insuperable mathematical difficulties for $\zeta > 150$ mV. The latter feature prevented Wiersema from establishing clearly whether $\varepsilon(\zeta)$ has a maximum or whether ε tends to a constant value as ζ increases. The data of Table 1 indicate that there is a maximum.

Conclusions

1. The theory of a thin polarized double layer has been applied to study the effects of double-layer deformation on the electrophoretic mobility. The formulation used in the Overbeek–Booth–Wiersema theory is retained.

2. A three-term formula has been derived for the electrophoretic mobility of a spherical solid particle, in which the first term coincides with Smoluchowski's formula, while the second and third terms reflect the effects of polarization. This formula agrees well with Overbeek's formula in the case of weakly charged particles, while it agrees well with numerical calculations for high ζ. The agreement shows that the theory of the thin polarized double layer has been developed correctly.

3. The formula gives additional information not available from Wiersema's treatment, since he did not perform any calculations for $\chi a > 50$, and the mathematical difficulties for $\zeta > 150$ mV were not overcome. The formula is applicable for any ζ.

References

1. P. Debye and E. Hückel, Phys. Z., 24:305 (1923).
2. M. W. Smoluchowski, Z. Phys. Chem., 92:129 (1918).
3. E. Hückel, Phys. Z., 25:204 (1924).
4. D. C. Henry, Proc. Roy. Soc., London, A133:106 (1931).
5. J. Th. G. Overbeek, Kolloid-Beihefte, 54:287 (1943).
6. L. Onsager, Phys. Z., 28:277 (1927).
7. M. Mooney, Z. Phys. Chem., 35:331 (1931); J. J. Bikerman, Z. Phys. Chem., Abt. A, 171:209 (1934); S. Komagata, Res. Electrotechn. Lab. (Tokyo), No. 287 (1935); J. J. Hermans, Phil. Mag., 26:650 (1938).
8. F. Booth, Proc. Roy. Soc., London, A203:514 (1950).
9. J. H. Gronwall, V. K. La Mer, and K. Sandved, Phys. Z., 29:358 (1928).
10. J. Th. G. Overbeek, Philips Research Rept., 1:135 (1956).
11. A. L. Loeb, P. H. Wiersema, and J. Th. G. Overbeek, The Electrical Double Layer around a Spherical Colloid Particle, MIT Press, Cambridge, Massachusets (1961).
12. P. H. Wiersema, On the Theory of Electrophoresis, Rijksuniversiteit, Utrecht (1964).
13. W. F. Picard, Kolloid-Z., 179:117 (1961).
14. B. V. Deryagin and S. S. Dukhin, this volume, p. 269.
15. J. Lyklema and J. Th. G. Overbeek, J. Coll. Sci., 16:501 (1961).
16. P. J. Anderson, Trans. Faraday Soc., 55:1421 (1959); D. A. Haydon, Proc. Roy. Soc., A258:319 (1960).
17. S. S. Dukhin, this volume, p. 287.

SECTION VI

TRANSPORT EFFECTS IN DISPERSE SYSTEMS
AND POROUS BODIES

A RADIOISOTOPE STUDY OF
THERMAL WATER TRANSPORT
IN FINELY DIVIDED MATERIALS

M. P. Volarovich and N. I. Gamayunov

Kalinin Polytechnical Institute

Radioactive tracers and γ-radiography have been used to examine thermal transport in peat, peaty soils, etc. The effects of such transport are reported for the drying of layers and small specimens of peat. Film transport is found to play a major part at low water contents.

Migration of water in response to temperature gradients is important in the water distribution in soils and building materials. Thermal transport is especially important in high-temperature drying and in hydrothermal treatment of materials.

There have been many studies [1-4] of such migration in granular materials; a major advance in the theory was made by Deryagin et al. [5, 6]. Various subsequent experimental papers [7-12] have confirmed the mechanism proposed in [5], but it has been observed that the phenomenon is very complicated, being accompanied by transport as liquid and as vapor, with marked effects from the temperature, temperature gradient, density, and initial water content.

An apparatus has been described [13] for examining the effects of these factors. Specimens of the material (peat) are placed in demountable cylindrical columns and are sealed before being placed between tanks in which water at controlled temperatures circulates. This provides control of the temperature difference along the column and also of the mean temperature. The peat contains Na_2S*O_4 as tracer, as this is not absorbed [14] by the solid phase of the peat or by the walls of the apparatus.

Figure 1 shows that the initial water content W_0 affects the distributions of water and label along the specimen at the end of the experiment; the proportion of water transferred towards the cold end increases as W_0 decreases, while there is hardly any transfer when W_0 approaches saturation. If the water content W is small (43% abs.), the water in the peat is in the sorbed state. Water evaporates at the hot end (curve 1) and migrates as vapor to the cold end, where it condenses. There are not yet continuous water films at $W_0 = 98\%$, but condensation at the cold end (curve 2) produces films of liquid there, and the water migrates against the heat flow in response to gradients in capillary pressure, as is clear from the change in N/N_0 (curve 2') on the right. Capillary film transport occurs throughout the length of the speci-

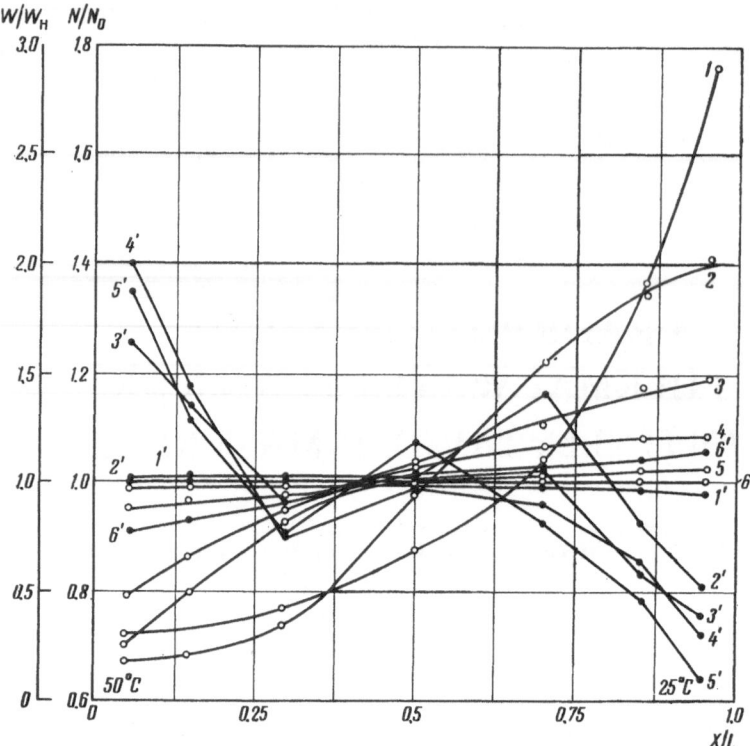

Fig. 1. Distribution of: 1-6) relative water content W/W_0;
1'-6') relative activity N/N_0 resulting from thermal transport in peat specimens at a mean temperature of 37.5 °C with
a temperature gradient Δt at the end of the experiment of
5 deg/cm for initial water contents W_0 (%) of: 1) 43; 2) 98;
3) 157; 4) 179; 5) 225; 6) 394.

men (curves 3-5) for W_0 of 150-300%; the water is transported as liquid and vapor along the
direction of heat flow, and as liquid in the reverse direction, i.e., circulation occurs [7-9], as
is clear from the S-shaped radioactivity curves 3'-5'.

Peat with $W_0 = 394\%$ is close to being a two-phase system; it has no pores through which vapor
can diffuse to the cold end, and migration in the liquid state is also hindered (curves 6 and 6').

The transport mechanism thus varies with W_0; for W_0 up to 150%, transfer in the direction
of heat flow occurs mainly in the form of vapor, while at higher levels there is film transport
in the liquid state. The transfer rate is inversely related to the density, i.e., is directly related
to the free porosity. The vapor diffuses to the cold end along the system of pore channels free
from water, though condensed water accumulates in these. The free porosity is inversely related to W_0, and so is the transport in vapor form. The result is that the mean rate of thermal
transport is inversely related to W_0.

The transport rate increases [15] with the temperature gradient and the mean temperature, mainly from increased evaporation and thermal diffusion of the vapor. There is also
some increase in migration in the liquid state from the reduction in the viscosity.

The above experiments do not define the transport kinetics. For this purpose, an apparatus
[16] was constructed for examining the water content along the specimen throughout the experiment by γ radiography. The specimen was placed in a rectangular container made of im-
permeable insulating material, which could be moved at right angles to the γ-ray beam. A

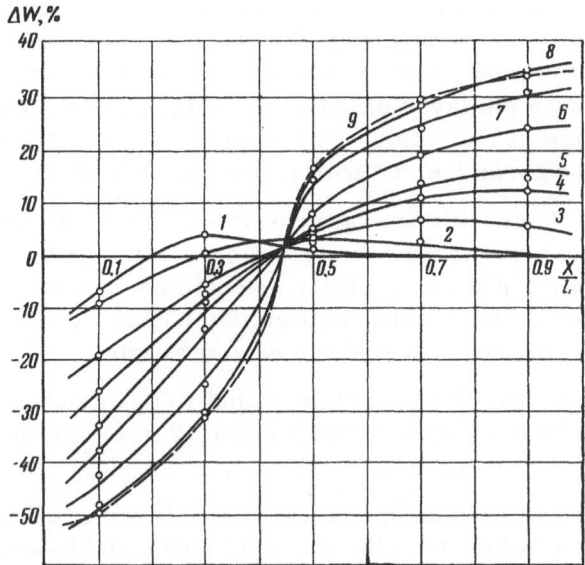

Fig. 2. Distribution of W along a peat specimen with $W_0 = 280\%$ and density of 0.135 g/cm^3
as measured after times (h) of: 1) 0.5; 2) 1; 3)
3; 4) 6; 5) 12; 6) 24; 7) 48; 8) 96; 9) 120.

heater was fitted at one end, while the other was kept at a constant temperature. The temperature was monitored by thermocouples at several points. We chose the strength and nature
of the γ-ray source, the counter type, the width of the collimator slit, and the disposition of
source and counter to be such as to minimize the statistical error of the measurement for the
optimal length of reading. The change ΔW in the water content is [14]

$$\Delta W = (1 + W_0) \ln N_1/N_2 / \ln N_0/N_1,$$

in which W_0 is initial water content measured in the usual way, N_1 is the corresponding count
rate, N_2 is the new count rate corresponding to the altered W, and N_0 is the count rate in the
absence of the material.

The experiments were done with peat and asbestos cement [17]; Fig. 2 shows distribution
curves for W, which resemble the ones for sand [11]. In every case, W is reduced at the hot
end, but the water migrates into the adjacent material, which produces a distribution of wave
type. The crest of the wave moves along the specimen. Then the peat, having greater free
porosity, shows the usual dehydration of the heated part and transfer of the water to the colder
half. A quasistationary distribution is attained after 3-4 days. The distribution at the end is
not linear, as has previously been demonstrated [8].

In the tests with asbestos cement, there were only these waves, whose crests moved in
the direction of heat flow, the transport rate being inversely related to W_0. No quasistationary
state was observed, evidently because the tests lasted only 24 h. There are two reasons why a
wave is produced in the first stage: 1) large temperature gradients are produced at the start
in the heated part, which cause thermocapillary flow down the temperature gradient; 2) water
evaporating near the heater diffuses to colder layers, where it condenses, which causes a local
increase in W and also tends to heat the material (the conductive transfer is very small for a
fibrous material, while the absorbed water produces a high specific heat). A steady-state temperature distribution is, however, produced much earlier than a steady-state water distribution
[12]. The net flow rate down the temperature gradient decreases as the test proceeds, since the
reverse flow increases with increase in the difference in W along the specimen.

The migration mechanism is even more complicated in an open system, where there is evaporation into the atmosphere. An apparatus has been built [18, 19] for research on the drying of peat contact with wet ground. The main item is an evaporator containing a block of peat, on which in boxes lies a 2 or 4 cm layer of cut upper peat of low degree of decomposition. These boxes had grids or foil at the bottom, to provide for drying studies with and without flow from the block. The peat had been treated with a solution of Na_2S*O_4.

The tests were done either with natural convection or with heating by an infrared heater above the peat. This heater was switched on periodically after 12 hours' drying, so radiation-convective (RC) and convective (C) modes of drying could be examined. The mean relative humidity φ and the air temperature t were respectively 55% and 22°C (RC drying) and 78% and 18°C (C drying). The radiation heat flux was 0.5 cal/cm$^2 \cdot$ min.

Figure 3 shows the vertical distribution of W and N/N_0 in a 2-cm layer. If convective drying is used for the first 12 h and then RC drying for 12 h, the top part of the layer dries out by evaporation (curves 1 and 2 of Fig. 3a).

The evaporation zone has reached the center of the layer at the end of the C period; the evaporation causes this to be colder than the material above and below, and this gives rise to thermocapillary flows towards the center. The effect is clearer in the layer without water exchange, as is clear from curve I' for the activity. There is also capillary film transport from the lower layers and the block into the evaporation zone. The result is a complicated distribution of the label.

In the subsequent RC period, curves 3 and 4 of Fig. 3a reveal a very different distribution of the water content. The water content in the top layer is actually less for the specimen having exchange with the block than it is for the one without exchange; drying of this layer thus occurs not only by evaporation but also by migration into the block. The heat absorbed by the isolated layer causes considerable evaporation and also redistribution within the layer. Thermocapillary transport and diffusion together [20] transport the label from the upper part of the specimen to the lower one (curves II and II' of Fig. 3a). The lower part of the layer having exchange has a higher water content than does that of the layer without exchange, but the final water contents of the layers are roughly equal. This indicates that radiative heating of moist ground (even with a fairly high groundwater level, say 25 cm [18]) does not appreciably influence the drying of a layer, since the water transfer from the base to the layer is balanced by reversal of thermal flow into the ground. If RC + C drying is used (Fig. 3b), the initial drying is more rapid than in C + RC, on account of rapid evaporation from the layer and transfer by thermocapillary action and thermal diffusion into the lower part or into the ground, as is clear from the activity curves I, I', II, and II' in Fig. 3b. In the second period (C drying), water (with label) enter the dried layer from the block, which leads to increases in water and radioactivity levels (curves 3 and III) in the lower part of the layer. Capillary film transport from underlying moist ground thus has a marked influence on convective drying of a layer.

The above transport mechanism applies also to a 4-cm layer, but drying occurs only in the upper half, while the lower half alters little in water content, since thermal transfer hinders drying by partly balancing out the upward flux of water under capillary forces. There is an increase in W only for the part in contact with the block.

Tests were also done with various levels of groundwater in the block: H of 25, 50, and 75 cm, which correspond respectively to capillary transport, capillary-film transport, and film transport to the surface of the block, so the rate of transfer decreases as H increases [21, 22], and there was a reduction in water exchange and thermal transport from the layer into the ground as the groundwater level fell. There was no observed exchange of a 4 cm layer with the block for H = 75 cm.

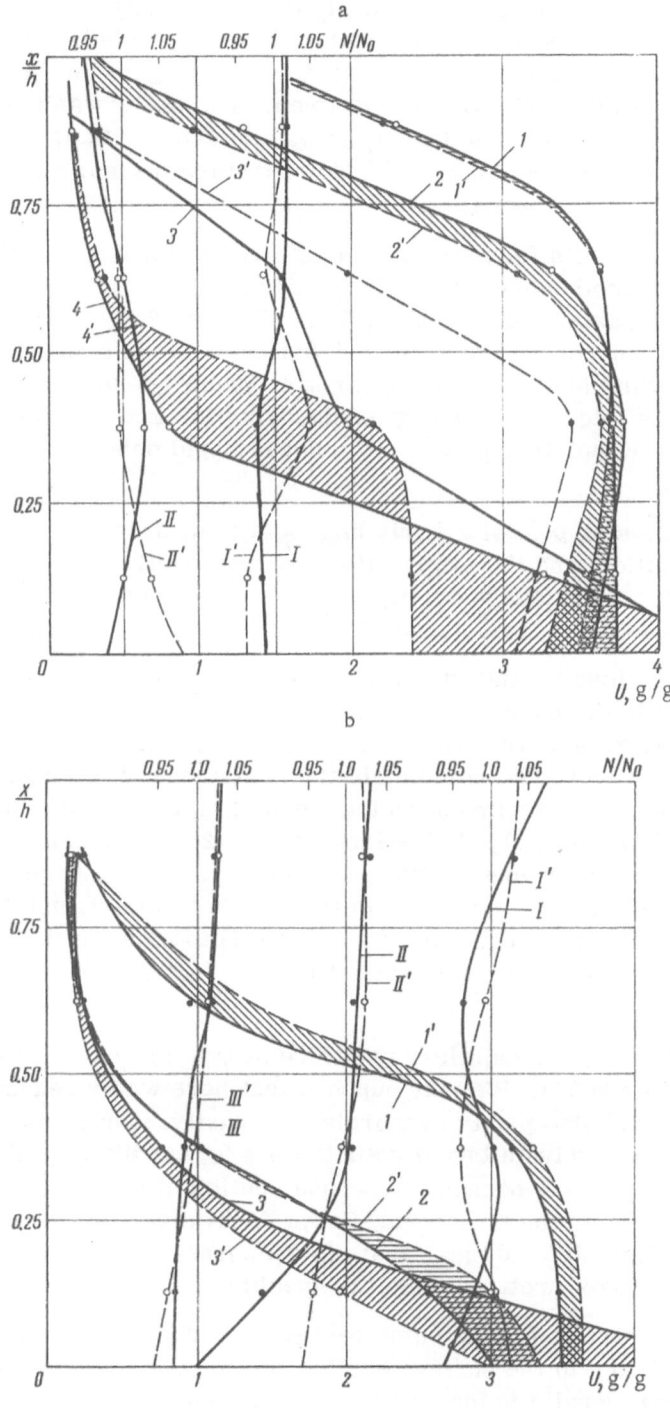

Fig. 3. Water content U and radioactivity N/N_0 (curves with roman numerals) for a 2-cm layer of cut peat: full lines, in presence of water exchange with the soil; broken lines, in the absence of exchange. Drying conditions: a) C + RC after time (h) of: 1) 6; 2) 12; 3) 18; 4) 24; b) RC + C after times (h) of: 1) 6; 2) 12; 3) 24.

It is also desirable to use radioactive traces to examine internal water transfer in denser peat specimens, as with an apparatus previously described [23], which consists of an air thermostat providing a constant temperature and relative humidity, a balance for determining the weight loss in the specimen (spheres of diameter 1-1.5 cm), and two photocells set at right angles to record change in the size of the specimen. The water was tagged with $Na_2S^*O_4$. The label migrated to the surface during drying, as was detected with an end-window counter in the thermostat directly under the specimen. Temperature measurements were also made at the surface and center of the sphere.

Figure 4a shows results for lower peat (30% decomposition) at 30°C with φ of 30, 50, 70, and 93%. The rate q of isothermal drying decreases as φ increases. The q(U) curves have parts of constant q and parts with variable q [24]. Curves 1", 3", and 4" show similar trends in the rate b of label transport to the surface,[†] but curve 2" is somewhat different. Other curves for b in other runs (not reproduced) for φ of 50-60% resemble those in Fig. 4a. The curves indicate that the internal transport mechanisms for $\varphi < 50\%$ and $\varphi > 50\%$ are different from that at $\varphi \approx 50\%$. Rapid drying ($\varphi = 30\%$) causes rapid dehydration of the large pores and then of the small ones.

Granular consolidated peat of a fairly high degree of decomposition has [14] a fairly narrow pore-size distribution, and the evaporation front gradually recedes into the specimen. However, there is pronounced shrinkage during drying, which produces an additional flow of water to the outside [24], as is clear from curves 1' (density) and 1‴ (coefficient β_0).

Here β_0 remains close to one[‡] down to U = 0.8 g/g, but it increases rapidly for lower U. The Laplace pressure in the menisci increases [1, 25] as φ is reduced, so the capillary pressure at $\varphi = 30\%$ will be greater than that at high φ. The capillary pressure at $\varphi = 93\%$ is so small as to be unable to overcome completely the resistance of the skeleton of the body, and hence $\beta_0 < 1$ (air enters the specimen), which means that evaporation occurs from the surface of the specimen and also internally, in the large pores [26]. The density of such a specimen is small (curve 4'). The considerable capillary and molecular shrinkage at $\varphi = 30\%$ causes a marked increase in density. The pressure is transmitted to the films and capillaries, and it forces the water in them with a high tracer content [25] to move to the surface. Also, the reduction in capillary size means that the water films are not lost at $\varphi = 0.3$ [26]. This explains the peak on curve 1".

There is a relatively high capillary pressure at $\varphi = 0.5$ (curve 2' indicates this indirectly), but much air enters (curve 2‴). We may suppose that here water reaches the surface along films and capillaries, and also partly evaporates internally. The films do not dry out (do not decrease in thickness to a solvate layer) until U $\approx$ 1 g/g, but the rate of capillary film transport falls rapidly for U < 1 g/g because the cross section of the films is reduced [26] and the evaporation front moves into the specimen. Curve 2° differs from the others in that t_s tends monotonically to t_c. This also indicates that the evaporation rate at the surface gradually falls because the evaporation front retreats into the specimen.

Figure 4b shows results for the same peat at $\varphi = 30\%$ with t_c of 30, 40, and 50°C. The radioactivity is brought out to the surface earlier and more rapidly the higher the temperature, whereas the final density is lower (curves 1'-3') but the drying rate increases (curves 1-3). One reason for this is that increase in t reduces the forces of capillary shrinkage because the surface tension of the water is lowered; but this is accompanied by increased capillary-film transport to the surface on account of increase in transport potential and decrease in viscosity, apart from other causes.

---

† See [24] for symbols.

‡ $\beta_0 = \Delta V \rho_b / \Delta p$ is the ratio of the volume change to the weight loss in the same time, in which ρ_b is the density of water; air enters for $\beta_0 < 1$, but is expelled for $\beta_0 > 1$.

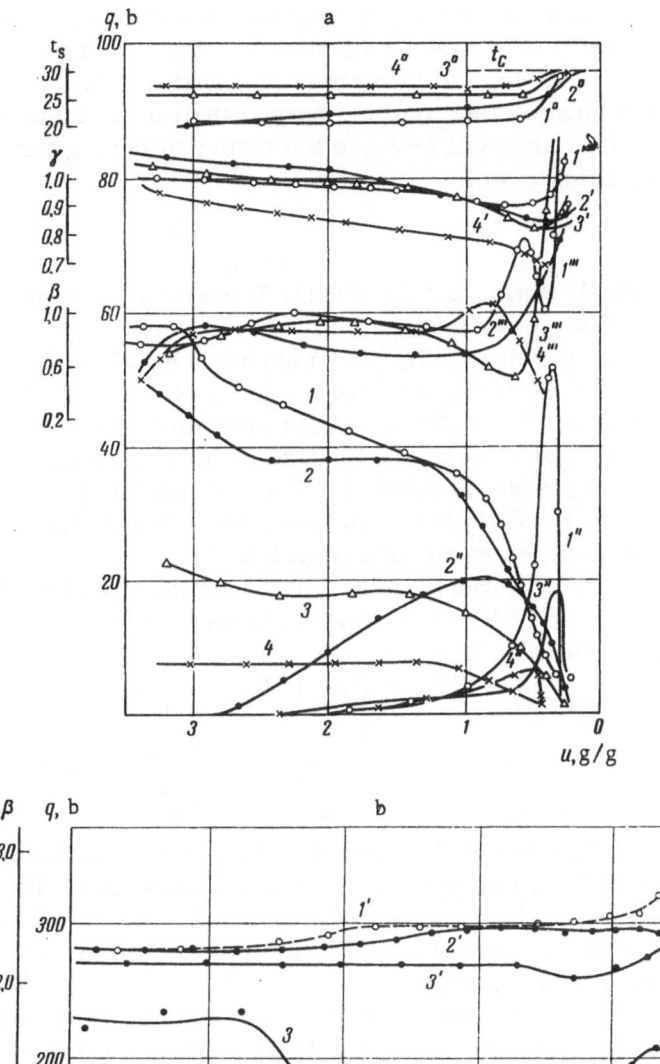

Fig. 4. Results on isothermal drying of peat: a) results of:
1-4) evaporation rate q (%/h); 1'-4') density γ (g/cm³); 1"-4")
activity brought to the surface (%/h); 1'''-4''') β_0, and 1°-4°)
surface temperature t_s, all as functions of water content U (g/g)
for a chamber temperature t_c = 30°C and relative humidities
φ (%) of: 1) 30; 2) 50; 3) 70; 4) 93; b) the same, but for φ = 30%
and t_c (°C) of: 1) 30; 2) 40; 3) 50.

Radioisotopes thus reveal the details of internal heat and water transport. The results show that Deryagin's theory of film water migration has important applications, but that the phenomena are much more complicated than those in model media [26-28] when there is shrinkage and also a local endothermic effect due to phase transitions. Future studies should provide the basis for better models, which will advance the theory of drying for natural granular materials and of water migration in soils.

References

1. A. V. Lykov, Zh. Prikl. Khim., 8:1354 (1935); Transport Phenomena in Capillary-Porous Media, Gostekhizdat, Moscow (1954).
2. N. A. Nasedkin and G. I. Pokrovskii, Zh. Tekh. Fiz., 9:1515 (1939).
3. A. E. Klyucharev and V. S. Shalygina, Torf. Prom., No. 5, p. 21 (1952).
4. G. P. Philip and D. A. de Vries, Trans. Amer. Geophys. Union, 38:222 (1957).
5. B. V. Deryagin, M. K. Mel'nikova, and S. V. Nerpin, Transactions of the Sixth International Soil Science Congress: Physics of Soils, Izd. AN SSSR, Moscow (1956), p. 101.
6. B. V. Deryagin and M. K. Mel'nikova, in: Aspects of Agronomic Physics, edited by A. F. Ioffe and N. I. Samoilov, Leningrad (1957), p. 30.
7. M. P. Volarovich, N. I. Gamayunov, and N. V. Churaev, Kolloidn. Zh., 22:535 (1960).
8. N. I. Gamayunov, Inzh. Fiz. Zh. Akad. Nauk Belorussk. SSR, 3:11 (1960).
9. N. V. Churaev, Kolloidn. Zh., 22:631 (1960).
10. A. M. Globus, Dokl. AN SSSR, 132:919 (1960); Pochvovedenie, No. 9, p. 103 (1961); No. 2, p. 7 (1962).
11. V. I. Oleinikov and M. F. Kazanskii, Inzh.-Fiz. Zh. Akad. Nauk Belorussk. SSR, 1:38 (1958).
12. D. N. Onchukov and V. P. Ostapchik, Pochvovedenie, 7:53 (1963).
13. M. P. Volarovich, P. N. Davidovskii, and N. I. Gamayunov, Kolloidn. Zh., 27:167 (1965).
14. N. P. Volarovich and N. V. Churaev, Radioisotope Research on Peat, Izd. AN SSSR, Moscow (1960).
15. M. P. Volarovich, N. I. Gamayunov, and P. N. Davidovskii, in: Combined Use of Peat, Izd. Nedra, Moscow (1965), p. 258.
16. M. P. Volarovich, N. I. Gamayunov, and P. N. Davidovskii, Kolloidn. Zh., 27:3 (1965).
17. M. P. Volarovich, N. I. Gamayunov, and É. S. Dolinskaya, Trudy NIIasbestotsementa, Stroiizdat, No. 19, p. 80 (1965).
18. M. P. Volarovich, N. I. Gamayunov, and A. P. Polyanicheva, in: Combined Use of Peat, Izd. Nedra, Moscow (1965), p. 265; Izv. VUZ, Gornyi Zhurnal, No. 10, p. 27 (1965).
19. M. P. Volarovich, N. I. Gamayunov, and A. P. Polyanicheva, Kolloidn. Zh., 27:505 (1965).
20. M. P. Volarovich, N. I. Gamayunov, and P. N. Davidovskii, Kolloidn. Zh., 26:139 (1964).
21. P. V. Panferov, Trudy Kalinin. Torf. Inst., No. 10, p. 90 (1959).
22. N. V. Churaev, in: New Physical Methods of Research on Peat, Gosénergoizdat, Moscow (1960), p. 180.
23. M. P. Volarovich, N. I. Gamayunov, and N. V. Churaev, Trudy Vses. Nauch.-Issled. Inst. Gidrotekh. i Melioratsii, No. 38, p. 97 (1962).
24. N. V. Churaev, Inzh.-Fiz. Zh. Akad. Nauk Belorussk. SSR, 5:41 (1962); 6:31 (1963).
25. P. A. Rebinder, All-Union Conference on Accelerated Drying and Quality Improvement in Dried Materials: Plenary Session, Profizdat, Moscow (1958).
26. N. V. Churaev, Dokl. AN SSSR, 148:1361 (1963); Kolloidn. Zh., 27:908 (1965).
27. S. V. Nerpin and N. V. Churaev, Inzh.-Fiz. Zh. Akad. Nauk Belorussk. SSR, 8:20 (1965).
28. B. V. Deryagin, S. V. Nerpin, and N. V. Churaev, Kolloidn. Zh., 26:301 (1964).

WATER TRANSPORT MECHANISM AND STRUCTURE
OF A POROUS BODY

N. V. Churaev, V. I. Lashnev, and M. M. Gorokhov

Division of Surface Phenomena, Institute of Physical Chemistry,
Academy of Sciences of the USSR;
Kalinin Polytechnical Institute

Tracer results are presented on the structures of porous media as deduced from convective diffusion.

The tracer method used here is based on the rate of replacement of pore water by a solution of an unadsorbed tracer [1]; it differs from displacement by gases or immiscible liquids in that there is no deformation by interfacial tension, which is especially important for readily deformed systems. Moreover, replacement of the natural dispersion medium by another one can cause structural changes on account of the shift in the dispersion equilibrium in the colloidal fraction. The composition of the dispersion medium is virtually unaltered in the tracer method, and this also eliminates osmotic effects.

The coefficient of convective diffusion D_* may be used to characterize the porous structure, which characterizes the distribution of the possible velocities of a tagged particle in the pore space [2-5]. Solutions have been derived [6] to the equation of convective diffusion for the boundary and initial conditions corresponding to the method of radiochromatography [1] for various values of the longitudinal Peclet number:

$$Pe = vl/D_*,$$

in which v is the filtration rate (constant during the experiment) and l is the length of the specimen.

These solutions allow one to find D_* via curves for $C = N/N_0$, the tracer concentration during displacement. The simplified solution for $Pe \geq 100$ is

$$C \simeq \frac{1}{2}\left\{ \mathrm{erfc}\left[\frac{\sqrt{Pe}}{2}\left(\frac{1}{\sqrt{T}} - \sqrt{T} \right) \right] + \frac{6}{T+1} \sqrt{\frac{T}{\pi Pe}} \exp\left[-\frac{Pe}{4}\left(\frac{1}{T} + T - 2 \right) \right] \right\}, \tag{1}$$

in which $T = V/V_0$ is the ratio of the volume of filtrate to the volume of liquid in the specimen, which characterizes (for v constant) the dimensionless time.

335

TABLE 1. Structural Parameters of Saturated Porous Bodies

Characteristics	Pe	l, cm	n	δ, μ	λ, μ
Quartz sand < 0.17 mm	186	3.2	0.41	—	172
The same	285	9.5	0.476	9.1	334
Quartz sand 0.1-0.25 mm	228	9	0.418	15	395
Quartz sand 0.2-0.3 mm	314	3.2	0.38	—	102
The same	276	5.6	0.38	12	202
" "	348	9.6	0.38	13.5	276
Quartz sand 0.25-0.5 mm	131	8.5	0.379	23.2	650
The same	170	13.6	0.37	21.1	801
" "	490	41.8	0.37	26.2	855
Quartz sand 0.5-1 mm	108	8.5	0.389	29.6	785
The same	300	52.0	0.374	49	1740
Glass fibers	57	9.5	0.76	—	1670
The same	53	9.3	0.912	—	1750

It is convenient to denote D_* from the slope of the linear part of the curve at T = 1. We get from (1) that

$$(dC/dT)_{T=1} = (vl/4\pi D_*)^{1/2}, \qquad (2)$$

so D_* can be calculated if v and l are known.

Equation (1) shows that plots of C against T for equal Pe should coincide. We have used quartz sand in various grain sizes (0.1 to 1.0 mm) with l = constant at v of 3×10^{-3} to 3×10^{-2} cm/sec; the experiments confirm this closely for the region of linear filtration. The invariance with respect to v arises from the linear relation [4, 5]

$$D_* = \lambda v, \qquad (3)$$

in which λ is some linear parameter of the porous medium that is responsible for the variance of the velocity due to inhomogeneity in the pore spaces.

It is usual [7, 8] to relate λ to the mixing length, so λ should increase with the particle size and the inhomogeneity of the packing. The tracer results of Table 1 confirm this conclusion: the λ increase as the degree of dispersion decreases, and there is a reduction for the systems more uniform in particle size (sand with grain size 0.2-0.3 mm). The inhomogeneity is greatest in the system of glass fibers. Parameter λ also tends to increase with l if the structural inhomogeneities are comparable with l [9], λ becoming independent of l if l is large enough, when there is a correlation between λ and the hydraulic radius of the pores $\delta = n/s$, in which n is porosity and s is specific surface. There is a linear relation of λ to δ for systems of the same type.

It was found that dispersion of the tracer in these systems was not due to molecular diffusion of the tracer (since the coefficient of molecular diffusion D is not dependent on v) and was due mainly to granulation effects. Numerical solutions to the unsimplified equation [6] must be used for Pe small.

Figure 1 shows tracer curves for Na_2SO_4 tagged with ^{35}S (ionic tracer) and HTO (molecular tracer) for quartz sand, glass fibres, and granulated silica gel. Curves for ^{35}S and HTO (a and b) agree well, which shows that there is no appreciable adsorption of ^{35}S on glass and quartz. The tritium partition coefficient for free and bound water is close to one [10], and the content of bound water is small,[†] so the experiment with HTO may be considered as free from tracer loss, which is

---

[†] The specific surfaces of quartz sand and glass fibers were respectively 524 and 263 m^2/kg.

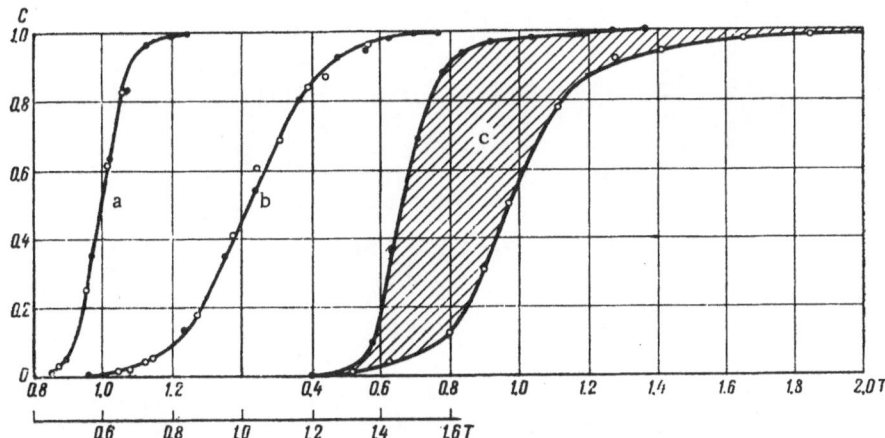

Fig. 1. Curves of C = f (T) for: ●) ^{35}S, ○) HTO for: a) quartz sand < 0.17 mm, l = 9.5 cm, n = 0.476; b) glass fibers, 0.25 × 5 mm, l = 9.3 cm, n = 0.912; c) 0.2-0.4 mm fraction of ground ASM-20 silica gel, l = 6.7 cm, n = 0.715.

confirmed by the agreement between the n calculated from the C curves by the method of [1] and from the volume of the specimen and the mass and density of the particles.

However, the curves for the two tracers do not agree for the more complicated system represented by silica gel (Fig. 1c); the C(T) curve for tritiated water differs from the other curve in being unsymmetrical, on account of mass exchange between the fine pores and the main channels. The equation in that case is [11, 12] usually formulated as

$$\frac{1}{\text{Pe}}\frac{\partial^2 C}{\partial X^2} - \frac{\partial C}{\partial X} = (1-f)\frac{\partial C}{\partial T} + f\frac{\partial C_\text{p}}{\partial T}; \; f\frac{\partial C_\text{p}}{\partial T} = K(C - C_\text{p}), \tag{4}$$

in which C_p is the mean tracer concentration in the blind pores, f is the fraction of the pore space accounted for by such pores, K is the mass-transfer factor, and X = x/l is the dimensionless coordinate.

Only numerical methods can be used to solve (4) subject to the boundary conditions previously used, so f, D_*, and K cannot be deduced from the shape and characteristic points of the curves. Analytic solutions have been obtained only for some special methods of introducing the label [13, 14].

We can find f and D_* from experiments with two tracers, one of which cannot enter the secondary pores, as in Fig. 1c, where the SO_4^{2-} cannot enter the fine pores, because of the negative charge on the surfaces, and so K ≈ 0, which means that (4) becomes as for convective diffusion. This failure of the anion to enter the pores was confirmed also by tests under static conditions. The hatched region in Fig. 1c represents the volume of water in the secondary pores, while the total porosity is characterized by the area between the ordinate axis and the C(T) curve for HTO. The results are n = 0.715 (total porosity) and m = 0.482 (intergranular porosity). For this system f = (n − m)/n = 0.326, which agrees well with the water content of the secondary pores as calculated from the porosity of the silica gel granules.

To find D_*, the C(T) curve for ^{35}S must be redrawn so that T = 1 corresponds to the volume of the intergranular pores. The results are D_* = 5.7 × 10^{-4} cm^2/sec and λ = 0.48 mm. The next step is calculation of K, which may have molecular and convective components.

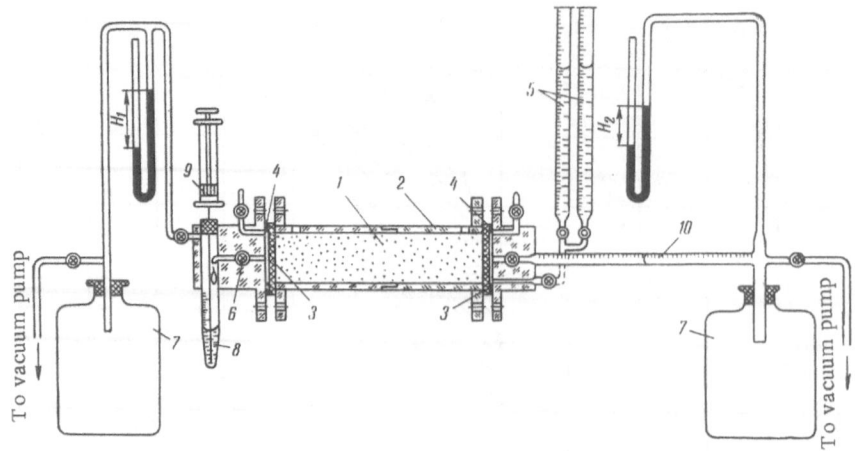

Fig. 2. Apparatus for radiochromatographic examination of the structure of unsaturated porous media.

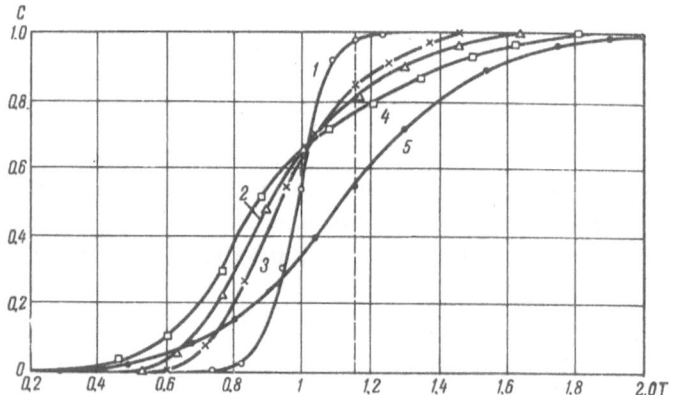

Fig. 3. C = f(T) curves for 0.2–0.3 mm quartz
sand (l = 3.2 cm, n = 0.38) for various degrees
of water saturation U: 1) 1; 2) 0.53; 3) 0.52; 4)
0.25; 5) 0.17.

In the latter case (K large) we can use simple solutions to the equation of convective
diffusion by including the volume of the blind pores in the total porosity.

It is of interest to apply radiochromatography to flow in incompletely saturated porous
media. The apparatus of Fig. 2 was used. The specimen 1 is mounted [15, 16] in the cylindrical body 2 between the sintered filters 3, which contact the small water-filled slots 4, which
connect the inlet to the calibrated capillary 10 and the outlet to the overflow 6. A reduced
pressure is set up in the capillary and measuring tube 8 by connection to the ballast vessels 7,
which allows the specimen to be drawn off to the required water content W. Gauges record the
corresponding reduced pressure P in the pore water.

When the required water content W(P) has been set up, an appropriate pressure difference[†] $\Delta P = H_2 - H_1$ is applied, which causes the water to move through the specimen. The flow

_____

[†] The pressure difference must not be large, in order not to set up substantial gradients in W;
usually, ΔP was 3–4 mm Hg.

TABLE 2. Structural Parameters of Unsaturated Porous Bodies

U	$v \cdot 10^3$, cm/sec	Pe	$D_* \cdot 10^4$, cm^2/sec	λ, μ	f
Quartz sand 0.2-0.3 mm; n = 0.38; l = 3.2 cm					
1	6.55	314	0.67	102	0
0.53	0.94	65	0.46	492	0.043
0.52	0.47	50	0.30	640	0.057
0.25	0.66	45	0.47	712	0.064
Quartz sand < 0.17 mm; n = 0.41; l = 3.2 cm					
1	3.7	186	0.64	172	0
0.38	0.48	46	0.33	695	0.060
0.21	0.25	39	0.21	813	0.075

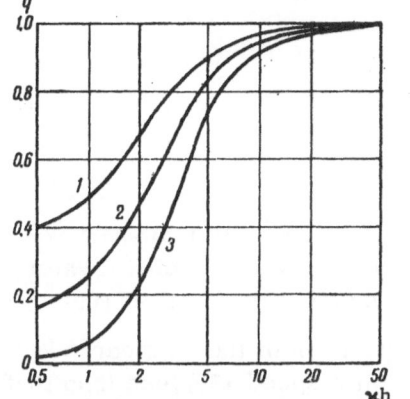

Fig. 4. Curves for q($\varkappa$h) obtained by numerical integration of (7) with zψ_s (mV) of: 1) 25; 2) 50; 3) 100.

rate is determined via the calibrated capillary and measuring tube. The steady-state flow rate provides determination of the water permeability and also the dependence on ΔP.

When the steady state has been reached, device 5 is operated to replace rapidly the pore water in the inlet chamber and capillary by the tracer solution. Burettes 5 also serve to supply the liquid to the capillary. Samples are taken with the syringe 9 via the rubber seal on tube 8 and are processed in the usual way [1] for tracer analysis. All parts are made of lucite and glass, which do not adsorb Na_2S*O_4.

Figure 3 shows C(T) curves for quartz sand with various degrees of water saturation

$$U = \overline{W}/W_0,$$

in which $\overline{W}$ is the mean water content and W_0 is the water content in the completely saturated state.

Table 2 shows that λ for quartz sand increases as U decreases, and so decrease in U increases the spread in the flow speed, on account of coexistence of film and capillary transport mechanisms. The symmetry in the C(T) curves shows that the equations of convective diffusion are applicable, since the content of immobile water, as calculated by the method of [1], is small for unsaturated quartz sand. The f = (n − m)/n (in which m is here the active porosity) increase as U decreases (Table 2). A similar trend in f with U has been observed [16] in experiments with a salt tracer.

Curve 5 of Fig. 3 (low U) corresponds to complete interruption of communication between capillary columns of film water, and this causes the C(T) curve to shift to the right relative to the axis T = 1, which occurs because thin films of liquid cause separation of anionic tracer [17, 18]. Of course, the results of such experiments cannot be interpreted in terms of the equations of convective diffusion, since the rate of transport of the tracer no longer corresponds to that of the water; but they can be used to examine the structure of diffuse ionic layers in thin water films.

The solution for the case of overlapping diffuse layers [19] gives the following concentration distribution for anions in thin films:

$$\frac{c_-}{c_0} = \exp\left[-\frac{zF\psi_s}{RT} \cdot \frac{\operatorname{ch} \varkappa (h-y)}{\operatorname{ch} \varkappa h}\right], \tag{5}$$

in which z is ion valency, F is the Faraday constant, ψ_s is surface potential, $1/\varkappa$ is the Debye thickness of the double layer, y is a coordinate reckoned from the surface of the particle, h is film thickness, and c_0 is the anion concentration in the bulk solution.

The reduction in the transport rate of an anion tracer in thin films is

$$q = \frac{Q_-}{Q_0} = \int_0^h cv\, dy \ \Big/ \int_0^h c_0 v\, dy. \tag{6}$$

If we simplify the treatment by neglecting the effect of the flow potential on the velocity profile (Poiseuille flow), we get

$$q = -3 \int_0^1 \exp\left[-\frac{zF\psi_s}{RT} \cdot \frac{\operatorname{ch} \varkappa h (1-Y)}{\operatorname{ch} \varkappa h}\right]\left(\frac{Y^2}{2} - Y\right) dY, \tag{7}$$

in which Y = y/h.

Figure 4 shows numerical results for q for three values of $z\psi_s$. The bulk transport velocity of the anions for $\varkappa h$ small becomes appreciably less than the flow velocity of the water, so an anion tracer can be used to examine the structure of porous bodies only for $\varkappa h \geq 10$.

The broken vertical line at T = 1.15 in Fig. 3 indicates the position of the axis corresponding to the mean transport velocity of the tracer, which moves at a speed 15% less than that of the water (T = 1), i.e., q = 0.85. Then z = 2 and $\psi_s = -25$ mV for quartz sand give $\varkappa h = 5$ from Fig. 4. Taking $\varkappa = 10^6$ cm^{-1} for pore water, we find that h $\approx$ 500 Å. Similar values are given by the Deryagin–Nerpin equation P = A/h^3; if we take A = 5×10^{-12} erg [20] with P = 48 cm Hg, we get h $\approx$ 470 Å. That is, study of anion transport for various h produced by variation of U can provide information on the thickness of the diffuse ion atmospheres in wetting films in pores.

In research on water transfer in media with porous particles it is desirable, as in research on filtration processes, to use double tracer. These results indicate possible tracer methods of research on the structure of porous bodies and on mass transfer.

References

1. M. P. Volarovich, N. V. Churaev, and B. M. Minkov, Dokl. AN SSSR, 114:964 (1957); N. V. Churaev, in: New Physical Methods of Research on Peat, Gosenergoizdat, Moscow (1960), p. 125.
2. L. V. Radushkevich, Dokl. AN SSSR, 57:471 (1947).
3. V. V. Rachinskii, Dokl. Timir. Sel'sk. Akad., No. 29, p. 89 (1957).
4. V. N. Nikolaevskii, Prikl. Matem. i Mekh., 23:1042 (1959).
5. A. É. Sheidegger, Physics of the Flow of Liquids through Porous Media, Gosstroiizdat, Moscow (1960).
6. N. V. Churaev and N. I. Gamayunov, Inzh.-Fiz. Zh. Akad. Nauk Belorussk. SSR, 4:106 (1961).
7. R. Aris and N. R. Amundson, Am. Inst. Chem. Eng. J., 3:280 (1957).

8. J. J. Carberry and R. H. Bretton, Am. Inst. Chem. Eng. J., 4:367 (1958).
9. N. V. Churaev and N. I. Il'in, Inzh.-Fiz. Zh. Akad. Nauk Belorussk. SSR, 4:44 (1961).
10. V. V. Rachinskii and L. A. Lenskii, Dokl. AN SSSR, 162:380 (1965).
11. R. C. Goodknight, W. A. Klikoff, and J. Fatt, J. Phys. Chem., 64:1162 (1960).
12. K. H. Coats and B. D. Smith, Soc. Petrol. Engrs. J., 4:73 (1964).
13. G. A. Turner, Chem. Eng. Science, 7:156 (1958).
14. V. G. Levich, V. S. Markin, and Yu. A. Chizmadzhev, Dokl. AN SSSR, 166:1401 (1966).
15. L. A. Richards and B. D. Wilson, J. Am. Soc. Agron., 28:427 (1936).
16. J. W. Biggar and D. R. Nielsen, J. Soil Sci., 12:188 (1961).
17. M. Tschapek, Z. Pflanz. Düng.; Bodenkunde, 102:193 (1963).
18. L. Dresner and K. A. Kraus, J. Phys. Chem., 67:990 (1963); L. Dresner, J. Phys. Chem., 69:2230 (1965).
19. B. V. Deryagin, Izv. AN SSSR, OMEN, Ser. Khim., No. 5, p. 1153 (1937).
20. S. V. Nerpin, Trudy Leningr. Inst. Inzh. Vodnogo. Transporta, No. 21, p. 126 (1954).

RADIOISOTOPE RESEARCH ON WATER FILM TRANSPORT IN POROUS BODIES

N. V. Churaev, A. E. Afanas'ev, and N. I. Gamayunov

Surface Phenomena Division, Institute of Physical Chemistry,
Academy of Sciences of the USSR;
Kalinin Polytechnical Institute

A description is given of tracer observations on the evaporation of water from porous bodies with various structures. It is shown that film flow and evaporation from the surfaces of films are the basic mechanisms of water transport in hydrophilic porous media.

The object of trace studies is to determine the rates and mechanisms of water transport (capillary and film flow, vapor diffusion) in relation to the structure of porous bodies and drying conditions. For this purpose, use is made of the distribution of the tracer within the body at various stages of the process [1-4] and also of measurements of tracer concentration at the outer (evaporation) surface [3, 5].

In the latter case, use is made of tracers emitting soft β-rays (^{35}S, ^{14}C) in order to record only the emission from a thin surface layer of liquid. The count rate N recorded by an end-window counter is dependent on the tracer and water contents of the surface layer. The effects of water content were examined by measuring N with a constant concentration of the tracer in the solution but with various water contents, which were adjusted by drawing off liquid from the specimen via a porous membrane. Results for quartz sand (various grain sizes) showed that N remains almost constant if the capillary pressure becomes less than P_k, the pressure corresponding to freeing of the smallest of the communicating pores. The reason is that the surface water content W_* is only slightly dependent on the pressure for $P < P_k$; it hardly varies throughout the first period of constant evaporation rate. A pressure $P < P_k$ in the pore water is set up from the start of drying,[†] so the measured N may be considered (during the first period of drying) as proportional to the tracer concentration in the surface layer.

Diffusion of the tracer also affects the results. If a solution of a nonvolatile radioisotope tracer is introduced into the liquid, the tracer accumulates at the surface during evaporation. The equation for convective diffusion may be used to obtain the following relation [6] in a linear

---

[†] As is evident from direct measurements of the capillary pressure.

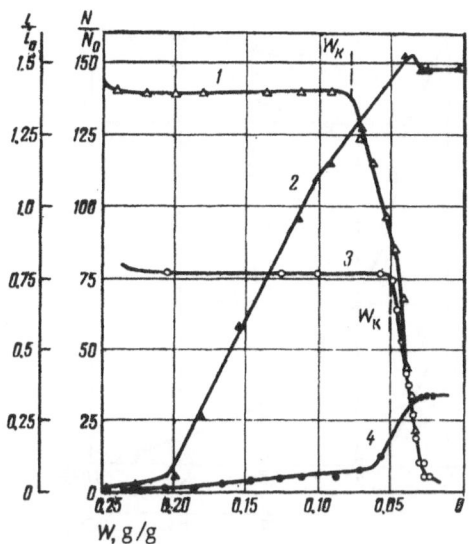

Fig. 1. 1, 3) Evaporation rate i/i_0; 2, 4) surface tracer concentration N/N_0 during drying of quartz sand at φ of: 1, 2) 0.5; 3, 4) 0.9.

approach for sufficiently large times τ and a constant rate v of movement of the liquid:

$$\frac{c_*}{c_0} \simeq 2 + \frac{v^2\tau}{D}, \qquad (1)$$

in which D is diffusion coefficient, c_* is the tracer concentration in the surface layer at time τ, and c_0 is the concentration at $\tau = 0$.

Then the following equation describes the distribution of the tracer concentration c in a porous body near the evaporation surface:

$$\frac{c}{c_0} \simeq 1 + \left(1 + \frac{v^2\tau}{D} - \frac{vx}{D}\right) \exp\left(-\frac{vx}{D}\right), \qquad (2)$$

in which x is distance reckoned from that surface.

Experiment shows that D is constant at $\sim 10^{-5}$ cm^2/sec (^{35}S-tagged Na$_2$SO$_4$) for water-saturated quartz sand with $v < 2 \times 10^{-5}$ cm/sec. Hence, only molecular diffusion of the tracer influences transport of the latter at low evaporation rates. Then D allows us to deduce the evaporation rate at the surface from the slope of c_*/c_0 as a function of τ.

However, the D for ions in thin films and pores can differ appreciably from the bulk values, so we must use D corresponding to various W_* in order to examine evaporation from an incompletely saturated porous medium. For instance, (1) gives for quartz sand that $D \approx 5 \times 10^{-8}$ cm^2/sec for $W_* \approx 1\%$ but $D \approx 5 \times 10^{-7}$ cm^2/sec for $W_* \approx 2\text{-}4\%$, which agree with D for these W_* as deduced by other methods [6].

Figures 1 and 2 show results for evaporation of water containing ^{35}S-tagged Na$_2$SO$_4$ from quartz sand (grain size $< 170\ \mu$) initially completely saturated. The specimens were 12 mm in diameter and 120 mm high, and were placed in test tubes in a chamber kept at a constant relative humidity φ and $26 \pm 0.1°$C. The specimens were weighed periodically, and measurements were made of the specific activity of the surface with an MST-17 end-window counter coupled to a B-2 radiometer. We used 3 or 4 identical specimens in each series, which were sampled by layers at various stages (at intervals of 2-5 mm in the upper parts and 10-20 mm in the lower ones), W_* and c in the pore water being determined for each layer by methods previously described [3, 7, 8].

The results gave the relative evaporation rate i/i_0 (curves 1 and 3 of Fig. 1) and the surface activity N/N_0 (curves 2 and 4) as functions of the mean water content $\overline{W}$ of the specimen. The evaporation rate i_0 for the surface of water was measured in the chamber under the same conditions. The count rate N_0 was measured before the start of drying. Figure 2 shows the distribution of the relative water content W/W_0 (W_0 corresponding to complete saturation) and of c/c_0 (c_0 being the initial tracer concentration in the pore water). The specimens were sampled at various $\overline{W}$.

Figures 1 and 2 show that the tracer accumulates in a relatively thin surface layer when the evaporation rate is constant (parts a and b of Fig. 2), and the inside concentration remains constant at c_0 within a dried-out specimen. This shows that there is no evaporation from the retreating meniscus front, whose position is clear from the curves for W/W_0. The part between the meniscus front

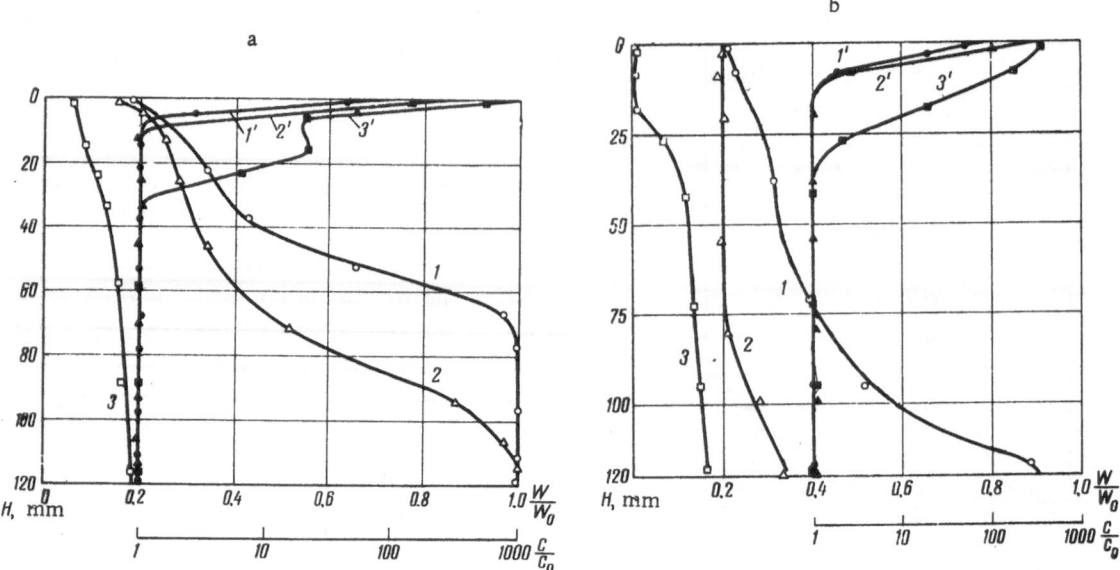

Fig. 2. Distribution of W/W_0 (curves 1-3) and c/c_0 (curves 1'-3') in the height H of specimens of quartz sand differing in $\overline{W}$: a) $\varphi = 0.5$; 1, 1') $\overline{W}$ of 0.18 g/g; 2, 2') $\overline{W}$ of 0.12; 3, 3') $\overline{W}$ of 0.034; b) $\varphi = 0.9$; 1, 1') $\overline{W}$ of 0.106 g/g; 2, 2') $\overline{W}$ of 0.051; 3, 3') $\overline{W}$ of 0.025.

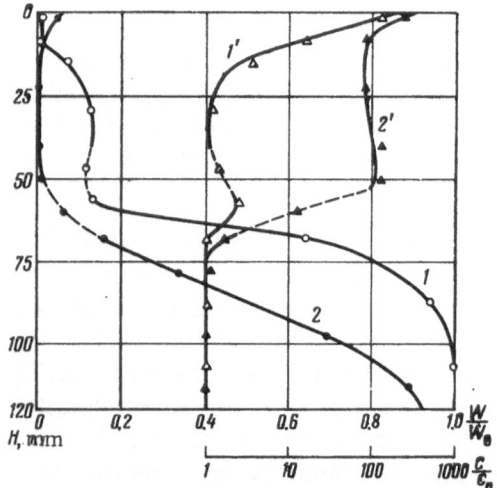

Fig. 3. 1, 2) W/W_0; 1', 2') c/c_0 as functions of height H in specimens of quartz sand of particle size 0.25-0.5 mm at $\varphi = 0.5$ and 26°C for $\overline{W}$: 1, 1') 0.1 g/g; 2, 2') 0.062 g/g. The broken lines indicate the positions of the hydrophobic bands.

and the surface has water transported not by diffusion (as was previously supposed) but as liquid. The isotherms $P(\overline{W})$ for the capillary pressure show that the water content of the surface layers corresponds to the state where the capillary droplets are separated by films, and the main transport mechanism is film flow. This mechanism is found to predominate in these granular systems of quartz sand type for mean particle sizes from 1 mm to 50 μ for φ of 0.2-0.95 and t of 20-30°C. This confirms the theory of evaporation from single capillaries [9, 10] and capillary-porous bodies [11], in which allowance is made for film flow of the liquid over the pore walls.

The tracer distribution for $\varphi = 0.5$ (Fig. 2a) is described satisfactorily by (2), so evaporation occurs from the outer surface of the specimen, the rate being determined by the conditions of external mass transfer and does not alter while the film mechanism maintains a sufficient flow of water to the evaporation surface.

Curves 1 and 2 of Fig. 1 show that the tracer concentration in the surface layer increases linearly during this period, which confirms that the film flow rate is constant. This continues until the meniscus front reaches the end of the specimen, after which the surface layers start to dry out. The films become thinner and so transport less readily, and there is a sharp fall in the drying rate. Evaporation begins to occur from the surfaces of the films and droplets within the surface layers. The evaporation zone, which corresponds to an elevated tracer concentration, is clear in Fig. 2a (curve 3'). Then the first critical point $\overline{W}_k$ on the drying-rate curves (Fig. 1) corresponds to transition from film transport to vapor diffusion in the surface layers.

In the second drying period, evaporation occurs from the receding front of the droplets and films. While this front is not far from the surface, vapor diffusion within the specimen is accompanied by film flow, as is clear from the continuing transfer of tracer to the outer surface. Then the film flow to the surface ceases, and the surface tracer concentration alters no further (curve 2 of Fig. 1). In the dried-out zone that occurs for $\overline{W} < \overline{W}_k$ (curves 3 of Fig. 2a), vapor diffusion is the predominant transport mechanism. Film flow continues below the film and droplet front, since the tracer concentration there remains at c_0.

Although the sand has only $\overline{W}$ of 0.03-0.04 g/g, and there are pores free from water, film flow is the main transport mechanism. Thin liquid films transport water over the particle surfaces in the parts between contiguous liquid columns (droplets), so the flow is not of pure film type; but the film parts produce the main resistance to the flow in such systems [12].

For $\varphi = 0.9$, the same general scheme applies (Fig. 2b), the only difference being that the water evaporates from a surface layer several mm thick instead of from the outer surface. This follows from the fact that the tracer concentration at the outer surface rises only slightly (curve 4 of Fig. 1), although the amount of tracer carried to the surface layer is as before (Fig. 2, a and b). The sharp rise in activity at the transition to the second period of drying (curves 3 and 4 of Fig. 1) occurs because the film flow brings to the outer surface the concentrated solution from the layers where the evaporation occurred. An extended evaporation zone at $\varphi = 0.9$ was observed also for sand of particle size 0.2-0.3 mm. No change in transport mechanism at $\varphi = 0.9$ was observed for coarse sand (0.5-1 mm).

Figure 3 shows results for the evaporation of tagged water from specimens containing 13-15 mm layers of sand made hydrophobic. There is a rise in tracer concentration under the hydrophobic band, in which film transport has been artificially impaired. Transfer through the hydrophobic layer occurs mainly by vapor diffusion, which causes $\overline{W}_k$ to shift to values higher than those for the unaltered sand. The rate of water loss is also reduced. Film transport from the lower layers cannot occur, so the upper layers dry out completely (curve 2) although the lower ones remain highly impregnated with water.

These results show that film flow must be taken into account in considering water transport in porous media, as it is a basic transfer mechanism. The film flow rate often determines the drying kinetics, and this rate is dependent on the hydrophilicity of the particle surfaces as well as on the pressure and composition of the solution. There is a need for further studies of the static and dynamic pressures exerted by wetting films as the basis for a quantitative theory of evaporation from hydrophilic porous bodies.

References

1. W. A. Headley and R. Eisenstadt, Heating, Piping, and Air Conditioning, 25:111 (1953).
2. A. I. Veinik and A. S. Shubin, Tr. Mosk. Tekhnol. Inst. Pishchevoi Prom., No. 8, p. 22. (1957).
3. N. V. Churaev, Kolloidn. Zh., 22:631 (1960).
4. A. M. Globus, Pochvovedenie, No. 9, p. 105 (1961).
5. N. V. Churaev, Inzh.-Fiz. Zh. Akad. Nauk Belorussk. SSR, 5:41 (1962); 6:31 (1963).
6. A. E. Afanas'ev, N. I. Gamayunov, and N. V. Churaev, The First Research Conference of Kalinin Young Scientists, Izd. Mosk. Rabochii, Kalinin (1967).
7. M. P. Volarovich and N. V. Churaev, in: Research in Surface Forces, Vol. 2, Consultants Bureau, New York (1966), p. 212.

8. B. V. Deryagin, S. V. Nerpin, and N. V. Churaev, Papers at the 8th International Soil Science Congress, Izd. Nauka, Moscow (1964), p. 43.

9. N. V. Churaev, Dokl. AN SSSR, 148:1361 (1963).

10. B. V. Deryagin, S. V. Nerpin, and N. V. Churaev, Kolloidn. Zh., 26:301 (1964).

11. N. V. Churaev, Kolloidn. Zh., 27:908 (1965).

12. S. V. Nerpin and M. K. Mel'nikova, in: Aspects of Agrophysics, Izd. VASKhNIL, Leningrad (1957), p. 41.

EFFECTS OF THE YIELD STRESS OF
A LIQUID ON SURFACE PHENOMENA

N. F. Bondarenko and S. V. Nerpin

Agrophysical Research Institute,
Lenin All-Union Agricultural Sciences Academy,
Leningrad

Glass and quartz capillaries have been used in measuring the yield stress for water, ethanol, and KCl solutions. The effects of shear strength are discussed for the equilibrium and movement of thin films and for the interaction of particles in a dispersion medium.

Introduction

Only the viscosity is usually considered in relation to particle interaction and liquid movement in a porous body, although effects' from the yield stress have been observed [1] in the filtration of water. The yield stress provides an explanation of deviations from Darcy's law and the nature of yield stress for suspensions has been discussed in [2, 3]. However, we have shown that the Buckingham—Reiner equation [4] describes the flow of a pure liquid through glass capillaries. The yield stress in that case cannot be ascribed to a quasi-crystalline lattice of colloidal particles.

Methods and Apparatus

The measurements were made in a thermostat with the systems shown schematically in Fig. 1. In the case of the water thermostat shown in Fig. 1a, the speed of the meniscus 4 in the capillary 3 (radius R = 0.9 mm) gives the volume Q of liquid flowing through the capillary 2 and the velocity V:

$$V = \frac{Q}{\omega t},$$

in which ω is the cross section of capillary 2 and t is time.

In the air thermostat of Fig. 1b, the liquid flows from vessel 1 under a set pressure via capillary 2 into vessel 3, which is supported on the balance 4. The cross sections of vessels 1 and 3 are so large that the difference in level between them remains constant to 0.1 mm during

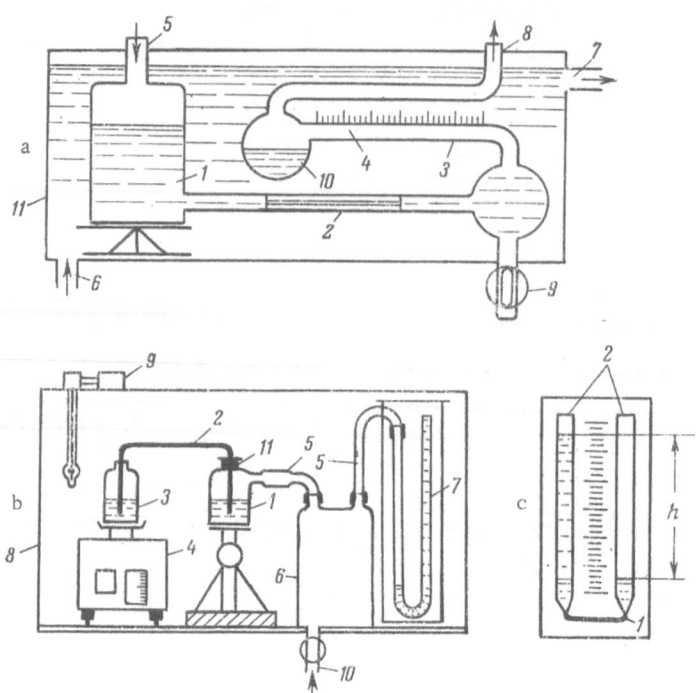

Fig. 1. Filtration systems: a: 1) glass vessel containing liquid; 2) capillary; 3) measuring capillary; 4) meniscus; 5) from pressure source and gauge; 6) to thermostat; 7) overflow; 8) to atmosphere; 9) drainage stopcock;
10) saturator; 11) thermostatic jacket; b: 1) vessel containing liquid; 2) capillary; 3) receiver; 4) VTK-500 balance; 5) rubber connecting tubes; 6) ballast space; 7)
liquid manometer; 8) thermostat; 9) regulator; 10) to
pressure source; 11) rubber bung; c: 1) capillary; 2)
quartz piezometers.

a measurement. The mass P of the transferred liquid gives

$$V = \frac{P}{\rho \omega t},$$

in which ρ is the density of the liquid at the set temperature.

The pressure differential H in systems 1a and 1b was measured to 0.2 mm of water with
a liquid manometer.

Figure 1c shows an apparatus made of quartz, which was suitable for use in air and
water thermostats; here V is defined by

$$V = \frac{(h_0 - h_1)\,\Omega}{(t_0 - t_1)\,2\omega},$$

in which Ω is the cross section of piezometers 2, while h_1 and h_2 are the level differences h at
instants t_0 and t_1 respectively.

The evaporation rate was monitored from the change in position of the meniscus (Fig.
1a and 1c) or the change in weight (Fig. 1b) at zero pressure difference [5]; in the same way,
it was shown that there were no capillary-osmotic and thermoosmotic flows. Various freshly

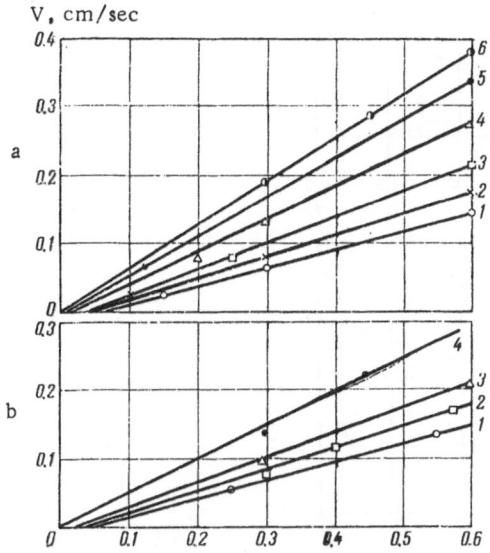

V, cm/sec

Fig. 2. Velocity V as a function of the hydraulic gradient I for system a of Fig. 1: a) water, r = 51 μ, glass No. 16 at temperatures (°C) of: 1) 15; 2) 20; 3) 30; 4) 40; 5) 50; 6) 55; b) 95% ethanol, r = 57 μ, glass No. 16 at temperatures (°C) of: 1) 25; 2) 30; 3) 40; 4) 50.

drawn capillaries were employed, in particular silica ones, with radii r from 51 to 265 μ and lengths L from 5 to 60 cm. High-purity double-distilled water was used, which was produced by a Polish Re-5 still. A specific conductivity less than 10^{-6} ohm$^{-1} \cdot$ cm^{-1} was produced by vacuum distillation (without boiling) of this water in a silica system. The instrument was filled with water directly from the condenser.

Results and Discussion

The observed V for small hydraulic gradients for water and ethanol are shown in Fig. 2 (apparatus a with trailing meniscus), Fig. 3 (apparatus b), and Fig. 4 (apparatus c).

The Buckingham−Reiner equation for V(I) may be put as

$$V = K_0 I \left[\frac{1}{3} \left(\frac{I_0}{I} \right)^4 - \frac{4}{3} \left(\frac{I_0}{I} \right) + 1 \right]. \tag{1}$$

Here K_0 is the coefficient for strictly Newtonian behavior, I is the hydraulic gradient, and I_0 is the threshold gradient.

The experimental I were such that $\frac{1}{3}(I_0/I)^4 \ll \frac{4}{3}(I_0/I) < 1$, and so (1) can be replaced by

$$V = K_0 \left(I - \frac{4}{3} I_0 \right). \tag{1'}$$

The observed V and the K_0 defined by $K_0 = r^2 \rho g / 8\eta$ (in which g is the acceleration due to gravity and η is viscosity) may be used to find I_0 from (1') for low values of I. In fact, I = H/L, in which H is the set pressure difference.

In the case of the apparatus of Fig. 1a,

$$H = H_M + H_R,$$

in which H_m is the pressure difference indicated by the manometer reading, $H_R = 2\sigma \cos \theta / R \rho g$ is the pressure difference under the surface of meniscus 4, σ is surface tension of the liquid, and θ is angle of contact.

We took θ as zero for a retreating meniscus. Also, H = H_m for the apparatus of Fig. 1b, while for Fig. 1c

$$I = \frac{h_0 + h_1}{2L}.$$

The full straight lines in Figs. 2-4 have slopes of K_0 and intercepts of $4I_0/3$. These figures show that the observed points lie close to the lines constructed from (1').

The yield stress τ_0 is defined by

$$\tau_0 = \frac{1}{2} r \rho g I_0. \tag{2}$$

Table 1 gives the results for τ_0 calculated for water from Figs. 2-4, while Table 2 gives results for 95% ethanol.

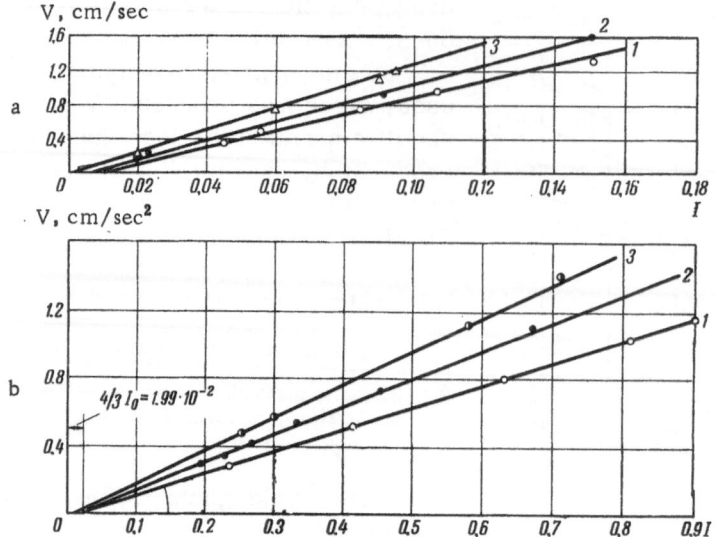

Fig. 3. Relation of V to I for apparatus of Fig. 1b: a) water, r = 265 μ, No. 16 glass at temperatures (°C) of: 1) 20; 2) 30; 3) 40; b) water, r = 103.5 μ, quartz at temperatures (°C) of: 1) 20; 2) 30; 3) 40.

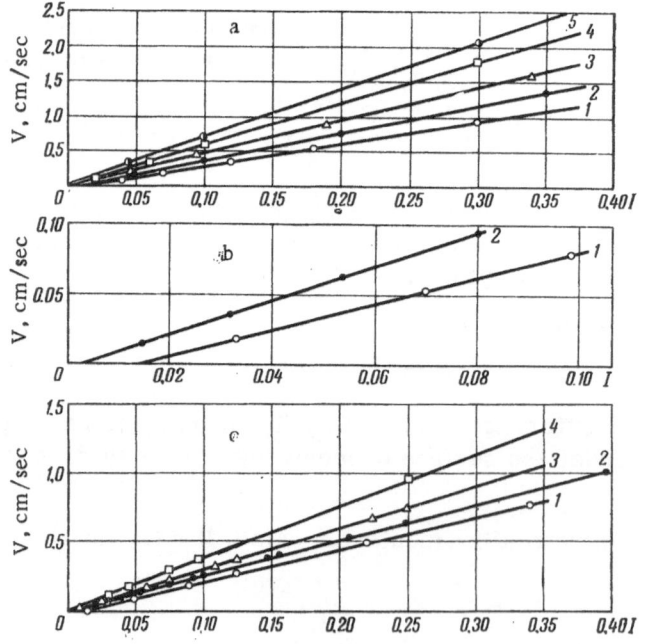

Fig. 4. Relation of V to I for apparatus of Fig. 1c: a) water, r = 165 μ, quartz, temperatures (°C) of: 1) 20; 2) 30; 3) 40; 4) 50; 5) 58; b) water, r = 94 μ, quartz, at temperatures (°C) of; 1) 20; 2) 30; c) 95% ethanol, r = 165 μ, quartz, at temperatures (°C) of: 1) 25; 2) 30; 3) 40; 4) 50.

TABLE 1. Yield Stress τ_0 for Water

Capillary radius and glass type	$\tau_0 \times 10^2$, dyn/cm^2					Apparatus
	15°C	20°C	30°C	40°C	50°C	
51 μ, No. 16 glass	11.5	9.5	8.0	3.5	1.0	1a
103.5 μ, quartz	–	7.6	6.4	4.9	–	1b
265 μ, No. 16 glass	–	7.0	4.0	3.2	–	1b
94 μ, quartz	–	8.6	2.6	–	–	1c*
165 μ, quartz	–	8.9	6.9	6.0	1.6	1c*
Mean τ_0	11.5	8.1	5.8	4.5	1.3	

*Water of specific conductivity less than 10^{-6} ohm$^{-1} \cdot$ cm^{-1} was used.

TABLE 2. Yield Stress τ_0 for Ethanol

Capillary radius and glass type	$\tau_0 \times 10^2$, dyn/cm^2				Apparatus
	25°C	30°C	40°C	50°C	
57 μ, No. 16 glass	10.5	7.6	5.5	2.0	1a
165 μ, quartz	7.0	3.0	1.2	1.0	1c
Mean τ_0	8.75	5.3	3.3	1.5	

Figure 5 shows $\tau_0(T)$ plotted from the mean values. There was no obvious dependence of τ_0 on r or on the grade of glass. Some experiments were done with aqueous KCl (0.01, 0.1, and 1 M) in the apparatus of Fig. 1c at 20°C with a capillary having r = 198 μ in order to establish the effects of an electrolyte.

The results show that the mean τ_0 (derived from three runs) were virtually independent of KCl concentration:

Electrolyte concentration	$\tau_0 \times 10^2$, dyn $\cdot$ cm^2
Double-distilled water	9.50
10^{-2} M	9.25
10^{-1} M	9.40
1 M	9.50

Similar results have been reported previously [1]. We may suppose that τ_0 for a pure liquid is related to the capacity to form molecular spatial structures via various types of bond. Intermolecular hydrogen bonds are responsible [8, 9] for the formation of polymer-type chains† in many compounds (water, alcohols, phenols, proteins, polypeptides, and polyhydroxy compounds, organic or inorganic. Some of these bonds tend to break as the temperature is raised, and the equilibrium between the free and associated species tends to shift towards the former, which is bound to affect the rheological behavior. This is confirmed by our results for $\tau_0(T)$ for pure water and ethanol.

---

† The elastoviscous properties of aqueous glycerol are explained [6, 7] by the quasi-polymer chain structure formed via hydrogen bonds.

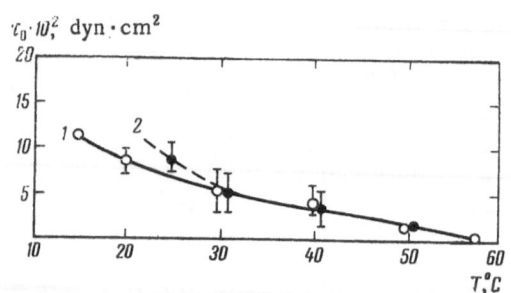

Fig. 5. Yield stress τ_0 vs temperature:
1) water; 2) 95% ethanol.

Effects of Yield Stress

on the Equilibrium and

Movement of Thin Liquid Films

and on Particle Interaction

in a Dispersion Medium

Under certain conditions, even the slightest traces of yield stress can play a major part in the equilibrum or movement of thin films of liquid, e.g., in filtration through a porous medium, or in the movement of water in the unsaturated zone of a soil.

We may take the equations of hydrodynamics (without quadratic terms) as being as follows for steady-state movement of a thin film along the y axis:

$$\frac{\partial P}{\partial y} - Y = \eta\left(\frac{\partial^2 V_y}{\partial x^2} + \frac{\partial^2 V_y}{\partial z^2}\right), \tag{3}$$

$$\frac{\partial V_y}{\partial y} = 0, \tag{4}$$

in which P is the pressure in the layer, xyz are the coordinate axes, Y is the projection of the bulk force on the y axis, V_y is the component of the velocity, and η is viscosity.

If we consider the motion at a point where the effects of the boundaries in the direction of one of the axes (e.g., z) are negligible, we neglect $\partial^2 V_y / \partial z^2$ to get

$$\frac{\partial P}{\partial y} - Y = \eta\frac{\partial^2 V_y}{\partial x^2}. \tag{5}$$

But $\eta \partial^2 V / \partial x^2 = \partial \tau / \partial x$, in which τ is shear stress, so (5) can be replaced by

$$\frac{\partial P}{\partial y} - Y = \frac{\partial \tau}{\partial x}. \tag{5'}$$

If the terms on the left are independent of x, the solution to (5') is

$$\tau = \left(\frac{\partial P}{\partial y} - Y\right) x + C. \tag{6}$$

If the liquid moves in a layer bounded by two parallel planes (separation h), we may place the origin half-way between the planes. Then, from symmetry, we have for x = 0 that

$$\tau = 0. \tag{7}$$

This gives C = 0, so (6) becomes

$$\tau = \left(\frac{\partial P}{\partial y} - Y\right) x.$$

At the boundary at x = h/2

$$\tau_b = \left(\frac{\partial P}{\partial y} - Y\right)\frac{h}{2}. \tag{8}$$

If the layer of liquid is bounded on one side by a solid and on the other by a gas, we place the origin at the gas–liquid interface and direct the x axis into the layer; then from condition (7) we again have C = 0, and so (6) is replaced by

$$\tau = \left(\frac{\partial P}{\partial y} - Y\right) x.$$

The boundary of the solid lies at x = h, and here

$$\tau_b = \left(\frac{\partial P}{\partial y} - Y\right) h. \tag{9}$$

If the liquid moves in a circular tube of radius r_t, with the y axis along the axis of the tube parallel to the flow, the Navier–Stokes equation takes the following form in cylindrical coordinates:

$$\frac{\partial P}{\partial y} - Y = \eta \left(\frac{\partial^2 V}{\partial r^2} + \frac{1}{r}\frac{\partial V}{\partial r}\right) \tag{10}$$

or, as

$$\eta \left(\frac{\partial^2 V}{\partial r^2} + \frac{1}{r}\frac{\partial V}{\partial r}\right) = \frac{'\partial \tau}{\partial r} + \frac{\tau}{r},$$

we may put

$$\frac{\partial P}{\partial y} - Y = \frac{\partial \tau}{\partial r} + \frac{\tau}{r}. \tag{10'}$$

Solution of (10') gives

$$\tau = \frac{1}{r}\left[\frac{r^2}{2}\left(\frac{\partial P}{\partial y} - Y\right) + C\right].$$

Now $\tau = 0$ at r = 0, so C = 0 and

$$\tau = \frac{r}{2}\left(\frac{\partial P}{\partial y} - Y\right).$$

At the wall, $r = r_t$, and

$$\tau_b = \frac{r_t}{2}\left(\frac{\partial P}{\partial y} - Y\right). \tag{11}$$

The Shvedov–Bingam law for a viscoplastic body gives

$$\tau = \tau_0 + \eta \frac{\partial V}{\partial n}$$

in which τ_0 is the yield stress and $\partial V/\partial n$ is the derivative of the velocity along the normal to this; then the liquid is motionless throughout the space between the bounding surfaces if $\tau_b = \tau_0$.

The limiting conditions for onset of motion are thus as follows:

1. For a symmetrical layer bounded by two fixed planes

$$\left(\frac{\partial P}{\partial y} - Y\right)\frac{h}{2} = \tau_0. \tag{12}$$

2. For an unsymmetrical layer (solid on one side, gas on the other)

$$\left(\frac{\partial P}{\partial y} - Y\right) h = \tau_0. \tag{13}$$

3. For a cylindrical capillary

$$\left(\frac{\partial P}{\partial y} - Y\right) \frac{r}{2} = \tau_0. \tag{14}$$

Conditions (12)-(14) show that the yield stress can have very considerable effects for thin layers. Consider, for example, the flow of a liquid down a vertical plane. Here[†] $\partial P/\partial y = 0$ ($P = P_g$ = constant at the boundary with the gas) and $Y = -\rho g$, so (13) is replaced by

$$\rho g h = \tau_0. \tag{15}$$

For instance, if $h = 10^{-6}$ cm and $\rho g \approx 10^3$ dyn/cm^3 it is sufficient to have $\tau_0 \approx 10^{-3}$ dyn/cm^2 for the film to remain motionless; but for this τ_0 the layer will move if there is a thickness gradient, even a very small one.

If the layer is horizontal but of uneven thickness, condition (13) may [11] be put as

$$\frac{\partial (\Delta P_h)}{\partial y} = -\frac{\partial R(h)}{\partial y} = -\frac{\partial R}{\partial h}\frac{\partial h}{\partial y} = \frac{\tau_0}{h} \tag{16}$$

or

$$\frac{\partial R}{\partial h} h = \frac{\tau_0}{\alpha}, \tag{17}$$

in which ΔP_h is the pressure difference between the phases, R is the shearing pressure, and α is the inclination of the liquid−gas interface to the solid−liquid interface.

If we take $R = \xi^3/h^3$, where ξ is[‡] a constant, then (17) is replaced by

$$\alpha = \frac{\tau_0}{3R}. \tag{18}$$

For instance, if R is 10^4-10^5 dyn/cm^2 and $\tau_0 = 10^{-3}$ dyn/cm^2, it is sufficient to have α of the order 10^{-7} to 10^{-8} in order to produce motion.

We may relate h to the distance y from a reference plane for τ_0 finite. Condition (13) and equation (16) give

$$\tau_0 - \rho g h + \frac{dR}{dh}\frac{dh}{dy} h = 0. \tag{19}$$

If $R = \xi^3/h^3$, (19) becomes

$$dy = 3\xi^3 \frac{dh}{h^3 (\tau_0 - \rho g h)}. \tag{20}$$

Integration of (20) from 0 to y and from $h = \infty$ (level of the bulk liquid) to h gives

---

[†] We assume that the layer is thin, so that there is no wave production, as occurs [10] at large thicknesses.

[‡] R (which has molecular and ionic components for aqueous films of electrolyte solutions) may be represented roughly as above for a wetting unsymmetrical film, in which ξ may be determined from the isotherms for R, e.g., from experiments with gas bubbles [12].

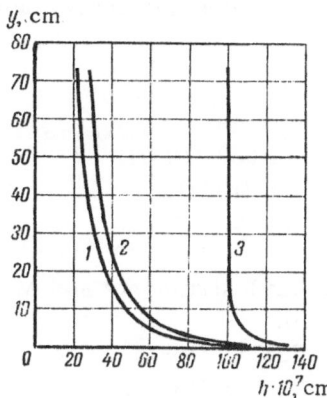

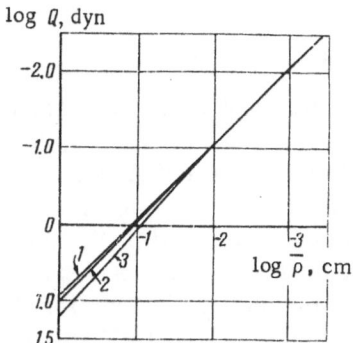

Fig. 6. Effects of τ_0 on the equilibrium distribution of a liquid on a vertical plane for τ_0 (dyn/cm^2) of: 1) 0; 2) 10^{-3}; 3) 10^{-2}.

Fig. 7. Effects of τ_0 on the state of equilibrium of two spherical particles of radius $\bar{\rho}$ for τ_0 (dyn/cm^2) of: 1) 0; 2) 10^{-1}; 3) 1.0.

$$y = -3\xi^3 \left[\frac{1}{2\tau_0 h^2} + \frac{\rho g}{\tau_0^2 h} + \frac{(\rho g)^3}{\tau_0^3} \ln\left(1 - \frac{\tau_0}{\rho g h}\right) \right]. \tag{21}$$

Integration of (20) for $\tau_0 = 0$ gives

$$y = \frac{\xi^3}{\rho g h^3}. \tag{22}$$

Comparison of (21) and (22) shows that, if $\tau_0 \neq 0$, h does not tend to zero as y increases without limit but tends to the finite limit $\tau_0/\rho g$.

Figure 6 shows curves for the layer thickness as a function of wall height for various τ_0 as deduced from (21) and (22), with ξ^3 taken as 10^{-12} dyn · cm, which corresponds to 10^{-3} M NaCl on glass [12]. It is clear that h(y) begins to be affected by the plasticity even for $\tau_0 = 10^{-3}$ dyn/cm^2 and that the equilibrium thickness of the film increases substantially with τ_0. Here we shall not consider the effects of τ_0 on the initial conditions for flow through narrow-pore filters when the pore space is completely filled by liquid, which has been examined in detail elsewhere [13].

Consider the equilibrium of two spherical particles (radius $\bar{\rho}$) separated by a thin layer of liquid. The force Q that must be exerted to bring the particles together is [14] given by

$$Q = \pi\bar{\rho}\left[\frac{\pi\varepsilon}{2\delta}\left(\frac{kT}{ze}\right)^2 - 2\sigma_{mi} + 2\tau_0\bar{\rho} \right], \tag{23}$$

in which ε is dielectric constant, δ is molecular size, k is Boltzmann's constant, T is absolute temperature, z is the electrovalency of an ion, e is electronic charge, and σ_{mi} is the coefficient for coverage of body i by liquid m.

Figure 7 shows the $Q(\bar{\rho})$ given by (23) for τ_0 of 0, 0.1, and 1 dyn/cm^2 with $\varepsilon = 80$, k = 1.38×10^{-16} erg/deg, e = 4.8×10^{-10} esu, T = 300 deg, z = 1, $\delta \approx 1.5 \times 10^{-8}$ cm, and $\sigma_{mi} \approx 30$.

It is clear from Fig. 7 that the elastoplastic resistance plays no appreciable part for $\bar{\rho} < 10^{-2}$ cm even for $\tau_0 = 1$ dyn/cm^2, and the equilibrium under these conditions is determined largely by thermodynamic factors. If $\bar{\rho} > 10^{-2}$ cm, the effects become appreciable for $\tau_0 > 1$ dyn/cm^2.

References

1. S. V. Nerpin and N. F. Bondarenko, Dokl. AN SSSR, 114:833 (1957).
2. I. F. Efremov and S. V. Nerpin, Dokl. AN SSSR, 113:846 (1957).
3. M. P. Volarovich and A. M. Gutkin, Dokl. AN SSSR, 143:896 (1962).
4. M. Reiner, Deformation, Strain, and Flow, H. K. Lewis and Co. Ltd., London (1960).
5. N. F. Bondarenko and S. V. Nerpin, in: New Measuring Instruments and Methods for Hydraulic Studies, Izd. AN SSSR, Moscow (1961), p. 49.
6. N. A. Dumanskii and L. V. Khailenko, Kolloidn. Zh., 22:277 (1960).
7. N. A. Dumanskii and L. V. Khailenko, Kolloidn. Zh., 23:684 (1961).
8. L. Pauling, Nature of the Chemical Bond, and the Structure of Molecules and Crystals; an Introduction to Modern Structural Chemistry, Cornell University Press, Ithaca, New York (1940).
9. G. Pimentel and A. McClellan, Hydrogen Bond, Freeman, San Francisco (1960).
10. P. L. Kapitsa, Zh. Éksp. Teor. Fiz., 18:3 (1948).
11. S. V. Nerpin, Tr. Leningrad. Inst. Inzh. Vodnogo Transporta, No. 25, p. 37 (1958).
12. B. V. Deryagin and M. M. Kusakov, Izv. AN SSSR, Ser. Khim., No. 5, p. 1119 (1937).
13. A. I. Kotov and S. V. Nerpin, Izv. AN SSSR, OTN, No. 9, p. 106 (1958); N. F. Bondarenko, S. V. Nerpin, and A. I. Kotov, Summaries of Papers at the Fourth All-Union Conference on Colloid Chemistry, Izd. AN SSSR, Moscow (1958), p. 35.
14. S. V. Nerpin, Dissertation, Leningrad. Inst. Inzh. Vodnogo Transporta (1956).

ION TRANSPORT IN GELATIN MEMBRANES

M. P. Sidorova, L. S. Avdeeva, and D. A. Fridrikhsberg

Zhdanov University, Leningrad

Concentration changes and polarization are reported for gelatin membranes carrying direct current. Results are reported for the effects of electrolyte agitation on the boundary polarization potential.

Passage of direct current through an electrochemically active membrane affects the electrolyte concentrations on the two sides and within the membrane. There have been numerous theoretical and experimental studies [1-5] of the behavior of homogeneous ion-exchange membranes. Although the general transport laws for homogeneous membranes have been extended to heterogeneous capillary systems of lower electrochemical activity, polarization has more effect on the properties of a capillary system because its electrochemical activity is very dependent on the concentration of the solution [6-8]. We have examined the effects of current strength, electroconvection, and agitation on ion transport through gelatin membranes.

Methods

Hardened gelatin membranes 10 mm thick and of area 3 cm^2 were made in the usual way [9] from 25% gelatin gel.

All the measurements were made in KCl solution, in which the membrane is negatively charged. We examined the diffusion, transport numbers, and electrolyte concentration in the membrane during current flow, as well as the fall in the potential when the current was switched off. All the measurements were made with the apparatus of Fig. 1, which allowed stirring of the solution near the membrane and also experiments with or without electroosmotic liquid transport. The transport numbers in the membrane were determined by an analytical method [9] and were calculated for each side of the membrane from

$$\Delta \bar{n} = \frac{\Delta m_0 \cdot F}{q},$$

in which Δm_0 is the change in electrolyte content in the cathode or anode space, $\Delta \bar{n}$ is the observed change in transport number on the corresponding side of the membrane, q is the quantity of charge passed, and F is Faraday's number.

357

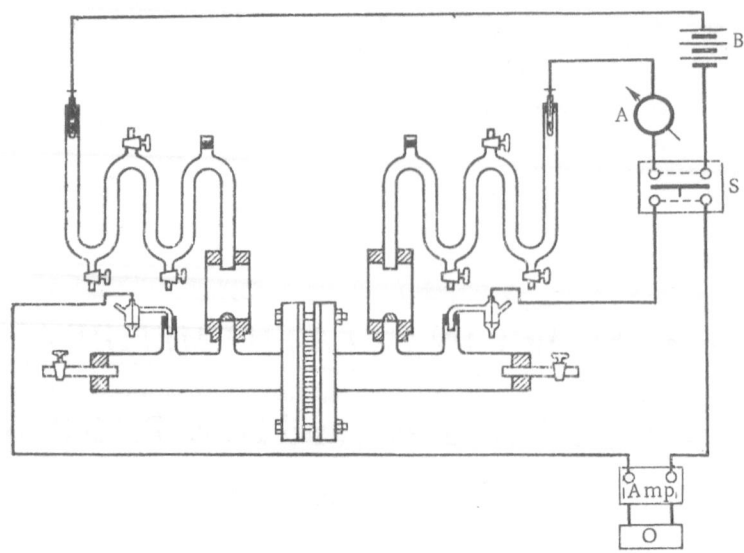

Fig. 1. Cell and measuring system: A = ELMA milliam-
meter, S = switch, Amp = amplifier (from LP-52 pH meter),
O = N-10 oscilloscope.

Sections of the membrane were analyzed layer by layer [8, 10] to determine the distribution of the concentration within the membrane. The polarization emf after switching off was measured with two calomel electrodes connected to the cell by agar bridges, the emf being measured by a loop oscilloscope.

Results and Discussion

a. Diffusion. One object of this work was to examine pure diffusion (ion transfer under a concentration gradient in the absence of an electric field). These tests were run for times of 15, 30, 60, and 120 min. Immediately afterwards, the membrane was cut into 4 or 5 layers, and the electrolyte concentration was deduced from the Cl content of each layer. The KCl concentrations on the two sides (in meq/liter) were 10:20, 10:40, 10:100, 5:100, 2:100, and 1:100. In every case, the membrane before use was kept in a 0.01 M KCl.

Figure 2 shows kinetic curves for the KCl in the membrane for the 10:100 case, in which x is the distance from the boundary between the membrane and the dilute solution.[†] This shows that the more concentrated solution gradually penetrates into the membrane and reaches the far side after about 1 h. This shows that allowance must be made for diffusion from the cathode space to the anode space in the analytical method of measuring transport numbers for membranes, since this decreases the change of the electrolyte content in each space, and hence also the observed $\Delta \bar{n}$.

Figure 3 shows the diffusion transport in relation to concentration ratio for a fixed time of 120 min.

The diffusion coefficients D (cm^2/sec) for the membrane were deduced from the accumulation of electrolyte via the formula [10]

$$D = \left(\frac{\Delta m}{2 c_0' S} \frac{\sqrt{\pi}}{\sqrt{t}} \right)^2,$$

[†] The layer thickness was determined by weighing.

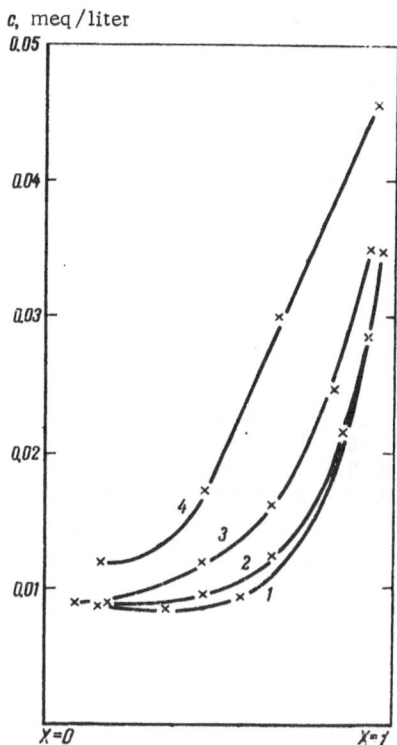

Fig. 2. Diffusion of KCl for a 10:100 meq/liter concentration ratio, showing distribution after times (min) of: 1) 15; 2) 30; 3) 60; 4) 120.

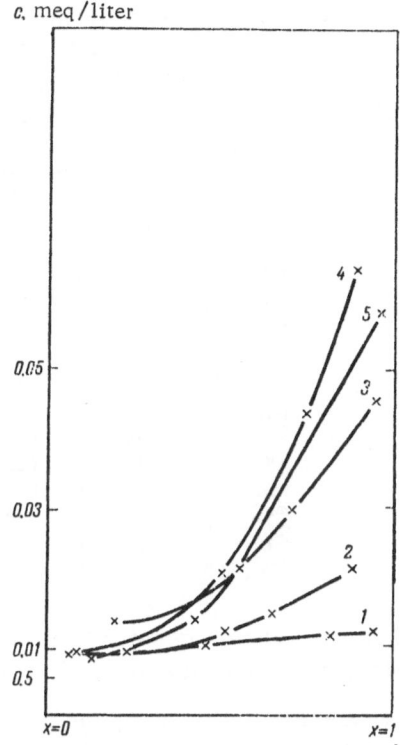

Fig. 3. Diffusion transport as concentration profiles for KCl concentration ratios (meq/liter) of: 1) 10:20; 2) 10:40; 3) 10:100; 4) 5:100; 5) 1:100.

with

$$c_0' = c_0 (1 - \Phi),$$

in which Δm is the change (in eq) in the amount of electrolyte in the membrane, c_0 is the concentration (eq/cm³) on the concentrated side, Φ is the fraction of the volume taken up by the skeleton, S is the area of the membrane, and t is diffusion time.

The resulting D (2-4 × 10⁻⁶ cm²/sec) are less than those for the free solution (D₀ = 2 × 10⁻⁵ cm²/sec) by nearly an order of magnitude and are less than D_0/β ($\beta = 2.1$, the coefficient of structural resistance for gelatin membranes). These results confirm the conclusion [11] that the electrochemical activity of a membrane affects the diffusion of electrolytes.

b. Concentration Polarization. The concentration changes on both sides of the membrane in response to passage of a current, and this causes unequal concentrations at the cathode and anode boundaries, as well as alterations in the transport numbers, with $c^C > c^A$ and $\Delta \bar{n}^C < \Delta \bar{n}^A$. This in turn leads to accumulation of electrolyte in the membrane, and hence to further deviation of the transport numbers from their initial values. This "poisoning" is characteristic of homogeneous ion-exchange membranes and of heterogeneous capillary systems, but it affects the electrochemical activity of the latter more extensively. Neglect of diffusion has the result that the observed $\Delta \bar{n}$ does not correspond to the true value.

Table 1 gives $\Delta \bar{n}^C$ and $\Delta \bar{n}^A$, while Fig. 4 gives concentration curves after passage of current under various conditions: no agitation (NA) and agitation in cathode (C) or anode (A) spaces by magnetic (M) or glass (G) stirrers. It is clear that the concentration profile always becomes

TABLE 1. Changes in the Transport Numbers of KCl on the Cathode and Anode Sides of a Membrane

	No electroosmosis			With electroosmosis		
	$\Delta \bar{n}^C$	$\Delta \bar{n}^A$	$\Delta m_0'$	$\Delta \bar{n}^C$	$\Delta \bar{n}^A$	$\Delta m_0'$
	$I = 9\,\text{mA}; q = 36\,\text{C}$					
NA	0.10	0.24	8.1	0.07	0.27	9.1
MCA	0.14	0.27	5.6	0.12	0.26	8.0
MC	0.13	0.22	6.3	0.15	0.25	7.0
MA	0.11	0.22	8.8	0.09	0.24	8.4
GC	0.19	0.25	3.2			
GA	0.08	0.26	10.9			
	$I = 5\,\text{mA}; q = 36\,\text{C}$					
NA	0.14	0.20	6.2	0.13	0.24	6.8
MCA	0.14	0.18	4.2	0.15	0.18	7.5
MC	0.11	0.23	5.3	0.15	0.20	2.7
MA	0.11	0.20	5.5	0.07	0.22	5.7
GC	0.19	0.22	0.4			
GA	0.10	0.21	5.7			

steeper when the current density is increased, which is due to larger concentration changes on the two sides, greater diffusion, and increase in the difference between $\Delta \bar{n}^C$ and $\Delta \bar{n}^A$.

Electroosmosis also steepens the concentration profile (Fig. 5). Preliminary tests on the electroosmosis rate u_{eo} gave a value of 26 μ/h, so runs lasting 1-2 h can show an appreciable effect only on the boundary conditions. On the anode, the boundary layer[†] of dilute solutions is drawn electroosmotically into the membrane, which increases $\Delta \bar{n}^A$. The electroosmotic flow at the cathode boundary is opposed to the diffusion flow and hinders the transfer of the concentrated solution from the boundary layer into the membrane, which reduces $\Delta \bar{n}^C$. The concentration profile steepens, on account of the increasing difference between $\Delta \bar{n}^C$ and $\Delta \bar{n}^A$.

The magnetic and glass stirrers operated respectively at 200 and 1000 rpm at distances of a few mm and 10 cm from the membrane. Table 1 shows that $\Delta \bar{n}^A$ is largely unaffected by stirring in the anode space. A constant electrolyte concentration is maintained in the anode boundary layer in the steady state by the diffusion fluxes from the solution and from the membrane, diffusion from the solution being the decisive process ($D_0 > D$). Diffusion from the membrane affects only $\Delta \bar{n}^A$, which is reduced relative to the true value Δn^A. The magnetic stirrer increases the flow of material from the solution to the boundary layer and so reduces the dilution of the latter and hence the true value Δn^A. At the same time, the reduced dc/dx reduces the diffusion from the membrane, so the observed value $\Delta \bar{n}^A$ may remain unchanged.

In agitation by the glass stirrer, we may consider that the layer of dilute solution on the anode side is eliminated. In that case, the transport numbers deduced from the change in electrolyte concentration in the space are close to the real ones corresponding to 0.01 M KCl. We mean neglect diffusion from the membrane into the anode space because there is only a small concentration gradient at the boundary in this case. The glass stirrer in the cathode space greatly reduces the poisoning and increases $\Delta \bar{n}^C$. The weaker agitation by the magnetic stirrer has little effect on $\Delta \bar{n}^C$; only the joint action of electroosmosis and agitation produces an appreciable increase in $\Delta \bar{n}^C$.

These explanations are confirmed by the measured polarization emfs.

[†] Boundary layer denotes the layer of solution with an altered concentration in contact with the outer surface of the membrane. We consider the term film inappropriate.

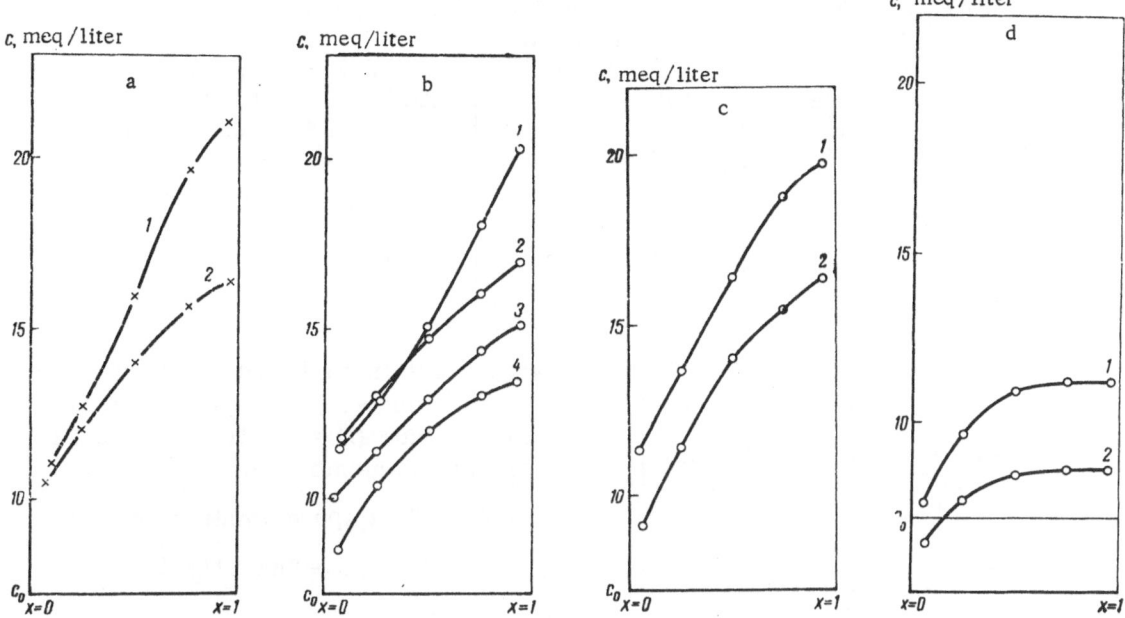

Fig. 4. Concentration profiles in gelatin membranes after passage of a current: a) no agitation at I (mA) of: 1) 9; 2) 5; b) agitation in both spaces by magnetic stirrers with I (mA) of: 1, 3) 9; 2, 4) 5; 1, 2) with electroosmosis; 3, 4) without electroosmosis; c) agitation in anode space by glass stirrer at I (mA) of: 1) 9; 2) 5; d) agitation in cathode space by glass stirrer at I (mA) of: 1) 9; 2) 5.

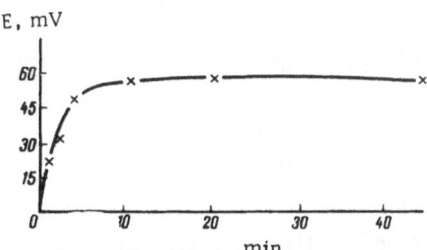

Fig. 5. Boundary polarization potential as a function of passage of polarizing current: I = 9 mA, no electroosmosis or agitation.

c. Polarization Emf. Concentration change in the boundary layers produces an emf of diffusion-potential type when the current is turned off. The sign of the emf coincides with the sign of the external field. We call this the boundary-polarization emf E_{bp}, to distinguish it from the emf due to polarization arising within a porous membrane in response to a current [12].

Figure 5 shows E_{bp} as a function of time. A steady value is reached after about 10 min, i.e., much earlier than the steady state in the system as a whole, which was not reached in the time used. Table 2 gives E_{bp} for various conditions, while Fig. 6 shows the decay of E_{bp}. Curves a and b in Fig. 6, and also the data of Table 2, show that agitation in the cathode space during passage of the current or during decay has no appreciable effect on the magnitude of E_{bp} or on the mode of decay, so the anode side is the one that determines this potential. In fact, stirring in the anode space during passage of the current results in very small E_{bp} on switching off (Fig. 6d), whereas we get large and rapidly decaying potentials in the absence of stirring (Fig. 6c). In this case, switching on the stirring (at t_1) during the decay rapidly reduces E_{bp}.

The rate of decay of E_{bp} on switching off is dependent on the rate of decay of the concentration changes in the boundary layers, i.e., is governed by diffusion. The decay time is several minutes in the absence of stirring, as against the few seconds needed for concentration equalization between the solution and a boundary layer 100 μ thick. This indicates that diffusion

TABLE 2. Boundary Polarization Potential E_{bp} (mV) for a Gelatin Membrane

	No electro-osmosis		With electro-osmosis			No electro-osmosis		With electro-osmosis	
	$I = 9mA$	$I = 5mA$	$I = 9mA$	$I = 5mA$		$I = 9mA$	$I = 5mA$	$I = 9mA$	$I = 5mA$
NM	85	60	94	90	MA	32	23	57	34
MCA	33	29	71	35	GC	60			
MC	88	65			GA	10			

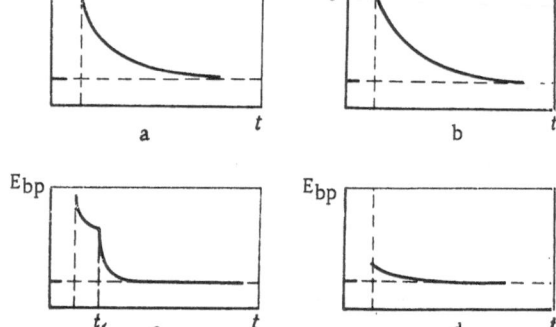

Fig. 6. Effects of electrolyte stirring on the decay of E_{bp}: a) no agitation; b) agitation in cathode space during polarization and decay; c) agitation in anode space during decay; d) agitation in anode space during polarization.

from the solution into the boundary layer ($D_0 = 1.7 \times 10^{-5}$ cm²/sec) is accompanied by diffusion at a lower rate ($D = 2 \times 10^{-6}$ cm²/sec) from this layer into the membrane.

As a first approximation we have

$$E_{bp} = 2\Delta n \cdot 58 \lg \frac{c_0^C}{c_0^A},$$

in which the superscripts to the concentrations denote respectively the anode and cathode sides of the membrane.

Agitation in the cathode space produces $c_0^C \approx c_0$, so

$$E_{bp} = 2\Delta n \cdot 58 \lg \frac{c_0}{c_0^A} = 60 \text{ mV}.$$

If $\Delta \bar{n}^A = 0.2$, then $c_0/c_0^A = 100$. Such small c_0^A means that the actual Δn^A is greater than the observed $\Delta \bar{n}^A$, which is reduced by diffusion transport into the anode space.

The effective coefficient for the membrane is $\alpha = 3.7$ for 10^{-3} M KCl; then the relation of α to Δn [9] gives $\Delta n = 0.4$, which is substituted into the expression for the potential to give $c_0/c_0^A = 15$, so $c_0^A = 0.6$ meq/liter.

The thickness δ (cm) of the boundary layers can be estimated from these results. In the steady state in unit time

$$\frac{j\Delta\bar{n}^C}{F} = D_0 \frac{c_0^C - c_0}{\delta}$$

because dc/dx = constant in the boundary layer; here j (A/cm²) is current density and c_0^C is the concentration in the boundary layer (eq/cm³). Then

$$\delta = \frac{D_0 F (c_0^C - c_0)}{j\overline{\Delta n}^K}.$$

The c = f(x) curves show that the profile can be extrapolated to the cathode boundary, so c_0^C is 20-30 μeq/cm³; with j = 4 mA/cm², $\Delta\bar{n}^C = 0.1$, and $c_0 = 10$ μeq/cm³, we get $\delta = 420$ μ. Similarly, $\delta = 105$ μ for j = 4 mA/cm² and $\Delta\bar{n}^A = 0.4$ at the anode boundary. These δ agree with published values [5], which confirms the arguments. The effects of current strength and electroosmosis on E_{bp} are similar to those on the concentration in the membrane and on the transport numbers. A future more detailed study of E_{bp} should probably provide a basis for estimating boundary concentration changes produced by dc.

Conclusions

1. A study has been made of concentration polarization of gelatin membranes produced by passage of dc. Diffusion coefficients for KCl solutions of various concentrations have been measured.

2. Electrolyte accumulates in the membrane on account of inequality of the ionic fluxes at the boundaries; this accumulation increases with the current density, and electroosmosis produces the same effect.

3. Vigorous stirring on the cathode side greatly reduces the polarization and increases the change in the transport numbers, whereas stirring on the anode side has no appreciable effect.

4. The boundary polarization potential is determined mainly by the anode side and is greatly reduced by stirring on that side.

References

1. R. Schlögl and V. Schodel, Z. Phys. Chem., 5:372 (1955).
2. T. R. E. Kressman and F. L. Tye, Trans. Faraday Soc., 8:1441 (1959)..
3. V. Subrahmanyan, Curr. Sci., 31:4 (1962).
4. Disc. Faraday Soc., No. 21 (1956).
5. F. Hellferich, Ion Exchange, McGraw-Hill, New York (1962).
6. L. Michaelis, R. Ellsevoth, and A. Weech, J. Gen. Physiol., 10:671 (1926).
7. O. N. Grigorov and E. I. Sutyagin, Vestnik LGU, Seriya Fiz. i Khim., No. 16, issue 3, p. 104 (1964).
8. O. N. Grigorov and E. I. Sutyagin, Vestnik LGU, Seriya Fiz. i Khim., No. 22, issue 4, p. 83 (1964).
9. P. A. Rebinder, ed., Electrokinetic Properties of Capillary Systems, Izd. AN SSSR, Moscow-Leningrad (1956).
10. M. A. Lauffer, Biophys. J., 1:3 (1961).
11. D. A. Fridrikhsberg and L. Pavlova, Kolloidn. Zh., 27(1):113 (1965).
12. D. A. Fridrikhsberg and M. P. Sidorova, Vestnik LGU, Seriya Fiz. i Khim., No. 4, issue 1, p. 57 (1961).

FLOW POTENTIAL AND CURRENT AS FUNCTIONS
OF TIME FOR NONPOLAR LIQUIDS

D. A. Fridrikhsberg and K. B. Shchiglovskii

Zhdanov University, Leningrad

A method is given for determining the convective and net currents for flow of a nonpolar liquid. It is found that the flow current has a maximum when a solution of Necal in petroleum ether flows through copper and PTFE tubes.

Flow currents in nonpolar liquids of very low conductivity (10^{-12}–10^{-16} ohm$^{-1} \cdot$ cm^{-1}) are of interest not only as regards electrification processes but also for practical purposes. Liquid hydrocarbons become electrified during pipeline transport, filtration, etc., and the vapors are capable of forming combustible mixtures with air, so the process can result in fire or explosion.

Theoretical and experimental relations have been published [1-4] for the charge generated as a function of the electrical and hydrodynamic parameters. The trends in the relationships indicate a resemblance to electrokinetic phenomena in aqueous solutions. The main difference lies in the enormous ratio of the electrical conductivities and hence in the ratio of the charge relaxation times τ, which are defined [1] by

$$\tau = \frac{\varepsilon \varepsilon_0}{\varkappa},\tag{1}$$

in which $\varkappa$ is specific electrical conductivity (ohm$^{-1} \cdot$ m^{-1}), ε is dielectric constant, and $\varepsilon_0 = 8.854 \times 10^{-12}$ F/m.

The liquids used in these experiments had τ of 0.2-200 sec, which indicate that the steady-state double layer during flow differs from the equilibrium layer on the same surface. This means that it is possible for the layer and flow current parameters to alter under fixed hydrodynamic conditions. These changes can be detected only by measuring the flow current as a function of time. Methods for measuring flow currents and potentials for nonpolar liquids differ from those for aqueous solutions mainly in the absence of macroscopic equipotential volumes of liquid whose potentials can be measured with a system of localized probes.

Previous studies of electrification in nonpolar liquids amount to measurement of the total charge or the time-averaged current reaching a receiving vessel, or else the current from a metal capillary to ground. A major problem was therefore to develop a method of measuring the electrification as a function of time, in particular for nonconducting capillaries.

The convective current (flow current) i_s in flow through a capillary gives the most information about charge separation and accumulation. A convenient measure of this current is the potential V taken up by a metal screen isolated from ground (Faraday cage) which contains the receiving vessel, into which the current flows along with the liquid. The current in the source vessesl (net current) i_r may be measured in the same way. It is also desirable to monitor the potentials of the metal capillary and of local electrodes placed in the vessels; although the results from these are not strictly quantitative, they do clearly indicate the polarity of the charges.

Methods and Materials

The choice of liquid was governed by the scope for varying $\varkappa$ within wide limits. Commercial vehicle fuels do not meet this requirement, so we used the purest direct-distillation petroleum product, namely the petroleum ether fraction boiling at 100-110°C, density 0.680 g/cm³, viscosity 0.42 cP, and $\varkappa \approx 10^{-15}$ ohm^{-1} · cm^{-1} (less than $\varkappa$ for aircraft gasoline by 2-3 orders of magnitude). This fraction has very little tendency to become electrified. This tendency was increased by adding Necal (sodium dibutylsulfonaphthalenate), which also increased $\varkappa$; 5 mg/liter gave $\varkappa = 2 \cdot 10^{-12}$, and 50 mg/liter gave 2.6×10^{-11} ohm^{-1} · cm^{-1}. Copper and PTFE ($\rho = 10^{17}$ ohm · cm) were used as the capillary materials in order to examine the effect of the resistance of the solid on the flow current.

The apparatus for measuring $i_s(t)$ and $i_r(t)$ consisted of two 10-liter vessels connected to a copper or PTFE capillary 3 mm in diameter and 50 cm long. Initially, each vessel was half full, while the capillary was full. Transfer was produced by supplying compressed air to one vessel.† The vessels stood on insulating supports in an electrically screened fume cupboard. A vessel was enclosed in a screen of insulated brass gauze for the purpose of measurement. Each vessel was also fitted with a platinum electrode dipping into the liquid. In some cases, the measuring instrument was connected to the metal tube. The measurements were made with an ÉMU-4 dc electrometer amplifier or an ÉM-61 electrometer voltmeter.

Figure 1 shows the circuit. Change in the charge within a cage,

$$\frac{dq}{dt} = i_s - i_1 = i_r ,$$

(2)

causes an equal change in the charge on the cage itself and hence the generation of a current in the RC circuit between cage and ground, where R consists of the insulation resistance R_c of the cage, the shunt resistance R_s, and the grid resistance R_g of the electrometer. As $R_c \gg R_s$ and R_g, we have

$$R = \frac{R_s R_g}{R_s + R_g} .$$

Also, C consists of the parallel capacities C_{cg} (cage to ground), C_1 (added capacitance), and C_g (grid capacitance of electrometer). As $C_1 \gg C_{cg}$ and C_g, we have $C \approx C_1$. Then the dependence of cage potential V on time t gives i_r as the difference between i_s and the leakage current i_1:

$$i_r = \frac{V}{R} + C \frac{dV}{dt} .$$

(3)

---

† The air pressure was adjusted so that the consequent difference in liquid levels had no effect on the flow.

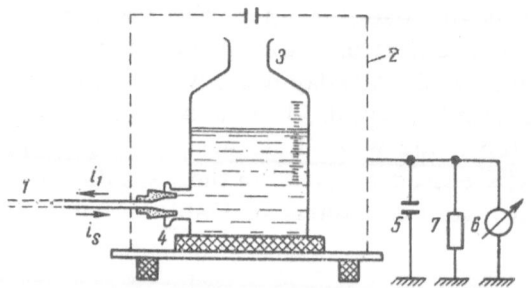

Fig. 1. The system: 1) tube carrying liquid; 2) Faraday cage; 3) vessel; 4) insulating support; 5) added capacitor; 6) measuring instrument; 7) shunt; i_s) flow current, i_1) leakage current (in liquid and tube).

This i_r is the basic practical quantity; the value of i_s can be deduced from the equivalent circuit (Fig. 2) if certain assumptions are made.

We can represent i_s as coming from an equivalent source whose internal shunt resistance is R_1, through which flows i_1. Taking the potential driving i_1 as equal to the potential of the cage,[†] we introduce into the circuit the equivalent elements R and C (see above), across which the potential difference is the measured V. Then

$$i_s = \frac{V}{R_1} + \frac{V}{R} + C\,\frac{dV}{dt}, \qquad (4)$$

in which we have the two unknowns $i_s(t)$ and R_1. All the same, $i_s(t)$ can be deduced from the measured V(t) as follows.

We make a series of measurements of V(t) with various R and C but with the other conditions unchanged, i.e., $V_1(t)$ with R = R' and C = C', and $V_2(t)$ with R = R" and C = C"; we assume that $i_s'(t) = i_s''(t)$ and $R_1' = R_2''$, which gives us a system of equations that may be solved to give

$$i_s(t) = \frac{V_1(t)\left[\dfrac{V_2(t)}{R''} + C''\,\dfrac{dV_2(t)}{dt}\right] - V_2(t)\left[\dfrac{V_1(t)}{R'} + C'\,\dfrac{dV_1(t)}{dt}\right]}{V_1(t) - V_2(t)}.$$

Figures 3 and 4 show i_s as calculated from $V_1(t)$, $V_2(t)$, and $V_3(t)$ in V_1-V_2 and V_2-V_3 combinations; these agree satisfactorily, which shows that the initial assumptions are reasonable.

More accurate results could be obtained without these assumptions by using an apparatus providing direct measurement of the potential causing the leakage; but in that case i_s should still be calculated in the same way with various parameters in the RC circuit, in view of the difficulty of measuring R_1.

Results and Discussion

For each Necal concentration we made many series of measurements of V as a function of t, each series corresponding to flow of 5-6 liters through the tube (one cycle), after which the system was restored to the initial condition in the following cycle.

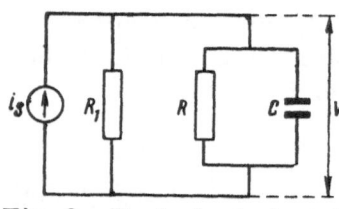

Fig. 2. Equivalent electrical circuit.

In prolonged tests with a copper tube we found a distinct fall from series to series in the positive potential recorded by the cage or electrode, with gradual attainment of a stable negative potential. Table 1 gives the results from one set, which shows only a few series, on account of lack of space (in every case, P = 1 cm Hg, R = R_g = 10^{11} ohm, C = 2.4×10^{-9} F, concentration 50 mg/liter). In analogous sets we also observed on the signal from the tube a transition from a negative potential to a positive one.

[†] On the assumption that the exit end of the tube intersects the equipotential V of the cage.

TABLE 1. Cage Potential V (volts) in 50 mg/liter Necal

Time	Copper								PTFE					
	Series													
	1	5	11	13	17 *	18	23	24 **	1	11	14	18	37	44
10″	0.04	0.15	0.10	0.05	—0.02	—0.04	—0.03	—0.02	0.18	0.03	0.06	0.07	0.03	0.07
20″	0.10	0.35	0.15	0.07	—0.02	—0.06	—0.06	—0.04	0.38	0.06	0.14	0.12	0.07	0.09
30″	0.15	0.50	0.20	0.08	—0.03	—0.09	—0.07	—0.07	0.62	0.11	0.21	0.17	0.12	0.12
40″	0.22	0.70	0.23	0.10	—0.03	—0.23	—0.10	—0.09	0.82	0.16	0.28	0.21	0.16	0.15
60″	1 5	0.88	0.30	0.14	—0.03	—0.20	—0.13	—0.13	1.26	0.25	0.42	0.30	0.24	0.24
2′	2.1	1.17	0.46	0.22	—0.03	—0.22	—0.14	—0.20	2.04	0.52	0.70	0.51	0.44	0.48
3′	2.0	1.35	0.59	0.35	—0.03	—0.23	—0.13	—0.21	—	0.78	0.86	0.72	0.61	0.70
4′	2.7	1.41	0.67	0.47	—0.03	—0.23	—0.12	—0.20	3.3	1.08	1.08	0.88	0.77	0.91
5′	2.8	1.47	0.75	0.53	—	—0.24	—0.11	—0.19	3.4	1.26	1.17	1.08	0.93	1.15
6′	2.8	1.50	0.75	—	—	—	—0.10	—0.16	3.4	1.44	1.20	1.14	1.11	1.18
7′	2.8	1.50	0.96	0.99	—	—	—0.08	—0.14	3.3	1.59	1.20	1.17	1.20	1.22
8′	3.2	—	—	1.05	—	—	—	—0.11	3.4	1.68	1.21	1.17	1.29	1.27

* Next day.
** Tube grounded.

This change of sign in V is ascribed to change in the composition and properties of the surface layer over a long period, since careful cleaning restored the original state, whereas use of a fresh batch of solution without cleaning did not affect the magnitude or sign (negative) of the potential.

Necal is a weak electrolyte and gives rise to a few Na^+ and A^- ions (the latter are organic anions). Initially, A^- is adsorbed by the free surface, which is accompanied by

$$A^- + Cu \rightarrow CuA + e.$$

The tube thereby acquires a negative charge, while the Na^+ produces a positive charge in the receiver ($i_s > 0$).

In the second stage, we suppose the formation a second layer with reversed orientation on the adsorbed CuA, the Na^+ being adsorbed and the A^- passing to the receiver, which acquires a negative charge. This mode of formation of a second layer has been demonstrated [5] for surfactants in aqueous media. It is possible that the two layers are formed together at a certain stage, according to the theory of multilayer adsorption [6]. The number of cycles needed for reversal decreases as the Necal concentration increases, which appears to confirm this view of the mechanism.

Table 1 gives results also for PTFE under the same conditions, which show that V (and hence i_r for the same R and C) is of the same order for both materials; but PTFE shows no decrease in V or change in sign, so no similar electrochemical process occurs on this inert and virtually nonconducting material. Perhaps there is continued adsorption of A^- over a very long period.[†] The following conclusion appears likely.

There are two distinct mechanisms for flow electrification: electrochemical for a conducting (grounded[‡]) solid and adsorption for a nonconducting one.

[†] A rough calculation shows that a monolayer of A^- (each of area 1000 Å²) would take 200 h to form on a tube of area 50 cm² at $i_s = 10^{-10}$ A with v = 1.5 m/sec.

[‡] The copper tube was grounded between series.

TABLE 2. Potentials Produced on a Copper Tube by 5 mg/liter Necal in Petroleum Ether ($C_1 = 15 \times 10^{-9}$ F, $R_s = 10^{10}$ ohm)

Time	RC*	RC**	SC	SC**	RC	T	SE	SC	SC**	SE**	SE	SC	SE	SC
	\multicolumn{14}{c}{Pressure, cm Hg}
	1	1	1	1	12	12	12	12	12	12	4	4	4	4
10″	0.03	0.02	—0.06	—0.07	—1.0	2.0	0.09	0.12	—0.03	—0.06	0.03	0.03	—0.02	—0.02
20″	0.05	0.05	—0.11	—0.10	—1.7	3.5	0.33	0.39	—0.06	—0.12	0.12	0.12	—0.05	—0.05
30″	0.07	0.07	—0.17	—0.15	—3.5	6.0	0.60	0.66	—0.09	—0.18	0.18	0.18	—0.08	—0.07
40″	0.09	0.10	—0.22	—0.19	—6.0	6.0	0.90	0.99	—0.12	—0.24	0.30	0.30	—0.10	—0.10
60″	0.13	0.13	—0.36	—0.27	—6	—	1.5	1.8	—0.21	—0.33	0.51	0.48	—0.13	—0.13
2′	0.22	0.58	—0.88	—0.66	—	—	3.1	2.8	—0.33	—0.45	0.93	0.84	—0.17	—0.18
3′	0.26	0.59	—0.80	—0.54	—	—	3.7	3.1	—0.36	—0.51	1.14	0.96	—0.17	—0.21
4′	0.26	0.66	—0.72	—0.60	—	—	—	—	—	—	—	—	—0.18	—0.21
5′	0.25	0.54	—0.78	—0.63	—	—	—	—	—	—	—	—	—0.18	—0.23
6′	0.26	0.45	—0.84	—0.63	—	—	—	—	—	—	—	—	—	—

*RC, SC, etc., are the signal sources.
**Tube grounded.

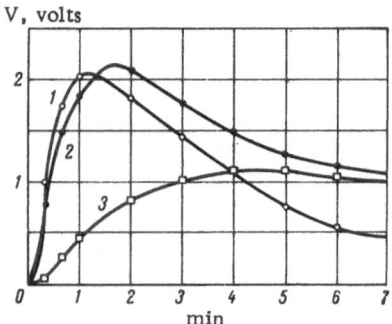

V, volts

Fig. 3. Cage potential V as a function of time: 1) $C \approx C_1 = 2.4 \times 10^{-9}$ F, $R_s = 10^{10}$ ohm, $R_g = 10^{11}$ ohm; 2) $C \approx C_1 = 2.4 \times 10^{-9}$ F, $R_s = 1.5 \times 10^{10}$ ohm, $R_g = 10^{11}$ ohm; 3) $C \approx C_1 = 15 \times 10^{-9}$ F, $R_s = 1.5 \times 10^{10}$ ohm, $R_g = 10^{11}$ ohm.

Various factors influence V and i_r. The electrification increases considerably with the flow speed v, as has previously been reported [1, 2, 7]. Also, i_r is proportional to the Necal concentration and has a maximum at a certain v.

After the liquid has been at rest for a long while, and especially after the surface of the tube has been cleaned mechanically, the V tend to be higher, which shows that the double layer comes to equilibrium slowly and that its state during flow is far from stationary. This leads to the practical suggestion that v should be restricted if new or cleaned tubes are being used, and also after halts in operation.

Grounding a copper tube during flow usually increased V by a factor 1.5-2 (series 24, Table 1), evidently because this facilitates charge separation at the interface. Grounding also greatly affects the magnitude and sign of the charge on the source vessel. (The potential at the source has not previously been discussed in the literature on nonpolar media.) We have found that it can be substantial (not less than on the receiving vessel) and so represents an explosion hazard.

If charge separation occurs at the surface of the metal or at the inlet to the tube, we would naturally expect that V for the source would be of the same sign as V for the tube (and opposite in sign to V for the receiver), on account of gradual leakage of charge from the tube to the source. In fact, the signs of V for source and tube were always the same when the tube was ungrounded; but grounding the tube, which prevents charge accumulation always produced a small negative V at the source (cage and Pt electrode), no matter what the sign of V for the tube.

The source and receiver thus acquire charges of opposite sign in the absence of grounding but of identical (negative) sign in the presence of grounding. Table 2 gives an example of this, in which C denotes the cage, E electrode, S source, R receiver, and T the tube (all as potential

TABLE 3. Potential V at Receiver Cage and Current i_r
for the Copper – Necal System

$C_1 \cdot 10^9$, F	R_S 10^{-9}, Ω				$C_1 \cdot 10^9$ F	R_S 10^{-9}, Ω			
	20	15	10	5		20	15	10	5
	V at receiver cage, volts					$i_r \cdot 10^{11}$, A			
0	6	2.8	1.7	1.3	0	6	21	18	28
2.4	6	2.9	1.7	0.9	2.4	8 (3)	22 (4)	18 (2)	18 (3)
4	6	2.2	—	0.6	4	12 (8)	16 (1)	—	13
15	2.2	1.2	1.3	—	15	10 (7)	11 (3)	17 (2)	—
100	0.5	0.4	0.4	—	100	11	13	14	

TABLE 4. Potential V at Receiver and Current i_r
for the Copper – Necal System

t	V, volts	$\frac{V}{R} \cdot 10^{10}$, A	$C \cdot \frac{dV}{dt} \cdot 10^{10}$, A	$i_r \cdot 10^{10}$, A
0″	0	0	—	0
10″	0.01	0.008	0.42	0.428
20″	0.06	0.04	1.08	1.12
30″	0.17	0.13	1.42	1.55
40″	0.28	0.21	1.40	1.61
50″	0.37	0.28	1.28	1.56
60″	0.45	0.34	1.12	1.46
2′	0.82	0.62	0.68	1.30
3′	1.01	0 76	0.36	1.12
4′	1.11	0.84	0.14	0.98
5′	1.11	0.84	—0.16	0.68

indicators). When the tube is grounded the charge on the interior leaks to ground, and the cause of the source potential may be a diffusion or convective mechanism. Theory [8] indicates that renewal of the double layer (if the diffusion coefficients are unequal) gives rise to an electric field that balances the ion fluxes; the bulk of the liquid then acquires a charge whose sign is that of the ion that moves more slowly (here the anion). The source electrification can thus be explained via the theory of the nonequilibrium double layer.

Consider now V and i_s as functions of time. It is clear that V is substantially dependent on the circuit parameters (R and C) and is only a qualitative characteristic of the electrification, but analysis of the V(t) curve gives us the data needed to find i_r and i_s via (3) and (5).

The V(t) curves (Fig. 3) are linear or S-shaped at the start, but the behavior for t large varies: 1) monotonic rise, 2) dV/dt = 0, 3) maximum followed by a fall, sometimes to a horizontal part. The shape of the curve gives some indication of $i_r(t)$; for instance, at the start (dV/dt > 0) it is clear that i_r rises, because

$$i_r = \frac{V}{R} + C \frac{dV}{dt}.$$

Subsequently, curves of the first type (which are observed for C_1 large) do not reach a steady state in the t used; a further passage of liquid would probably give rise to a curve of the second or third type. Curves of the second type represent attainment of a steady state not only

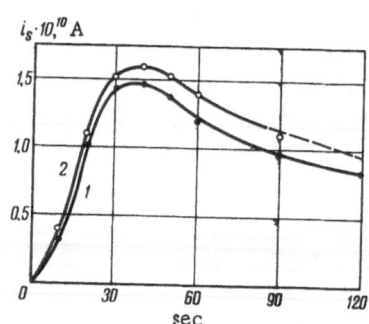

Fig. 4. Dependence of i_s on time: 1) $i_{V_1 - V_3}$; 2) $i_{V_2 - V_3}$.

at the cage $(dV/dt = 0)$ but also in the liquid flow.[†] Curves of the third type have $dV/dt = 0$, i_r = constant at the maximum; later, $dV/dt < 0$, and C_1 discharges through R_g and R_s, so i_r also has a maximum (since $di_r/dt > 0$, for t small). If $V(t)$ subsequently becomes constant, the flow attains a steady state in which i_r = constant $< i_{r\,max}$. Curves of the third type occur for high P, so the steady state in the flow deviates further from the equilibrium state as v increases.

Tables 3 and 4 give some examples illustrating these relations. Table 3 gives V measured at the receiver cage and the i_r deduced from these for copper with 5 mg/liter Necal at P = 1 cm Hg and various R_s and C_1 ($R_g = 10^{11}$ ohm). The underscored values correspond to the steady state $(dV/dt = 0)$. The V and i_r for the largest t are given for curves of the first type, and the initial i_r ($= CdV/dt$ for $t \to 0$, $V \to 0$) when the curve has an initial linear part. These results show that, in spite of considerable variation in R_s and C_1 (and hence in V), the i_r are relatively constant under steady conditions.[‡]

We have calculated $i_r(t)$ from (3) via the $V_3(t)$ of Fig. 3 for the system PTFE-Necal (50 mg/liter) at 4 cm Hg (v $\approx$ 0.6 cm/sec). The results show that i_r increases slowly at first, reaches a maximum after somewhat less than a minute, and then slowly falls. Similar results were obtained in other cases. The time taken to reach the steady state very greatly exceeds the time of passage through the tube (0.5 sec) and the charge relaxation time (0.2 sec).

Calculations on i_s from (5) were made for $V_1 - V_3$ and $V_2 - V_3$ combinations (Fig. 3) and are shown in Fig. 4. It is clear that i_s rises rapidly to a maximum and then falls slowly.

These results show that i_r and i_s, and hence all the parameters of the double layer, change considerably during the flow. This time dependence appears not to have been discussed in the literature. The present method allows one to examine the details of the charge generation. The peaks in $i_r(t)$ and $i_s(t)$ are significant to future theoretical developments, since they show that electrokinetic processes can be time-dependent.

Conclusions

1. A method is proposed for determining the net current $i_r(t)$ entering the receiving vessel from a flow of nonpolar liquid, and also for deducing approximately the flow current i_s.

2. Necal in petroleum ether produces i_r and i_s that pass through maximum in times of under 1 min on flow through PTFE and copper tubes 3 mm in diameter.

3. The results indicate the need to incorporate the time coordinate in the description of electrokinetic processes in nonpolar liquids.

References

1.   A. Klinkenberg and J. Van der Minne, Electrostatics in the Petroleum Industry, Elsevier, Amsterdam (1958).

[†] But not in the receiver, where the potential U increases linearly with t for i_r = constant and V = constant. This rise should subsequently become slower as a steady state in the receiver is approached (U = constant, i_r = 0, V = 0).

[‡] There is a slow change in the electrification in experiments with copper tubes (Table 1).

2. B. Hampel and H. Luther, Chem. Ing. Techn., 29:323 (1957); H. Luther and B. Hampel, Erdöl und Kohle, 10:297 (1957).

3. J. Gavis and J. Koszman, J. Coll. Sci., 16:375 (1961); J. Koszman and J. Gavis, Chem. Eng. Sci., 17:1013, 1023 (1962); J. Gavis, Chem. Eng. Sci., 19:237 (1964).

4. D. G. Rogers and C. E. Schleckser, Riv. Combust., 14:291 (1960).

5. O. N. Grigorov, Z. P. Koz'mina, A. V. Markovich, and D. A. Fridrikhsberg, in: Electrokinetic Properties of Capillary Systems, Izd. AN SSSR, Moscow-Leningrad (1966), p. 148.

6. S. Brunauer, The Adsorption of Gases and Vapors, Vol. I: Physical Adsorption, Princeton University Press, New Jersey (1943).

7. D. A. Fridrikhsberg, A. N. Zhukov, K. F. Kulikova, and M. I. Tarabanova, Summaries of Papers at the Fifth All-Union Conference on Colloid Chemistry, Odessa (June 1962), Izd. AN Ukr. SSR (1962), p. 29.

8. B. V. Deryagin, Kolloidn. Zh., 22(2):148 (1960); S. S. Dukhin and B. V. Deryagin, Kolloidn. Zh., 20(6):705 (1958); S. S. Dukhin, in: Research in Surface Forces, Vol. 2, Consultants Bureau, New York (1966), p. 54.

TANGENTIAL ACTION OF A FLOW OF LIQUID IN
THE REMOVAL OF THIN OIL FILMS FROM FLAT SURFACES

E. S. Klepikov and E. A. Kamener

It is found that an oil film less than $0.6\,\mu$ thick has an anomalous viscosity when acted on by a tangential water flow. The effects of surfactants on oil film removal are examined.

Flushing of oil contaminants from surfaces is an important practical problem. Oil removal by droplet water jets from large surfaces requires the use of surfactants and substantial forces. The forces usually needed are not proportional to the energy of adhesion of the contaminant, so the role of the surfactant is of some interest. Also, the forces on the contaminant are partly tangential, and other modes (impact of continuous or droplet jets) consume more energy.

We have examined the mechanism of removal of oil films by a tangential flow containing a surfactant. Deryagin and others have shown that many interfacial effects require three-dimensional consideration of the boundary layer.

It is well known that a solid surface has an orienting action on boundary layers of liquid. Deryagin and Kusakov [1] have shown that the radius of action of surface molecular forces is at least $0.2\text{-}0.3\,\mu$, and they consider that the effect is probably due to molecular orientation or polarization extending from the solid into the liquid.

Kusakov and Titievskaya [2] concluded that a solid surface has an action out to about $0.1\,\mu$ in a nonpolar liquid; their results imply that the effects of the substrate vary with the oil. Deryagin and Zakhavaeva [3] obtained similar results in tests on the stability of lubricant films [4]. Deryagin et al. [5-7] examined the boundary viscosity changing abruptly as the surface is approached. Traces of polar substances (fatty acids, esters) also resulted in boundary layers with distinctive viscosities.

Zakhaeva and Andreev [8] have shown that a solid surface affects the viscosity of a macromolecular compound out to $10\,\mu$. Akhmatov et al. [9] considered in detail the mechanical and physicochemical properties of boundary layers of polar liquids. Fuks [10] examined the structures of thin films of lubricants.

We have examined how the structure of the boundary layer affects the removal of thin oil films, the conditions for complete removal of oil from surfaces, and the mode of action of tangential water flow containing surfactants.

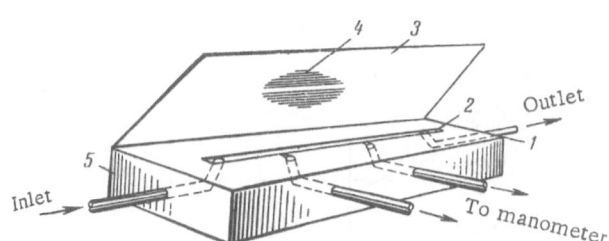

Fig. 1. Apparatus for flushing off oil films:
1) spacer; 2) channel for liquid; 3) plate
(cover); 4) oil film; 5) body.

We used thin films of automobile oil (GOST 5303-50), which were produced on steel and duralumin surfaces of class 9 finish by deposition of a 10% solution in petroleum ether to give a film of known thickness. The oil specimen contained a small amount (2 degrees) of naphthenic acids, and so it can be considered as a weak solution of polar compounds in a nonpolar hydrocarbon. The viscosity at 20°C was 230 cP; the acid number was 3, and the density was 0.920 g/cm³. It is thus to be expected that adsorbed layers of the polar impurities will influence the boundary viscosity and removal [9]. Although the material is nonpolar and interacts relatively weakly with metal surfaces, removal was not simple. The surfactants were aqueous solutions of sulfonol.

The approach was much as in Deryagin's experiments [3-7] on an air-blown film of oil on a smooth surface, where the viscosity change was deduced from the profile of the film as recorded optically. We used a water flow instead of air and judged the removal mechanism from the reduction in the thickness of the oil film.

The apparatus (Fig. 1) included a spacer of thickness H = 0.04 cm. The liquid flowed through a plane-parallel slot, one wall of which was coated with the oil film. The thickness of the residual oil film was determined after exposure to the flow, together with the tangential force of the flow. The following conditions must be met for the flow to exert only tangential forces on the film.

1. The effects of the pressure gradient along the flow on the motion of the oil film should be negligible relative to the tangential action of the flow, which is so if h/H ≤ 0.01, in which h is film thickness and H is channel height.

2. The tangential pressure F must be constant along the length of the channel, which is so if Δp/p ≤ 0.05, in which Δp is the fall in pressure along the channel and p is the pressure in the flow.

3. The liquid must flow laminarly in the channel.

If these conditions are obeyed, F can [4] be calculated from

$$F = \frac{H}{2} \frac{\Delta p}{\Delta l} \ (\text{dyn/cm}^2).$$

All of the above conditions were obeyed in our experiments. The channel had a width 0.5 cm; H = 0.04 cm and h = 10^{-4} cm. Also, Δp was indicated by a mercury differential manometer over a length Δl = 3.5 cm. Reynolds number was much less than the critical value of 1200. The flow was tested for laminar behavior by another method (agreement between the observed resistance coefficient and that calculated for laminar flow).

The fluorescence method [11] was used to determine quantitatively the thickness of the residual oil film. The oil film is exposed to UV and is photographed through a yellow filter, the thickness being calculated from the optical density by reference to calibration curves. A correction for variations in processing was made by photographing a standard specimen on the same film. The optical density was measured with a DNF-10 densitometer, which reads the mean optical density over a circle of diameter 3 mm. The method allows h to be measured to 0.01 μ.

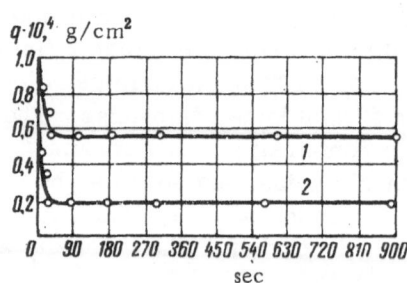

Fig. 2. Removal of oil from a steel plate at F (dyn/cm^2) of: 1) 190; 2) 1640.

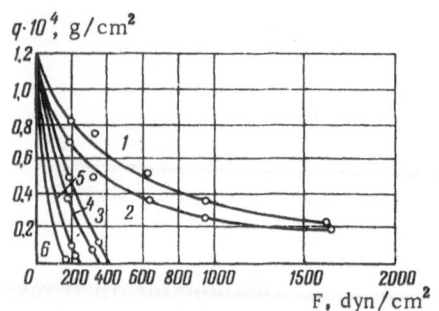

Fig. 3. Residual oil q on steel as a function of F for aqueous sulfonol solutions of concentrations (%): 1) 0.1; 2) 0; 3) 0.3; 4) 0.5; 5) 0.75; 6) 1.

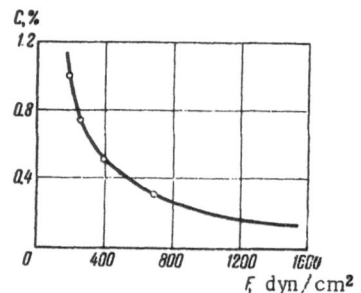

Fig. 4. Relation between concentration C and tangential pressure F for complete removal of oil from a steel surface.

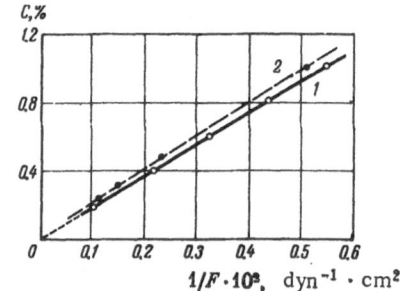

Fig. 5. Relation between concentration C and tangential pressure F for removal of oil to a thickness of 0.01 μ from: 1) steel; 2) duralumin.

Removal by Pure Water

Calculation of the removal rate (Fig. 2) showed that the residual film attained a constant thickness after 30 sec, so all subsequent tests were done with an exposure of 60 sec. The tangential pressures ranged from 190 through 1640 dyn/cm^2; each F corresponded to a definite residual thickness q which was constant along the length and which was virtually independent of the exposure time (Fig. 3).

It was found that q for a given F was larger than that deduced from the viscosity of the bulk liquid, which confirms the altered rheology of thin films on solids. The oil is not removed by ordinary viscous shear but by detachment from the surface. The effect may be characterized by the F dependence of q (Fig. 3) in terms of a limiting shear stress. Curve 2 of Fig. 3 shows that the F needed to remove the film increases rapidly for q < 0.1 μ.

Removal by Surfactant Solutions

Figure 3 shows the results; 0.1% sulfonol has less effect than pure water, but concentrations above 0.2-0.3% produce complete removal, and F for complete removal decreases as the surfactant concentration increases. The deterioration at the lowest concentration can be ascribed to adsorption of the oil, which serves to screen the oil film, but higher concentrations overcome the screening effect.

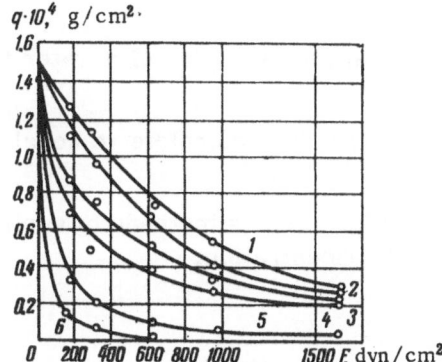

Fig. 6. Relation of q to F for steel
surfaces previously treated with solutions having sulfonol concentrations (%) of: 1) 0.1; 2) 0.5; 3) 1.0;
4) 0; 5) 2; 6) 3.

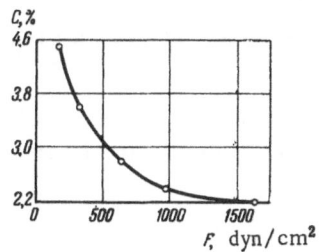

Fig. 7. Relation between concentration C and tangential
pressure F for previously
treated steel plates exposed
to water.

Figure 4 shows the relation of sulfonol concentration to the F needed for complete removal. The critical concentration for micelle formation is 0.15% for sulfonol, but this is not high enough to give the most effective removal, and higher concentrations produce better performance under these dynamic conditions. This may be because adsorbed layers at the interface are flushed away by the flow, and high concentrations are required to reduce the surface tension sufficiently for the effect to be pronounced. Figure 5 illustrates the effect of concentration even more clearly; it also shows that steel and duralumin plates (class 9 finish) are roughly equal in this respect. Curves of this type may be used to evaluate the effects of surfactants of surfaces.

Removal of Oil by Water after Previous Treatment

of Plates with Surfactant

Equilibration with a detergent solution takes a certain time. Our measurements indicated that the surface tension in sulfonol solutions reached a steady value in 40-60 sec. The plates before coating with oil were exposed to the sulfonol solution for 10-20 min; after coating with oil, they were exposed to a water flow.

It was found that the oil film did not break at any sulfonol concentration between 0.1 and 3%; pinpoints occurred in certain instances, but these did not expand. However, the film broke at concentrations above 3%. Figure 6 shows the results on exposure to water. In each case, an observed point is the mean from 4-6 determinations.

Solutions up to 1% produced firmer adhesion, evidently because an adsorbed layer is produced at the phase interface, which has two effects. Firstly, the surfactant tends to break up the oil film and detach it from the plate; secondly, the structured adsorbed layer of water tends to prevent the flow from attacking the film. The first effect predominates at high concentrations and removal is facilitated.

Figure 7 shows the relation of C to the F for complete removal by water from plates previously treated with sulfonol; here the concentrations needed are larger than those for the previous case by an order of magnitude, which shows that the tangential flow removes the film layer by layer, starting with the adsorbed layers; as these are not subsequently replaced, the flushing as a whole deteriorates.

Conclusions

1. A method has been developed for examining the tangential action of a water flow on thin oil films.

2. Tangential stresses of 190 through 1640 dyn/cm^2 produce complete removal by successive detachment of layers.

3. The mode of action of surfactants is considered for various methods of application, and it is found that the mechanical force is related to the concentration.

References

1. B. V. Deryagin, M. M. Kusakov, and L. S. Lebedeva, Dokl. AN SSSR, 23:670 (1939).
2. M. M. Kusakov and A. S. Titievskaya, Dokl. AN SSSR, 28:333 (1940).
3. B. V. Deryagin and N. N. Zakhavaeva, Kolloidn. Zh., 4(4):230 (1949).
4. B. V. Deryagin, G. M. Strakhovskii, and D. S. Malysheva, Zh. Éksp. Teor. Fiz., 16(2):172 (1946).
5. B. V. Deryagin and E. F. Pichugin, Transactions of the First All-Union Conference on Friction and Wear in Machines, Vol. 1 (1947), p. 103; Vol. 3 (1949), p. 101.
6. B. V. Deryagin and E. F. Pichugin, Dokl. AN SSSR, 63(1):53 (1948).
7. V. V. Karasev and B. V. Deryagin, Kolloidn. Zh., 15(5):365 (1953).
8. N. N. Zakhavaeva and S. V. Andreev, Summaries of Papers at the Fifth All-Union Conference on Colloid Chemistry, Izd. AN SSSR, Moscow (1962), p. 91.
9. A. S. Akhmatov, Molecular Physics of Boundary Friction, Gosfizmatizdat, Moscow (1963).
10. G. I. Fuks, Research in Surface Forces, Vol. 1, Consultants Bureau, New York (1963), p. 79.
11. V. N. Kochetov and E. S. Klepikov, Byull. Izobret. i Tov. Znakov, No. 9, p. 19 (1963).

SECTION VII

SURFACE FORCES IN ADHESION, COHESION, AND FRICTION

CHEMICAL ACTIVITY OF THE SURFACE OF A POLYMER ON DISRUPTION OF AN ADHESION BOND

L. P. Morozova and N. A. Krotova

Surface Phenomena Division, Institute of Physical Chemistry,
Academy of Sciences of the USSR

IR methods are used to show that a polymer surface (nitrocellulose or acetylcellulose) freshly formed by detachment from a substrate shows a very great increase in grafting of acrylonitrile. It is considered that electron emission from the freshly formed surface is responsible for this.

A polymer surface freshly mechanically generated has an elevated chemical activity, which is ascribed [1, 2] to free radicals, which can be detected by electron spin resonance and IR spectroscopy.

We have observed the emission of fast electrons from a polymer surface freshly formed by detachment [3]. It may be supposed that the emitting surface will affect reactions near it and also gas adsorption.

We have tested this supposition by reference to the surface compounds formed on initial and freshly prepared surfaces of films of cellulose esters in contact with the vapor of acrylonitrile.

We constructed an apparatus (Fig. 1) allowing the film to be detached from a substrate under vacuum, the surface then being exposed to the vapor of a reagent. The system is evacuated to 10^{-4} mm Hg before the film is detached from the glass roller by a weight released by an electromagnet, the saturated vapor of acrylonitrile being present at the time of detachment. The device was mounted in the beam of an IKS-14 IR spectrophotometer. The LiF windows allowed recording over the range 1500–3000 cm^{-1}, which corresponds to the characteristic frequencies of the functional groups in these polymers.

In the first series of tests we used films on glass that had previously been treated with trimethylchlorosilane to render it hydrophobic, in order to prevent sticking. The film removed from this surface was stored for two days over CaCl$_2$ and was then placed in the cell.

Figure 2 shows the IR spectra of acrylonitrile adsorbed by the various celluslose esters. The heights of the CN peaks decrease in the following order: nitrocellulose > acetylcellulose > ben-

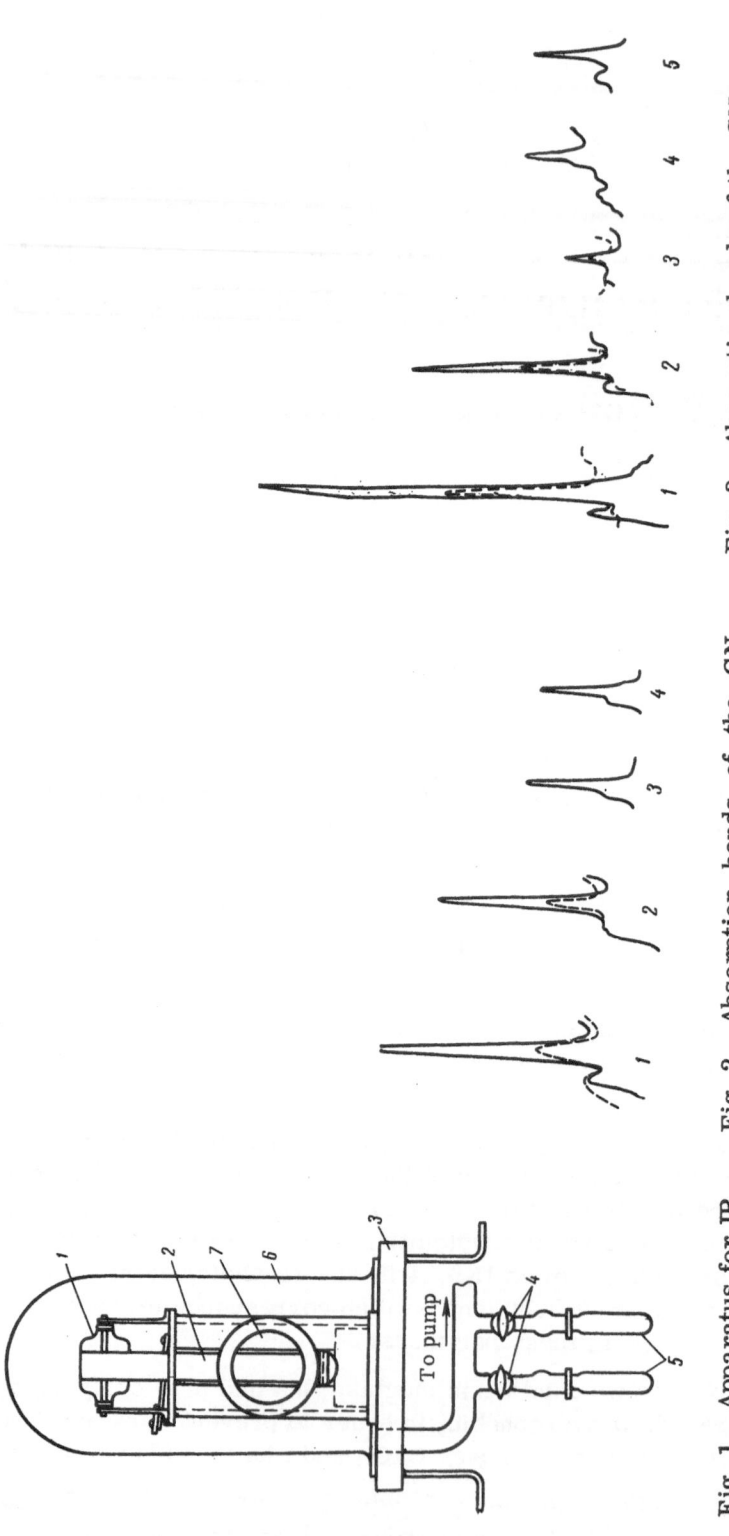

Fig. 1. Apparatus for IR study of the interaction between a monomer vapor and a freshly formed surface: 1) roller; 2) film; 3) support; 4) valves; 5) tubes; 6) chamber; 7) window.

Fig. 2. Absorption bands of the CN groups of acrylonitrile on interacting with the initial surfaces of cellulose esters. The broken line indicates the spectrum after evacuation for 20 h: 1) nitrocellulose; 2) acetylcellulose; 3) benzylcellulose; 4) ethylcellulose.

Fig. 3. Absorption bands of the CN groups of acrylonitrile on interacting with freshly prepared surfaces of cellulose esters. The broken line indicates the spectrum after evacuation for 20 h: 1) nitrocellulose; 2) acetylcellulose; 3) acetylcellulose–methacrylic acid copolymer; 4) benzylcellulose; 5) ethylcellulose.

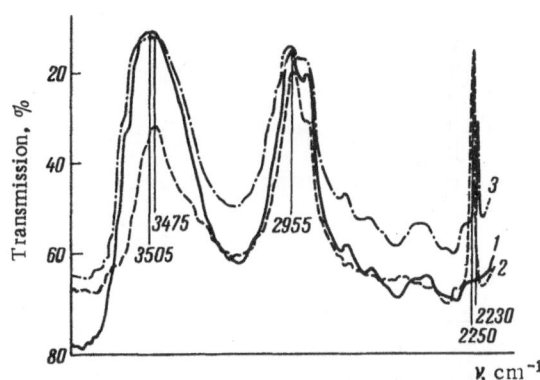

Fig. 4. IR spectrum from: 1) acetylcellulose; 2) copolymer of acetylcellulose with acrylonitrile; 3) product from interaction of acrylonitrile with a freshly prepared surface of acetylcellulose.

zylcellulose > ethylcellulose. Evacuation for 20 h caused the spectrum of the CN group to vanish completely from the last two. The residual peak for nitrocellulose was much larger than that for acetylcellulose.

Chemosorption by fresh surfaces revealed the same trends as in the first series (Fig. 3), i.e., complete loss by the latter two compounds. The peaks for the other two show that these bind irreversibly a substantial quantity of acrylonitrile.

Curve 3 of Fig. 4 shows the IR spectrum of the surface compound of acrylonitrile with acetylcellulose. Curve 2 is the IR spectrum of a copolymer of these substances (provided by the Vladimir Paint Industry Institute). These two curves show that the CN group is present and that there are certain changes in the frequency, shape, and strength of the OH absorption band, which are more marked for the copolymer; these show that acrylonitrile interacts chemically with our films. The spectra for the 1300-1800 cm^{-1} range indicate the following features of the surface compounds.

Prolonged pumping suppresses the 1880 cm^{-1} band (overtone of the out-of-plane deformation mode of the $>C=C$ group, with the H deviating from the CCH plane) and the 1580 cm^{-1} band ($-CH=CH_2$ stretching), which are characteristic only of the liquid monomer and which do not occur for polyacrylonitrile. From this we conclude that the film bears a layer of polyacrylonitrile (Fig. 5). It is reasonable to suppose that acrylonitrile is grafted to the cellulose, but formation of homopolymer cannot be ruled out, because this cannot be dissolved away, since it is very difficult to find a solvent that will dissolve the homopolymer butnot attack the cellulose.

The functional groups of the cellulose ester thus have a specific effect on the interaction with acrylonitrile. The reactive groups in cellulose esters are the OH groups of glucoside residues, or groups replacing these (nitro, acetyl, etc.). The acrylonitrile is most probably attached at the OH groups, as is clear from the alterations in the OH band. The substituents have a specific effect by influencing the reactivity of OH on account of an altered electron-density distribution in the macromolecules. The OH groups in nitro- and acetylcellulose are less screened, so they can interact with acrylonitrile [4].

It has been shown [5] that a freshly vacuum-detached cellulose ester film emits electrons at 10^4-10^5 sec^{-1} with energies of the order of 10 keV, the emission falling exponentially at a rate governed by the functional groups.

The IR results [6] indicate the following explanation. The OH groups in the polymer most probably form hydrogen bonds to the glass as below:

$$\begin{array}{l} \quad\quad -\overset{|}{\underset{|}{Si}}-O-H \ldots OH-R \\ glass \quad -\overset{|}{\underset{|}{Si}}-O-H \ldots OH-R \quad polymer \\ \quad\quad -\overset{|}{\underset{|}{Si}}-O-H \ldots OH-R \end{array}$$

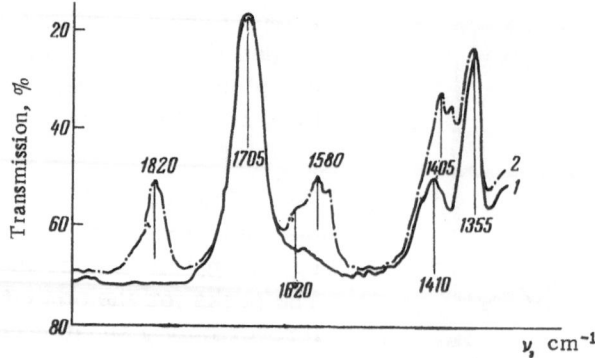

Fig. 5. IR spectra (1300–1800 cm^{-1}) of acetyl-cellulose: 1) after prolonged evacuation; 2) in saturated vapor of acrylonitrile.

The glass becomes positively charged on detaching the polymer film, so the potential well for the proton lies near the oxygen in the glass, while the electron cloud is drawn off towards the polymer oxygen. This unsymmetrical charge distribution represents a double electrical layer, the glass on detachment being protonated, while the polymer is induced by the strong fields in the gap to emit electrons in accordance with the following schema:

Glass surface	Polymer surface
—Si—OH$^+$	$e \leftarrow$ OH—R
—Si—OH$^+$	$e \leftarrow$ OH—R
—Si—OH$^+$	$e \leftarrow$ OH—R

The polymer emits electrons of energy ~10 keV, which is more than sufficient to break the double bond $CH_2=CH$ in the monomer to give a free radical via

$$CN$$

$$\text{Radiation} + CH_2=CH \rightarrow -CH_2-\dot{C}H$$
$$\quad\quad CN \quad\quad\quad CN$$

The subsequent reactions may take two paths:

1. Growth of the polyacrylonitrile chain and formation of homopolymer:

$$-CH_2-\dot{C}H + CH_2=CH \rightarrow CH_2-CH-CH_2-\dot{C}H \text{ etc.}$$
$$\quad CN \quad\quad CN \quad\quad\quad CN \quad\quad CN$$

2. Grafting of acrylonitrile by chain transfer to the cellulose:

$$-CH_2-CH-CH_2-\dot{C}H + \begin{matrix} OH- \\ OH- \\ OH- \end{matrix} \rightarrow -C_2-CH-CH_2-CH_2 \begin{matrix} \dot{O}- \\ OH- \\ OH- \end{matrix} + \begin{matrix} -\dot{O} \\ -OH \\ -OH \end{matrix} \rightarrow \begin{matrix} -\dot{O} \\ -OH \\ -OH \end{matrix} CH_2=CH \rightarrow \begin{matrix} -O-CH_2-CH' \\ -OH \\ -OH \end{matrix}$$

Not only do the electrons generate radicals from the monomer, but also the surface itself has emission centers, which may also be active adsorption centers. An emitting center is at a high energy level before its electrons have reached the surface, so it also may break double bonds and bind acrylonitrile by producing a free radical, which subsequently generates a chain attached to the surface.

Conclusions

The IR spectra of freshly formed polymer surfaces indicate that electron emission produces rupture of double bonds in nearby molecules, which leads to production of homopolymer and to the formation of polymer grafted to the surface.

These reactions at the surface may also involve active centers, because the surface of the polymer contains electrons at high energy levels.

References

1. N. K. Baramboim, Polymer Mechanochemistry, Gostekhizdat, Moscow (1961).
2. S. N. Zhurkov, I. I. Novak, and V. I. Vettegren', Dokl. AN SSSR, 157:1431 (1964).
3. V. V. Karasev, N. A. Krotova, and B. V. Deryagin, Dokl. AN SSSR, 88(5):777 (1953); L. P. Morozova and N. A. Krotova, Dokl. AN SSSR, 115:747 (1957).
4. N. V. Ivanova and R. G. Zhbankov, in: The Hydrogen Bond, Nauka, Moscow (1964), p. 149.
5. A. M. Polyakov and N. A. Krotova, in: Research in Surface Forces, Vol. 2, Consultants Bureau, New York (1966), p. 280.
6. N. A. Krotova and L. P. Morozova, in: Research in Surface Forces, Vol. 1, Consultants Bureau, New York (1963), p. 64.

CHEMICAL MODIFICATION OF THE SURFACE
OF PTFE AND PRODUCTION OF ADHESION

N. N. Stefanovich and N. A. Krotova

Surface Phenomena Division, Institute of Physical Chemistry,
Academy of Sciences of the USSR

A study is reported of surface grafting of polystyrene and polymethylmethacrylate to polytetrafluoro-ethylene films after treatment in a glow discharge. It is considered that the grafting occurs via a radical mechanism. The adhesion of the resulting modified films is substantially increased and is related to the extent of the grafting and the nature of the functional groups.

Attachment of polytetrafluoroethylene (PTFE) is an important but difficult problem. PTFE has properties (chemical inertness and dielectric strength) that make it indispensable in some chemical and electronic applications, but the difficulties of attachment restrict its use. The importance of the problem is demonstrated by the large number of papers on the modification of the surface of PTFE to produce adhesion.

Methods have been described involving the use of vigorous chemical reagents, such as molten potassium acetate [1], sodium-naphthalene complex [2], sodium in liquid ammonia [3], and alkoxy derivatives of metals of groups 1, 2, and 3 in the periodic system [4]. IR studies [5] of the surface after such treatments show that these introduce conjugated double bonds and functional groups such as OH, CH_3, CO, NH, and CH_2. These groups result in a considerable increase in the adhesion.

Another approach is to graft on a polymer showing good adhesion, for which purpose the surface of the PTFE must be activated, e.g., by γ irradiation. The grafting is done either directly in the beam [6] or after the irradiation [7]. A silent electrical discharge has been used for the same purpose [8], subsequent grafting of polymethylmethacrylate leading to some increase (5-22 kg/cm^2) in the adhesion. The surface properties of PTFE were unaffected by treatment with a plasma gun [9]. We have modified the surface of PTFE by activation in a glow discharge, with subsequent grafting of vinyl polymers [10].

Methods

The PFTE film is placed between the electrodes 25-30 mm from the cathode. The discharge is struck at 600-700 V and 0.4-0.6 mm Hg, with a discharge current of 30 mA. The

TABLE 1. Surface Properties of Modified PTFE Films

PTFE film	Grafted polymer		Angle, deg	Adhesion, kg/cm^2
	mg/cm^2	%		
Initial	–	–	108	No adhesion
Treated in glow discharge	–	–	83-45	109
Grafted with:				
PS	2.4×10^{-1}	1.1	75	49
PMMA	2.6×10^{-1}	1.2	67	77
PMMA	1.2	5.7	67	128
PMMA	1.7	8.1	67	190

subsequent grafting of the polymer (polymethylmethacrylate PMMA or polystyrene PS) is performed by heating the film in a sealed tube with the appropriate monomer at 80°C for 4-6 h. Most of the tests were done with methyl methacrylate (MMA).

In the early tests, the heating with the monomer was done in an atmosphere of nitrogen, but it was found that equally good grafting was produced in air. The amount of grafted polymer was determined by weighing after the film had been freed from homopolymer by washing to constant weight. It was found that boiling for 14 h in a solvent removed the homopolymer completely. The amount of grafted polymer was expressed as percent of the weight of the film after the glow-discharge treatment or as mg/cm^2. As the grafting occurs at the surface, the latter provides a better basis for comparing fibers, films of various thicknesses, etc.; the per cent expression was used only for tests with a single material.

Table 1 gives the characteristics of the films (wetting angle for water and strength of attachment of ÉD-6 epoxide resin with polyethylene polyamide as hardening agent.[†]

The angles were measured to ±4-5° and the adhesion to 15%.

These properties showed a large variation over the surface of the specimen and from specimen to specimen after the glow-discharge treatment. The performance of the treatment may be judged from the weight loss, which may be expressed as per cent of the initial weight. Figure 1 shows this loss as a function of treatment time. The loss for a given treatment time is dependent on the variation of current; the full line denotes the result for a current of 25-30 mA. The variations over the surface of the specimen are related to the position of the specimen in the discharge; it is likely that not all points are equally affected by the discharge. The weight loss Δq on treatment of one side was 1-1.5%.

The amount G of grafted polymer is related to Δq (Fig. 2); these results are for treatment with pure MMA at 80°C for 6 h without initiator, and they show about 3% grafting or 1.3 mg/cm^2. Up to 15% can be obtained by longer treatment in the monomer. Also, G for a given Δq is related to the molecular weight of the grafted chains. Solvent or initiator substantially reduces G, other conditions being the same.

Figure 3 shows G as a function of benzoyl peroxide concentration in pure MMA. The presence of 0.3% initator accelerates the reaction but reduces G, which is due to shortening of the grafted chains.

[†] The adhesion tests were done by Yu. M. Emmanuilov in the Polymer Coatings Division of this institute.

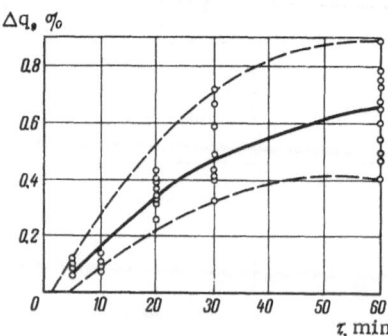

Fig. 1. Weight loss Δq of a
PTFE film as a function of
time τ of treatment in a glow
discharge.

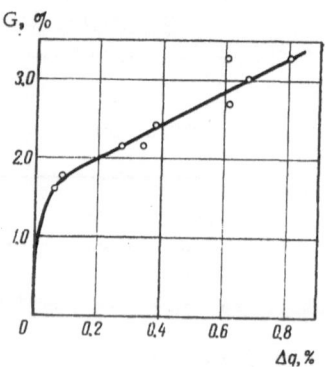

Fig. 2. Grafting G of PMMA as
a function of weight loss Δq for
PTFE treated in a glow discharge.

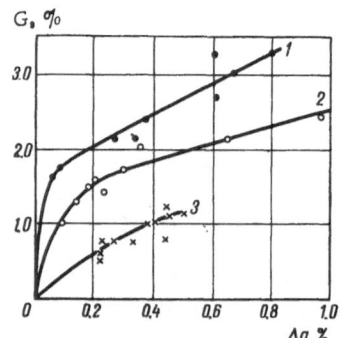

Fig. 3. Grafting G of PMMA
as a function of weight loss Δq
for PTFE treated in a glow discharge in relation to initiator
concentration (benzoyl peroxide):
1) 0; 2) 0.05%; 3) 0.3%

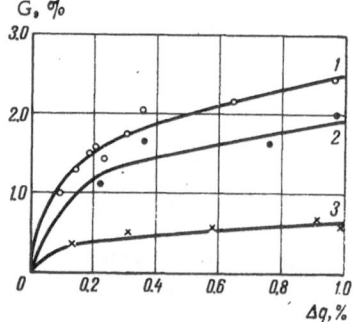

Fig. 4. Grafting G of PMMA as
a function of dilution of MMA
with benzene in the presence of
0.05% benzoyl peroxide; MMA
concentration (%): 1) 100; 2) 50;
3) 20.

Dilution of the monomer also affects G; Fig. 4 shows the effects of dilution with benzene.
Dilution with CCl_4 (a good chain-transfer agent) produces an even greater fall in G.

The glow discharge would appear to produce hydroperoxide groups on the surface, but
these are undetectable by the methods of chemical analysis, perhaps because the concentration
is too low. The film does not initiate polymerization of the monomer; instead, grafting occurs when the monomer polymerizes in response to heat. Also, grafting in air is as effective
as grafting in an inert medium, so it may occur by a simple mechanism of chain transfer from
a radical of the monomer to the activated film. An untreated film, or a treated one that has
lost no weight, does not gain in weight on heating with the monomer.

No conclusion can be drawn at present on the surface groups responsible for the grafting.
The weight loss in the discharge indicates that oxidation or surface destruction occurs, the
products of low molecular weight being lost.

A plasma torch produces activated oxygen (atomic oxygen), and this also causes a loss in
weight. There is no weight loss if helium replaces oxygen [9]. However, treatment of PTFE

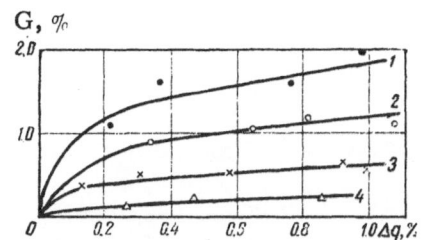

Fig. 5. Effect of nature of solvent on grafting G of PMMA in the presence of 0.05% benzoyl peroxide: 1) 50% in benzene; 2) 50% in CCl_4; 3) 20% in benzene; 4) 20% in CCl_4.

with atomic oxygen does not produce a change in surface properties, in spite of the weight loss [9, 11]. In this case, the material is not in a discharge but is merely acted on by atomic oxygen. A discharge exposes the surface to electrons as well as atoms. The weight loss indicates that oxidation or destruction products are lost, but presumably some oxidized molecules remain, which participate in grafting.

This method of modification is of considerable interest when adhesion is to be produced only on one side, the unaltered side retaining the initial properties. In principle, the method allows a vinyl polymer with any functional groups to be grafted. A glow discharge as a means of producing active centers is a promising method of surface modification.

References

1. A. Ya. Korolev, in: Adhesives and Adhesion Technology, edited by D. A. Kardashev, Oborongiz, Moscow (1960), p. 35.
2. E. R. Nelson, T. J. Kilduff, and A. A. Benderly, Ind. Eng. Chem., 50:329 (1958); A. A. Benderly, J. Appl. Polymer Sci., 6:221 (1962).
3. J. Skeist, Plastics Technol., 2:458 (1956); A. P. Curd, British Patent 924833 (1963).
4. D. G. Gluck, U. S. Patent 3067078 (1960).
5. F. K. Borisova, G. A. Galkin, A. V. Kiselev, A. Ya. Korolev, and V. I. Lygin, Kolloidn. Zh., 27:320 (1965).
6. A. Chapiro, J. Polymer Sci., 34:481 (1959).
7. J. Dobo, A. Somogyi, and T. Czvikovszky, J. Polymer Sci., C2:1173 (1964).
8. V. V. Korshak, K. K. Mozgova, and T. M. Babchinitser, Dokl. AN SSSR, 151:1332 (1963).
9. R. Mantell and W. L. Ormand, Ind. Eng. Chem. Prod. Res. Develop., 3:300 (1964).
10. N. N. Stefanovich, USSR N201652 (Ce C08f) 3/13/65, Appl. 9/8/67.
11. R. H. Harsen, J. V. Pascale, T. Benedictis, and P. M. Reutzepis, J. Polymer Sci., A3:2205 (1965).

ADHESION OF GERMANIUM

V. I. Anisimova, N. A. Krotova, and L. A. Sukhareva

Surface Phenomena Division, Institute of Physical Chemistry,
Academy of Sciences of the USSR

It has been found that the functional groups in polymer films on germanium affect the electrophysical properties of the latter. The surface conductivity is altered on detaching the film. Measurements have been made on the shrinkage stresses in these films.

Adhesion of polymers to germanium is of substantial interest, but little has been published on adhesion to semiconductors. Allowance must be made for effects on the electrophysical properties in choosing protective coatings. Evidence on the adhesion mechanism can be provided by parallel measurement of adhesion and change in electrical parameters. We have examined the effects of various cellulose esters on the electrophysical properties of germanium, the conductivity change $\Delta\sigma$ on detaching the film, and the adhesion.

Methods

The adhesion was evaluated from the work of detachment as measured with an adhesiometer providing a controlled rate in an adjustable atmosphere (Fig. 1). The fixed plate 5 bears the vacuum disc 3. Bellows 4 allows vertical displacement of the specimen 2 without breaking the vacuum. The speed of the specimen was set by a gearbox, which has a magnetic clutch, and could be varied from 3×10^{-4} to 2.5 cm/sec. The force of detachment was measured with a dynamometer, which consisted of strain gauges connected in a bridge [1]. The bridge signal was passed to a UD-3 amplifier and thence to an N102 loop oscillograph. The dynamometer was calibrated with weights. Allowance was made for the internal stresses produced in making the films, which were measured via the birefringence [2]. It is true that these measurements are made with the films mounted on glass prisms, but it has been shown [3] that the internal stresses are independent of the substrate and are governed by the mechanical properties of the films.

The electrophysical properties of the surface were examined via the field effect at constant voltage [4] and for large sinusoidal signals [5] as well as via photoconduction [6]. The conductivity change $\Delta\sigma$ during rupture of the contact was measured via the bridge circuit used in the dc measurements on the field effect. These measurements were made at 10^{-5} mm Hg and in air.

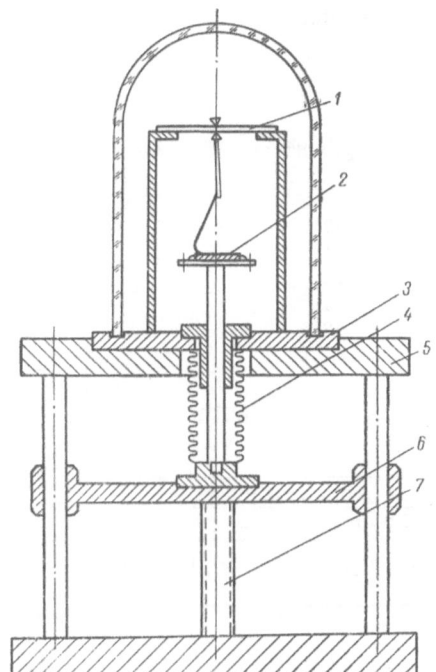

Fig. 1. The adhesiometer: 1) dynamometer; 2) specimen bearing film; 3) baseplate; 4) bellows; 5) plate; 6) clutch; 7) lead screw.

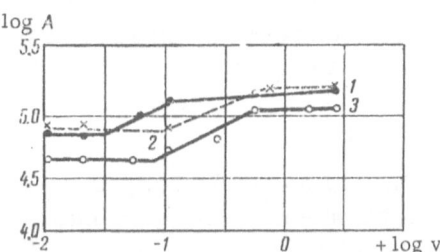

Fig. 2. Velocity dependence of A for Ge bearing: 1) nitrocellulose; 2) methylcellulose under vacuum; 3) the same in air.

We used n-type germanium of specific resistance 40 ohm · cm as plates $20 \times 10 \times 1$ mm cut parallel to the 111 axis. The specimens were ground with boron carbide powder, etched in boiling 30% H_2O_2 containing a few drops of 10% KOH, and repeatedly washed with hot double-distilled water. The etching was monitored via S, the rate of surface recombination; specimens with $S \le 30$ cm/sec were chosen. The polymer film was deposited from 4% solution; in some instances, the film was deposited on both sides, which allowed the adhesion and electrophysical properties to be measured on the same specimen. The solutions were poured onto the plates in a chamber having 5-7% relative humidity and were left there for 2-3 days. The film thickness was usually 55-60 μ. We used cellulose derivatives: nitrocellulose, benzylcellulose, acetylcellulose, ethylcellulose, and methylcellulose.

Experimental

It has been shown [7, 8] for metal−polymer and glass−polymer systems that the following features occur in most instances:

1) Velocity dependence of the work of detachment A;

2) Effects from the medium on the work of detachment;

3) Electron emission on breaking the adhesion bond.

It was of interest to establish whether these occur here. The velocity dependence of A was present for all the materials except acetylcellulose, for which the log A = f (log v) relation was not very clear.

Figure 2 shows curves with the three parts characteristic of other systems [7].

The velocity dependence of A was largely unaffected by the use of air instead of vacuum (curve 2 of Fig. 2); the three parts persist, but the adhesion under vacuum is usually stronger, as Table 1 shows by reference to the limiting A for v = 2.5 cm/sec.

The compounds may be arranged in order of increasing σ_{in} as follows: benzylcellulose, methylcellulose, nitrocellulose, acetylcellulose, ethylcellulose (Table 1). It is difficult to say whether the adhesion is dependent on the chemical structure of the polymer or on the magnitude of σ_{in}.

All the compounds were found to affect the relation of surface conductivity in the field effect to the induced charge, $\Delta\sigma = f$ (Q). The $\Delta\sigma = f$ (Q) curves may be compared with theoretical curves to give the charge trapped in surface states as a function of band bending, $Q_{ss} = f$ (Y_s) (Fig. 3). The $\Delta\sigma = f$ (Q) and $Q_{ss} = f(Y_s)$ curves gave the potential and state of charge of the surface.

TABLE 1. Work of Detachment A and Internal Stress σ_{in} for Polymers on Ge

Polymer	Vacuum	Air	σ_{in}, kg/cm²
	A, erg/cm²	A, erg/cm²	
Benzylcellulose	$3.2 \cdot 10^5$	$2.1 \cdot 10^5$	12
Nitrocellulose	$2.3 \cdot 10^5$	$1.8 \cdot 10^5$	21
Acetylcellulose	$2.67 \cdot 10^5$	$1.5 \cdot 10^5$	24
Ethylcellulose	$1.94 \cdot 10^5$	$1.3 \cdot 10^5$	34
Methylcellulose	$1.7 \cdot 10^5$	$1.1 \cdot 10^5$	18

TABLE 2. Electrical Parameters of Germanium Surfaces with and without Films

Film	Before detachment		After detachment		
			v = 2.5 cm/sec		v = 10⁻² cm/sec
	Y_s, kT/e	$Q_{ss} \cdot 10^{-8}$, C/cm²	Y_s, kT/e	$Q_{ss} \cdot 10^{-8}$, C/cm²	Y_s, kT/e
None	+1.3	+0.14	—	—	
Benzylcellulose	−4.1	−1.34	—	—	
Nitrocellulose	+1.5	+0.14	−0.6	−0.1	−0.4
Acetylcellulose	−2	−0.35	−0.1	−0.1	—
Ethylcellulose	−2.3	−0.44	−4.2	−0.8	—
Methylcellulose	−2.6	−0.6	+1.5	+0.15	+1.4

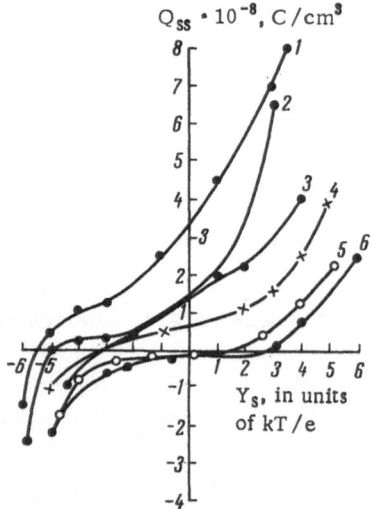

Fig. 3. Charge Q_{ss} in surface states as a function of Y_s for Ge bearing films of: 1) benzylcellulose; 2) methylcellulose; 3) ethylcellulose; 4) acetylcellulose; 5) nitrocellulose; 6) no polymer.

Table 2 gives the results, which show that the films alter the surface properties, which points to changes in the surface levels. All of the present compounds, except nitrocellulose, are donors with respect to germanium, since the surface acquires a negative charge.

It has been shown [9] that film detachment produces a conductivity change and electron emission whose intensity is dependent on the rate of detachment.

Figure 4 shows typical curves for the detachment of ethylcellulose from germanium; benzylcellulose, nitrocellulose, and acetylcellulose gave similar curves.

The following trends occur:

1. $\Delta\sigma$ increases with v.

2. The conductivity after detachment relaxes towards some stationary value in a way dependent on the film. The curve is exponential in the case of benzylcellulose but logarithmic for ethylcellulose and nitrocellulose.

3. There is limiting $\Delta\sigma$ attained at v = 10⁻² cm/sec for all compounds except benzylcellulose, while $\Delta\sigma$ = 0 for v = 10⁻³ cm/sec.

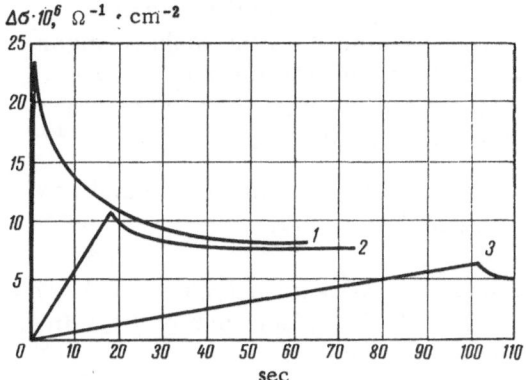

Fig. 4. $\Delta\sigma$ as a function of time in the detachment of ethylcellulose at a rate (cm/sec) of: 1) 1; 2) 10^{-1}; 3) 10^{-2}.

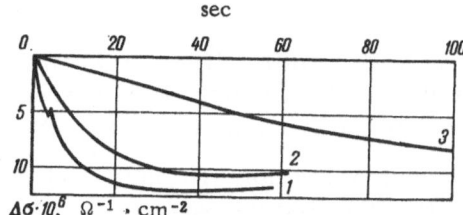

Fig. 5. $\Delta\sigma$ as a function of time in the detachment of methylcellulose at a rate (cm/sec) of: 1) 1; 2) 10^{-1}; 3) 10^{-2}.

4. Usually, $\Delta\sigma$ is less for detachment in air.

5. The relaxation of $\Delta\sigma$ in air differs from that under vacuum.

Methylcellulose shows a different mode of variation in $\Delta\sigma$, which here is negative (Fig. 5), the trends being as follows:

1. $\Delta\sigma$ is dependent on v, increasing as v decreases.

2. $\Delta\sigma$ relaxes after detachment.

3. The relaxation time and mode of relaxation are almost independent of the thickness of the oxide film on the germanium.

4. The medium affects the magnitude of $\Delta\sigma$ and the mode of relaxation.

The surface properties of the Ge were examined after detachment (Table 2). Only methylcellulose gave complete recovery of the potential and charge to the values before the film was applied. Table 2 also shows that there is no effect of v on the properties (in particular, the equilibrium conductivity) in the cases of nitrocellulose and methylcellulose.

Discussion

Ge-polymer contacts show the trends characteristic of glass—polymer contacts for cellulose esters [7]; in particular, the work of detachment is affected by v and by the surrounding medium (Fig. 2 and Table 1), but comparison with the results for glass [7] shows that Ge has a very much narrower transition region, and there is no great difference between ranges I and III, evidently because a semiconductor has an additional mechanism for charge recombination from the surface by band-level exchange [9].

These trends can be explained on the basis of a double electrical layer at the interface [7, 10].

It has been shown [9] that the various trends for Ge can be explained if we assume that such a layer is present and makes itself felt on breaking the contact, because the semiconductor surface retains an uncompensated charge, which affects the conductivity in the space-charge layer, as in the field effect.

The $\Delta\sigma = f(t)$ curves for these cellulose derivatives can be explained via a single mechanism for the change in $\Delta\sigma$ on account of the uncompensated charge after detachment. The films (other than methylcellulose) have a negative charge after detachment, so the Ge has a positive charge, which increases the electron concentration in the space-charge layer and so increases the conductivity of the Ge.

Methylcellulose also causes the space-charge layer to become enriched in electrons (Table 2), but it leaves behind an uncompensated negative charge, which reduces the conductivity.

TABLE 3. Work of Detachment A and
Uncompensated Charge $\varkappa_0$

Polymer	A, erg/cm^2	$\varkappa_0$, cgse/cm^2
Benzylcellulose	3.2×10^5	100
Acetylcellulose	2.6×10^5	88
Ethylcellulose	1.84×10^5	95
Nitrocellulose	2.3×10^5	50
Methylcellulose	1.6×10^5	30

Relaxation in $\Delta\sigma$ shows that the charge leaks away. If we assume that the leakage mechanism is always the same, the different trends in $\Delta\sigma$ would mean that the surface after detachment has different electronic structures.

The medium affects the magnitude and relaxation of $\Delta\sigma$, which also indicates that a double layer is present; $\Delta\sigma$ is reduced because the atmosphere provides an additional leakage path (gas discharge). Thus, all the effects can be explained in terms of the double layer.

Table 1 shows that the adhesion and internal stress σ_{in} vary with the polymer. As regards σ_{in}, it should be borne in mind that the solvent has a plasticizing effect. The solvents for methylcellulose, ethylcellulose, and nitrocellulose were of low volatility (water, butyl acetate), and so the effects of these may persist for a long time and reduce σ_{in}. The volatile solvents acetone and dichloroethane were used for acetylcellulose and benzylcellulose, which should result in higher σ_{in}.

The σ_{in} series does not agree with the adhesion series, so it is difficult to say whether the difference in adhesion is due to σ_{in} or is related to the chemical structure of the polymer. Table 1 indicates, however, that the structure does not have a pronounced effect on the adhesion.

It has been supposed [11] that the OH groups in the glucoside link play the main part in the adhesion of cellulose derivatives to glass, substituents affecting the result in so far as they block the OH groups. The surface of Ge resembles that of glass in bearing OH groups capable of forming hydrogen bonds to the OH groups of cellulose derivatives, while the functional groups serve only to block the OH.

Table 3 gives results on the space charge produced in Ge by detachment of films of cellulose derivatives. The surface charge densities $\varkappa_0$ are low because the charge starts to be neutralized even during detachment via tunneling recombination, leakage to the point of detachment, band-level exchange, gas discharge (detachment in a gas), and electron emission (detachment under vacuum).

Information on the type of detachment is obtained [9] by measuring the electrophysical properties of the surface before coating, when the film is present, and after detachment. If the Ge does not recover the parameters found before coating, we may speak of mixed detachment; if recovery does occur, we have adhesion detachment. Table 2 shows that only methylcellulose shows the latter, and the mode of detachment does not alter when the detachment rate is varied from 10^{-2} to 2.5 cm/sec. The other derivatives show mixed detachment.

References

1.   N. A. Krotova, L. P. Morozova, and G. A. Sokolina, Fiz. Tverd. Tela, 3:2000 (1961).
2.   P. I. Zubov and L. A. Lepilkina, Kolloidn. Zh., 24:174 (1962).
3.   A. T. Sanzharovskii and G. I. Epifanov, Vysokomolekul. Soedin, 2:1709 (1960); 3:1641 (1961).
4.   S. R. Morrison, in: Physics of Semiconductor Surfaces, IL, Moscow (1959).
5.   A. V. Rzhanov, Yu. F. Novototskii-Vlasov, and I. G. Neizvestnyi, Zh. Tekh. Fiz., 27:2448 (1957).
6.   A. B. Engler and C. J. Kavane, Rev. Sci. Inst., 28:548 (1957).
7.   N. A. Krotova, Yu. M. Kirillova, and B. V. Deryagin, Zh. Fiz. Khim., 30:1921 (1956).

8.   L. P. Morozova, Dissertation, Institute of Physical Chemistry, AN SSSR (1958).
9.   G. A. Sokolina, Dissertation, Institute of Physical Chemistry, AN SSSR (1966).
10. B. V. Deryagin and N. A. Krotova, Adhesion, Izd. AN SSSR, Moscow (1949).
11. L. P. Morozova and N. A. Krotova, this volume, p. 377.

EFFECTS OF SURFACE CHEMICAL STRUCTURE AND CONTACT DESTRUCTION RATE ON ELECTRON EMISSION

A. M. Polyakov and N. A. Krotova

Surface Phenomena Division, Institute of Physical Chemistry,
Academy of Sciences of the USSR

A novel adhesiometer is used to make simultaneous measurements of adhesion and electron emission rate. This is applied to study the effects of rate of detachment and of type of functional group in the polymer. It is found that the adhesion is correlated with the emission rate at various rates of detachment.

The electrical theory of adhesion treats the double electrical layer at an adhesion as a consequence of the donor−acceptor bond [1], and it is naturally assumed that formation of the double layer is dominated by the specific properties of the surfaces, especially the nature of the functional groups. In particular, if one surface bears many donor groups, and the other has acceptor ones, we may have donor−acceptor interaction in the simplest sense, and so the role of the electrical forces in adhesion is directly related to the chemical nature of the containing materials. The activity of the functional groups should affect the strength of the bond via the surface charge density, and the latter should influence the electron emission on breaking the contact, though this emission should also be dependent on the conditions of contact rupture, in particular the rate of detachment. These features are considered here.

A glass roller was coated with polymer from a 4% solution, and the force of detachment was measured with a strain-gauge system at rates of detachment from 10^{-3} to $1 \, cm/sec$, as selected with a four-speed gearbox. The electron emission was recorded by a secondary-electron multiplier. A detailed description of the apparatus has been given elsewhere [2].

The electron transitions involved in the donor−acceptor interaction cause the surfaces to become charged, the charges being retained after the contact is broken. Functional groups naturally differ in electron affinity and so may acquire charges of different signs.

We have shown [3] that some functional groups may be arranged in the following order as regards tendency to produce a positive charge:

$$\rightarrow NH_2 > OH > OCOR > C_6H_5 > Cl > COOH > CN > -C = C-$$

The electron−acceptor behavior increases from left to right in this series. The greater the difference in electron affinity between the contacting surfaces, the stronger the adhesion

TABLE 1. Emission Rate N in the Detachment of Polymers from Glass

	Cellulose derivative			SKS-30 + 15% MMA	Gutta-percha	SKN-40
	acetyl-	ethyl-	nitro-			
$N \times 10^{-3}$, sec^{-1}	1.4	4.0	5.5	7.2	55	80

TABLE 2. Emission Rate N and Adhesion A
in Detachment of Polymers from Glass at 0.1 cm/sec

	Cellulose derivative		
	acetyl-	ethyl-	nitro-
$A \times 10^{-4}$, erg/cm^2	1.7	3.3	5.3
$N \times 10^{-3}$, sec^{-1}	1.4	4.0	5.5

bond. Smigla [4] drew an analogous conclusion from a theoretical study of the dependence of the number of reacting donor−acceptor pairs on the difference between the electron levels in the two types of center.

Increase in double-layer density should be accompanied by increase in electron emission rate on detachment, as we actually find (Table 1).

Untreated glass bears −OH groups, and detachment of polymer films gives rise to the following sequence in increasing emission rate N: cellulose derivatives with −OCOR groups, butadiene−styrene rubber (−C$_6$H$_5$) containing methacrylic acid (−COOH, very active), and SKN-40 nitrile rubber (−CN groups). Guttapercha does not fit properly into the series, as it comes after nitrile rubber in the donor−acceptor series but before it as regards N.

The differences between cellulose derivatives are determined by the activity of radical R in the −OCOR functional group. Table 2 shows that acetylcellulose (−OCOCH$_3$ group) gives the lowest N and the worst adhesion, while the more active −OC$_2$H$_5$ and −ONO$_2$ groups produce better adhesion and higher N.

Morozova [5] examined the effects of functional groups and found that the adhesion to glass increased with the proportion of methacrylic acid (MAA) in copolymers of MAA with butadiene−styrene rubber. We have used an analogous series of carboxylate rubbers to trace the effect on N (Fig. 1); the very good agreement between the shapes of the curves shows that the adhesion and N are determined primarily by the functional groups in the polymer.

We have previously observed [2, 3] that an electron multiplier provides reliable observation of the afteremission from freshly prepared polymer films, which is of the same nature as the emission in film removal (field emission). The gradually decaying emission from a fresh film occurs under a high vacuum, in complete darkness, and without any external energy supply. The effect is clearly due to cold emission caused by distortion of the potential barrier by the large surface charge left behind by the double layer.†

Most instances of exoemission are ones of afteremission due to previous treatment of the solids, and such treatments often produce a surface charge [6, 7], with an exponential decay, as in our experiments. In most cases of rapid detachment from glass, a logarithmic plot (Fig. 2) gives a straight line representing

$$N = at^{-b},$$

_____

† See [13] on the direct relation of film charge to afteremission rate.

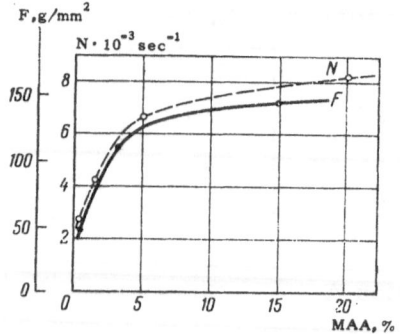

Fig. 1. Electron emission rate N and detachment force F as a function of % methacrylic acid (MAA) in SKS-30 copolymer.

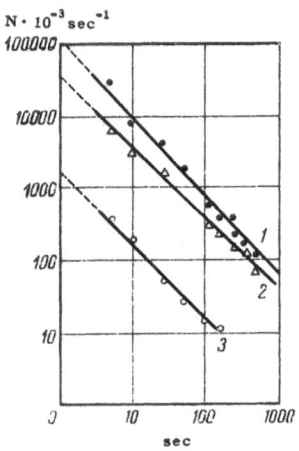

Fig. 2. Decay of emission from polymer films detached from glass: 1) SKN-40; 2) guttapercha; 3) SKS-30-1.

in which a is a constant corresponding to the initial rate and characterizing the adhesion activity, while b is a constant close to one.

This decay law does not always apply. For instance, acetylcellulose after detachment from glass often gives a high rate persisting for a long time, whereas nitrocellulose gives virtually no afteremission. This pure emission during contact rupture only gives a clear indication of the change in rate during detachment itself. The processes in rupture of an adhesive contact provide a test of adhesion theories, quite apart from their practical importance.

In the electrical theory, the elementary act of rupture of a double electrical layer is considered as the separation of the plates of planar capacitor, whose energy is taken as that of adhesion (detachment). This is confirmed by the quasi-periodic voltage variations during detachment in a Polyani-type tester with a fixed detachment rate [8], and also by the discrete flashes of light given by the glass during spontaneous detachment of polymer films in response to cooling [9]. However, the clearest signs of stepwise detachment are seen by recording electron emission during the detachment of nitrocellulose from glass. Even at 10^{-2} cm/sec, the emission occurs in regular bursts (Fig. 3a), and slower detachment reveals the structure of these bursts (Fig. 3b, 10^{-3} cm/sec). The emission gradually rises before the onset of a burst, which corresponds to complete detachment of an elementary area.

The potential difference between the plates of the condenser gradually rises during detachment, which is accompanied by a slow increase in emission, until it reaches a critical value such that the electrons accelerated by the field acquire energy sufficient to initiate an avalanche, which greatly reduces the charge density in the double layer and hence the force. The strain gauge recording the detachment force simultaneously reveals a marked fall in the force. The number of bursts does not vary greatly from run to run, which allows one to estimate the length of an elementary detachment area as ~300 μ.

The electron emission from a freshly detached polymer substantially determines the chemical activity of the surface. Electron emission favors vapor-state polymerization of the monomer [10], and rapid grafting occurs at fresh mechanically produced breakage surfaces [11]. These effects have been ascribed to the free radicals always present on such surfaces. However, the following detailed study (by L. A. Morozova) of the above surfaces shows that this is not the only feature in the grafting of acrylonitrile vapor.

IR spectra show that the chemical activity is directly proportional to the emissivity, an important factor being the emission after detachment. The peak representing acrylonitrile grafted to the surface of nitrocellulose, for example, increases at once to its maximal value and does not thereafter alter, whereas acetylcellulose shows a gradual increase, i.e., grafting continues, which is related to the substantial afteremission of this material. This indicates that the emission centers participate actively in adsorption, grafting, and polymerization.

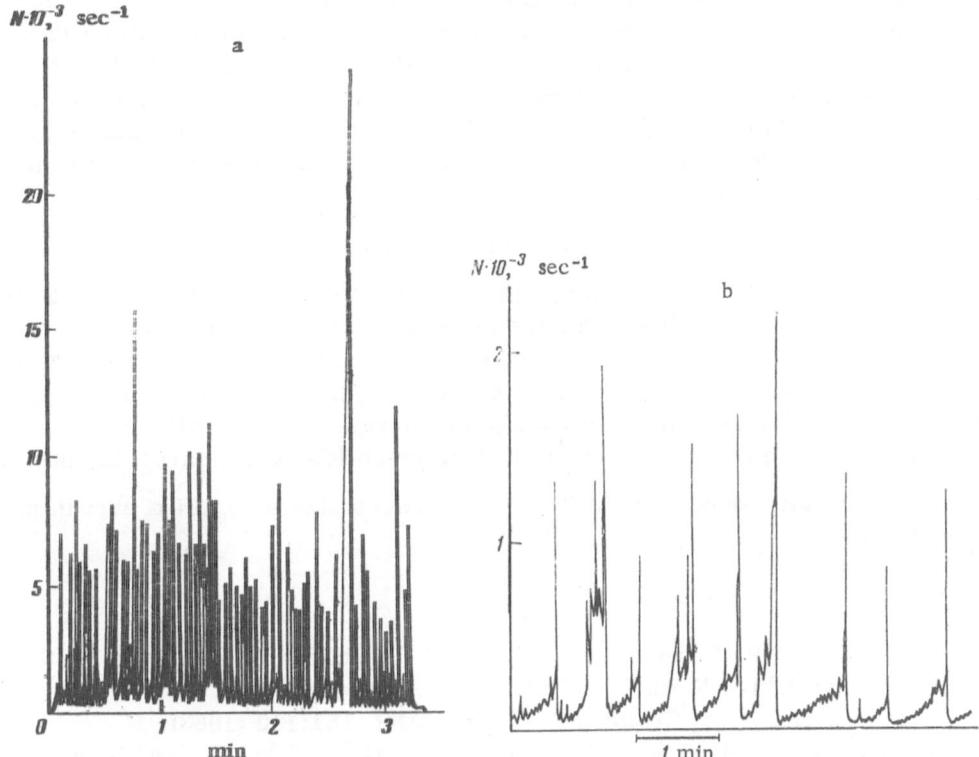

Fig. 3. Rate of electron emission in detachment of nitrocellulose films from glass at a rate (cm/sec) of: a) 10^{-2}; b) 10^{-3}.

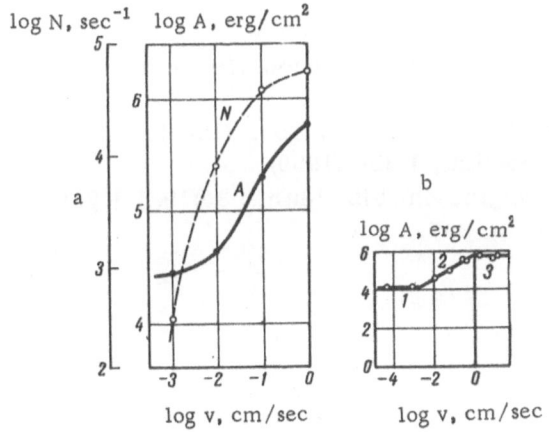

Fig. 4. a) Work of adhesion A and electron emission rate N as functions of detachment rate v for guttapercha on glass; b) adhesogram for guttapercha on glass [6].

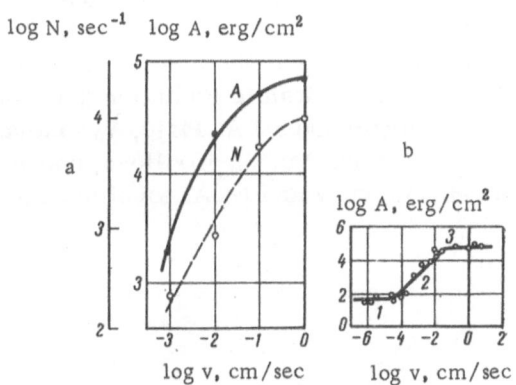

Fig. 5. a) Work of adhesion A and electron emission rate N as functions of detachment rate v for nitrocellulose on glass; b) adhesogram for nitrocellulose on glass [12].

The relation of work of adhesion to detachment rate $A = f(v)$ is an important aspect of any adhesion theory; Figs. 4b and 5b show that this relation has three distinct parts, and only the electrical theory explains these. At low v (part 1), the charges can leak away, and A is governed mainly by the intermolecular forces. Part 3 represents high v, where most of the work is done against the electrostatic forces, when the charges have no time to leak away; all the effects due to disruption of the double layer are observed. Part 2 corresponds to a substantial loss from charge leakage or recombination.

This relation naturally implies a relation of N to v.

We used the range 10^{-3} to 1 cm/sec, which corresponds generally to part 2, though this varies from one polymer to another. For instance, 10^{-3} cm/sec corresponds to the start of part 2 for guttapercha (Fig. 4), and this is accompanied by the rapid onset of emission, which thereafter increases in the same way as A. On the other hand, 10^{-3} cm/sec for nitrocellulose on glass corresponds to the straight-line section of part 2 (Fig. 5), with an analogous correlation between A and N. The rate of increase in N is much less when part 3 is entered.

The present results show that the functional groups play a decisive part in producing the electrical double layer.

References

1. B. V. Deryagin and V. P. Smigla, Dokl. AN SSSR, 121(5):877 (1958); V. P. Smigla and B. V. Deryagin, Dokl. AN SSSR, 122(6):1049 (1958).
2. A. M. Polyakov and N. A. Krotova, Dokl. AN SSSR, 151:130 (1963).
3. N. A. Krotova, L. P. Morozova, A. M. Polyakov, G. A. Sokolina, and N. N. Stefanovich, Kolloidn. Zh., 26:207 (1964).
4. V. P. Smigla, in: Research in Surface Forces, Vol. 1, Consultants Bureau, New York (1963), p. 58.
5. L. P. Morozova, Dissertation, Institute of Physical Chemistry, Moscow (1958).
6. J. Kramer, Z. Phys., 129:34 (1951).
7. W. Hanle and G. Gourge, Phys. Blät., 14:499 (1958).
8. B. V. Deryagin and N. A. Krotova, Adhesion, Izd. AN SSSR, Moscow (1949).
9. R. C. Williams, J. Appl. Phys., 28:1043 (1957).
10. A. Charlesby, Atomic Radiation and Polymers, Pergamon, New York (1960).
11. V. A. Kargin and N. A. Platé, Vysokomolekul. Soedin., 1:330 (1959).
12. N. A. Krotova, Yu. M. Kirillova, and B. V. Deryagin, Zh. Fiz. Khim., 30:1921 (1956).
13. A. M. Polyakov and G. A. Sokolina, this volume, p. 397.

ELECTRICAL PHENOMENA ON BREAKAGE
OF A POLYMER-SEMICONDUCTOR CONTACT

A. M. Polyakov and G. A. Sokolina

Surface Phenomena Division, Institute of Physical Chemistry,
Academy of Sciences of the USSR

Electron emission and change in surface conductivity have been observed for germanium on detachment of films of guttapercha, nitrile rubber, etc. The surface conductivity change is dependent on the thickness of the oxide film. The effects are explained in terms of the electrical theory of adhesion.

The basic concept in the electrical theory of adhesion is that the latter is due to electrostatic attraction in the parts of the double electrical layer arising at the interface between the solids. We have examined the adhesion of various polymers to single-crystal germanium. A semiconductor was chosen as providing much scope for examining the electrophysical parameters of the contact as well as the adhesion. If there is this double layer, the charge localized at the surface of the semiconductor should produce an altered surface conductivity during detachment. Figure 1a shows the type of charge distribution near the contact before detachment. The total charge is zero taken over regions I (polymer), II (Ge surface), and III (space-charge layer). When the film is detached (plates of the microcapacitor moved apart), the surface of the semiconductor has an uncompensated charge (Fig. 1b). The resulting field causes carrier injection or extraction; the altered carrier distribution as between the bulk and the space-charge region alters the electron and hole concentrations near the surface, which is reflected in the conductivity.

Separation of the parts of the double layer should [1] give rise to electron emission, whose rate is dependent on the charge density at the contact and on the conditions of bond rupture. We have made simultaneous measurements of the work of adhesion, the electron emission, and the change in surface conductivity.

We would expect that the electrical emission and the change in surface conductivity should be correlated with the adhesion when the electrostatic forces are responsible for the adhesion. Factors that affect the work of adhesion (detachment rate, surrounding medium) should affect also the conductivity change. The best experiment is one in which measurement of this change is accompanied by measurement of the electrophysical properties of the surface before and after detachment, which will indicate whether the detachment is of adhesion, cohesion, or mixed type. In the pure adhesion type, the surface properties after detachment should

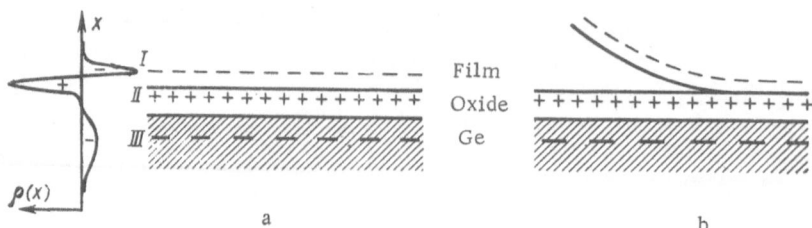

Fig. 1. Charge distribution at the contact: a) before rupture;
b) during rupture.

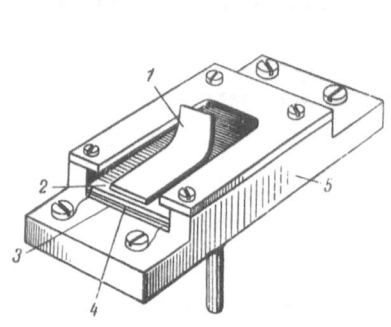

Fig. 2. Holder for Ge bearing a
polymer film: 1) film; 2) Ge; 3)
field electrode; 4) film; 5) Teflon
body.

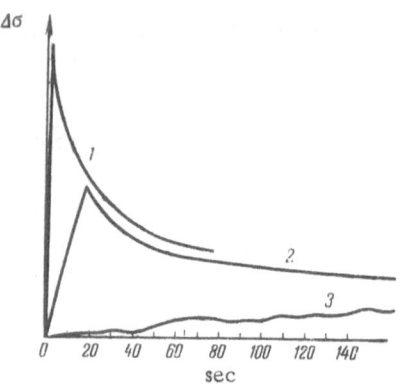

Fig. 3. $\Delta\sigma$ for Ge during detachment of films of nitrile rubber at
rates (cm/sec) of: 1) 1; 2) 10^{-1};
3) 10^{-2}.

be as before application of the film; in the cohesion type, the properties will be those of the
coated surface; and in the mixed type, some intermediate values will be found.

A bridge circuit [2] was used to measure the conductivity change $\Delta\sigma$ during detachment,
the out-of-balance voltage being recorded by an ÉPP-09 potentiometric recorder (10 mV). The
adhesion strength was expressed as the work of adhesion, as measured with an evacuable instrument capable of operating at $3-5 \times 10^{-6}$ mm Hg with detachment rates from 10^{-3} to 1 cm ·
$\sec^{-1}$ [3].

The electron emission was recorded by an open-ended secondary-electron multiplier [9].

We determined the effects of the polymer on the electronic structure of the surface via
the dc field effect [2] and from the decay of photoconductivity [5]. All of the experiments were
done with n-type Ge as wafers $10 \times 20 \times 0.8$ mm cut from a single crystal. The films were
deposited on both sides after the samples had been hatched in H_2O_2; this made it possible to make
all the measurements on a single specimen. The holder (Fig. 2) provided a capacitor for measurement of the field effect.

We have previously shown [3] that all the results from such measurements have a logical
explanation in terms of the double layer. In the present case we used SKN-40 nitrile rubber and
polyvinylchloride. The field-effect measurements with SKN-40 show that this rubber greatly increases the number of slow surface states in Ge without altering the surface charge.

Figure 3 shows results on $\Delta\sigma$ on detaching SKN-40 films at various rates. Here the
changes were similar to those for guttapercha and are due to removal of the negatively charged

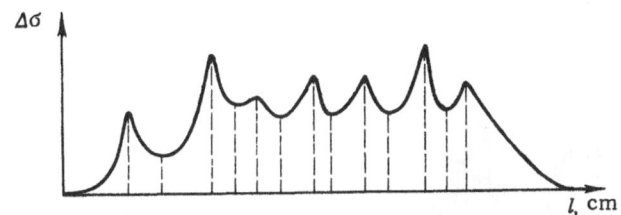

Fig. 4. $\Delta\sigma$ during detachment of an SKN–40 film at 10^{-2} cm/sec.

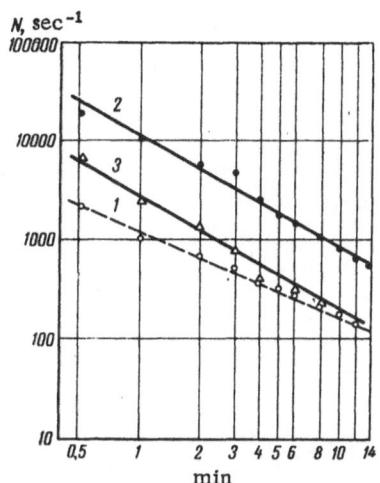

Fig. 5. Decay of afteremission after breaking contact of Ge with: 1) guttapercha; 2) nitrile rubber; 3) PVC.

rubber film. The magnitude of $\Delta\sigma$ is dependent on the detachment rate v on account of processes of charge loss from the surface.

Figure 4 shows a recording of $\Delta\sigma$ during adhesion detachment of SKN–40 from Ge in sections, which was responsible for the fluctuations in $\Delta\sigma$. The $\Delta\sigma$ were zero during cohesion detachment, as would be expected, and no electron emission then occurred.

SKN–40 on Ge provided proof of an important relation between the rate N of electron emission and the charge on the freshly detached polymer film. The theory would indicate that the afteremission is due to residual charge, and this receives indirect confirmation. On the one hand, it has been found that, as for insulators [6] and metals [7], the residual charge increases with v; on the other, N increased with v in our tests.

If the film is of very low conductivity (e.g., guttapercha, cellulose esters, 10^{-15} ohm · cm), the residual charge is lost almost entirely by electron emission. In fact, the decay after rapid detachment was logarithmic (Fig. 5). Nitrile rubber has a conductivity higher by 10^4, and here neutralization can occur by leakage along the film. In that case, there is a marked effect from partial detachment.

Figure 6 shows emission curves during and after detachment of SKN–40 films from Ge; curve 1 is for a film left touching the Ge after detachment, and here the emission virtually stops at the end of detachment, on account of leakage. If the detachment is complete and the film is insulated, there is a considerable afteremission (curve 2 of Fig. 6), whose decay is logarithmic, as for the other polymers. If this film is allowed to touch the Ge or the metallic holder containing it, the emission rate falls rapidly, as for an incompletely detached film.

PVC produced negative $\Delta\sigma$ (Fig. 7); the field-effect data show that the space-charge layer is enriched in electrons, so the negative $\Delta\sigma$ certainly implies† that a negative charge is produced on the Ge. Here the conductivity after detachment did not relax to the initial value, either because detachment alters the system of surface levels or because the change in surface conductivity occurs near the minimum surface conductivity. The second possibility is supported by the dc field effect from the rate of entry of carriers in the surface levels (Fig. 8).

PVC gave rapid initial emission ($\sim 10^4$ sec^{-1}) and slow decay, as for the other materials.

† Negative $\Delta\sigma$ have been reported [8] for the detachment of methyl cellulose films.

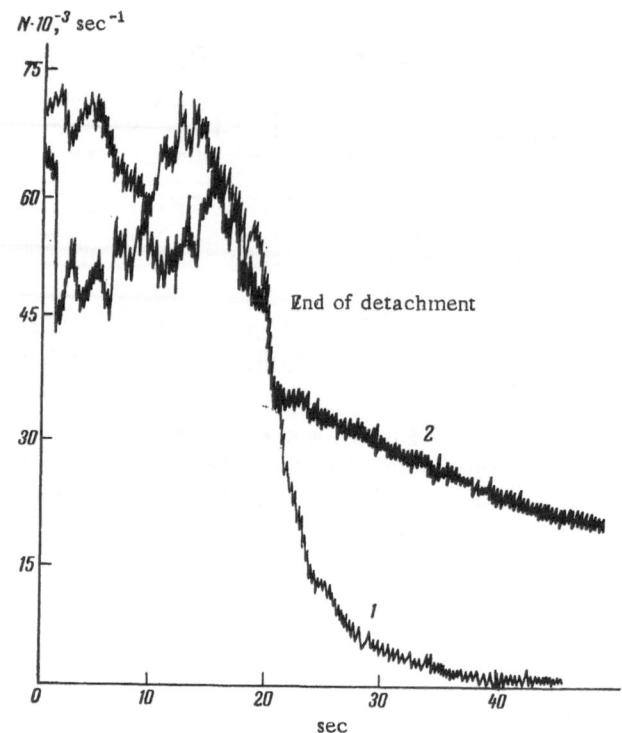

Fig. 6. Emission and afteremission of SKN-40 films:
1) touching substrate after detachment; 2) removed
completely.

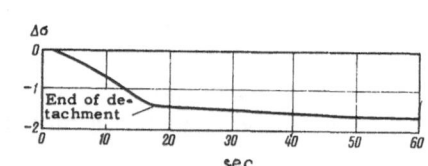

Fig. 7. Conductivity change on detachment of PVC.

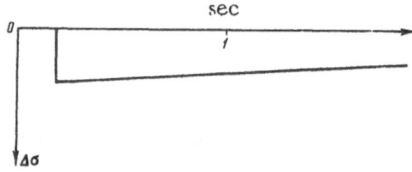

Fig. 8. Response in the field effect for Ge coated with PVC.

These results show that electrical effects due to a double electrical layer occur when
adhesion contacts of Ge with nitrile rubber and PVC are broken.

References

1.   V. V. Karasev, N. A. Krotova, and B. V. Deryagin, Dokl. AN SSSR, 88(5):777 (1953).
2.   G. A. Sokolina, N. A. Krotova, and Yu. A. Khrustalev, Dokl. AN SSSR, 147:1409 (1962).
3.   G. A. Sokolina, Dissertation, Institute of Physical Chemistry, AN SSSR (1966).
4.   A. M. Polyakov and N. A. Krotova, Dokl. AN SSSR, 151:130 (1963).
5.   A. B. Engler and C. J. Kavane, Rev. Sci. Inst., 28:548 (1957).
6.   B. V. Deryagin and N. A. Krotova, Adhesion, Izd. AN SSSR, Moscow (1949).
7.   M. M. Bredov and I. Z. Kshemenskaya, Zh. Tekh. Fiz., 27:921 (1957).
8.   V. I. Anisimova, N. A. Krotova, and L. A. Sukhareva, this volume, p. 386.

EFFECTS OF SURFACE AND BULK DEFECTS
ON THE STRENGTH OF GLASS

V. A. Ryabov and D. V. Fedoseev

Surface Phenomena Division, Institute of Physical Chemistry,
Academy of Sciences of the USSR

The distribution curve for the strength of glass specimens has two peaks, the first due to surface defects and the second to bulk defects. It is shown that the breaking strength of glass fibers is related to the strength of a plastic impregnated with many noninteracting glass fibers.

1. The strength found for specimens of glass varies widely, so statistical analysis is used. A reliable experimental test of a statistical theory requires numerous measurements, and the statistical theory of the strength tests on glass rods 2 mm in diameter [1] has two peaks. There are three distinct techniques in the making of glass of a given composition: stirring of the mass to produce homogenization, surface cleaning (skimming) before use, and preparation without either of these techniques. Three types of rods differing in quality were made in this way. The most probable strength for the first was 300 kg/mm² and that for the third was 30 kg/mm², while the curve for the second had peaks at 30 and 300 kg/mm². We have obtained analogous results.

We did the following tests. We used ordinary window glass made by vertical drawing, from which we made seven types of specimen with natural and strengthened surfaces (total 11,000 specimens) as squares of side 60 mm with various thicknesses. These specimens were diamond-cut from 500 × 500 mm sheets, the deviations on the sides being less than 1%, while the deviations in thickness were less than 10%.

The chemical composition was roughly 72% SiO_2, 15% Na_2O, 3% MgO, 8% CaO, and 1.5-2% Al_2O_3. The strengthening was produced by removing the defective surface layer by spraying with hydrofluoric acid in a laboratory system. The strength in symmetrical bending was measured with type RM machines with limiting loads of 100, 250, 500, 1000, 5000, and 10,000 kg. The specimens were tested on a square support having a square hole 50 × 50 mm, the disc plunger being 6 mm in diameter. Local overloading of the glass was avoided by the use of a soft lining.

Figure 1 shows some results for specimens 2.2 ± 0.2 mm thick; curve 1 is for unstrengthened glass (800 specimens), while curve 2 is for strengthened glass (4000 specimens).

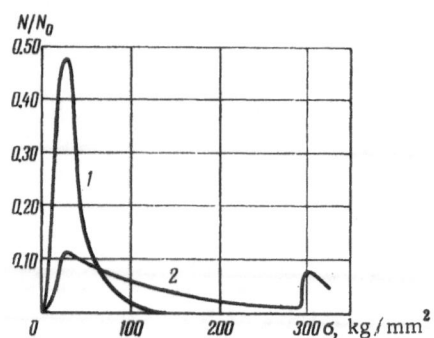

Fig. 1. Strength distribution curves for: 1) unstrengthened; 2) strengthened glass.

The latter gave two peaks on the plot of N/N_0 against σ, in which σ is breaking stress and N/N_0 is the ratio of the number N of specimens failing in the range σ to $\sigma + \Delta\sigma$ to the total number N_0. The curves are envelopes of the corresponding histograms.

Greene's statistical theory [2] for the strength of glass has been extended [1], the two peaks on the curve being explained in terms of two types of surface defect: type A is responsible for the first peak, while type B is responsible for the peak at $\sigma = 300$ kg/mm^2.

It is impossible to explain why there should be two types of defects (or cracks) on the surface, so we suppose that the second peak is due to all types of internal defects, i.e., the second peak is of bulk origin. On the basis we can explain the behavior of glass under load.

Vitman et al. [3] have shown that removal of a surface layer displaces the first peak. If a layer is removed in hydrofluoric acid (with measures to prevent subsequent surface damage), the surface has fewer and smaller cracks, so a higher load is required to break the glass. On the other hand, the internal defects are unaffected by the surface chemical treatment. If there were the type B surface defects proposed previously [1], these would be the first to be eliminated by chemical treatment, because they would be much smaller in size.

We assume that there is one crack on the surface, whose position is random. Let $F_1(\sigma_1)$ be the probability that the glass will break under a load σ_1. If the surface has n_1 cracks whose distribution is $P_1(n_1)$, the probability that the specimen will fall under a load σ_1 will be

$$\Phi_1 = 1 - \sum_{n_1=0}^{\infty} P_1(n_1)\,[1 - F_1(\sigma_1)]^{n_1}. \tag{1}$$

Similarly we find for internal defects that

$$\Phi_2 = 1 - \sum_{n_2=0}^{\infty} P_2(n_2)\,[1 - F_2(\sigma_2)]^{n_2}. \tag{2}$$

The corresponding distribution functions are

$$\varphi_1 = \frac{d\Phi_1}{d\sigma_1}\,; \qquad \varphi_2 = \frac{d\Phi_2}{d\sigma_2}\,.$$

In a system having surface and bulk defects,

$$\varphi(\sigma) = \varphi_1(\sigma) + \varphi_2(\sigma).$$

Then the probability of failure at a load σ will be

$$\Phi(\sigma) = \int_0^{\sigma} \varphi(\sigma)\,d\sigma. \tag{3}$$

We assume that $P_1(n_1)$ and $P_2(n_2)$ are Poisson distributions:

$$P_1(n_1) = \frac{(\mu A)^{n_1} e^{-\mu A}}{n_1!}\,,$$

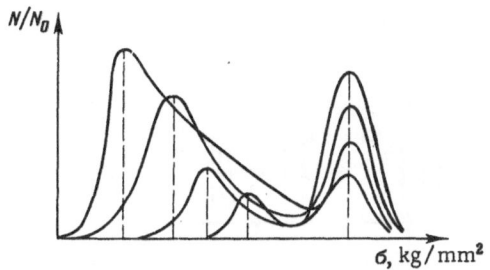

Fig. 2. Change in the strength distribution curve for glass on gradual removal of the defective surface layer.

$$P_2(n_2) = \frac{(\nu V)^{n_2} e^{-\nu V}}{n_2!},$$

in which μ and ν are respectively the mean numbers of defects per unit surface A and volume V. Then (1) and (2) are replaced by

$$\Phi_1(\sigma_1) = 1 - e^{-\mu A F_1(\sigma_1)}, \tag{4}$$

$$\Phi_2(\sigma_2) = 1 - e^{-\nu V F_2(\sigma_2)}. \tag{5}$$

As $F_1(0) \to 0$ and $F_2(0) \to 0$, we have $\Phi_1 \to 0$ and $\Phi_2 \to 0$, and from (4) we get

$$\frac{\partial \Phi_1}{\partial \mu} = A F_1(\sigma_1) e^{-\mu A F_1(\sigma_1)}.$$

Then the first peak on the distribution will be displaced towards larger σ (Fig. 2). Naturally, this is so only when $h \ll V^{1/3}$, in which h is the thickness of the layer of glass removed. If this inequality is not obeyed, we expect chemical treatment to affect the second peak.

The ratio of the logarithms of the probabilities of failure due to bulk and surface defects is proportional to the smallest characteristic dimension of the specimen (thickness, radius).

These experiments and calculations show that one of the simplest ways to increase the strength of glass is to preserve the original surface; this alone is sufficient to reduce breakage considerably. As regards improvement in the region of the second peak, the aim should be to preserve a liquid structure [4].

2. Statistical methods can also be applied with advantage in relating the strength of individual glass fibers to the strength of glass-fiber materials. Components reinforced with glass fiber have strengths of glass-fiber materials. Components reinforced with glass fiber have strengths much less than those of the elementary fibers. Continuous-drawn glass fibers show a considerable spread in strength even at a single diameter, and this spread is described by an error curve. Glass-fiber moldings are made by drum winding of individual fibers with coatings applied during drawing; here the fibers are parallel and are bonded by materials applied to the moving glass [5]. Such moldings are used in materials of density around 2 g/cm^3. The individual fibers may be several meters long, with mean diameters of 10-100 μ or so.

For simplicity, we consider such a component as a simple strip consisting of numerous noninteracting fibers of total area S. As the fibers are very numerous, we will use everywhere integrals in place of sums. If the component tears at a load P, the failure stress is usually calculated as

$$\sigma_0 = \frac{P}{S}. \tag{6}$$

The process may be represented as follows. The load P is gradually increased, and hence also the stress σ on each glass fiber. Fibers that cannot withstand this stress will break, so the load P is applied to a gradually decreasing area. The entire component fails at some stress σ_n, at which time a fraction $1/n$ of the area S has failed, so

$$\sigma_n = \frac{P}{(1-n)S}. \tag{7}$$

Clearly, $\sigma_n \geq \sigma_0$; equality applies if the distribution curve has a variance of zero.

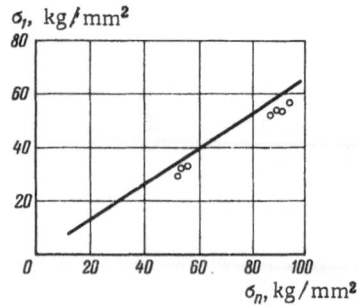

Fig. 3. Relation of strength σ_n of spun glass to σ_1, most probable strength of glass fibers.

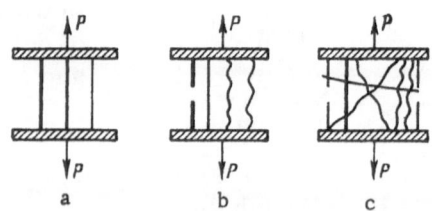

Fig. 4. Scheme for rupture of a set of glass fibers.

We assume the following law for the strength distribution of the fibers:

$$d\left(\frac{N}{N_0}\right) = \frac{2}{\sqrt{2\pi}} e^{-\frac{1}{2}\left(\frac{\sigma}{\sigma_1}-1\right)^2} d\left(\frac{\sigma}{\sigma_1}\right), \tag{8}$$

in which N_0 is the total number of glass fibers, whose most probable strength is σ_1.

From (3) we get $(1-n)$ as

$$1-n = \frac{2}{\sqrt{2\pi}} \int_{\frac{\sigma_n}{\sigma_1}}^{\infty} \exp\left[-\frac{1}{2}\left(\frac{\sigma}{\sigma_1}-1\right)^2\right] d\left(\frac{\sigma}{\sigma_1}\right). \tag{9}$$

The true breaking stress σ_n is

$$\sigma_n = \frac{P/S}{\dfrac{2}{\sqrt{2\pi}} \displaystyle\int_{\frac{\sigma_n}{\sigma_1}}^{\infty} \exp\left[-\frac{1}{2}\left(\frac{\sigma}{\sigma_1}-1\right)^2\right] d\left(\frac{\sigma}{\sigma_1}\right)}.$$

There is the following relation between σ_n, σ_n/σ_1, and σ_0:

$$\frac{\sigma_0}{\sigma_n} = \frac{2}{\sqrt{2\pi}} \int_{\frac{\sigma_n}{\sigma_1}}^{\infty} \exp\left[-\frac{1}{2}\left(\frac{\sigma}{\sigma_1}-1\right)^2\right] d\left(\frac{\sigma}{\sigma_1}\right) \leqslant 1.$$

The condition for failure of a spun-glass component consisting of noninteracting fibers is

$$S\sigma_n \int_{\frac{\sigma_n}{\sigma_1}}^{\infty} \exp\left[-\frac{1}{2}\left(\frac{\sigma}{\sigma_1}-1\right)^2\right] d\left(\frac{\sigma}{\sigma_1}\right) - p \leqslant 0.$$

We can determine σ_n from the condition for a maximum in

$$K(y) = y_n \int_{y_n}^{\infty} \exp\left[-\frac{1}{2}\left(\frac{\sigma}{\sigma_1}-1\right)^2\right] d\left(\frac{\sigma}{\sigma_n}\right),$$

in which

$$y_n = \frac{\sigma_n}{\sigma_1}, \qquad y = \frac{\sigma}{\sigma_1}.$$

From $dK(y)/dy_n = 0$, we get

$$y_n e^{-\frac{1}{2}(y_n-1)^2} = \int_{y_n}^{\infty} e^{-\frac{1}{2}(y-1)^2} dy. \tag{10}$$

Equation (10) may be solved graphically via tables [6]. In the present case, $y_n \approx 0.63$, i.e., $\sigma_n = 0.63\sigma_1$, or the ratio is constant. Figure 3 shows the relationship, where the experimental values [7] fit satisfactorily to the theoretical straight line.

It was assumed above that the load is applied to all the fibers uniformly, which all act as struts in tension (Fig. 4a); but if we envisage the successive application of the load of different fibers (Fig. 4b), the resulting strength is much lower. It has long been known that the bending of a fiber in spun-glass components increases as the fiber diameter decreases, so a component made of small-diameter glass fibers shows a marked loss of strength because not all fibers are tensioned simultaneously, and this may balance out the high strength of the small-diameter fibers. This is especially so for materials obtained by drawing from a die with a flow of air, where the fibers are short, variable in cross section, and often not cylindrical (Fig. 4c).

It has been shown in research at this institute and at the State Glass Research Institute that thick fibers (30-100 μ diameter), which are much cheaper than thin ones, can be used to advantage in oriented glass-fiber materials and in unwoven filters.

References

1. J. Cornelissen, H. Meyer, A. Kruithof, and H. Hamaker, Technical Papers of the Sixth International Congress on Glass, Washington (July 8-14, 1962).
2. C. H. Greene, J. Am. Ceram. Soc., 39:66 (1956).
3. F. F. Vitman, V. P. Pukh, and N. N. Shekberg, Steklo i keramika, No. 8, p. 9 (1964).
4. E. D. Shchukin, L. A. Kachanova, and Z. M. Zanozina, Dokl. AN SSSR, 160:1061 (1965).
5. B. V. Deryagin and S. M. Levi, Physical Chemistry of Deposition of Thin Films on Moving Substrates, Izd. AN SSSR, Moscow (1959); Film Coating Theory, Focal Press, New York (1964).
6. E. Jahnke and F. Emde, Tables of Functions [Russian translation], Gostekhizdat, Moscow-Leningrad (1949).
7. V. A. Ryabov, V. G. Yartsev, B. N. Aleshin, et al., Byull. Vses. Nauch.-Issled. Inst. Stekla, No. 3(112), p. 23 (1961).

RELATIVE ATMOSPHERIC HUMIDITY
AND THE FRICTION OF SOLIDS

Yu. M. Luzhnov

Surface Phenomena Division, Institute of Physical Chemistry,
Academy of Sciences of the USSR

Results are presented on the effects of atmospheric humidity on the static friction of clean and dusty bodies.

It has been found [1-4] for the friction of dusty bodies under conditions of controlled humidity that the friction between solids is dependent on the humidity of the contacting surfaces. Additional studies on clean and dusty surfaces have been made in order to elucidate the relationship.

The measurements were made with the apparatus previously described [5], which allows control of the relative humidity of the air in the space surrounding the bodies whose static friction is to be measured. The surfaces were a flat plate with a slide resting on three balls of radius 0.5 mm. The slide provided a constant mean pressure of 65 kg/mm^2 at the contacting surfaces. The method of measurement allowed the relative humidity to be measured to ±2-3% and the frictional force at the dusty surfaces to ±4%.

All the measurements were made on surfaces previously cleaned in a glow discharge by Karasev's method [6]. The cleanliness was checked via measurements of the coefficient of static friction μ, specimens with values below 0.65 being discarded.

It was found that μ for silver and glass surfaces with silver and glass contacts was virtually independent of the relative humidity p/p_s at low and medium values, but that there was a sharp fall in μ for p/p_s near saturation (0.8 and 0.92 respectively) (Fig. 1a). The adsorption isotherms for water on these surfaces (Fig. 1b) show that the largest reduction in μ occurs when a nearly monomolecular layer of water is formed; subsequent increase in the thickness of the water film does not cause any marked change in μ. Similar results for μ as a function of p/p_s were obtained (Fig. 2) with steel surfaces and with nickel-plated surfaces.

The clean surfaces show a marked fall in μ as p/p_s increases, but the dusty ones showed a different relation (Fig. 3), μ being much smaller at low p/p_s and rising substantially as saturation is approached. Another characteristic feature was that μ was further reduced at low p/p_s, and the rise was displaced to higher p/p_s, as the content of organic matter in the dust layer was increased.

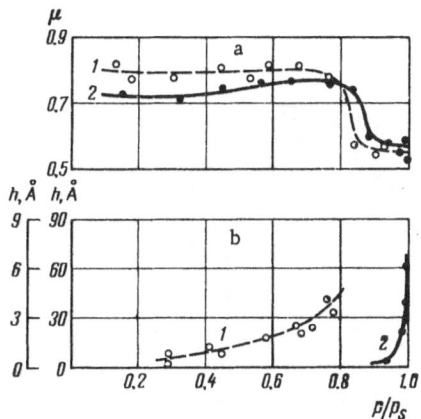

Fig. 1. a) Coefficient of static friction μ as a function of relative humidity p/p_s; b) adsorption isotherms for water vapor on: 1) silver surfaces [8]; 2) glass surfaces [7].

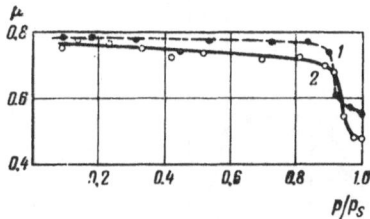

Fig. 2. Relation of μ to p/p_s for: 1) nickel-coated surfaces; 2) clean steel surfaces.

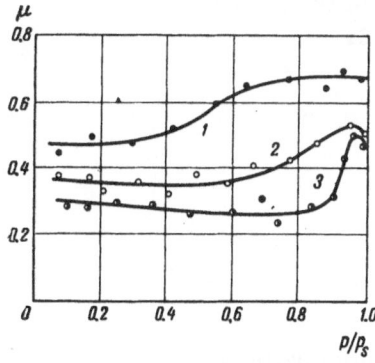

Fig. 3. Relation of μ to p/p_s for surfaces coated (thickness 10-12 μ, particle size > 0.1 μ) with powdered silicon oxide containing various amounts of hydrocarbon oil: 1) pure powder; 2) powder left in contact with oil vapor for a day; 3) powder as previous, but five days.

Tests were done with silicon oxide powder, for which the surface treatment was identical but the mean grain size varied; higher μ went with smaller grain size at all p/p_s (Fig. 4a), and the region of rapid rise in μ shifted to higher p/p_s as the grain size increased, which represents a diminution in the influence of capillary forces. This is especially clear from the curves for $[(\mu/\mu_{min})-1]$ as a function of p/p_s (Fig. 4b). Here we took as μ_{min} the mean value at the point preceding the onset of capillary condensation in the powder layer, while μ was the measured value.

The normal pressure in the friction zone has (Fig. 5) a marked influence on the relation of μ to p/p_s for dusty bodies, the effect being the most pronounced for small normal pressures, where μ may increase by 60% during capillary condensation. Increase in normal pressure tends [9] to diminish the effects of capillary forces on the friction between dusty bodies.

Experiments were also done with bentonite powders, as bentonite has a layered lattice with only weak bonds between layers; it tends to swell as p/p_s increases.

These experiments (Fig. 6) were done with natural Kokaita bentonite (which swells considerably), Na-substituted Keles bentonite (medium swelling), and H-substituted Keles bentonite (little swelling). In all cases there was a monotonic fall in μ as p/p_s increased. Thin films of liquid were formed between the particles when the internal crystal layers had become largely saturated with water or had completely ceased to take up water from the surface. Concave water menisci appeared within the powder pores at about p/p_s = 0.95, which substantially increased the forces of interaction between the particles, altered the mechanical parameters, and increased μ by 30-40% (for a pressure of 65 kg/mm^2 in the friction zone).

The action of the capillary menisci ceased when the pores of the powder became completely filled with water, and μ fell to the level found before the appearance of the menisci. Similar relationships were observed with montmorillonite powder either clean or containing up to 25-30% quartz sand with particles of size from 0.1 to 100 μ (Fig. 7).

Bentonite and montmorillonite on clean metal surfaces showed no extrusion from the contact zone even when highly diluted with water. Under these

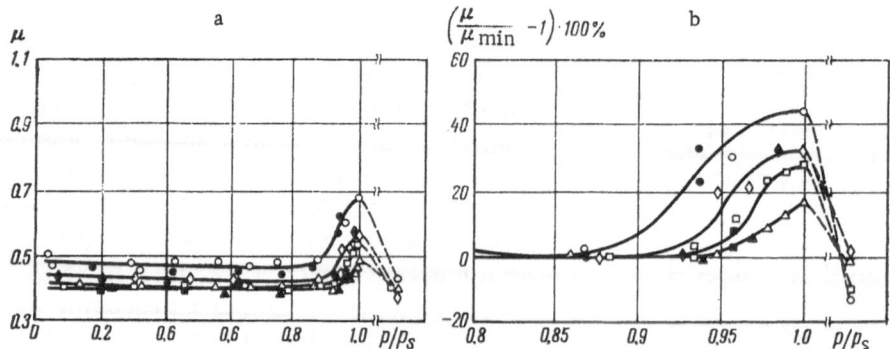

Fig. 4. a) Relation of μ to p/p_s for surfaces coated with silicon oxide powder having a particle size (cm) of: ○, ●) 10^{-4} (increasing and decreasing respectively); ◇) 1.3×10^{-4}; □) 1.8×10^{-4}; △) 2.5×10^{-4}; b) relative change in μ with p/p_s. Values of $p/p_s > 1$ represent surfaces coated with layers of water.

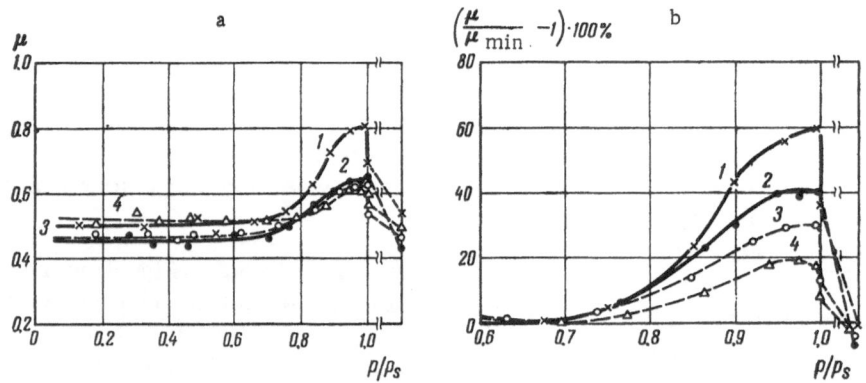

Fig. 5. a) Relation of μ to p/p_s for surfaces coated (thickness 10-12 μ, particle size < 0.1 μ) with powdered silicon oxide; b) relative change in μ with p/p_s for various normal pressures in the friction zone. Slide mass (g): 1) 5.6; 2) 16.5; 3) 30.6; 4) 41.6.

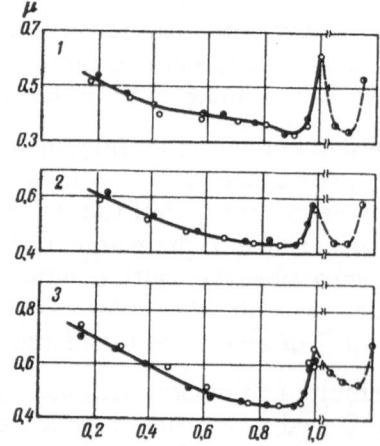

Fig. 6. Relation of μ to p/p_s for surfaces coated with bentonite powder: 1) natural Kokaita bentonite; 2) Na-substituted Keles bentonite; 3) H-substituted Keles.

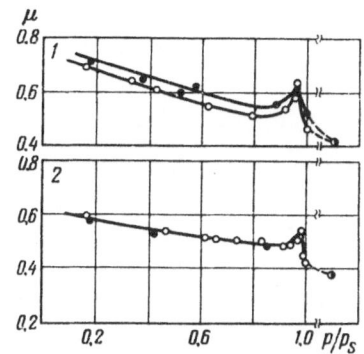

Fig. 7. Relation of μ to p/p_s for surfaces coated with: 1) montmorillonite powder; 2) montmorillonite containing quartz sand for p/p_s: ○) increasing; ●) decreasing.

conditions, μ was always low and close to the value observed before the production of menisci. Increase in μ on moist surfaces coated with bentonite was seen only when the powder was flushed away from the contact spots by the water.

The following conclusions may be drawn:

1. The coefficient of static friction between clean solid surfaces remains constant up to the point where a monolayer of condensed water starts to appear; thereafter, the coefficient falls rapidly, but further increase in the thickness of the layer leaves the coefficient unaffected.

2. The friction between dusty bodies is affected mainly by processes within the layer of powder separating the bodies as the humidity varies. The details of these processes are dependent on the normal pressure in the contact zone, on the surface treatment of the powder, on the particle size, and on the degree of contamination with organic substances.

I am indebted to B. V. Deryagin, Associate Member of the Academy of Sciences of the USSR, for direction in this work.

References

1. S. I. Kosikov, Izv. AN SSSR, OTN, No. 8, p. 63 (1957).
2. Yu. P. Toporov, Inzh.-Fiz. Zh. Acad. Nauk Belorussk.SSR, 3(4):44 (1960).
3. Y. M. Lujnov (Yu. M. Luzhnov) and S. J. Kosikov, Convention on Adhesion, Paper 3, London (1963).
4. Yu. M. Luzhnov, in: Research in Surface Forces, Vol. 2, Consultants Bureau, New York (1966), p. 317.
5. Yu. M. Luzhnov, TsITÉIN, PNTO, No. p-62-72/10, Moscow (1962), p. 24.
6. V. V. Karasev and G. I. Izmailova, Zh. Tekh. Fiz., 24:871 (1954).
7. B. V. Deryagin and Z. M. Zorin, Zh. Fiz. Khim., 29(6):1010 (1955); No. 10, p. 1755 (1955).
8. F. P. Bowden and W. R. Throssell, Proc. Roy. Soc., A209:297 (1951).
9. Yu. M. Luzhnov, Dissertation, Institute of Physical Chemistry, AN SSSR, Moscow (1966).

SECTION VIII

SURFACE FORCES IN GASES

KINETICS OF CONDENSATION
ON SMALL INSOLUBLE NUCLEI

V. A. Tereshin and L. M. Shcherbakov

Tula Polytechnical Institute

Thermodynamic perturbation theory based on capillary effects of the second kind is used to calculate the excess free energy of a thin layer of liquid on an insoluble solid core. The Frenkel−Zel'dovich scheme is used for the nucleation, which allows nucleation to be considered as Brownian motion along the axis of nucleus size.

Nucleation occurs mainly at the surfaces of solid particles (condensation nuclei), whose activity is dependent on the size and on the physicochemical properties of the surface. The parts of the surface of a real nucleus may differ in activity, on account of irregular shape and various inhomogeneities. It is usual to simplify the problem by considering ideal homogeneous spherical nuclei.

There are two possibilities in condensation on insoluble spherical solid nuclei: 1) condensation on large nuclei, which begins with the production of nuclear liquid lenses,[†] 2) condensation on small nuclei, for which the most probable form of nucleation is as a layer of liquid uniformly enclosing the particle. The latter was proposed by Kristanow [2], but he neglected the excess free energy of the layer.

Various difficulties arise in relation to the thermodynamics of nucleation, on account of the small number of particles in a nucleus. One of these is that the fluctuations in the thermodynamic quantities for the nucleus become too large for the quantities themselves to be considered as macroscopic. This difficulty is removed to some extent if we introduce [3] an ensemble of nucleus + medium type. Then the thermodynamic quantities for the nucleus (including the surface) are defined as statistical averages over the ensemble. Another difficulty is that it is impossible to distinguish bulk and surface parts in the nucleus. This is overcome by the concept of capillary effects of the second kind, in which the excess free energy is not related solely to the interface but is assigned to the volume as a whole. If the small object has a composite structure (e.g., a heterogeneously formed nucleus), the excess free energy may reasonably be assigned to the entire system of small object + medium.

[†] Fletcher [1] has considered this type in detail on the basis of the classical theory of surface phases.

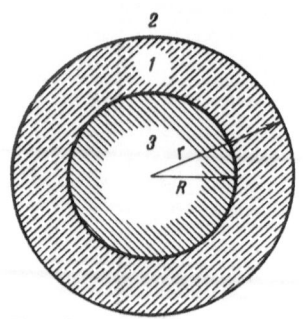

Fig. 1. Derivation of the formula for the excess free energy of a nucleation layer on a spherical solid core.

Thermodynamic perturbation theory[†] allows one to reduce the calculation of the excess free energy of such a system to calculation of the mean perturbation energy due to deposition of the individual parts of the systems from the corresponding massive phases [6, 7].

Consider the following system (Fig. 1): spherical liquid layer 1 on an insoluble solid core 3 in the presence of vapor 2, in which all parts are assumed to be bounded by equimolecular separating surfaces corresponding to the condition for absence of autoabsorption. Then the perturbation energy ΔU has three components: 1) the excess interaction energy $U_{12}^{lsu} + U_{13}^{lso} - U_{1,2+3}^{ll}$, which is due to isolation of the layer from the massive liquid and transfer to the core,[‡] 2) the change in the interaction energy, $U_{31}^{sol} + U_{32}^{sosu} - U_{3,2+1}^{soso}$, which is due to isolation of the core from the massive solid phase and transfer within the liquid layer, 3) the change in the interaction energy arising from insertion of the core bearing the liquid into the vapor. Then, far from the critical point, the excess free energy of this system is

$$\Psi = \langle U_{1,2+3}^{11} \rangle - \langle U_{3,2+1}^{soso} \rangle + 2 \langle U_{13}^{1so} \rangle. \tag{1}$$

This relation implies that calculation of Ψ for a system including a small object reduces to calculation of the mean interaction energy of the regions into which the space is dissected by the bonding surfaces. Such a calculation is most simply performed via the binary correlation function $F^{II}(r_{ij}, \xi)$ for the inhomogeneous (two-phase) system.[§] This function defines the probability of finding particle j in a volume element of region 2 at a distance r_{ij} from particle i in the corresponding volume element of region 1. Assuming as a first approximation that the particles j are uniformly distributed in region 2, we can [7, 8] represent F^{II} as

$$F^{II}(r_{ij}, \xi) = \delta(\xi) \rho(r_{ij}), \tag{2}$$

in which $\rho(r_{ij})$ is the usual radial distribution function for the massive phase and $\delta(\xi)$ is a delta function in ξ, which characterizes the distance along the normal from particle i to the boundary of region 1:

$$\delta(\xi) = 0 \quad (\xi \leqslant a); \qquad \delta(\xi) = 1 \quad (\xi > a). \tag{3}$$

The linear parameter a may be called the cutoff parameter.

The Lennard–Jones potential [6–12] may be used to describe the interaction between particles; as a first approximation, we put $\rho(r_{ij}) = 0$ for $r_{ij} < a$ and $\rho(r_{ij}) = 1$ for $r_{ij} > a$, which gives[¶] for Ψ after identification[‖] of the energy constants with the corresponding macroscopic quantities:

$$\Psi = 4\pi r^2 \sigma_\infty \left(1 - \frac{a}{r} - \frac{R}{r}\frac{a^2}{h^2}\right) + 4\pi R^2 \left[\sigma_\infty^{so}\left(1 - \frac{a}{R}\right) - (\sigma_\infty^{so} - \sigma_\infty^{sol})\left(1 + \frac{a}{R} - \frac{r}{R}\frac{a^2}{h^2}\right)\right], \tag{4}$$

[†] See [5].

[‡] Here the subscripts indicate the regions of interaction and the superscripts indicate the phases filling these.

[§] In accordance with the basic idea of the perturbation method, we envisage the correlation function for the massive system.

[¶] See [8] for the details of the calculation.

[‖] By passage to the limits r, R → ∞.

in which r is the outside radius of the liquid layer, R is the radius of the core, h = r − R is the thickness of the layer, and σ_∞, σ_∞^{so}, and σ_∞^{sol} are the specific free surface energies for the following massive interfaces: liquid−vapor, solid−vapor, and solid−liquid.

This Ψ gives us the following expression for the thermodynamic potential of the system 1-3 (Fig. 1):

$$\Phi = \mu_1(p, T)\cdot g + \mu_2(p, T)(N - g) + \Psi, \tag{5}$$

in which μ_1 and μ_2 are the chemical potentials (per particle) in the massive liquid and vapor (as referred to a vapor pressure p), g is the number of particles in the layer, and N is the total number of particles in the system.

The final expression for thermodynamic equilibrium is obtained in the following form by variation subject to the additional conditions p, N, T = constant:

$$nkT \ln p/p_s = \frac{2\sigma_\infty}{r} - \frac{\sigma_\infty a}{r^2} - (m - 1)\sigma_\infty \frac{R(r + R)}{r^2} \frac{a^2}{h^3}, \tag{6}$$

in which n is the particle concentration in the liquid, k is Boltzmann's constant, p is saturation vapor pressure, and m = $(\sigma_\infty^{so} - \sigma_\infty^{sol})/\sigma_\infty$ is the wetting coefficient.†

This equation is a generalization of the isotherm for polymolecular adsorption and of Kelvin's equation for the vapor pressure of a small drop; if r, R → ∞ it becomes Hill's equation [9] for polymolecular adsorption on a planar adsorbent. If R = 0, we may use the feature [6] that $\sigma = \sigma_\infty[1 - (a/r)]$ for a homogeneous drop of liquid, and (6) coincides with the corrected form of Kelvin's equation previously considered by one of us [10].

Figure 2 shows a series of isotherms from (6) for benzene at 10°C on particles of various sizes for m = 1.5, and also the curve for a homogeneous drop. The abscissa is r, not the layer thickness, for convenience in comparing the isotherms.

It is clear from (6) that the isotherm has two branches corresponding to physically distinct cases. The left ascending branch (Fig. 2) is the adsorption branch, which corresponds to a thermodynamically stable polymolecular adsorption layer, while the right descending part is the condensation branch, which corresponds to a labile condensation equilibrium. Therefore, (6) describes not only polymolecular adsorption on readily wetted particles (i.e., equilibrium of a polymolecular layer with a vapor, which occurs before saturation is reached) but also condensation proper. It may therefore be called the equation for the isotherm for polymolecular adsorption and condensation.

We can now give the following scheme for nucleation on small readily wetted insoluble particles. Before saturation is reached (for p < p_s), a polymolecular adsorbed layer appears uniformly on the particle. This layer grows as p increases, but the thickness at saturation remains finite (a difference from adsorption on a planar readily wetted surface), and the thickness continues to increase in the supersaturation region until some critical value is reached (the value corresponding to the peak in Fig. 2). The layer is stable in this range of thicknesses, i.e., any change in p leads to a corresponding change in thickness. The layer becomes labile beyond the critical region, and any increase in thickness due to fluctuation leads to spontaneous condensation and formation of a macroscopic liquid drop.

The general features of the above scheme apply also to nucleation on not completely wetted small particles, the difference being that here the adsorption and condensation branches relate

---

† In the case of incomplete wetting, m = cos θ.

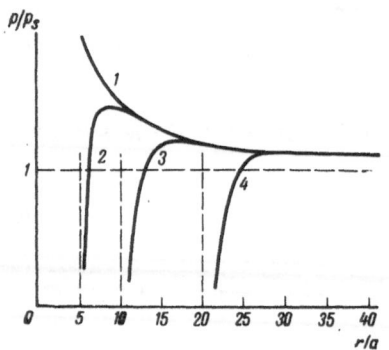

Fig. 2. Nucleation isotherms for benzene at 10°C on particles with m = 1.5 for R/a of: 1) 0; 2) 5; 3) 10; 4) 20.

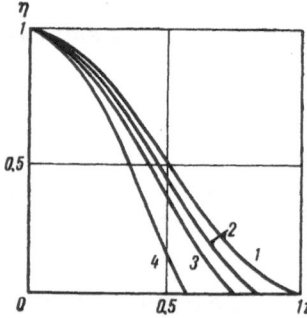

Fig. 3. Variation of η with R/r* for several values of the supersaturation parameter M: 1) 0; 2) 0.066; 3) 0.097; 4) 0.190.

to layers in distinct phases.[†] Adsorption on the particle thus first produces a thermodynamically stable adsorption sublayer, which serves as basis for a labile layer of ordinary liquid on passing to the condensation branch.

If a nucleated layer is to survive, it must attain the critical size r* (or h*) corresponding to transition from the adsorption branch of the isotherm to the condensation one. This requires a certain amount of energy, which is a basic parameter of the condensation. The performance (effectiveness) of a particle as a condensation center may be defined as

$$\eta = W^*/W_0^*, \tag{7}$$

in which W^* and W_0^* represent the energies of formation for heterogeneously and homogeneously produced nuclei in labile equilibrium with the supersaturated vapor. This ratio may also be called the performance coefficient; $\eta < 1$ means activity in respect of condensation, while $\eta > 1$ means passivity. We have previously [8] considered the relation of η to particle size with allowance for capillary effects of the second kind, and so we merely give the results (Fig. 3) for η as a function of z = R/r* for several values of the supersaturation parameter

$$M = \frac{nkTa}{\sigma_\infty} \ln p/p_s.$$

We use the Frenkel'–Zel'dovich scheme for nucleation kinetics, in which nucleation is interpreted as Brownian motion along the size axis. Here we assume that the concentration ν of active particles in the supersaturated vapor is sufficient to provide heterogeneous condensation. These particles are coated with stable adsorption films before supersaturation is reached, and the layers of liquid begin to arise on these on passage to the condensation branch.

The size distribution (number of particles g) for the nuclei at a time t is a nonequilibrium distribution function $f(g, t)$, which is defined, as for homogeneous nucleation, by an equation of diffusion type:

$$\frac{\partial f}{\partial t} = \frac{\partial}{\partial g}\left[D \cdot f^0 \frac{\partial}{\partial g}(f/f^0)\right] \tag{8}$$

subject to the same boundary and initial conditions[‡]:

$$
\begin{aligned}
f(g, t) &= f^0(g) &&\text{(for } g = 1\text{)},\\
f(g, t) &= 0 &&\text{(for } g = \infty\text{)},\\
f(g, t) &= 0 &&\text{(for } t = 0\text{)}.
\end{aligned}
\tag{9}
$$

[†] Equation (6) in this case applies only to the condensation branch.

[‡] Somewhat different edge conditions were introduced by Reiss [11] in a treatment of the kinetics of heterogeneous nucleation, but these lead to logical contradictions.

Here D is the coefficient for diffusion of the nuclei along the size axis and $f^0(g)$ is the equilibrium† size distribution:

$$f^0(g) = C \exp[-W(g)/kT]. \tag{10}$$

Constant C here differs from the case of homogeneous nucleation in that it cannot be identified with the concentration of single molecules. The normalization condition now takes the form

$$\int_1^\infty f^0(g)\, dg = \nu,$$

which implies that C may be equated to the particle concentration ν as a first approximation.

In steady-state nucleation, there is a constant flux I of nuclei through all sizes less than some value $G > g^*$; in this case, $\partial f/\partial t = 0$ and (8) reduces to quadratures, with

$$I^{-1} = \int_1^\infty \frac{dg}{D(g) f^0(g)}.$$

This integral may be evaluated by the standard method [12, 13] used in nucleation kinetics; then we use the expression [14] for the rate I_0 of homogeneous nucleation with allowance for capillary effects of the second kind to get I/I_0, the ratio of the rates of stationary heterogeneous and homogeneous nucleation, in a close approximation as

$$I/I_0 = \frac{\nu}{n_2} \exp\left[(1-\eta)\frac{W_0^*}{kT}\right]. \tag{11}$$

Here n_2 is the concentration of single molecules in the vapor, the other symbols being as before.

We are indebted to B. V. Deryagin, Associate Member, Academy of Sciences of the USSR, for a useful discussion.

References

1. N. H. Fletcher, J. Chem. Phys., 29:572 (1953); 31:1136 (1953).
2. L. Kristanow, Idojaras, Budapest, 61:333 (1957).
3. T. L. Hill, J. Chem. Phys., 36:3182 (1962).
4. L. M. Shcherbakov, in: Research in Surface Forces, Vol. 1, Consultants Bureau, New York (1963), p. 19.
5. L. D. Landau and E. M. Lifshitz, Statistical Physics, Addison-Wesley, Reading, Mass. (1958).
6. L. M. Shcherbakov, Research in Surface Forces, Vol. 2, Consultants Bureau, New York (1966), p. 26.
7. L. M. Shcherbakov, Dokl. AN SSSR, 168:388 (1966).
8. L. M. Shcherbakov and V. A. Tereshin, Kolloidn. Zh., 28:758 (1966).
9. T. L. Hill, J. Phys. Coll. Chem., 34:1381 (1950).
10. L. M. Shcherbakov, P. P. Ryazantsev, and N. P. Filippova, Kolloidn. Zh., 23:338 (1961).
11. H. Reiss, J. Chem. Phys., 18:529 (1950).
12. Ya. B. Zel'dovich, Zh. Éksp. Teor. Fiz., 12:525 (1942).
13. D. Stachoroska, J. Chem. Phys., 42:1887 (1965).
14. L. M. Shcherbakov and V. A. Matveev, this volume, p. 418.

† Equilibrium in the sense of labile equilibrium.

CAPILLARY EFFECTS OF THE SECOND KIND
IN THE KINETICS OF HOMOGENEOUS CONDENSATION

L. M. Shcherbakov and V. A. Matveev

Tula Polytechnical Institute

Capillary effects of the secondary kind are examined for the Frenkel—Zel'dovich scheme, in which
nucleation is interpreted as Brownian motion along the axis of nucleus size. This provides calculation of the steady-state nucleation rate. The difficulties are overcome via Hill's nucleus + medium
ensemble.

Formation of an aerosol by condensation can be divided into two states: 1) nucleation,
i.e., production of liquid nuclei in the supersaturated vapor, 2) subsequent behavior. Kinetic
effects are important, especially in the second stage; in the first stage, they act mainly by
producing thermodynamic conditions favorable to heterophase fluctuations in microvolumes.
We now consider the thermodynamics of nucleation.

1. Thermodynamics of Nucleation

Formation of a surviving nucleus involves an energy barrier, whose magnitude is usually
deduced thermodynamically (Gibbs—Volmer approach), the nucleus being considered as part
of the thermodynamic system. However, here there are major difficulties, on account of the
small number of particles in a nucleus. One of these is that the fluctuations in the thermodynamic parameters applicable to a nucleus are too large for these parameters to be considered
as macroscopic. This difficulty can be overcome [1] by introducing an ensemble of nucleus +
medium type. Then the thermodynamic quantities that characterize a nucleus are deduced as
statistical averages over the ensemble.[†] Examples are the excess free energy of a nucleus
and the same for the surface of a nucleus. The spatial location of the nucleus is determined
by introducing the geometrical separating surface as the boundary.

Choice of the Gibbs tension surface as the separating surface enables us to avoid correction terms of $\partial\sigma/\partial r$ type, but simultaneously we introduce the effect of autoadsorption on the

---

[†] This approach is justified from the physical viewpoint by the fact that a condensing supersaturated vapor always contains a reasonably large number of nuclei, each of which (with part
of the surrounding medium) may be considered as one of the systems in the ensemble.

surface tension, which is not accessible to measurement. It is more convenient to locate the separating surface in such a way that the autoadsorption is zero. Such an equimolecular separating surface may be defined on the basis of the virtual incompressibility of the liquid far from the critical point:

$$v'_\infty g + v''_\infty (N - g) = V, \tag{1.1}$$

in which v'_∞ and v''_∞ are the volumes per particle in the bulk phases, a single prime refers to the liquid nucleus (g particles), a double prime refers to the vapor, N is the total number of particles, and V is the volume of the system.

This boundary allows us to determine the radius r of the nucleus via

$$r = \left(\frac{3}{4\pi} v'_\infty\right)^{1/3} g^{1/3} = 0.239 \left(\frac{g}{n'_\infty}\right)^{1/3}, \tag{1.2}$$

in which n'_∞ is the particle concentration in the bulk liquid.

A second difficulty arises because it is impossible to distinguish volume and surface parts in a nucleus, since the number of particles is small. To avoid this difficulty, one of us [2] introduced the concept of capillary effects of the second kind, in which the excess free energy Ψ (or other potential) of the system including the small object is not associated with the separating surface but is assigned to the system small object + medium. The standard state for the definition of Ψ is taken as the state corresponding to a system of one of the bulk phases.

As the excess free energy refers to the nucleus as a whole, it can be considered as a function of the size or of g. For g large, $\Psi(g)$ becomes the usual expression for the surface free energy of a drop, which can be represented formally[†] as

$$\Psi = \sigma(r) \cdot 4\pi r^2 = \sigma_\infty \varphi(r) \cdot 4\pi r^2, \tag{1.3}$$

in which the dimensionless function $\varphi(r) = \sigma(r)/\sigma_\infty$ (σ_∞ is the specific surface free energy of a large drop). This representation of $\Psi(g)$ facilitates comparison of this theory with the classical theory of surface phases.

These arguments allow us to give a reasonably complete thermodynamic description of systems including small objects [3].

The thermodynamic theory of perturbations allows one to show [4] that calculation of the excess free energy for a system including a small object reduces to calculation of the mean interaction energy for regions corresponding to the given and standard states, which are separated by this geometrical surface. In accordance with the basis of the perturbation method, the averaging is performed via the unperturbed distribution corresponding to the bulk phase.[‡] We have performed calculations [5] of the excess free energy for a small drop + vapor via various model interaction potentials (the Lennard–Jones potential, Buckingham's potential, model of a rectangular well), and these lead in the first approximation to the same expression for the specific surface free energy of a small drop:

$$\sigma(r) = \sigma_\infty \left(1 - \frac{a}{r}\right), \tag{1.4}$$

in which a is a linear parameter (the cutoff parameter).

[†] We should not overlook the formal character of this concept of specific surface free energy $\sigma(r)$ of a nuclear drop.

[‡] See [4, 5] for details.

We can express the work W needed to produce a liquid nucleus of g particles in a supersaturated vapor (pressure p) in terms of the difference between the thermodynamic potentials in the final and initial states. Including capillary effects of the second kind, we have

$$W = \Delta\Phi = (\mu'_\infty - \mu''_\infty) g + \Psi(g),$$

in which μ'_∞ is the chemical potential (per particle) in the bulk liquid and μ''_∞ is the same for the vapor, as referred to the vapor pressure p. We use (1.2) to pass from g to the size of the nucleus, with $\Delta\mu_\infty = \mu'_\infty - \mu''_\infty = -kT \ln p/p_\infty$ (in which p_∞ is the saturation vapor pressure for a planar surface), and rewrite the last result as

$$W = n'_\infty kT \ln s \cdot \frac{4}{3}\pi r^3 + \sigma(r) \cdot 4\pi r^2, \tag{1.5}$$

in which $s = p/p_\infty$ is the supersaturation.

Taking the turning point in (1.5), we get the corrected Kelvin equation[†]

$$n'_\infty kT \ln s = \frac{2\sigma}{r^*} + \frac{\partial\sigma}{\partial r^*}, \tag{1.6}$$

which defines (in inexplicit form) the radius r* of a critical nucleus in labile equilibrium with the supersaturated vapor.

Then (1.6) gives us W* (work of formation of a critical nucleus) as

$$W^* = \frac{4}{3}\pi r^{*2}\left(\sigma - r\frac{\partial\sigma}{\partial r}\right). \tag{1.7}$$

This implies that W* becomes zero not for r = 0 but for some finite value r_0, which is defined by $\sigma(r_0)/r_0 = (\partial\sigma/\partial r)_{r=r_0}$; the supersaturation s_m corresponding to this size being [7] the upper limit to the attainable supersaturations (absolute critical supersaturation), while r_0 (or the corresponding g_0) may [7] be considered as the conditional boundary separating the region of associated molecular complexes ($r < r_0$) from that of liquid nuclei ($r > r_0$).

If $r \gg r_0$, (1.6) and (1.7) become the classical Gibbs-Volmer equations:

$$n'_\infty kT \ln s = \frac{2\sigma_\infty}{r_m} \tag{1.8}$$

and

$$W^*_c = \frac{4}{3}\pi r_c^2 \cdot \sigma_\infty = \frac{16\pi}{3} \cdot \frac{\sigma_\infty^3}{(n'_\infty kT \ln s)^2}. \tag{1.9}$$

It is best to convert (1.6) and (1.7) to dimensionless variables in order to characterize the deviations of this new approach from the classical theory, e.g., the dimensionless radius $x = r/a$ and dimensionless surface tension $\varphi(x) = \sigma/\sigma_\infty$. Then, assuming s to be given, we use (1.6) and (1.8) to get the ratio of r* to the classical radius r_c of a nucleus as

$$r^*/r_c = \varphi + \frac{1}{2}x \cdot \frac{\partial\varphi}{\partial x}. \tag{1.10}$$

Similarly, the ratio of the W in the critical case is found from (1.7), (1.9), and (1.10) as

$$W^*/W^*_c = \left(\varphi + \frac{1}{2}x \cdot \frac{\partial\varphi}{\partial x}\right)^2\left(\varphi - x \cdot \frac{\partial\varphi}{\partial x}\right) = \zeta(x). \tag{1.11}$$

_____
[†] See [6] for details.

Table 1 illustrates the results for a nonpolar liquid, for which $\sigma(r)$ may be represented to a first approximation by (1.4).

2. Growth and Evaporation of Nuclear Drops

If $g \approx g*$, evaporation and growth may be considered as quasi-stationary, since the equilibrium vapor concentrations are close to the actual value, and so Fuks's [8] theory of a two-stage process can be applied. The molecules are transported by molecular flow through a region of width l directly adjoining the surface of the drop, and then subsequently by diffusion.[†] Knudsen's expression [9] may then be applied for the time dependence of g in the first region:

$$\dot{g} = 4\pi r^2 vc \, (n_l'' - n_r''). \tag{2.1}$$

Here $c = \sqrt{kT/2\pi m}$ is the mean speed of a vapor molecule in a given direction (m is the mass of a molecule), ν is the condensation coefficient, n_r'' is the concentration of molecules directly at the surface of the drop, and n_l'' is the same at the surface of the region.

To determine n_l'' we solve the diffusion equation for the second region and link the result up with that from (2.1) for the first region.[‡] Then we get the growth rate as

$$\dot{g} = \theta \cdot 4\pi r^2 vc \, (n'' - n_r''), \tag{2.2}$$

in which n" is the concentration of vapor molecules far from the drop and θ is a correction factor, which for the present small drops ($r < l$) is virtually 1.

We convert from the concentration to the vapor pressure via $p = n''kT$, substitute for c, and incorporate the value of θ to get

$$\dot{g} = \frac{4\pi r^2 v}{(2\pi mkT)^{1/2}} \, [p - p(r)]. \tag{2.3}$$

Here p(r) is the vapor pressure in equilibrium with a drop of radius r and p is the actual pressure of the supersaturated vapor (at the periphery), which can also be put as $p = p(r*)$, in which r* is the radius of the critical nucleus corresponding to the existing vapor pressure p_0.

From (1.2) and (1.6), the g, p, and p(r) of (2.3) are given by

$$p = p_\infty \exp\left[\left(\frac{2\sigma}{r*} + \frac{\partial\sigma}{\partial r*}\right) \Big/ n_\infty' kT\right], \quad p(r) = p_\infty \exp\left[\left(\frac{2\sigma}{r} + \frac{\partial\sigma}{\partial r}\right) \Big/ n_\infty' kT\right],$$
$$g = \frac{4}{3}\pi r^3 n_\infty'. \tag{2.4}$$

3. Steady-State Nucleation Rate in a Supersaturated Vapor

Following Frenkel' [10] and Zel'dovich [11], we can interpret nucleation as Brownian motion along the phase axis of size in an external field. The flux of nuclei along the size axis g in this case consists of two components:

$$I = -\left[u^0 \frac{\partial W}{\partial g} \cdot f(g, t) + D(g) \frac{\partial f}{\partial g}\right]. \tag{3.1}$$

The first defines the average motion (ordinary growth), while the second is the diffusion component, which describes transfer between sizes by fluctuation; also, $f(g, t)$ is the nonequilibrium

---

[†] Here l is of the order of the mean free path for vapor molecules.

[‡] Allowance must be made for equality of the molecular and diffusion fluxes at the boundary of the two regions.

TABLE 1. Dimensionless Thermodynamic
Characteristics as Functions of Relative Size
for a Drop of Nonpolar Liquid

$x = r/a$	$\varphi(x) = \sigma/\sigma_\infty$	$r*/r_C$	$\zeta = W*/W_C^\bullet$
∞	1	1	1
1000	0.9990	0.9995	0.997
500	0.9980	0.9990	0.994
100	0.9900	0.9950	0.970
50	0.9800	0.9900	0.941
40	0.9750	0.9875	0.928
30	0.9667	0.9833	0.908
20	0.9500	0.9750	0.855
10	0.9000	0.9500	0.722
9	0.8888	0.9444	0.702
8	0.8750	0.9375	0.659
7	0.8571	0.9285	0.615
6	0.8333	0.9161	0.560
5	0.8000	0.9000	0.486
4	0.7500	0.8750	0.383
3	0.6667	0.8333	0.231
2	0.5000	0.7500	0

size distribution at time t, and u^0 is the mobility of a nucleus, which is related to the coefficient D(g) for diffusion along the g axis by Einstein's relation $u^0 = D(g)/kT$.

Then (3.1) can be rewritten as

$$I = -De^{-W/kT} \cdot \frac{\partial}{\partial g}[fe^{W/kT}].$$ (3.2)

The labile equilibrium distribution is defined by (3.2), if we put I = 0, in accordance with the principle of detailed balancing. Then[†]

$$f^0(g) = Be^{-W/kT}.$$ (3.3)

Frenkel' [10] showed that constant B corresponds to the molecule concentration in the supersaturated vapor, $n'' = p/kT$; Courtney [12] states that this is roughly the concentration in the saturated vapor, $n_s'' = p_\infty/kT$.

We consider the steady-state nucleation via Szillard's scheme, i.e., we assume that nuclei of size G > g* are lost from the volume and are replaced by the corresponding quantity of single vapor molecules to leave the concentration of the latter unchanged. Then the nucleation rate (the number of nuclei of size G formed per second in unit volume) coincides with the flux I, which in this case is constant. Then (3.2) with I = constant reduces to quadratures; use of (3.3) gives

$$f(g)/f^0(g) = \frac{I}{B}\int_{g_\bullet}^{G}\frac{1}{D(g)}\exp\left[\frac{W(g)}{kT}\right]dg.$$

[†] This distribution has a physical meaning only for nuclei not too large relative to the critical size, but this is no obstacle to its use in discussion of steady-state nucleation, since the scheme (see below) involves removal of nuclei that exceed the critical size.

Also, $f(g_0)$ and $f^0(g_0)$ are virtually the same for $g = g_0$, so

$$I = B \left/ \int_{g_0}^{G} \frac{1}{D(g)} \exp\left[\frac{W_g}{kT}\right] dg. \right.$$

(3.4)

The integral may be evaluated from the feature [11] that $W(g)$ has a sharp peak at $g = g*$.

We perform a series expansion with respect to $(g - g*)$ in the integrand to get[†]

$$I \cong B \cdot D(g*) e^{-W(g*)/kT}.$$

(3.5)

$D = u^0 kT$ (Einstein's equation), while u^0 is defined by $\dot{g} = -u^0 \partial \dot{g} / \partial g$. The two relations together give

$$D(g) = - kT \frac{\dot{g}}{\partial W/\partial g}.$$

(3.6)

This relation becomes indeterminate for $g = g*$, and the usual technique for resolving the indeterminacy gives

$$D(g*) = - kT \left(\frac{\partial \dot{g}/\partial g}{\partial^2 W/\partial g^2}\right)_{g=g*}.$$

(3.7)

We now use (1.2), (2.3), and (2.4) to calculate $\partial \dot{g}/\partial g$, and (1.2), (1.3), and (1.5) to find $\partial^2 W/\partial g^2$, which gives $D(g*)$, which is substituted into (3.5), followed by the value of B. We express n'_∞ in terms of the density ρ'_∞ and molecular weight M (with use of the universal constants), and also use (1.11), which gives the steady-state nucleation rate as

$$\ln I = \ln K - A_c \zeta.$$

(3.8)

The kinetic factor K, the exponent A_c, and the correction factor ζ are defined[‡] by

$$K = 9.57 \cdot 10^{25} \cdot \nu M^{1/4} \frac{\sigma_\infty^{1/2}}{\rho_\infty} (p_\infty/T)^2 \cdot s \,,$$

$$A_c = 4.392 \frac{\sigma_\infty^3 M^2}{T^3 \rho_\infty^2 (\ln s)^2},$$

$$\zeta = \left(\varphi + \frac{1}{2} x \cdot \frac{\partial \varphi}{\partial x}\right)^2 \left(\varphi - x \cdot \frac{\partial \varphi}{\partial x}\right),$$

(3.9)

in which all quantities apart from p_∞ are expressed in cgs units, while p_∞ is in mm Hg, in order to facilitate the use of tabulated data.

These equations differ from the classical ones only in the correction factor ζ (the deviation coefficient). Table 1 shows that $\zeta \approx 1$ at moderate supersaturations, for which $x = r/a$ is relatively large, and so the result is essentially the classical one in that case. For high supersaturations (roughly $x < 50$), the two theories give different results; the new theory indicates a lower critical supersaturation, i.e., the supersaturation corresponding to a given I.

This prediction of lower critical supersaturations is, in general, in better agreement with experiment. Unfortunately, there are few published data on critical supersaturations for non-polar liquids, and we are able to use only Scharrer's results [14] for benzene. The quantities used were as follows: T = 251°K, M = 78 g/mole, ν = 0.9, p_∞ = 4.26 mm Hg, σ_∞ = 34.7 erg/cm² (extrapolated), ρ_∞ = 0.92 g/cm³ (extrapolated), and a = 4.879 Å.

[†] See [13] for details of the steps involved.

[‡] The exponent to s in the expression for K corresponds to Courtney's [12] estimate for B; if Frenkel's [10] estimate for B is used, the expression for K contains s^2.

We put $I = 10$ in (3.8) and (3.9) to get $s_{theor} = 4.18$, while Scharrer found $s_{exp} = 5.32$, whereas the classical theory gives $s_c = 9.3$.

References

1. T. L. Hill, J. Chem. Phys., 36:3182 (1962).
2. L. M. Shcherbakov, Tr. Tul'sk Mekhan. Inst., No. 7, p. 117 (1955); Research in Surface Forces, Vol. 1, Consultants Bureau, New York (1963), p. 19.
3. L. M. Shcherbakov and V. I. Rykov, Kolloidn. Zh., 23:221 (1961); L. M. Shcherbakov, in: Surface Effects in Melts in Powder Metallurgy, Izd. AN Ukr. SSR, Kiev (1963), p. 38; L. M. Shcherbakov and S. A. Pilyus, Ibid., p. 74.
4. L. M. Shcherbakov, in: Research in Surface Forces, Vol. 2, Consultants Bureau, New York (1966), p. 26.
5. L. M. Shcherbakov and V. A. Tereshin, in: Surface Effects in Melts and Solid Phases, Nal'chik (1965), p. 157.
6. L. M. Shcherbakov, Uch. Zap. Kishinevsk. Gos. Univ., 5:117 (1952); Kolloidn. Zh., 24:502 (1962).
7. L. M. Shcherbakov, Kolloidn. Zh., 20:759 (1958); L. M. Shcherbakov, P. P. Ryazantsev, and N. P. Filippova, Kolloidn. Zh., 23:338 (1961).
8. N. A. Fuks, Zh. Éksp. Teor. Fiz., 4:747 (1934).
9. M. Knudsen, Ann. Phys., 29:179 (1909); 47:697 (1915).
10. Ya. I. Frenkel', Zh. Éksp. Teor. Fiz., 9:192, 952 (1939).
11. Ya. B. Zel'dovich, Zh. Éksp. Teor. Fiz., 12:525 (1942).
12. W. G. Courtney, J. Chem. Phys., 35:2249 (1961).
13. A. J. Barnard, Proc. Roy. Soc., A220:132 (1953).
14. L. Scharrer, Ann. Phys., 35:619 (1939).
15. V. A. Tereshin and L. M. Shcherbakov, this volume, p. 413.

THEORY OF DIFFUSOPHORESIS
OF LARGE NONVOLATILE AEROSOL PARTICLES

B. V. Deryagin and Yu. I. Yalamov

Surface Phenomena Division, Institute of Physical Chemistry,
Academy of Sciences of the USSR

A formula is deduced for the diffusophoresis of nonvolatile particles whose size is much greater than the mean free path of the gas molecules. This is derived from the kinetic equations for the transport of gases through a partition consisting of spheres rigidly fixed in space. The Chapman—Enskog theory is used. The theoretical diffusophoresis rate is compared with experiment.

Introduction

The theory of aerosol motion in a diffusing gas mixture takes two different forms for particles whose radii R satisfy the conditions

$$R \ll \lambda_i, \tag{1}$$

$$R \gg \lambda_i, \tag{2}$$

in which λ_i is the mean free path for the molecules of component i of the gas mixture.

One of us has developed [1-2] the theory of diffusophoresis for small particles on the assumption that the velocity distribution of the molecules incident on a particle is not perturbed by the presence of the particle. Chapman and Enskog [3] derived the diffusophoresis rate V as

$$V = - D_{12} \frac{n}{n_2} \frac{\sqrt{m_1}}{n_1 \sqrt{m_1} + n_2 \sqrt{m_2}} \operatorname{grad} n_1, \tag{3}$$

in which that rate is reckoned in a coordinate system linked to the gas at rest in the case of a vapor—gas mixture, m_1 and m_2 are the masses of the components of the binary mixture, n_1 and n_2 are the corresponding mean numbers of molecules per unit volume, $n = n_1 + n_2$, and D_{12} is the mutual diffusion coefficient.

Brock [4] has considered the diffusophoresis rate for large aerosol particles, and he incorporated the coefficient of diffusion slip at the surface, though in deriving this he made the unfounded assumption that the interface does not affect the velocity distribution of the incident molecules. Here we derive the diffusophoresis rate for large nonvolatile aerosol particles via

427

the kinetic equations for gas transport through an aerosol partition, as in the calculation of thermophoresis for large particles [5, 6] and of diffusophoresis in solution [7].

Diffusophoresis Rate

Consider a system consisting of two vessels containing a binary gas mixture, which are joined via a wide capillary (diameter $2R \gg \lambda$), and which are both at a temperature T. Between the vessels there is a constant pressure difference Δp and also constant concentration differences ΔC_1 and ΔC_2, with $C_1 = n_1/n$ and $C_2 = n_2/n$.

The rate of production of entropy $\Delta \dot{S}$ is [16, 17]

$$\Delta \dot{S} = - I_1 \frac{\Delta \mu_1}{T} - I_2 \frac{\Delta \mu_2}{T}, \tag{4}$$

in which I_1 and I_2 are the mass flow rates through the capillary and $\Delta \mu_1$ and $\Delta \mu_2$ are the differences in the chemical potentials corresponding to ΔC_1 and ΔC_2:

$$\Delta \mu_{1,2} = \left(\frac{\partial \mu_{1,2}}{\partial p} \right)_{C_1; C_2} \Delta p + \left(\frac{\partial \mu_{1,2}}{\partial C_1} \right)_p \Delta C_1 + \left(\frac{\partial \mu_{1,2}}{\partial C_2} \right)_p \Delta C_2. \tag{5}$$

The volume flow rate is

$$V = - I_1 \left(\frac{\partial \mu_1}{\partial p} \right)_{C_1; C_2} - I_2 \left(\frac{\partial \mu_2}{\partial p} \right)_{C_1; C_2}, \tag{6}$$

in which $(\partial \mu_{1,2}/\partial p)_{C_1; C_2}$ are the partial specific volumes; then (4) becomes

$$\Delta \dot{S} = V \Delta p - I_1 \Delta_p \mu_1 - I_2 \Delta_p \mu_2, \tag{7}$$

in which

$$\Delta_p \mu_{1,2} = \left(\frac{\partial \mu_{1,2}}{\partial C_1} \right)_p \Delta C_1 + \left(\frac{\partial \mu_{1,2}}{\partial C_2} \right)_p \Delta C_2. \tag{8}$$

We introduce the flow

$$I_k^{\bullet} = I_k - \rho_k V, \tag{9}$$

(k = 1, 2), which defines the volume transfer of the component less the amount due to transfer as a whole. The V of (9) is the mean linear barocentric speed of the flow.

Gibbs's theorem gives

$$\rho_1 \Delta_p \mu_1 + \rho_2 \Delta_p \mu_2 = 0, \tag{10}$$

from which we get

$$I_1 \Delta_p \mu_1 + I_2 \Delta_p \mu_2 = I_1^{\bullet} \Delta_p \mu_1 + I_2^{\bullet} \Delta_p \mu_2. \tag{11}$$

Substitution of (11) into (7) gives us that

$$\Delta \dot{S} = V \Delta p - I_1^{\bullet} \Delta_p \mu_1 - I_2^{\bullet} \Delta_p \mu_2, \tag{12}$$

in which the generalized flow rates take the form

$$I_1^{\bullet} = - \alpha_{11} \Delta_p \mu_1 - \alpha_{12} \Delta_p \mu_2 + \alpha_{13} \Delta p, \tag{13}$$

$$I_2^{\bullet} = - \alpha_{21} \Delta_p \mu_1 - \alpha_{22} \Delta_p \mu_2 + \alpha_{23} \Delta p, \tag{14}$$

$$V = - \alpha_{31} \Delta_p \mu_1 - \alpha_{32} \Delta_p \mu_2 + \alpha_{33} \Delta p. \tag{15}$$

The Onsager relations are

$$\alpha_{12} = \alpha_{21}; \; \alpha_{13} = \alpha_{31}; \; \alpha_{23} = \alpha_{32}. \tag{16}$$

The V for $\Delta p = 0$ characterizes the diffusion osmosis due to the gradients in $\Delta_p \mu_1$ and $\Delta_p \mu_2$:

$$V|_{\Delta p=0} = -(\alpha_{31}\Delta_p\mu_1 + \alpha_{32}\Delta_p\mu_2). \tag{17}$$

The coefficients α_{31} and α_{32} may be derived from (13) and (14) with $\Delta_p\mu_1 = \Delta_p\mu_2 = 0$, i.e., when there are no gradients ΔC_1 and ΔC_2:

$$\alpha_{31} = \alpha_{13} = I_1^*|_{C_1; \, C_2}/\Delta p; \tag{18}$$

$$\alpha_{32} = \alpha_{23} = I_2^*|_{C_1; \, C_2}/\Delta p. \tag{19}$$

From (17)-(19) we get

$$V|_{\Delta p=0} = -\left(\frac{I_1^*|_{C_1; \, C_2}}{\Delta p} \Delta_p\mu_1 + \frac{I_2^*|_{C_1; \, C_2}}{\Delta p} \Delta_p\mu_2 \right) \tag{20}$$

or via (11)

$$V|_{\Delta p=0} = -\left(\frac{I_1|_{C_1; \, C_2}}{\Delta p} \Delta_p\mu_1 + \frac{I_2|_{C_1; \, C_2}}{\Delta p} \Delta_p\mu_2 \right). \tag{21}$$

In (21) we have the flows

$$I_1 = \rho_1 \vec{v}_1, \tag{22}$$

$$I_2 = \rho_2 \vec{v}_2, \tag{23}$$

in which $\vec{v}_1$ and $\vec{v}_2$ are the mean linear velocities along the normal to the partition.

If the Gibbs relation is put as

$$n_1\Delta_p\mu_1 + n_2\Delta_p\mu_2 = 0, \tag{24}$$

we get from (21)-(24) that

$$V|_{\Delta p=0} = \frac{n_1 (\vec{v}_2 - \vec{v}_1)|_{C_1; \, C_2}}{\Delta p} \Delta_p\mu_1. \tag{25}$$

The rate $\vec{v}_2 - \vec{v}_1$ of mutual diffusion may be derived from the general diffusion equation for a binary mixture for the case of isothermal flow in the absence of external forces [3]:

$$\vec{v}_2 - \vec{v}_1 = \frac{n^2}{n_1 n_2} D_{12} \left\{ \mathrm{grad}\, C_1 + \frac{n_1 n_2 (m_2 - m_1)}{n\rho} \mathrm{grad}\, (\ln p) \right\}. \tag{26}$$

Now grad $C_1 = 0$ if there are no concentration gradients, and (26) averaged over the width of the partition becomes

$$(\vec{v}_2 - \vec{v}_1)|_{C_1; \, C_2} = D_{12} \frac{n (m_2 - m_1)}{p\rho} \overline{\mathrm{grad}\, p}. \tag{27}$$

The symbols are as follows in (26) and (27): ρ mean density of mixture, D_{12} mixture diffusion coefficient.

From (25) and (27) we get

$$V = \frac{n n_1}{p\rho} D_{12} (m_2 - m_1) \frac{\Delta_p\mu_1}{H}, \tag{28}$$

where the gradient in the mean chemical potential is

$$\frac{\Delta_p \mu_1}{H} = kT \frac{\Delta C_1}{C_1 H}, \tag{29}$$

in which k is Boltzmann's constant.

Substitution of (29) into (28) and use of p = nkT gives

$$V = \frac{n D_{12} (m_2 - m_1)}{\rho} \frac{\Delta C_1}{H}. \tag{30}$$

The thermodynamic derivation of (30) does not allow one to establish whether the volume or surface mechanism determines the transport through the capillary. Kucherov and Rikenglaz [24] were the first to consider the rate due to the surface mechanism with allowance for the perturbation in the distribution function. Independently, Brock [4] made a more general study and gave an analogous but more convenient final formula for diffusion slip of a binary gas mixture at a solid surface. However, in both cases [4, 24] the distribution function for the gas molecules at the wall was derived without consideration of the collision integral. A correct approach to this problem must be based on the Boltzmann kinetic equation in conjunction with the collision integrals at the solid, which must cause the distribution function to become dependent on the distance from the wall, whereas these studies [4, 24] allow for this effect only via the incident and reflected molecules on the assumption that the distribution for the incident molecules retains its bulk character right up to the wall, while the distribution for the reflected molecules was expressed as the sum of the diffusely reflected molecules (with an equilibrium Maxwellian distribution) and the specularly reflected ones (with the same distribution as the incident molecules).

However, the incident molecules (and correspondingly the specularly reflected ones) must undergo at least one collision with reflected molecules on penetrating to a distance of the order of λ into the Knudsen layer, and so the bulk (but nonuniform) Chapman–Enskog distribution is perturbed.

Exact consideration of this problem is extremely complicated, and no solution has yet been obtained, so we restrict our consideration to estimation of the effect from Brock's [4] value for the diffusion slip rate:

$$V_g = \left\{ \frac{\frac{n m_1 m_2}{m} \left(\frac{\beta_1}{m_1^{1/2}} - \frac{\beta_2}{m_2^{1/2}} \right)}{(\beta_1 n_1 m_1^{1/2} + \beta_2 n_2 m_2^{1/2})} - \frac{1}{2} \left(\frac{2kT}{\pi} \right)^{1/2} \frac{n^2}{d_0 n_1 n_2} \frac{\left(\beta_1 m_1^{1/2} n_1 \frac{d_1}{m_1^{1/2}} + \beta_2 n_2 m_2^{1/2} \frac{d_{-1}}{m_2^{1/2}} \right)}{\left[\beta_1 \rho_1 \left(\frac{2kT}{\pi m_1} \right)^{1/2} + \beta_2 \rho_2 \left(\frac{2kT}{\pi m_2} \right)^{1/2} \right]} \right\} D_{12} \cdot \text{grad } C_1. \tag{31}$$

in which ρ_1 and ρ_2 are the densities of the two components of the mixture, β_1 and β_2 are the coefficients of diffuse reflection for these, T is absolute temperature, $m = (n_1/n)m_1 + (n_2/n)m_2$, and the coefficients d_{-1}, d_0, and d_1 take the form

$$d_{-1} = \frac{\delta_p B_{-1}}{A}, \qquad d_0 = \frac{\delta_p B_0}{A}, \qquad d_1 = \frac{\delta_p B_1}{A}, \tag{32}$$

in which

$$\delta_p = \frac{3}{2}\left(\frac{2kT}{n_1 n_2}\right)^{1/2}; \qquad A = \begin{vmatrix} a_{-1-1} & a_{0-1} & a_{1-1} \\ a_{0-1} & a_{00} & a_{01} \\ a_{1-1} & a_{01} & a_{11} \end{vmatrix}; \qquad (33)$$

$$B_{-1} = \begin{vmatrix} a_{01} & a_{0-1} \\ a_{11} & a_{1-1} \end{vmatrix}; \qquad B_0 = \begin{vmatrix} a_{-1-1} & a_{1-1} \\ a_{1-1} & a_{11} \end{vmatrix}; \qquad B_1 = \begin{vmatrix} a_{1-1} & a_{01} \\ a_{-1-1} & a_{0-1} \end{vmatrix}. \qquad (34)$$

The a_{pq} [3] are expressed via the collision integrals, which contain the Chapman–Enskog distribution functions.

The coefficient within the braces consists of two terms K_1 and K_2, of which the first is as follows for $\beta_1 = \beta_2 = 1$:

$$K_1 = \frac{m_1 m_2 \left(\sqrt{m_2} - \sqrt{m_1}\right)}{m \left(n_1 \sqrt{m_1} + n_2 \sqrt{m_2}\right)\left(\sqrt{m_1 m_2}\right)}, \qquad (35)$$

while second under those conditions is

$$K_2 = \frac{n^2 \left(n_1 d_1 + n_2 d_{-1}\right)\sqrt{m_1 m_2}}{2 d_0 n_1 n_2 \left(\rho_1 \sqrt{m_2} + \rho_2 \sqrt{m_1}\right)}. \qquad (36)$$

K_2 arises from the nonequilibrium character of the bulk distribution function and can be as much as 50% of K_1 for $n_1 \approx n_2$ for the hard-spheres model for a vapor–air mixture, i.e., this part substantially reduces the slip. It is to be expected that a more precise calculation via the collision integrals at the surface would produce a further substantial reduction, and so we will henceforth neglect the diffusion slip near the wall, the more so since even (31) indicates that this contributes no more than 15-20% of the mean transport rate given by (30).

We now assume that these vessels are separated by a highly porous partition formed of randomly placed spheres of radius $R \gg \lambda_i$, which are rigidly fixed in space at separations much greater than R. The concentration differences ΔC_1 and ΔC_2 are maintained at the ends of the partition. Then the mean linear barocentric rate of passage of the binary mixture through the partition equals in magnitude the diffusophoresis rate V_D for $\Delta p = 0$ and $\Delta T = 0$, being defined by (25), which shows that this rate is determined by the mutual diffusion rate $(\vec{v}_2 - \vec{v}_1)_{C_1; C_2}$, which differs from (27) when there are spheres in the partition on account of the local diffusion field around each sphere. We derive these fields by considering the velocities and concentrations of the components around the spheres. This involves solving a system of equations for convective diffusion and hydrodynamic effects in a viscous binary mixture, on the assumption that the process is isothermal and quasi-stationary. It has been shown [8, 9] that, for low relative density of one component ($\rho_1/\rho_1 + \rho_2 \ll 1$) and low velocities, this system splits up into purely hydrodynamic Navier–Stokes equations

$$\eta \Delta \vec{v} = \operatorname{grad} p, \qquad (37)$$

$$\operatorname{div} \rho \vec{v} = 0 \qquad (38)$$

and a pure diffusion equation for C_1'

$$\Delta C_1' = 0, \qquad (39)$$

in which C_1' is the deviation of the concentration from the mean value. The $\vec{v}$ of (37) and (38) is the total momentum of unit mass of mixture, while η is the viscosity.

Equations (37) and (38) are to be solved subject to the boundary conditions

$$v_r = 0 \quad \text{for} \quad r = R,$$

$$v_\theta = C_m \lambda \left[r \frac{\partial}{\partial r} \left(\frac{v_\theta}{r} \right) + \frac{1}{r} \frac{\partial v_r}{\partial \theta} \right]_{r=R}. \tag{40}$$

In (40), v_r and v_θ are the radial and angular components of the velocity of the incident gas flow, while r and θ are polar coordinates whose OZ axis is parallel to the flow velocity $\vec{u}$ far from the sphere. Also, C_m is [10] 1.09 for reflection by diffusion from the sphere and λ is the molecular mean free path. A detailed analysis has been given [11, 12] for the boundary conditions of (40). The solution to (37) and (38) subject to (40) gives [14, 15] the pressure gradient as

$$\text{grad } p = 2A\eta \left[\frac{3 (\vec{u} \cdot \vec{r}) \vec{r}}{r^5} - \frac{\vec{u}}{r^3} \right], \tag{41}$$

in which

$$A = \frac{3}{4} R \left(\frac{1 + 2C_m \dfrac{\lambda}{R}}{1 + 3C_m \dfrac{\lambda}{R}} \right). \tag{42}$$

On account of grad p, closed local barodiffusion flows arise around each sphere, these starting and ending at the surfaces, and passing through the partition in the sense opposite to that of the flow of (27).

The radial local barodiffusion isoconcentric flow at the surface of a sphere is [13]

$$n_1 (\vec{v}_2 - \vec{v}_1)_{r=R} |_{C_1; C_2} = \frac{n_1 n (m_2 - m_1)}{p \rho} D_{12} (\text{grad}_r p)_{r=R}, \tag{43}$$

in which grad $p|_{r=R}$ is defined by (41) as

$$\text{grad}_r p |_{r=R} = \frac{4 A \eta | \vec{u} | \cos \theta}{R^3}. \tag{44}$$

We substitute the grad$_r$ p of (44) into (43); also, the total flux through the part of an intersecting plane external to a sphere is equal to the total flux beginning from the surface of a sphere on one side of that plane, so

$$\Delta \left[n_1 (\vec{v}_2 - \vec{v}_1) \right]_{r=R} |_{C_1; C_2} = \frac{n_1 n (m_2 - m_1)}{p \rho} D_{12} \frac{4 A \eta | \vec{u} |}{R^3} \int_{\Delta S_i} \cos \theta \, dS. \tag{45}$$

Integration over the surface ΔS_i gives

$$\Delta \left[n_1 (\vec{v}_2 - \vec{v}_1) \right]_{r=R} |_{C_1; C_2} = \frac{n_1 n (m_2 - m_1)}{p \rho} D_{12} \frac{4 A \eta | \vec{u} |}{R^3} \Delta S_i. \tag{46}$$

Clearly, $S = \Sigma \Delta S_i$ is the proportion of the plane within the spheres; the randomness means that this is always the same (no matter what the position of the plane) and equals $4\pi R^3 N/3$, in which N is the number of spheres per unit volume.

Summation of (46) over all spheres in a given cross-section gives

$$n_1 (\vec{v}_2 - \vec{v}_1)_S |_{C_1; C_2} = \frac{16 \pi A \eta | \vec{u} | N n_1 n (m_2 - m_1)}{3 p \rho} D_{12}. \tag{47}$$

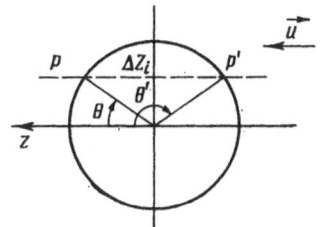

Fig. 1. Calculation of the pressure differential at the ends of the part of a straight line within a sphere.

from (41) and (42), the flow resistance of a sphere is

$$F \equiv \frac{\Delta p}{NH} = 6\pi\eta R \, |\vec{u}| \left(\frac{1 + 2C_m \frac{\lambda}{R}}{1 + 3C_m \frac{\lambda}{R}} \right), \tag{48}$$

so we get from (47) that

$$n_1 (\vec{v_2} - \vec{v_1})_S \mid_{C_1; C_2} = \frac{2}{3} \frac{n_1 n (m_2 - m_1)}{p\rho} D_{12} \frac{\Delta p}{H}. \tag{49}$$

H is the length of the partition in (48) and (49).

This flux is directed against the external flux of (27). Before we can compare (27) and (49), we must distinguish between the $\overline{\text{grad p}}$ of (27) and the grad $\overline{p}$ of (49), the latter being equal to $\Delta p/H$.

The pressure differential across the partition is as follows along a straight line normal to the partition and parallel to the flow:

$$\Delta p = \sum \Delta p_i + H \overline{\text{grad } p}, \tag{50}$$

in which Δp_i is the pressure differential at the ends of the part of a straight line within a sphere. Here we have used the fact that the mean values $\overline{\text{grad p}}$ on parts of a straight line and of a plane section of the partition outside the spheres are equal to the mean $\overline{\text{grad p}}$ over the volume and so are mutually equal. Division of (50) by H gives

$$\frac{\Delta p}{H} = \sum \frac{\Delta p_i}{H} + \overline{\text{grad } p}. \tag{51}$$

The pressure differential at the ends of the part of a straight line ΔZ_i within a sphere is Δp_i (Fig. 1):

$$\Delta p_i = p - p' = 2A\eta \frac{|\vec{u}|}{R^2} (\cos \theta' - \cos \theta). \tag{52}$$

Expressions for p and p' are derived from the solutions to (37) and (38) subject to (40). As $\cos \theta' = -\cos \theta$, we get

$$\Delta p_i = 4A\eta \frac{|\vec{u}|}{R^2} \cos \theta = 2A\eta \frac{|\vec{u}|}{R^3} \Delta Z_i. \tag{53}$$

Summation over all spheres intersected by the straight line gives

$$\frac{\sum \Delta p_i}{H} = 2A\eta \frac{|\vec{u}|}{R^3} \frac{\sum \Delta Z_i}{H}, \tag{54}$$

in which $\Delta Z_i/H$ is the proportion of the line within the spheres, which also is independent of the position on account of the disorder and is also $4\pi R^3 N/3$. Then

$$\sum \frac{\Delta p_i}{H} = \frac{8}{3} \pi A N \eta |\vec{u}|. \tag{55}$$

From (48) and (55)

$$\frac{\sum \Delta p_i}{H} = \frac{1}{3} \cdot \frac{\Delta p}{H}. \tag{56}$$

Then from (51)

$$\overline{\operatorname{grad} p} = \frac{2}{3} \cdot \frac{\Delta p}{H}.$$

(57)

From (27) and (57)

$$n_1 (\vec{v_2} - \vec{v_1}) |_{C_1; C_2} = \frac{2}{3} D_{12} \frac{n_1 n (m_2 - m_1)}{p\rho} \frac{\Delta p}{H}.$$

(58)

The flows of (49) and (58) then sum to zero.

However, the sum of (49) and (58) is not equal to the total true flux from mutual diffusion when

$$C_1 = \text{const}, \ C_2 = \text{const}$$

(59)

in view of the perturbation caused by the concentration distributions around the spheres, which violate (59). As in the theory of thermophoresis, we must take into account the thermal polarization of each sphere and of the partition as a whole [5, 6].

To determine the concentration distribution at low speeds, we integrate (39) subject to the boundary conditions

$$C_1'(r) \to 0 \quad \text{for} \quad r \to \infty,$$

(60)

$$-n D_{12} \frac{\partial C_1'}{\partial r} \bigg|_{r=R} + \frac{n^2 C_1 C_2 (m_1 - m_2)}{p\rho} D_{12} \operatorname{grad}_r p |_{r=R} = 0.$$

(61)

The solution to (39) subject to (44), (60), and (61) is

$$C_1'(r) = \frac{2n C_1 C_2 (m_2 - m_1) A\eta |\vec{u}| \cos \theta}{p\rho r^3}.$$

(62)

The analogy with electrostatics indicates that the concentration distribution of (62) is produced by a dipole whose moment is

$$|\vec{K}_0| = \frac{2n C_1 C_2 (m_2 - m_1) A\eta |\vec{u}|}{p\rho}.$$

(63)

The concentration distributions of the individual spheres sum if the spheres are widely separated, and the entire partition resembles a polarized dielectric in that it can be represented via a moment for concentration polarization:

$$|\vec{K}| = N |\vec{K}_0|,$$

(64)

the ends of the partition having a concentration difference

$$\Delta C_1' = -4\pi |\vec{K}| H.$$

(65)

The concentration difference for the second component, $\Delta C_2'$, is calculated similarly. Condition (59) is violated by $\Delta C_1'$ and $\Delta C_2'$, and this condition is necessary in order to find the fluxes of (43) and (58). To eliminate this violation, we have to introduce mass sources and sinks at the ends of the partition to balance out $\Delta C_1'$ and $\Delta C_2'$. The rate of mutual diffusion due to these sources is calculated via the diffusion equation, in which the gradient $\Delta C_1'/H$ in the mean concentration is determined via (62)-(65). Then this contribution to the mutual diffusion rate takes the form

$$n_1 (\vec{v_2} - \vec{v_1})' |_{C_1; C_2} = D_{12} \frac{n_1 n (m_2 - m_1)}{p\rho} \frac{\Delta p}{H}.$$

(66)

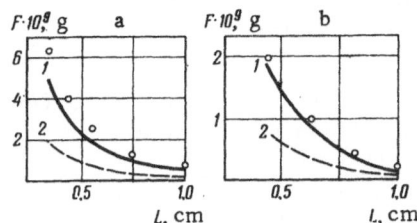

Fig. 2. Relation of repulsion force F between a silvered glass sphere and an evaporating water droplet to distance L between the centers as given by: 1) formula (78); 2) theory of [8, 9]. The points are from experiment [12]: a) 0% humidity; b) 40% humidity.

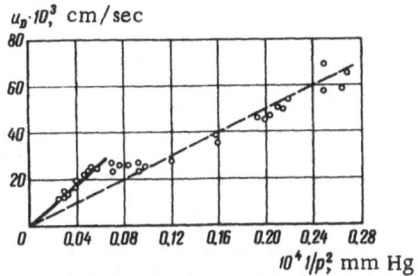

Fig. 3. Diffusophoresis rate U_D of an oil aerosol as a function of $1/p^2$: full line from (77), broken line from (3).

The barocentric transport rate relative to the partition is found by substituting (66) into (25) and using (28) and (29):

$$V = D_{12} \frac{n(m_2 - m_1)}{\rho} \frac{\Delta C_1}{H}. \tag{67}$$

The diffusophoresis rate for the spheres is found on transferring to a coordinate system in which the center of inertia of the gas is at rest, i.e., by reversing the sign in (67):

$$V_D = -D_{12} \frac{n(m_2 - m_1)}{\rho} \cdot \frac{\Delta C_1}{H}. \tag{68}$$

Consider now the case of a vapor–gas mixture in a vessel in which one wall acts as vapor source, while the parallel opposite wall acts as an absorber. The ordered motion of the gas has a zero velocity component normal to these walls. Here we must transfer to the laboratory coordinate system, in which the center of inertia is in motion. The speed of the center of inertia of a binary gas mixture is

$$V_{ci} = \frac{\rho_1 \vec{v}_1 + \rho_2 \vec{v}_2}{\rho_1 + \rho_2}, \tag{69}$$

where the densities are

$$\rho_1 = n_1 m_1, \tag{70}$$

$$\rho_2 = n_2 m_2, \tag{71}$$

and since the above component is zero,

$$\vec{v}_2 = 0 \tag{72}$$

we get

$$V_{ci} = \frac{\rho_1}{\rho_1 + \rho_2} \vec{v}_1 = \frac{\rho_1}{\rho} \vec{v}_1. \tag{73}$$

The kinetic theory of gases gives [14, 18] the flux of the first component in the binary mixture relative to the laboratory coordinate system as

$$n_1 \vec{v}_1 = -n D_{12} \frac{\Delta C_1}{H} + \frac{n^3 C_1 G_2}{\rho}(m_1 - m_2) D_{12} \overline{\text{grad} \ln p} + n_1 \frac{\vec{n_1 v_1} + \vec{n_2 v_2}}{n}. \tag{74}$$

For isobaric conditions, we get via (72) and (74) that

$$\vec{v}_1 = -D_{12} \frac{n^2}{n_1 n_2} \frac{\Delta C_1}{H}. \tag{75}$$

Substitution of (75) into (73) gives us, on considering the case $n_2 \approx n$ (vapor concentration very low) that

$$V_{ci} \approx -D_{12} \frac{m_1 n}{\rho} \frac{\Delta C_1}{H}. \tag{76}$$

Summation of (68) and (76) then gives the diffusophoresis rate for large nonvolatile aerosol particles relative to the walls of the vessel in a diffusing vapor−gas mixture:

$$V_D^{(\text{lab})} = -D_{12} \frac{n m_2}{\rho} \frac{\Delta C_1}{H} .\tag{77}$$

Formula (77) gives the force acting on a particle:

$$F = 6\pi\eta R \left(\frac{1 + 2C_m \dfrac{\lambda}{R}}{1 + 3C_m \dfrac{\lambda}{R}} \right) V_D^{(\text{lab})}.\tag{78}$$

Formula (77) differs by about 12% from our previous formula [25] for the diffusophoresis rate in an analogous case, since here we have taken account of the difference between the mean value of the pressure gradient over the partition and the gradient of the mean pressure [see (57)].

Comparison with Experiment

Formula (78) can be compared with measurements [19] of the force between a silvered glass sphere and an evaporating drop of water (Fig. 2). The pressure in these experiments was close to atmospheric, while the sphere had a radius of 0.1 cm, and hence the slip correction in (67) is unimportant (Knudsen's number $\text{Kn} \equiv \lambda/R \ll 1$), and the formula is

$$F = 6\pi\eta R V_D^{(\text{lab})}.\tag{79}$$

In this case, theory agrees well with experiment.

Consider now the comparison of (77) with direct measurements of diffusophoresis rate for large aerosol particles. In one case [20], the error of measurement was very high (about 50%), and satisfactory agreement with experiment in this case was obtained for small particles by the baseless introduction of an "effective length" for the connecting tube, which differed by 44% from the actual length. It would also seem that there were convection currents, but these were not examined. It is therefore impossible to compare these results precisely with experiment, the more so since convective effects may vary greatly with the pressure.

Another paper [21] gives results for water vapor in air, in which the air was at rest relative to the walls [formula (77)], and here again this unjustified effective length was introduced, so comparison with our theory is difficult. However, the theoretical ratio of the reduced speeds for large and small particles is about 1.4, whereas Fig. 3 of [21] indicates a value close to one. However, the error of experiment at high pressures was about 40%, so it cannot be said that the results conflict with our theory.

Storozhilova's [22] results appear the most accurate, as the error of experiment was about 10%. Here the diffusophoresis rate was determined from the deflection of a thin aerosol jet by the method used in thermophoresis [23]. An oil aerosol made by condensation was used, which had a mean particle radius of about 0.25×10^{-4} cm. Figure 3 shows V_D as a function of $1/p^2$, where the full line is from (77), while the broken line is from (3) (small particles). It is clear from Fig. 3 that our theory for large particles [formula (77)] applies closely for $\text{Kn} < 0.5$, while formula (3) for small particles applies for $\text{Kn} > 0.7$. The transition region between small and large particles is thus extremely narrow.

Conclusions

1. It is not possible to calculate directly the diffusophoresis rate for large particles via Onsager's principle in thermodynamics, which contrasts with the direct development of the theory for particles small relative to the mean free path of the gas molecules.

2. A theory has been developed for the motion of a gas through a porous partition or cylindrical channels under a concentration gradient.

3. Detailed consideration is given to a partition in the form of randomly placed spheres, which provides a formula for diffusophoresis of particles much larger than the mean free path of the gas molecules.

4. Allowance must be made for barodiffusion in considering the mass transfer in order to obtain the correct result.

5. The result for the diffusophoretic force is in good quantitative agreement with measurements [19] of the force between a silvered glass sphere (Kn very small) and an evaporating water drop.

6. Formula (77) for large particles agrees satisfactorily with results [22] from the deflection of a thin jet at Kn < 0.5, whereas agreement with formula (3) for small particles is found for Kn > 0.7. The transition region thus covers only a narrow range in Kn.

References

1. B. V. Deryagin and S. P. Bakanov, Dokl. AN SSSR, 117(6):959 (1957).
2. S. P. Bakanov and B. V. Deryagin, Kolloidn. Zh., 21(4):377 (1959).
3. S. Chapman and T. G. Cowling, Mathematical Theory of Nonuniform Gases, Cambridge University Press, New York (1952).
4. J. R. Brock, J. Coll. Sci., 18:489 (1963).
5. B. V. Deryagin and S. P. Bakanov, Dokl. AN SSSR, 147(1):139 (1962).
6. B. V. Deryagin and Yu. I. Yalamov, Dokl. AN SSSR, 155(4):886 (1964).
7. S. S. Dukhin and B. V. Deryagin, Dokl. AN SSSR, 159(2):401 (1964).
8. B. V. Deryagin and S. S. Dukhin, Dokl. AN SSSR, 106(5):851 (1956).
9. B. V. Deryagin and S. S. Dukhin, Dokl. AN SSSR, 111(3):613 (1956).
10. S. P. Bakanov and B. V. Deryagin, Dokl. AN SSSR, 139(1):71 (1961).
11. R. E. Street, Rarefied Gas Dynamics, New York (1960), p. 276.
12. E. N. Kennard, Kinetic Theory of Gases, New York (1938), pp. 295, 327.
13. B. V. Deryagin and S. P. Bakanov, Dokl. AN SSSR, 141(2):284 (1961).
14. J. O. Hirschfelder, C. F. Curtiss, and R. S. Bird, Molecular Theory of Gases and Liquids, New York (1954), Chapter 8.
15. L. D. Landau and E. M. Lifshits, Mechanics of Continuous Media, Gosfizmatizdat, Moscow (1953).
16. S. R. de Groot, Thermodynamics of Irreversible Processes, Wiley, New York (1951).
17. S. R. de Groot and P. Mazur, Nonequilibrium Thermodynamics (Interscience), Wiley, New York (1962).
18. E. A. Mason and S. Chapman, J. Chem. Phys., 36:627 (1962).
19. P. S. Prokhorov and L. F. Leonov, Kolloidn. Zh., 23:464 (1961); Disc. Faraday Soc., No. 30, p. 124 (1960).
20. K. H. Schmitt and L. Waldmann, Z. Naturforsch., 15a:843 (1960).
21. K. H. Schmitt, Z. Naturforsch., 16a:144 (1961).
22. A. I. Storozhilova, Dokl. AN SSSR, 155:426 (1964).
23. B. V. Deryagin and A. I. Storozhilova, Kolloidn. Zh., 26(5):583 (1964).
24. R. Ya. Kucherov and L. É. Rikenglaz, Zh. Éksp. Teor. Fiz., 36:1758 (1959).
25. Yu. I. Yalamov and B. V. Deryagin, Dokl. AN SSSR, 165(2):364 (1965).

THE BARO-EFFECT AND DIFFUSION SLIP

P. E. Suetin and P. V. Volobuev

Ural Polytechnical Institute, Sverdlovsk

A theoretical relation is derived between the velocity of diffusion slip and the potential parameters. The solution is tested via the baro-effect.

Three ranges in the Knudsen number $Kn = \lambda/d$ are distinguished in the motion of a gas with respect to a solid, in which λ is mean free path and d is a characteristic dimension of the body. The range $10^{-2} \leq Kn \leq 10^2$ is termed the intermediate region, because the flow is then dependent on collisions with the surface as well as by collisions between molecules, and so it is impracticable to apply either the mechanics of continuous media or the methods of elementary kinetic theory. If the hydrodynamic approach is to be applied for Kn small, we have to alter the boundary conditions for the macroscopic equations.

Maxwell [1] was the first to use kinetic methods in discussing the boundary conditions for the macroscopic equations of motion of a gas, which provided an explanation of the observed [2] viscous slip. Of course, the same approach applies if a mixture of gases is involved.

On the other hand, it is possible to apply rigorous kinetic theory and not resort to the hydrodynamic equations, as via the linearized Boltzmann equation and a perturbation function [3, 4]. These results show that the method is a promising one for all intermediate values of Kn. Here we consider only a binary gas mixture in the hydrodynamic approach.

It has been shown [5, 6] that here the boundary conditions include diffusion slip (macroscopic transport of a gas mixture along a wall on account of concentration inhomogeneity). The rate of this slip can be determined precisely by the Chapman—Enskog method, but experimental test involves various difficulties.

We consider that the baro-effect [7] in mutual diffusion provides the best method of examining this phenomenon. This effect is the production of a pressure difference in the isothermal mutual diffusion of gases. The essence of the effect may be explained by the methods of nonequilibrium thermodynamics [8], but kinetic methods have to be invoked in order to obtain satisfactory agreement between theory and experiment for the pressure difference. Naturally, diffusion slip only governs the baro-effect for a mixture of two gases having the same molecular weight but different collision diameters.

438

Theory of the Baro-Effect

Consider two fixed closed volumes V_I and V_{II} filled with different gases under isothermal conditions. V_I initially contains only the light gas (molecules of weight m_1, density $n_I(0) = n_1$), while V_{II} contains only the heavy gas (m_2 and $n_{II}(0) = n_2$).

We take the initial pressures as equal. The two volumes are then connected via a broad capillary (length L and radius R). The lighter molecules have the higher velocity, which leads to a pressure rise in V_{II} and vice versa, and hence to a pressure difference ΔP, which gives rise to compensating hydrodynamic transport. The resulting mean numerical velocity $\vec{\omega}_0$ expressed in terms of the mean mass velocity $\vec{v}_0$ and the diffusion velocity $\vec{V}_\alpha$ of component α [9] is:

$$\vec{\omega}_0 = \vec{v}_0 + \frac{m_2 - m_1}{m_2 \cdot m_1} \frac{\rho_1}{n} \vec{V}_1,$$ (1)

in which $n = n_1 + n_2$ is the density of the gas mixture.

This shows that the diffusion transport is exactly balanced by viscous transport when $\vec{\omega}_0 = 0$, and the resulting ΔP is entirely determined by the concentration difference.

The time dependence of the effect can be deduced from the condition of conservation of the number of molecules in the system:

$$\frac{dn_{\alpha\,I}}{dt} = \frac{s_k}{V_I} g_\alpha; \qquad \frac{dn_{\alpha\,II}}{dt} = -\frac{s_k}{V_{II}} g_\alpha,$$ (2)

in which s_k is the cross-sectional area of the capillary.

The density of the total flux of component α can be put as

$$g_\alpha = \omega_0 n_\alpha - n D_{12} \frac{dC_\alpha}{dx},$$ (3)

in which $C_\alpha = n_\alpha/n$ is the molar concentration of the α component, and where the term describing the barodiffusion has been omitted.

Solution of (2) in the quasi-stationary approximation [10] shows that ΔP increases to a maximum ΔP_{max} determined by the state in which $\omega_0 = 0$, and then slowly falls as the concentrations become equal. The rate of increase in ΔP is characterized by a relaxation time τ_P, while the concentration equalization is characterized by a relaxation time τ_C. These times may be deduced by assuming that the concentration is constant throughout each of the volumes separately. It follows from Fick's equation for diffusion within the volumes [11] that this approximation is true for viscous flow of the mixture through the capillary during the rise to ΔP_{max}. Here

$$\tau_P = \frac{8\pi\bar{h}VL}{Ps_k^2(1 + 4l/R)}, \qquad \tau_C = \frac{VL}{D_{12}s_k},$$ (4)

in which $1/V = (1/V_1) + (1/V_2)$ and

$$4\bar{l}/R = 4\sqrt{2\pi kT}\,\bar{\eta}/P(\sqrt{m_1} + \sqrt{m_2})R \approx \lambda/R$$

is a correction to the hydrodynamic transport due to viscous slip, η is the mean viscosity of the mixture, and k is Boltzmann's constant.

The relation between the times is such that the change in concentration up to the attainment of ΔP_{max} may be neglected even for Kn very large; ΔP tends exponentially to ΔP_{max}. In the terminology of nonequilibrium thermodynamics, the state corresponding to ΔP_{max} may be classifed as a first-order stationary state [12].

Thermodynamic Solution

Relaxation phenomena are characterized via the change in the entropy S of the system from the viewpoint of nonequilibrium thermodynamics. The rate of increase in S in this system under isothermal conditions is given [13] by

$$\frac{\partial}{\partial t} \int \rho S \, dV = \int \frac{1}{T} \sigma_{ik}' \frac{\partial v_{0i}}{\partial x_k} \, dV - \int j \frac{\nabla \mu}{T} \, dV, \tag{5}$$

in which the first integral represents the increase in S from viscous transport and the second represents that from diffusion. Here σ_{ik}' is the tensor for the viscous stresses, j is the mass diffusion flux, and μ is the chemical potential.

The first term on the right in (5) may be put as

$$\int \frac{1}{T} \sigma_{ik}' \frac{\partial v_{0i}}{\partial x_k} \, dV = \int \frac{1}{T} \operatorname{div} (v_0 \sigma_{ik}') \, dV - \int \frac{1}{T} v_{0i} \frac{\partial \sigma_{ik}'}{\partial x_k} \, dV.$$

The first integral on the right may be transformed to an integral over the fixed surface of the system, and in this way it may be shown to be zero. Then

$$\frac{\partial}{\partial t} \int \rho S \, dV = - \int \frac{1}{T} v_{0i} \frac{\partial \sigma_{ik}'}{\partial x_k} \, dV - \int j \frac{\nabla \mu}{T} \, dV. \tag{6}$$

The Navier–Stokes equation gives for steady-state flow in the capillary that

$$\frac{\partial \sigma_{ik}'}{\partial x_k} = \nabla_i P \tag{7}$$

and

$$\frac{\partial}{\partial t} \int \rho S \, dV = - \int \vec{v}_0 n \frac{\nabla P}{nT} \, dV - \int \vec{j} \frac{\nabla \mu}{T} \, dV. \tag{8}$$

The expression for the increase in S allows us to determine the forces and hence the flow rates, since by definition

$$\dot{S} = \vec{I}_1 \vec{X}_1 + \vec{I}_2 \vec{X}_2. \tag{9}$$

Then comparison with (8) gives

$$\vec{I}_1 = \vec{v}_0 n; \qquad \vec{I}_2 = \vec{j}; \tag{10}$$

$$\vec{X}_1 = \frac{\nabla P}{nT}; \qquad \vec{X}_2 = \frac{\nabla \mu}{T}. \tag{11}$$

The following linear relations apply for these forces and flows:

$$\vec{I}_1 = - L_{11} \vec{X}_1 - L_{12} \vec{X}_2,$$
$$\vec{I}_2 = - L_{21} \vec{X}_1 - L_{22} \vec{X}_2. \tag{12}$$

When $\omega_0 = 0$, it is convenient to perform the following linear transformations on the flows:

$$\vec{I}_1 = \vec{I}_1 + a\vec{j}; \qquad \vec{I}_2 = \vec{I}_1 C + \frac{1}{m_1} \vec{j}; \tag{13}$$

in which

$$a = \frac{m_2 - m_1}{m_1 m_2}.$$

In these expressions, I_1' and I_2' are the total fluxes through the capillary in the laboratory coordinate system for the mixture and the light component respectively.

The forces corresponding to these fluxes are defined by (9), which still applies when we use the linear transformations

$$\vec{X}_1' = \frac{\vec{X}_2 C - \vec{X}_1 \frac{1}{m_1}}{b}, \qquad \vec{X}_2' = \frac{\vec{X}_1 a - \vec{X}_2}{b}, \tag{14}$$

in which

$$b = Ca - \frac{1}{m_1}.$$

For the transformed fluxes we have

$$\begin{aligned}
\vec{I}_1' &= -L_{11}'\vec{X}_1' - L_{12}'\vec{X}_2', \\
\vec{I}_2' &= -L_{21}'\vec{X}_1' - L_{22}'\vec{X}_2'.
\end{aligned} \tag{15}$$

Now $\vec{I}_1' = 0$ in the state we envisage, so (15) implies that

$$\frac{\vec{X}_2'}{\vec{X}_1'} = -\frac{L_{11}'}{L_{12}'}. \tag{16}$$

The ratio of the coefficients may be deduced via Onsager's relation: for $X_2' = 0$ we have from (15) that

$$\frac{L_{11}'}{L_{21}'} = \frac{\vec{I}_1'}{\vec{I}_2'}\bigg|_{X_2'=0}, \tag{17}$$

but

$$L_{12}' = L_{21}'. \tag{18}$$

Then

$$\frac{\vec{X}_2'}{\vec{X}_1'} = \frac{a\frac{\vec{X}_1}{\vec{X}_2} - 1}{C - \frac{\vec{X}_1}{\vec{X}_2}\frac{1}{m_1}}\bigg|_{X_2'=0}. \tag{19}$$

Reverting to the former symbols,

$$\frac{\vec{X}_1}{\vec{X}_2} = -\frac{\vec{I}m}{\vec{I}_1}\bigg|_{X_2'=0}. \tag{20}$$

Then from (10) and (11)

$$\frac{\nabla P}{n\nabla\mu} = -\frac{\vec{j}}{nv_0}\bigg|_{X_2'=0}. \tag{21}$$

But the definition of X_2' implies that $X_2' = 0$ when

$$\frac{\nabla P}{n\nabla\mu} = \frac{1}{a}. \tag{22}$$

Then in this steady state

$$\frac{1}{a} = -\frac{j}{n v_0}.$$ (23)

The problem then reduces to determination of the explicit form of $\vec{v}_0$ and the diffusion rate.

There are two methods of interpreting transport in low-pressure gases. The first is macroscopic (via nonequilibrium thermodynamics), while the second is microscopic (via kinetic theory and molecular interactions). The two methods give, to a first approximation, equivalent expressions for the macroscopic fluxes of physical quantities, but the phenomenological approach does not allow one to determine the explicit form of the transport coefficients that appear in these. This applies also to the momentum transfer in a gas mixture, especially the effects of boundary conditions on $\vec{v}_0$.

Boundary Conditions and Diffusion Slip

The following is the result for $\vec{v}_0$ from integrating the Navier–Stokes equations for steady-state flow through a capillary whose axis lies along the X axis:

$$\vec{v}_0 = -\frac{r^2}{4}\frac{1}{\eta}\frac{dP}{dx} + a.$$ (24)

The constant of integration a can be deduced from the boundary conditions at the wall. The elementary kinetic solution [5] for a binary gas mixture shows that the concentration gradient causes diffusion slip along the wall with a speed

$$V_{0d} = \sigma_{12} D_{12}\frac{dC}{dx}.$$ (25)

The coefficients σ_{12} for this slip is derived via the equilibrium distribution function as

$$\sigma_{12} = \frac{m_2 - m_1}{m_1 C + m_2 (1-C)} - \frac{\sqrt{m_2} - \sqrt{m_1}}{\sqrt{m_1} C + \sqrt{m_2}(1-C)}.$$ (26)

Grade's method allows us to determine the boundary conditions directly for the nonequilibrium distribution function, via the reflected half of the distribution if the half corresponding to incident molecules is known. V_{0d} for pure diffuse scattering at the wall has been derived [6] by Grade's method [14] via the moments of the nonequilibrium distribution. In the 13-moment approximation

$$V_{0d} = -\frac{kT\,(R_{1x}/\sqrt{m_1} + R_{2x}/\sqrt{m_2})}{2\,(P_1\sqrt{m_1} + P_2\sqrt{m_2})} - \frac{S_{1x}\sqrt{m_1} + S_{2x}\sqrt{m_2}}{5\,(P_1\sqrt{m_1} + P_2\sqrt{m_2})},$$ (27)

in which P_α is the partial pressure. If f_α is taken as the distribution for the α component,

$$R_{\alpha x} = m_\alpha \int V_{\alpha x} f_\alpha\, dv_\alpha,$$
$$S_{\alpha x} = \frac{m_\alpha}{2} \int V_\alpha^2 V_{\alpha x} f_\alpha\, dv_\alpha,$$ (28)

in which the integration is with respect to the total velocity v_α, while V_α is the corresponding thermal velocity. The physical significance of the moments $R_{\alpha x}$ and $S_{\alpha x}$ is obvious.

Analogy with previous results [15] allows us to transform the first term on the right in (27) to

$$- \frac{kT\,(R_{1x}/\sqrt{m_1} + R_{2x}/\sqrt{m_2})}{2\,(P_1\sqrt{m_1} + P_2\sqrt{m_2})} = \frac{1}{2}\,\sigma_{12}D_{12}\,\frac{dC}{\partial x}\,. \tag{29}$$

We need to know the explicit form of the distribution function for a component in order to determine S_α.

The problem may be solved in a first approximation on the basis that $\Delta n_\alpha \gg \Delta n$ in the capillary; then f_α is determined as the distribution function for component α in the diffusion field of the binary mixture under isothermal conditions. Enskog's method gives this as having a form similar to that previously derived [16]:

$$f_1 = f_1^{(0)}\left\{1 - \left[\left(\frac{\rho_1\rho_2}{\rho n_1 m_1^{1/2}}\,d_0 + \frac{5}{2}\,d_1\right)\vec{W}_1 - d_1 W_1^2\vec{W}_1\right]\operatorname{grad} n_1\right\},$$

$$f_2 = f_2^{(0)}\left\{1 - \left[\left(-\frac{\rho_1\rho_2}{\rho n_2 m_2^{1/2}}\,d_0 + \frac{5}{2}\,d_{-1}\right)\vec{W}_2 - d_{-1} W_2^2\vec{W}_2\right]\operatorname{grad} n_2\right\}, \tag{30}$$

in which $\vec{W}_\alpha = \left(\frac{m_\alpha}{2kT}\right)^{1/2}\vec{v}_\alpha$ is the dimensionless velocity of the molecules of the α component and

$$d_{-1} = \frac{\delta_p B_{-1}}{A}\,; \quad d_0 = \frac{\delta_p B_0}{A}\,; \quad d_1 = \frac{\delta_p B_1}{A}\,;$$

$$\delta_p = \frac{3}{2}\left(\frac{2kT}{n_1 n_2}\right)^{1/2}\,;$$

$$A = \begin{vmatrix} a_{-1-1} & a_{0-1} & a_{1-1} \\ a_{0-1} & a_{00} & a_{01} \\ a_{1-1} & a_{01} & a_{11} \end{vmatrix}\,;$$

$$B_{-1} = \begin{vmatrix} a_{01} & a_{0-1} \\ a_{11} & a_{1-1} \end{vmatrix}\,; \quad B_0 = \begin{vmatrix} a_{-1-1} & a_{1-1} \\ a_{1-1} & a_{11} \end{vmatrix}\,; \quad B_1 = \begin{vmatrix} a_{1-1} & a_{01} \\ a_{-1-1} & a_{0-1} \end{vmatrix}\,.$$

The elements of the determinants are expressed via the collision integrals and the values have been given [15].

These expressions allow us to determine the $S_{\alpha x}$ explicitly, and hence also V_{0d}. We get from (28) and (30) that

$$S_{1x} = \frac{5}{8}\left(\frac{m_1}{2kT}\right)^{-3/2}\left(d_1 - \frac{\rho_1\rho_2}{\rho n_1 m_1^{1/2}}\,d_0\right)\rho_1\,\frac{dn_1}{dx}\,,$$

$$S_{2x} = \frac{5}{8}\left(\frac{m_2}{2kT}\right)^{-3/2}\left(d_{-1} + \frac{\rho_1\rho_2}{\rho m_2^{1/2} n_2}\,d_0\right)\rho_2\,\frac{dn_2}{dx}\,. \tag{31}$$

Then the kinetic derivation of the first approximation for the mutual diffusion coefficient gives

$$D_{12} = \frac{3}{2}\,\frac{kT}{na_{00}}\,. \tag{32}$$

The second term in (27) may be put in the Kihara approximation† as

$$- \frac{S_{1x}\sqrt{m_1} + S_{2x}\sqrt{m_2}}{5\,(P_1\sqrt{m_1} + P_2\sqrt{m_2})} = \left(\frac{1}{2}\,\sigma_{12} - K\right)D_{12}\,\frac{dC}{dx}\,, \tag{33}$$

---

† Brock [18] has given the formula for the general case, where some molecules are specularly reflected and the others diffusely (Editor).

in which

$$K = \frac{(a_{1-1}a_{0-1} - a_{01}a_{-1-1})\, C + (a_{01}a_{1-1} - a_{0-1}a_{11})\,(1-C)}{2C\,(1-C)\,(a_{-1-1}a_{11} - a_{-1-1}^2)\,[\sqrt{m_1}\,C + \sqrt{m_2}\,(1-C)]}. \tag{34}$$

Then

$$V_{0d} = (\sigma_{12} - K)\, D_{12} \frac{dC}{dx}. \tag{35}$$

The solution obtained by the methods of rigorous kinetic theory not only gives a revised value for V_{0d} but also elucidates the effects of the potential parameters on V_{0d}. If the masses of the molecules are similar, K is close to α_T, the coefficient of thermal diffusion; in the limit $m_1 = m_2$ we have for the hard spheres model that

$$V_{0d} = 0.068 \frac{\sigma_2 - \sigma_1}{\sigma_1 + \sigma_2}, \tag{36}$$

in which σ_α is the effective collisional diameter for molecules of component α. This result cannot be obtained within the framework of elementary kinetic theory.

Integration of the Initial Equations

Starting from (24), the mean of v_0 over the capillary cross section area s may be found via (35) as

$$\bar{v}_0 = -\frac{s\,(1 + 4\bar{l}/R)}{8\pi\eta} \frac{dP}{dx} + (\sigma_{12} - K)\, D_{12} \frac{dc}{dx}, \tag{37}$$

in which $4\bar{l}/R$ is the correction due to viscous slip.

The ΔP for steady-state mutual diffusion is found by integrating (23), where flux j for one-dimensional diffusion is given by kinetic theory as

$$j = -\frac{n^2}{\rho}\, m_1 m_2 D_{12} \frac{dC}{dx}. \tag{38}$$

Then (23), (26), (37), and (38) together give explicitly the relation between dP/dx and dC/dx:

$$-\frac{s\,(1 + 4\bar{l}/R)}{8\pi\eta} \frac{dP}{dx} = \left[-\frac{\sqrt{m_2} - \sqrt{m_1}}{\sqrt{m_2}\,(1 - C) + \sqrt{m_1}C)} + K\right] D_{12} \frac{dC}{dx}. \tag{39}$$

But V_I contains only light gas; then the concentration change up to the attainment of ΔP_{max} may be neglected, so

$$\Delta P_{max} = P_{II} - P_I = \frac{8\pi\bar{\eta}D_{12}}{s\,(1 + 4\bar{l}/R)} \left(\frac{1}{2}\ln \frac{m_2}{m_1} + \int_1^0 K\,dC\right). \tag{40}$$

Integration of K for the general case is rather difficult, but the problem simplifies considerably if we use the feature that $a_{-1-1}a_{11} \gg a_{1-1}^2$. In the limit $m_1 = m_2$, we get for the hard spheres model that

$$\Delta P_{max} = 1.70 \frac{\bar{\eta}D_{12}}{s\,(1 + 4\bar{l}/R)} \frac{\sigma_2 - \sigma_1}{\sigma_2 + \sigma_1}. \tag{41}$$

Barodiffusion has been neglected here; its effect may be evaluated to a first approximation by solving for the baroeffect without considering the slip effects, whereupon (39) becomes

$$- \frac{dP}{dx} = \frac{8\pi\eta}{s} \, \frac{m_2 - m_1}{m_2 (1 - C) + m_1 C} \, D_{12} \left[\frac{dC}{dx} + \alpha_p C \, (1 - C) \, \frac{1}{P} \, \frac{dp}{dx} \right], \qquad (42)$$

in which $\alpha_p = (m_2 - m_1)/[m_2(1 - C) + m_1 C]$ is the barodiffusion coefficient. The correction term for barodiffusion may be put as

$$\frac{8\pi\eta D_{12}}{sP} \ln \frac{m_2}{m_1} \cdot \frac{\alpha_p \, (m_2 - m_1) \, C \, (1 - C)}{[m_2 (1 - C) + m_1 C]} \ln \frac{m_2}{m_1} \approx \frac{\Delta P}{P},$$

in which ΔP is the baro-effect found by neglecting barodiffusion; but $\Delta P \ll P$, in which P is the pressure in the system, so

$$\frac{\Delta P}{P} \ll 1 \qquad (43)$$

for all pressures corresponding to viscous flow of the mixture through the capillary.

Measurement of the Baro-Effect

Apparatus. The measurements were made on a system as above, which was such that no corrections for concentration change were required. However, measurements on a capillary in the region of viscous flow involved major technical difficulties, in view of the smallness of the effect. ΔP_{max} does not exceed 1 N/m² for Kn $\approx 10^{-3}$ (P = 760 mm Hg and s_k = 0.1 mm), and no standard instrument is of adequate sensitivity for the purpose, so we had to develop a suitable differential manometer.

The apparatus [17] had the following main parts: the volumes containing the gases, a system for measuring the pressure difference, an air thermostat, and a gas-handling system. Two brass vessels each of volume ~500 cm³ were joined via a calibrated glass capillary, whose end was closed by a rubber cap during filling. The pressures were equalized before the measurement via a bypass tube about 4 mm in diameter having a Teflon-sealed stopcock.

ΔP was measured with a capacity transducer having two beryllium-bronze corrugated membranes 64 mm in diameter and about 0.1 mm thick, which constituted a paparallel-plate condenser, one of the membranes separating the regions at the different pressures. This transducer was easy to make and gave stable readings; also, its response could be adjusted via the initial gap. The limit of detection for an initial gap of 0.2 mm was 4.25×10^2 N/m².

A IPE-2 instrument was used as secondary indicator. This employs the comparison of the frequencies of two high-frequency oscillators, the micromanometer being connected in the circuit of one of them. This instrument has a response constant of 5.5×10^{-4} pF/Hz. The rectified voltage representing the frequency difference is recorded by a millivoltmeter and pen recorder.

Gravitational errors are avoided by placing the capillary, bypass tube, vessels, and micromanometer all in the same horizontal plane. Pressure differences due to valve operation were avoided by eliminating bellows valves and using stopcocks with Teflon seals.

Particular attention was given to temperature errors. Temperature gradients were minimized by making the parts of the vacuum system of materials of high conductivity, all parts being mounted on a thick aluminum plate in an air thermostat.

A calibration device was included in the system. A pressure difference to calibrate the micromanometer was produced by altering a known volume in the right half of the system by displacing a rod with d = 5.89 ± 0.01 mm via a Teflon seal, the position of the rod being set with a micrometer. The initial volume of the system was determined by comparison with a standard system.

The gases were admitted simultaneously to the two vessels via needle valves and silica-gel filters with fabric sealing. The pressure difference during filling was recorded by an oil manometer and did not exceed 2-3 mm in that gauge ($\rho \approx 0.8$ g/cm^3). The absolute pressure was measured with a mercury manometer or (at low pressures) an OM-30 gauge. The gases were pumped out with a rotary pump, the residual pressure being recorded with a VIT-1 conductivity gauge fitted with an LT-4M head.

After filling, the working part of the system was isolated from the gauge and filling system. The stability of the working conditions was tested from the readings of the micromanometer. The pressures were equalized via the bypass valve, and then the capillary was opened. The zero of the millivoltmeter was set when the pressures were equal. Oscillator pulling near zero frequency difference was avoided by setting the frequency difference at ~40 Hz at zero pressure difference. On closing the bypass valve, a pressure difference arises by diffusion through the capillary. ΔP was recorded by the millivoltmeter and pen recorder. When ΔP_{max} had been reached, the pressures were equalized via the bypass valve, and repeat measurements were made without changing the gases. Check measurements were made by interchanging the gases between the vessels.

Results.   Measurements were made at ~293°K on 10 pairs of gases with various capillaries at pressures ranging from 1 to 700 mm Hg. The spread in the ΔP_{max} under fixed conditions did not exceed 3%.

ΔP tended exponentially to ΔP_{max}, and the latter value persisted unchanged throughout the period of measurement, except when $Kn \geq 10^{-2}$, when there was a slight fall after ΔP had been attained. It was found that ΔP_{max} was inversely proportional to P, in accordance with theoretical predictions.

Table 1 gives ΔP_{max} for $H_2 - D_2$ at various pressures.

The τ_p / τ_C calculated as a function of Kn show that the concentration change up to the attainment of ΔP_{max} is small when there is viscous flow through the capillary.

$He - CO_2$ was used in examining the effect of capillary length L. Results for $s_k = 5.68 \times 10^{-4}$ cm^2 and $L_1 = 5.85$ cm, $L_2 = 4.36$ cm, and $L_3 = 2.30$ cm show that ΔP_{max} is independent of L, which confirms the assumption that the concentrations are uniformly distributed over the volumes. The effects of capillary diameter were examined with $H_2 - He$ for $s_1 = 6.90 \times 10^{-4}$ cm^2, $s_2 = 4.12 \times 10^{-4}$ cm^2, and $s_3 = 2.47 \times 10^{-4}$ cm^2. The results confirm that correct allowance has been made for the effects of viscous slip on the baro-effect; in fact, $s(1 + 4\bar{l}/R)\Delta P_{max}$ is independent of capillary diameter for pressures corresponding to viscous flow through the capillary.

The results also show that the boundary conditions have been correctly determined for hydrodynamic flow of a diffusing binary mixture.

Table 2 compares the observed ΔP_{max} with the values calculated via the Lennard-Jones potential for $P \approx 200$ mm Hg. Results have been given for a wide range of pressures [11], which confirm the above theoretical conclusions within the error of experiment.

$Ar - CO_2$ gives an especially good illustration of the effects of diffusion slip on the baro-effect. Here the difference in the effective diameters of the molecules leads to pressure rise in the volume filled initially by the lighter gas. A further test is provided by $N_2 - CO$, since

TABLE 1. The Baro-Effect for H_2-D_2 Diffusing
through a Capillary ($s = 2.21 \times 10^{-4}$ cm^2, L = 5.5 cm)

P, mm Hg	Kn	τ_P/τ_C	τ_P, min	t_{max}, min	$\overline{\Delta P}_{max}$, N/m^2
702	$1.4 \cdot 10^{-3}$	$9.15 \cdot 10^{-6}$	0,497	3	0.506
402	$2.5 \cdot 10^{-3}$	$2.75 \cdot 10^{-5}$	0.867	6	1.03
208	$4.7 \cdot 10^{-3}$	$1.03 \cdot 10^{-4}$	1.66	12	1.77
103	$9.8 \cdot 10^{-3}$	$4.15 \cdot 10^{-3}$	3,31	15	3.59
55.0	$1.8 \cdot 10^{-2}$	$1.42 \cdot 10^{-3}$	6,04	27	6.63
20.6	$4.8 \cdot 10^{-2}$	$9.26 \cdot 10^{-3}$	14.8	54	15.8
10.6	$9.3 \cdot 10^{-2}$	$3.11 \cdot 10^{-2}$	25,4	63	22.3

TABLE 2. Comparison of Observed and Theoretical
Values for the Baro-Effect

Gases	$S \cdot 10^4$, cm^2	L, cm	P, mm Hg	$\Delta P_{max\ meas}$, N/m^2	$\Delta P_{max\ theor}$, N/m^2
He — CO$_2$	5.68	5.85	200	2.26	2,21
H$_2$ — He	6.90	4.6	205	1.53	1.23
H$_2$ — D$_2$	2.21	5.5	208	1.77	1.91
H$_2$ — N$_2$	4.12	5.5	206	4.52	3.94
He — Ar	4.12	5.5	482	2.16	1.92
H$_2$ — SF$_6$	6.90	4.6	195	2.37	2.30
Ar — N$_2$	2.21	5.5	202	0,316	0.298
He — N$_2$	2.21	4.40	210	6.36	6.83
Ar — CO$_2$	1.10	2.50	14,6	—0.454	—0.425
N$_2$ — CO	4.12	5.5		Not observed	

here the masses and effective diameters are virtually equal; no effect was observed within the error of the measurements.

Conclusions

1. The rigorous kinetic theory has been applied to give the velocity of diffusion slip for a binary gas mixture at a solid wall. It is found that the boundary conditions affect the pressure difference arising on mutual diffusion in a closed system.

2. The baro-effect is deduced via the general methods of nonequilibrium thermodynamics

3. The baro-effect has been measured with a capacity differential micromanometer to ~3% for 10 pairs of gases.

4. The experimental results confirm the theoretical conclusions.

References

1. J. C. Maxwell, Scientific Papers, 2:704 (1892).
2. A. Kundt and E. Warburg, Phil. Mag., 50:53 (1875).
3. E. Gross and S. Ziring, Mekhanika, No. 6, p. 37 (1959).
4. S. P. Bakanov and B. V. Deryagin, Dokl. AN SSSR, 139(1):71 (1961).
5. H. A. Kramer and J. Kistemaker, Physica, 10:699 (1943).
6. R. Ya. Kucherov and L. É. Rikenglaz, Zh. Éksp. Teor. Fiz., 36(6):547 (1959).
7. B. V. Deryagin and G. A. Batova, Dokl. AN SSSR, 128(2):323 (1959).
8. P. E. Suetin and P. V. Volobuev, Zh. Tekh. Fiz., 35:1689 (1965).

9.   J. Hirschfelder, C. Curtiss, and R. Bird, Molecular Theory of Gases and Liquids, John Wiley and Sons, New York (1954).
10. P. V. Volobuev and P. E. Suetin, Zh. Tekh. Fiz., 35:336 (1965).
11. P. V. Volobuev, Dissertation, Ural Polytechnical Institute (1965).
12. S. R. de Groot and P. Mazur, Non-Equilibrium Thermodynamics (Interscience), Wiley, New York (1962).
13. L. D. Landau and E. M. Lifshits, Mechanics of Continuous Media, GITTL, Moscow (1953).
14. H. Grade, Mekhanika, No. 4, p. 23 (1952); No. 5, p. 19 (1952).
15. S. Chapman and T. G. Cowling, Mathematical Theory of Nonuniform Gases, Cambridge University Press, New York (1952).
16. B. V. Deryagin and S. P. Bakanov, Dokl. AN SSSR, 117(6):959 (1957).
17. P. E. Suetin and P. V. Volobuev, Zh. Tekh. Fiz., 34:1107 (1964).
18. J. R. Brock, J. Coll. Sci., 18:489 (1963).